Control and Regulation of Transport Phenomena in the Cardiac System

Samuel (Sam) Sideman
1929–2008

Professor Samuel Sideman just finished drafting the near-final program for the 6th Fairberg Workshop on Analysis of Cardiac Development to be held in Jerusalem in 2009, anticipating his 80th birthday. He was happily spending time with his loving family, when at the age of 79, still as young and enthusiastic as ever, he died suddenly leaving a huge gap behind him.

Born in Israel in 1929, he had been at the Technion since 1957. Following a successful career in chemical engineering he was appointed as head of Biomedical Engineering in 1982. He was the incumbent of the R.J. Matas Winnipeg Chair in Biomedical Engineering from 1982 until 1997 and established and co-chaired the Heart System Research Center since 1982.

The outstanding significance of Dr. Sideman's studies lies in his integrated approach to the multivariant cardiovascular system, a pioneering approach that has contributed to the basic understanding of cardiac physiology. His studies were described in over 400 publications, exposed at multiple international conferences, and brought out in 20 books. Together with Rafael Beyar and later with Amir Landesberg, he was the driving force and initiator of the sequence of prestigious workshops that he organized and chaired. These meetings focused on multidisciplinary cardiac research bringing in premier international collaborations. This scholarly activity catalyzed the formation of the "cardiome" and the "physiome," the newly formed international program of modeling human physiology and cardiology.

Professor Sideman was founder and past president of the Israel Institute of Chemical Engineering, and president of the International Assembly of Heat Transfer. He was also active in other key international positions. His long list of awards reflects the respect he earned from both the engineering and medical communities. Samuel Sideman is known for his total involvement in science and his inspiration to students and colleagues. His enthusiasm and excitement in the pursuit of science have been contagious, and he has shared the fruits of his creative mind freely and joyfully. His world-class achievements, his contributions to Israel as a distinguished scientist and an inspiring intellectual educator, combined with his unassuming and generous personality, will stay with us forever.

ANNALS OF THE NEW YORK ACADEMY OF SCIENCES
Volume 1123

Control and Regulation of Transport Phenomena in the Cardiac System

Edited by
SAMUEL SIDEMAN, RAFAEL BEYAR, AND AMIR LANDESBERG

Published by Blackwell Publishing on behalf of the New York Academy of Sciences
Boston, Massachusetts
2008

Library of Congress Cataloging-in-Publication Data

Larry & Horti Fairberg Workshop on Control and Regulation of Transport Phenomena in Biological Systems with Special Emphasis on the Cardiac System (5th : 2007 : Antalya, Turkey)
   Control and regulation of transport phenomena in the cardiac system / editors, Samuel Sideman, Rafael Beyar, and Amir Landesberg.
      p.; cm. – (Annals of the New York Academy of Sciences, ISSN 0077-8923)
   Includes bibliographical references.
   ISBN-13: 978-1-57331-706-1 (alk. paper)
   ISBN-10: 1-57331-706-3 (alk. paper)
   1. Cardiovascular system–Physiology–Congresses. 2. Biological transport–Congresses. I. Sideman, S. II. Beyar, Rafael. III. Landesberg, Amir. IV. Tekhniyon, Makhon tekhnologi le-Yisra'el. V. International Center for Heat and Mass Transfer. VI. Title. VII. Series.
   [DNLM: 1. Cardiovascular Physiology–Congresses. 2. Biological Transport–physiology–Congresses. 3. Cardiovascular Diseases–physiopathology–Congresses. W1 AN626YL v.1123 2007 / WG 102 L334c 2007]

   QP101.L334 2007
   612.1–dc22

                                          2007044705

The *Annals of the New York Academy of Sciences* (ISSN: 0077-8923 [print]; ISSN: 1749-6632 [online]) is published 28 times a year on behalf of the New York Academy of Sciences by Blackwell Publishing with offices at (US) 350 Main St., Malden, MA 02148-5020, (UK) 9600 Garsington Road, Oxford, OX4 2ZG, and (Asia) 165 Cremorne St., Richmond VIC 3121, Australia. Blackwell Publishing was acquired by John Wiley & Sons in February 2007. Blackwell's program has been merged with Wiley's global Scientific, Technical, and Medical business to form Wiley-Blackwell.

MAILING: *Annals* is mailed Standard Rate. Mailing to rest of world by IMEX (International Mail Express). Canadian mail is sent by Canadian publications mail agreement number 40573520. POSTMASTER: Send all address changes to *Annals of the New York Academy of Sciences*, Blackwell Publishing Inc., Journals Subscription Department, 350 Main St., Malden, MA 02148-5020.

Disclaimer: The Publisher, the New York Academy of Sciences and Editors cannot be held responsible for errors or any consequences arising from the use of information contained in this publication; the views and opinions expressed do not necessarily reflect those of the Publisher, the New York Academy of Sciences and Editors.

Blackwell Publishing is now part of Wiley-Blackwell.

Information for subscribers: For ordering information, claims, and any inquiry concerning your subscription please contact your nearest office:

UK: Tel: +44 (0)1865 778315; Fax: +44 (0) 1865 471775
USA: Tel: +1 781 388 8599 or 1 800 835 6770 (toll free in the USA & Canada); Fax: +1 781 388 8232 or Fax: +44 (0) 1865 471775
Asia: Tel: +65 6511 8000; Fax: +44 (0)1865 471775,
Email: customerservices@blackwellpublishing.com

Subscription prices for 2008 are: Premium Institutional: US$4265 (The Americas), £2370 (Rest of World). Customers in the UK should add VAT at 7%; customers in the EU should also add VAT at 7%, or provide a VAT registration number or evidence of entitlement to exemption. Customers in Canada should add 5% GST or provide evidence of entitlement to exemption. The Premium institutional price also includes online access to the current and all online back files to January 1, 1997, where available. For other pricing options, including access information and terms and conditions, please visit www.blackwellpublishing.com/nyas.

Delivery Terms and Legal Title: Prices include delivery of print publications to the recipient's address. Delivery terms are Delivered Duty Unpaid (DDU); the recipient is responsible for paying any import duty or taxes. Legal title passes to the customer on despatch by our distributors.

Membership information: Members may order copies of *Annals* volumes directly from the Academy by visiting www.nyas.org/annals, emailing membership@nyas.org, faxing +1 212 298 3650, or calling 1 800 843 6927 (toll free in the USA), or +1 212 298 8640. For more information on becoming a member of the New York Academy of Sciences, please visit www.nyas.org/membership. Claims and inquiries on member orders should be directed to the Academy at email: membership@nyas.org or Tel: 1 800 843 6927 (toll free in the USA) or +1 212 298 8640.

Printed in the USA. Printed on acid-free paper.

Annals is available to subscribers online at Blackwell Synergy and the New York Academy of Sciences Web site. Visit www.blackwell-synergy.com or www.annalsnyas.org to search the articles and register for table of contents e-mail alerts.

The paper used in this publication meets the minimum requirements of the National Standard for Information Sciences Permanence of Paper for Printed Library Materials, ANSI Z39.48 1984.

ISSN: 0077-8923 (print); 1749-6632 (online)
ISBN-10: 1-57331-706-3 (paper); ISBN-13: 978-1-57331-706-1 (paper)

A catalogue record for this title is available from the British Library.

ANNALS OF THE NEW YORK ACADEMY OF SCIENCES

Volume 1123
March 2008

Control and Regulation of Transport Phenomena in the Cardiac System

Editors
SAMUEL SIDEMAN, RAFAEL BEYAR, AND AMIR LANDESBERG

Conference Chairs
SAMUEL SIDEMAN AND RAFAEL BEYAR

International Scientific Advisory Committee
CHARLES ANTZELEVITCH, JAMES B. BASSINGTHWAIGHTE, RAFAEL BEYAR,
GIANLUIGI CONDORELLI, JAMES DOWNEY AMIR LANDESBERG, ANDREW MARKS,
ROGER MARKWALD, ANDREW D. McCULLOCH, MARTIN MORAD, DENIS NOBLE,
YORAM RUDY, SAMUEL SIDEMAN, RAIMOND WINSLOW

International Organizing Committee
SAMUEL SIDEMAN, FARUK ARINC, AND GIANLUIGI CONDERELLI

This volume is the product of the 5th Larry & Horti Fairberg Workshop on CONTROL AND REGULA-
TION OF TRANSPORT PHENOMENA IN BIOLOGICAL SYSTEMS, WITH SPECIAL EMPHASIS
ON THE CARDIAC SYSTEM, held September 16–20, 2007, in Antalya, Turkey, as part of the Interna-
tional Centre of Heat & Mass Transfer series of conferences on Transport Phenomena in Biological Systems.
The meeting was organized by the Technion-Israel Institute of Technology, Haifa, Israel and the Interna-
tional Centre Heat & Mass Transfer and sponsored by the Larry and Horti Fairberg Memorial Fund, ATS,
NY, the MultiMedica Hospital Research Center, Milan, Italy, the Middle East Technical University, Ankara,
Turkey and TUBITAK, the Scientific Technological Research Council, Turkey.

CONTENTS

Part V. Transport Models and Hierarchical Analysis

Part VI. Clinical Aspects

Financial support for holding the conference was received from:

- Technion-Israel Institute of Technology, Haifa, Israel
- The New York Academy of Sciences
- The Larry and Horti Fairberg Memorial Fund, American Technion Society,
  NY, USA
- IRCCS MultiMedica, The MultiMedica Hospital Research Center,
  Milan, Italy
- Turbitack, The Scientific and Technological Research Council and of Turkey

Prologue

Mind Over Molecule: Activating Biological Demons

DENIS NOBLE

Department of Physiology, Anatomy and Genetics, Oxford University,
Oxford, United Kingdom

The new vogue for systems biology is an important development. It is time to complement reductionist molecular biology by integrative approaches. But this welcome development is in danger of losing its way. Many of the early implementations of the approach are very low level, in some cases hardly more than an extension of genomics and bioinformatics. In this paper, I outline some general principles that could form the basis of systems biology as a truly multilevel approach. We need the insights obtained from a higher level analysis in order to succeed at the lower levels. Higher levels in biological systems impose boundary conditions on the lower levels. Without understanding those conditions and their effects, we will be seriously restricted in understanding the logic of living systems. Sydney Brenner has insisted that "the cell is the correct level of abstraction." I would go further and insist on the value of abstraction at even higher levels than the cell, while recognizing the cell as a landmark level of biological organization. The principles outlined are illustrated with examples from cardiac and other aspects of physiology and biochemistry.

Key words: systems biology; computational physiology; human physiome project; genetic determinism

Introduction: Mind and Molecule Meet

Imagine a world in some m^{th} dimension in which Mind and Molecule meet. Molecule, the little David, surveys Mind, the big Goliath, and teases him: "You think yourself so big. In fact you have no dimensions at all. And when you do, you just turn out to be a bunch of molecules like me!" Mind recalls this taunt from their days in the world of four dimensions. "You jumped-up tiny midget! Where do you get this idea that your level of causality is privileged?" Molecule likes this challenge. "It's obvious isn't it? I am what the whole of life reduces to. In my most extended form, as DNA, I am the great Genie. As one of our Earthlings put it, *I* created *you*, body and mind." Mind: "But he also called you selfish!" Molecule: "That's why I always win!"

Mind is tired of this tautologous nonsense, so he climbs Meditation Mountain and thinks. He thinks so deeply that he becomes empty of form. Even in *m* dimensions he finds that when he meditates profoundly he seems to disappear. As he returns from Nirvana he has his Eureka moment. "I've got it," he exclaims, "we are *all* empty, even that little Molecule. He's just . . . he's just a bunch of tangled strings. He is a molecule precisely because he interacts with other molecules; he is nothing on his own."

Meditation Mountain has 10 large tablets of stone. Mind carves into them his 10 commandments: the principles of systems biology. In this paper I will explain what he wrote and use the heart and some other organs and systems to illustrate his principles.

First Principle: Biological Functionality Is Multilevel

I start with this principle because it is obviously true, all the other principles follow from it, and it is the basis on which a physiological understanding of the phenomenon of life must be based. It is obviously true in the field represented at this meeting because, to take just one example, functions, such as cardiac pacemaker activity, simply do not exist below the cellular level, while many forms of arrhythmia, re-entrant arrhythmias, and fibrillation do not exist below the multicellular level of tissues and even the organ as a whole. Similar statements could be made about functionality in all the organs and systems of the body.

Yet, the language of modern reductionist biology often seems to deny this obvious truth. The enticing metaphor of the "book of life" made the genome into

Address for correspondence: Professor Denis Noble, PhD, CBE, FRS, FMedSci, Balliol College and Department of Physiology, Anatomy and Genetics, Oxford University, Parks Road, Oxford OX1 3PT UK. Fax: +44-1865-272554.

denis.noble@physiol.ox.ac.uk

Ann. N.Y. Acad. Sci. 1123: xi–xix (2008). © 2008 New York Academy of Sciences.
doi: 10.1196/annals.1420.000

the modern equivalent of the "embryo homunculus," the old idea that each fertilized egg contains within it a complete organism in miniature. That the miniature is conceived as a digital "map" or "genetic program" (yet another reductionist metaphor against which I will argue later in this article) does not avoid the error to which I am drawing attention, which is the idea that the living organism is simply the unfolding of an already existing program, fine-tuned by its interaction with its environment, to be sure, but in all essentials, already there in principle as a kind of zipped-up organism. In its strongest form, this view of life leads to gene selectionism and to gene determinism: "They [genes] created us body and mind."[1]

I don't suppose that Dawkins himself believes that. In a more recent book he entitles one chapter "Genes aren't us"[2] and, even in *The Selfish Gene*, the bold simple message of the early chapters is qualified at the end. My reservations, however, go much further than his. For, in truth, the stretches of DNA that we call genes do nothing on their own. They are simply databases used by the organism as a whole. This is the reason why I replace the metaphor of the "selfish" gene by genes as "prisoners."[3(Chapter 1)] From the viewpoint of the organism, genes are captured entities, no longer having a life of their own independent of the organism. They are forced to cooperate with many other genes to stand any chance of survival. As Maynard Smith and Szathmáry[4] express it, "co-ordinated replication prevents competition between genes within a compartment, and forces co-operation on them. They are all in the same boat."

Second Principle: Transmission of Information Is Not One Way

The standard biological dogma is that information flows from DNA to RNA, from RNA to proteins, which then form protein networks, and so on up through the biological levels to that of the whole organism. It does not flow the other way. This is the dogma that is thought to safeguard modern neo-Darwinian theory from the specter of Lamarckism, the inheritance of acquired characteristics.

There are two respects in which the dogma fails. The first is that it defines the relevant information uniquely in terms of the DNA code, the sequence of C, G, A, T bases. But the most that this information can tell us is which protein (or proteins in the case of genes with multiple exons) will be made. It does not tell us how much of each protein will be made. Yet, this is one of the most important characteristics of any

protein-producing machinery. I first encountered this type of problem when working on the mathematics of excitable cell conduction.[5] The speed of conduction of a nerve or muscle impulse depends on the density of rapidly activated sodium channels: the larger the density, the greater the ionic current, and the faster the conduction. But this rule applies only up to a certain optimum density. Because the channel gating also contributes to the cell capacitance, which itself slows conduction, there is a point at which adding more channel proteins is counterproductive.[5(p. 432)] This optimum density was also shown theoretically in the same year by Alan Hodgkin who noted that the actual density in an unmyelinated nerve, such as the squid giant axon, is close to this optimum.[6] This means that a feedback mechanism must operate between the electrical properties of the nerve and the expression levels of the sodium-channel protein. We now refer to such feedback mechanisms in the nervous system, which take many forms, as electro-transcription coupling.[7]

Similar processes must occur in the heart. One of the lessons I have learned from many attempts to model cardiac electrophysiology[8] is that, during the slow phases of repolarization and pacemaker activity, the ionic currents are so finely balanced that it is inconceivable that nature arrives at the correct expression levels of the relevant proteins without some kind of feedback control. We do not yet know what that control might be, but we can say that it must exist. Nature cannot be as fragile as our computer models are! Robustness is an essential feature of successful biological systems.

There is nothing new in the idea that such feedback control of gene expression must exist. It is, after all, the basis of cell differentiation. All nucleated cells in the body contain exactly the same genome (with the exception, of course, of the germ cells, which contain only half the DNA). Yet the expression patterns of, for example, cardiac, hepatic, or bone cells, are completely different. Moreover, whatever is determining those expression levels is inherited when more cells of a certain type are produced via cell division. This cellular inheritance process is robust and it must depend on some form of gene marking. It is this information on relative gene expression levels that is critical in determining each cell type.

By what principle could we possibly say that this is not relevant information? It is just as relevant as the raw DNA. Yet, it is also clear that this information *does* travel the other way. The genes are told by the cells and tissues what to do, how frequently they should be transcribed, and when to stop. That is why I have developed the metaphor of the genome as an organ of 25,000 pipes (there are pipe organs this big!) with the organ player being the cells, tissues, organs, and systems of the body.[3(Chapter 2)] In a word, there is "downward causation" from those higher levels to determine how the genome is "played" in each cell.

The dogma also fails even at the level of the code. Sure, the C, G, A, T code itself is never altered by the organism. Such alterations occur as evolutionary mutations but not as inheritance of acquired characteristics. Nor can protein sequences be back-translated to form DNA sequences. But the DNA code is nevertheless marked by the organism. This is the focus of the rapidly growing field of epigenetics.[9] At least two such mechanisms are now known at the molecular level: methylation of cytosine bases and control by interaction with the tails of histones around which the DNA is wound. Both of these processes modulate gene expression. The terminological question then arises: do we regard this as a form of code modification? Is a cytosine, the C of the code, a kind of C* when it is methylated? That is a matter of definition of the code and one which I will deal with in the next section, but what is certain is that it is relevant information determining levels of gene expression and that this information does flow against the direction of the central dogma.

Another way to look at this principle is to say that a form of inheritance of acquired characteristics (those of specific cell types) is rampant within all multicellular organisms with very different specialized cell types.[3(Chapter 7)] At the least we have to say that, during the lifetime of the individual organism, transmission of information is far from being one way.

Third Principle: DNA Is Not the Sole Transmitter of Inheritance

The defenders of the central dogma of biology would argue that, while my conclusions regarding the second principle are correct, transmission of information to the next generation nevertheless involves wiping the slate clean of epigenetic effects. Methylation of cytosine bases and other forms of genome marking are removed. The genome is reset so that Lamarckism is impossible.

But this is to put the matter the wrong way round. We need to explain *why* the genome reverts to an unmarked state. We do not explain that by appealing to the central dogma for that dogma is simply a restatement of the same idea. We are in danger of circular logic here. Later, I will suggest a plausible reason why, at least most of the time, the resetting is complete, or

nearly so. In order to do that, we first need to analyze the idea that genetics, as originally understood, is just about DNA. The modern view is that only the DNA contains genes. That is why we can count the number of genes by determining which sequences of DNA form protein-coding sequences.

But this is not the original biological meaning of *gene*. The concept of a gene has changed.[10–13] The original biological meaning of gene was an inheritable phenotype characteristic, such as eye or hair or skin color, body shape and weight, number of legs or arms—to which we could perhaps add more complex traits, such as intelligence, personality, and sexuality. Unless we subscribe to the view that the inheritance of all such characteristics is attributable entirely to DNA sequences (which is just false), then genes, as originally conceived, are not the same as stretches of DNA.

The main reason why it is just false to say that all transmitted "nature" characteristics are attributable to DNA sequences is that, by itself, DNA does nothing at all. We also inherit the egg cell, together with any epigenetic characteristics transmitted by sperm,[14] perhaps via RNA in addition to its DNA, and all the epigenetic influences of the mother and environment. Of course, the latter begins to be about nurture rather than nature, but one of my points is that this distinction is fuzzy. The proteins that initiate gene transcription in the egg cell and impose an expression pattern on the genome are initially from the mother, and other such influences continue throughout development in the womb. Where we draw the line between nature and nurture is not at all obvious. There is an almost seamless transition from one to the other. Lamarckism, the inheritance of acquired characteristics, lurks in this fuzzy crack to a degree yet to be defined.[15,16] As the evolutionary geneticist Maynard Smith says, "It [Lamarckism] is not so obviously false as is sometimes made out."[17] He also gives some examples of unicellular organisms where forms of Lamarckism clearly occur.

This inheritance of the egg cell is important for two reasons. First, it is the egg-cell gene reading machinery (a set of around 100 proteins and the associated cellular ribosome architecture) that enables the DNA to be used as a template to make more proteins. Second, the set of other cellular elements, mitochondria, endoplasmic reticulum, microtubules, nuclear and other membranes, and a host (billions) of chemicals arranged specifically in cellular compartments, is also inherited. Most of this is not coded for by DNA sequences. Lipids certainly are not so coded, but they are absolutely essential to the cell architecture. There would be no cells, nuclei, mitochondria, endoplasmic reticulum, ribosomes, and all the other cellular machinery and compartments without the lipids.

Could epigenetic inheritance and its exclusion from the germ-cell line be a requirement of multicellular harmony? The exact number of cell types in a human is debatable. It is partly a question of definition. Findings from a project that seeks to model all the cell types in the body, the Human Physiome Project,[18] provide us with an estimate that there are around 200 cell types, all with completely different gene-expression patterns. There would be even more if one took account of finer variations, such as those that occur in various regions of the heart and which are thought to protect the heart against fatal arrhythmias. The precise number is not too important. The important fact is that the number of cell types is large and that the range of patterns of gene expression is, therefore, also large and varied. Their patterns must also be harmonious in the context of the organism as a whole. They are all in the same boat; they sink or swim together. Disturbing their harmony would have serious consequences because this harmony was arrived at after more than 2 billion years of experimentation. It may not be perfect, but most of the time it works.

Each cell type is so complex that the great majority of genes are expressed in many cell types. So it makes sense that all the cells in the body have the same gene complement and that the coding for cell type is transmitted by gene marking rather than by gene complement. I think that this gives the clue to the purpose of resetting in germline inheritance. Consider what would happen if germline inheritance reflected adaptive changes in individual cell types. Given that all cell types derive ultimately from the fused germline cells, what would the effect be? Clearly, it would be to alter the patterns of expression in nearly all the cell types. There would be no way to transmit an improvement in, say, heart function to the next generation via gene marking of the germ cells without *also* influencing the gene-expression patterns in many other types of cell in the body. And, of course, there is no guarantee that what is beneficial for a heart cell will be so in, say, a bone cell or a liver cell. On the contrary, the chances are that an adaptation beneficial in one cell type would be likely to be deleterious in another.

We frequently encounter this problem in the pharmaceutical industry. We develop drugs for an action on one type of cell in the body, such as a kidney, bone, or liver cell. Unfortunately, these drugs also influence many other cells. Cells that express the same genes cannot avoid also being sensitive to the same drugs. The result can be a serious disturbance of the delicate harmony between the cells of the body. In the case of a drug, we call this a side effect, a problem that is particularly serious in the case of cardiac arrhythmia.

Side effects present a serious problem. Indeed, they are often debilitating and sometimes fatal. The same would apply to acquired-characteristic genetic changes. Much better, therefore, to let the genetic influences of natural selection be exerted on undifferentiated cells, leaving the process of differentiation to deal with the fine-tuning required to code for the pattern of gene expression appropriate to each type of cell. If this explanation is correct, we would not necessarily expect it to be 100% effective. It is conceivable that some germline changes in gene-expression patterns might be so beneficial for the organism as a whole, despite deleterious effects on a few cell lines, that the result would favor selection. This could explain the few cases where germline Lamarckian inheritance seems to have occurred and also motivates the search for other cases. The prediction would be that germline changes in gene-expression patterns will occur in multicellular species only when beneficial to overall intercellular harmony. Such changes might be more likely to occur in simpler species. That makes sense in terms of the few examples that we have so far found.[17]

Finally, in this section, I will comment on the concept of code. Applied to DNA, this is clearly metaphorical. It is also a very useful metaphor, but we should beware of its limitations. One of these is to imply that only information that is coded is important, as in talk of the genome as the book of life. The rest of cellular inheritance is not so coded; in fact, it is not even digital. The reason is very simple. The rest of the cellular machinery does not code for or get translated into anything else. It does not need to because it represents itself; cells divide to form more cells, to form more cells, and so on. In this sense germline cells are just as "immortal" as DNA, but a lot of this information is transmitted directly without having to be encoded. We should beware of thinking that only digitally coded information is what matters in genetic inheritance.

Fourth Principle: The Theory of Biological Relativity

A fundamental property of systems involving multiple levels between which there are feedback control mechanisms is that there is no privileged level of causality. Consider, as an example, the cardiac pacemaker mechanism. This depends on ionic current generated by a number of protein channels carrying sodium, calcium, potassium, and other ions. The activation, deactivation, and inactivation of these channels proceed in a rhythmic fashion in synchrony with the pacemaker frequency. We might, therefore, be tempted to say that their oscillations generate the oscillations of the cell electrical potential (i.e., the higher level functionality since it is this potential that drives the heart rhythm). But this is not the case. The kinetics of these channels varies with the electrical potential. There is, therefore, feedback between the higher level property, the overall cell potential, and the lower level property, the channel kinetics.[3(Chapter 5)] This form of feedback was originally identified by Alan Hodgkin working on the nerve impulse, so it is sometimes called the Hodgkin cycle.[6] If we remove the feedback (e.g., by holding the potential constant, as in a voltage clamp experiment), the channels no longer oscillate. The oscillation is, therefore, a property of the system as a whole, not of the individual channels or even of a set of channels unless they are arranged in a particular way in the right kind of cell.

Nor can we establish any priority in causality by asking which comes first, the channel kinetics or the cell potential. This fact is also evident in the differential equations we use to model such a process. The physical laws represented in the equations themselves and the initial and boundary conditions, operate *at the same time* (i.e., during every integration step, however infinitesimal), not sequentially. It is simply a prejudice that inclines us to give some causal priority to lower level molecular events. The concept of level in biology is itself metaphorical. There is no literal sense in which genes and proteins lie *underneath* cells, tissue, and organs. It is a convenient form of biological classification to refer to different levels, and we would find it very hard to do without the concept. But we should not be fooled by the metaphor into thinking that *high* and *low* here have their normal meanings. From the metaphor itself, we can derive no justification for referring to one level of causality as privileged over others. That would be a misuse of the metaphor of level.

One of the aims of my book, *The Music of Life*,[3] is to explore the limitations of biological metaphors. This is a form of linguistic analysis that is rarely applied in science, although a notable exception is Steven J. Gould's monumental work on the theory of evolution[19] in which he analyzes the arguments for the multiplicity of levels at which natural selection operates. These points can be generalized to any biological function. The only sense in which a particular level might be said to be privileged is that, in the case of each function, there is a level at which the function is integrated, and it is one of our jobs as scientists to determine what that level may be. These will generally be higher, not lower, levels. We have found that cardiac pacemaker activity is integrated at the cellular level. Cardiac pumping and the generation of blood flow are integrated at the level of the whole organ.

The idea that there is no privileged level of causality partly resembles scale theories of relativity proposed by some theoretical physicists,[20] the fourth principle is itself a theory of biological relativity.

Fifth Principle: Gene Ontology Will Fail without Higher Level Insight

Genes, as defined by molecular genetics to be the coding regions of DNA, code for proteins. Biological function then arises as a consequence of multiple interactions between different proteins in the context of the rest of the cell machinery. Each function, therefore, depends on many genes, while many genes play roles in multiple functions. What then does it mean to give genes names in terms of functions? The only unambiguous labeling of genes is in terms of the proteins they code for. Thus, the gene for the sodium–calcium-exchange protein is usually referred to as *ncx*. Ion-channel genes are also often labeled in this way, as in the case of sodium-channel genes being labeled *scn*. This approach, however, naturally appears unsatisfactory from the viewpoint of a geneticist, since the original question in genetics was not which proteins are coded for by which stretches of DNA (in fact, early ideas on where the genetic information might be found[21] favored the proteins), but rather what is responsible for higher level phenotype characteristics. There is no one-to-one correspondence between genes or proteins and higher level biological functions. Thus, there is no pacemaker gene. Cardiac rhythm depends on many proteins interacting within the context of feedback from the cell electrical potential.

Let's do a thought experiment. Suppose we could knock out the gene responsible for L-type calcium channels and still have a living organism (perhaps because a secondary pacemaker takes over and keeps the organism viable—and something else would have to kick in to enable excitation–contraction coupling and so on throughout the body because L-type calcium channels are ubiquitous!). Since L-type calcium current is necessary for the upstroke of the action potential in the sinoatrial node of most species, we would find that we had abolished normal pacemaker rhythm. Do we then call the gene for L-type calcium channels the pacemaker gene? The reason why this is unsatisfactory, even misleading, to a systems-level biologist is obvious. Yet, it is the process by which we label many genes with high-level functions. The steadily growing list of "cancer genes" have been identified in this way, by determining which mutations (including deletions) change the probability of cancer occurring. We can be fairly sure, though, that this characteristic is not why these genes were selected during the evolutionary process.

Another good example of this approach is the discovery of what are termed *clock* genes, which are involved in circadian rhythm. Mutations in a single gene (now called the *period* gene) are sufficient to abolish the circadian period of fruit flies.[22] This discovery of the first clock gene was a landmark since it was the first time that a single gene had been identified as playing such a key role in a high-level biological rhythm. The expression levels of this gene are clearly part of the rhythm generator. They vary (in a daily cycle) in advance of the variations in the protein that they code for. The reason is that the protein is involved in a negative feedback loop with the gene that codes for it.[23] The idea is very simple. The protein levels build up in the cell as the period gene is read to produce more protein. The protein then diffuses into the nucleus where it inhibits further production of itself by binding to the promoter part of the gene sequence. With a time delay, the protein production falls off and the inhibition is removed so that the whole cycle can start again. So, we not only have a single gene capable of regulating the biological clockwork that generates circadian rhythm, it is itself a key component in the feedback loop that forms the rhythm generator. However, such rhythmic mechanisms do not work in isolation. There has to be some connection with light-sensitive receptors (including the eyes). Only then will the mechanism lock on to a proper 24-h cycle rather than free running at, say, 23 h or 25 h. In the mouse, for example, many other factors play a role. Moreover, the clock gene itself is involved in other functions. That is why Foster and Kreitzman have written, "what we call a clock gene may have an important function within the system, but it could be involved in other systems as well. Without a complete picture of all the components and their interactions, it is impossible to tell what is part of an oscillator generating rhythmicity, what is part of an input, and what is part of an output. In a phrase, it ain't that simple!"[24]

Indeed not. The period gene has also been found to be implicated in embryonic development as the adult fly is formed over several days. And it is deeply involved in coding for the male love songs generated by wing-beat oscillations, which are specific to each of around 5000 species of fruit fly and ensure that courtship is with the right species. Perhaps it should be renamed the *fruitfly love gene*!

The point is obvious. We should not be misled by gene ontology. The first function a gene is found to be involved in is rarely, if ever, the only one and may not even be the most important one. Gene ontology will require higher level insight to be successful in its mission.

Sixth Principle: There Is No Genetic Program

I can already sense that, out there in my imagined m^{th} dimension, Molecule is getting impatient with Mind. We are halfway through the tablets of stone. Mind has been running all over him like a steamroller flattening everything in its path. Surely, Molecule thinks, I must win on the question of genetic programs. They are all over the place. They are the crown jewels of the molecular genetic revolution, invented by none other than the famous French Nobel Prize winners Monod and Jacob.[25,26] Their enticing idea was born during the early days of electronic computing when computers were fed with paper tape or punched cards coded with sequences of instructions. Those instructions were clearly separate from the machine itself that performed the operations. They dictated those operations. Moreover, the coding is digital. The analogy with the digital code of DNA is obvious. So, are the DNA sequences comparable to the instructions of a computer program?

An important feature of such computer programs is that the program is separate from the activities of the machine that it controls. Originally, the separation was physically complete, with the program on the tape or cards only loaded temporarily into the machine. Nowadays, the programs are stored within the memory of the machine, and the strict distinction between the program, the data, and the processes controlled may be breaking down. Perhaps computers are becoming more like living systems, but in any case the concept of a genetic program was born in the days when programs were separate identifiable sets of instructions.

So, what do we find when we look for genetic programs in an organism? We find no genetic programs! There are no sequences of instructions in the genome that could possibly play a role similar to that of a computer program. The reason is very simple. A database is not a program. To find anything comparable to a program we have to extend our search well beyond the genome itself. Thus, as we have seen above, the sequence of events that generates circadian rhythm includes the period gene, but it necessarily also includes the protein it codes for, the cell in which its concentration changes, the nuclear membrane across which it is transported with the correct speed to effect its inhibition of transcription. This is a gene–protein–lipid–cell network, not simply a gene network. The nomenclature matters. Calling it a gene network fuels the misconception of genetic determinism. In the generation of a 24-h rhythm, none of these events in the feedback loop is privileged over any other. Remove any of them, not just the gene, and you no longer have circadian rhythm. Moreover, it would be strange to call this network of interactions a program. The network of interactions is itself the circadian rhythm process. As Enrico Coen, the distinguished plant geneticist, put it, "Organisms are not simply manufactured according to a set of instructions. There is no easy way to separate instructions from the process of carrying them out, to distinguish plan from execution."[27] In short, the concept of a program here is completely redundant.

Seventh Principle: There Are No Programs at Any Other Level

I have introduced the analogy of the genome as a database and the metaphor of genes as prisoners in order to provoke the change in mindset that is necessary for a complete systems approach to biology to be appreciated. The higher levels of the organism use the database and play the genome to produce functionality. If the genome can be likened to a huge pipe organ, then it seems correct to ask who is the player, who was the composer? If we cannot find the program of life at the level of the genome, at what level do we find it?

We should view all such metaphors simply as ladders of understanding. They enable us to escape more easily from the errors arising from the limitations of reductionist metaphors, but they also have important limitations. All metaphorical language behaves in this way since metaphorical language highlights some features of the target to which they are applied at the expense of downplaying others.[28,29] We may need to use multiple, even incompatible, metaphors in order to capture all that we want to say. Then, once we have used such ladders, we can, as it were, throw them away. This way of thinking can seem strange to some scientists for whom there must be just one correct answer to any scientific question. I explore this important issue in *The Music of Life* by analyzing the selfish-gene and prisoner-gene metaphors linguistically to reveal that no conceivable experiment could decide which is correct.[3(Chapter 1)] These metaphors highlight totally different aspects of the properties of genes. This philosophy is applied throughout the book as it answers questions, such as "where is the program of life?" It would take too long to repeat all the arguments here

in this paper. The conclusion is simply that there are no such programs at any level. At all levels the concept of a program is redundant since, as with the circadian rhythm network, the networks of events that might be interpreted as programs are themselves the functions we are seeking to understand. Thus, there is no program for the heart's pacemaker separate from the pacemaker network itself.

While causality operates within and between all levels of biological systems, there are certain levels at which so many functions are integrated that we can refer to them as important levels of abstraction. Sydney Brenner[30] wrote, "I believe very strongly that the fundamental unit, the correct level of abstraction, is the cell and not the genome." He is correct since the development of the eukaryotic cell was a fundamental stage in evolutionary development, doubtless requiring at least a billion years to be achieved. To systems physiologists, though, there are other important levels of abstraction, including whole organs and systems.

Eighth Principle: There Are No Programs in the Brain

By the eighth tablet, Molecule is getting more than impatient. We are moving onto what ought to be Mind's territory, the brain. Suspecting that Mind might have an easy ride at this level, in desperation Molecule calls in help from his brilliant discoverer, the Nobel Prize winner Francis Crick. In his book, *The Astonishing Hypothesis*, Crick proclaimed, "You, your joys and your sorrows, your memories and your ambitions, your sense of personal identity and free will, are in fact no more than the behaviour of a vast assembly of nerve cells and their associated molecules."[31] As Molecule savors the implications of this sweeping statement, he does a *pas de deux* with his alter ego, the other strand of the double helix.

Meanwhile, Mind is at last in some kind of trouble of his own. Many biologists tell him that the solution to the old mind–brain problem is simple. In some sense or other, the mind is just a function of the brain. The pancreas secretes insulin, endocrine glands secrete hormones . . . and the brain "secretes" consciousness! All that is left is to find out how and where that happens in the brain. In one of his last statements Crick has even hinted at where that may be: "I think the secret of consciousness lies in the claustrum"[31] (quoted by V.S. Ramachanran,[32] www.edge.org). This structure is a thin layer of nerve cells in the brain. It is very small and it has many connections to other parts of the brain, but the details are of no importance to the argument. The choice of brain location for the secret of consciousness varies greatly according to the author. Descartes even thought that it was in the pineal gland. The mistake is always the same, which is to think that in some way or other the brain is a kind of performance space in which the world of perceptions is reconstructed inside our heads and presented to us as a kind of Cartesian theatre. But that way of looking at the brain leaves open the question: where is the "I", the conscious self that sees these reconstructions? Must that be another part of the brain that views these representations of the outside world?

We are faced here with a mistake similar to that of imagining that there must be programs in the genomes, cells, tissues, and organs of the body. There are no such programs, even in the brain. The activity of the brain and of the rest of the body simply *is* the activity of the person, the self. Once again, the concept of a program is superfluous. When I play my guitar and my fingers rapidly caress the strings at an automatic speed that comes from frequent practice, there is no separate program that is making me carry out this activity. The patterns and processes in my nervous system and the associated activities of the rest of my body simply *are* me playing the guitar. Similarly, when we deliberate intentionally, there is no nervous network forcing us to a particular deliberation. The nervous networks, the chemistry of our bodies, together with all their interactions within the social context in which any intentional deliberation makes sense, *are* us acting intentionally.

Ninth Principle: The Self Is Not an Object

In brief, Mind wins yet again, not because Mind is a separate object competing for activity and influence with the molecules of the body. Thinking in that way was originally the mistake of the dualists, like Sherrington and Eccles, led by the philosophy of Descartes. Modern biologists have abandoned the separate substance idea, but many still cling to a materialist version of the same mistake[33] based on the idea that somewhere in the brain the self is to be found as some neuronal process. The reason why that level of integration is too low is that the brain, and the rest of our bodies which are essential for attributes like consciousness to make sense,[3(Chapter 9)] are tools (back to the database idea again) in an integrative process that occurs at a higher level involving social interactions. We cannot attribute the concept of *selfness* to ourselves without also doing so to others.[34] Contrary to Crick's view, our selves are indeed much "more than the behaviour

of a vast assembly of nerve cells and their associated molecules" precisely because the social interactions are essential even to understanding what something like an intention might be. I analyze an example of this point in much more detail in *The Music of Life* (Chapter 9). This philosophical point is easier to understand when we take a systems view of biology since it is in many ways an extension of that view to the highest level of integration in the organism.

Tenth Principle: A Genuine Theory of Biology Does Not Yet Exist

Well, of course, choosing just 10 tablets of stone was too limiting. There are many more to be discovered. This last one points the way to many others of whose existence we have only vague ideas. We do not yet have a genuine theory of biology. The theory of evolution is not a theory in the sense in which I am using the term. It is more a historical truth, itself standing in need of explanation. We do not even know yet whether that explanation will itself be primarily historical, depending on events that are difficult, if not impossible, to analyze fully from a scientific perspective or whether it was a process that would have homed in to the organisms we have regardless of the conditions. My own suspicion is that it is most unlikely that, if we could turn the clock back and let the process run again, we would end up with anything like the range of species we have today on earth. But, whichever side of this particular debate you may prefer, the search for general principles that could form the basis of a genuine theory of biology is an important aim of systems biology. Can we identify the logic by which the organisms we find today have succeeded in the competition for survival? In searching for that logic, we should not restrict ourselves to the lower levels. Much of the logic of living systems is to be found at the higher levels since these are the levels at which selection has operated and determined whether organisms live or die.

It is time to leave Mind and Molecule arguing it out in their m^{th} dimension. Perhaps I have convinced you that in our world, at least, the argument is sterile. It will be an interesting subject for future historians of science to determine why we were misled for so long by the idea of molecular determinism. But we can move on. We need to understand all those molecules, but we need even more to understand their interactions. The parameter space in which those interactions occur is mind-bogglingly vast. "There wouldn't be enough material in the whole universe for nature to have tried out all the possible interactions even over the long period of billions of years of the evolutionary process."[3](Chapter 2)

Summary

The systems approach works well together with another important insight, which is that it is somewhat artificial to view objects and their relationships as separate aspects of reality. An object that had no relationship to any other would be undetectable. That is why, when Mind cuts himself off from the world as he meditates, he literally becomes empty in the sense in which I used the term in the Introduction. Everything, even Molecule, becomes empty. Nothing and everything meet out there in the mth dimension.

Conflicts of Interest

The author declares no conflicts of interest.

References

1. DAWKINS, R. 1976. The Selfish Gene. Oxford University Press. Oxford.
2. DAWKINS, R. 2003. A Devil's Chaplain. Weidenfeld and Nicolson. London.
3. NOBLE, D. 2006. The Music of Life. Oxford University Press. Oxford.
4. MAYNARD SMITH, J. & E. SZATHMÁRY. 1999. The Origins of Life. Oxford University Press. New York.
5. JACK, J.J.B., D. NOBLE & R.W. TSIEN. 1975. Electric Current Flow in Excitable Cells. Oxford University Press. Oxford.
6. HODGKIN, A.L. 1975. The optimum density of sodium channels in an unmyelinated nerve. Proc Royal Society B **270:** 297–300.
7. DEISSEROTH, K., P.G. MERMELSTEIN, H. XIA, *et al.* 2003. Signaling from synapse to nucleus: the logic behind the mechanisms. Current Opinion in Neurobiology **13:** 354–365.
8. NOBLE, D. 2002. Modelling the heart: insights, failures and progress. BioEssays **24:** 1155–1163.
9. QIU, J. 2006. Unfinished Symphony. Nature **441:** 143–145.
10. KITCHER, P. 1982. Genes. British J Philosophy of Science **33:** 337–359.
11. DUPRÉ, J. 1993. The Disorder of Things. Harvard. Cambridge, Mass.
12. MAYR, E. 1982. The Growth of Biological Thought. Harvard. Cambridge, Mass.
13. PICHOT, A. 1999. Histoire de la notion de gène. Flammarion. Paris.
14. ANWAY, M.D., A.S. CUPP, M. UZUMCU & M.K. SKINNER. 2005. Epigenetic transgenerational actions of endocrine disruptors and male fertility. Science **308:** 1466–1469.
15. JABLONKA, E. & M. LAMB. 2005. Evolution in Four Dimensions. MIT Press. Boston.
16. JABLONKA, E. & M. LAMB. 1995. Epigenetic Inheritance and Evolution. The Lamarckian Dimension. Oxford University Press. Oxford.
17. MAYNARD SMITH, J. 1998. Evolutionary Genetics. Oxford University Press. New York.
18. CRAMPIN, E.J., M. HALSTEAD, P.J. HUNTER, P. NIELSEN, D. NOBLE, N. SMITH & M. TAWHAI. 2004. Computational physiology and the physiome project. Experimental Physiology **89:** 1–26.
19. GOULD, S.J. 2002. The Structure of Evolutionary Theory. Harvard. Cambridge, Mass.
20. NOTTALE, L. 2000. La relativité dans tous ses états. Du mouvements aux changements d'échelle. Hachette. Paris.
21. SCHRÖDINGER, E. 1944. What is Life? Cambridge University Press. Cambridge, Mass.
22. KONOPKA, R.J. & S. BENZER. 1971. Clock mutants of *Drosophila melanogaster*. Proc Nat Acad Sciences **68:** 2112–2116.
23. HARDIN, P.E., J.C. HALL, & M. ROSBASH. 1990. Feedback of the *Drosophila* period gene product on circadian cycling of its messenger RNA levels. Nature **343:** 536–540.
24. FOSTER, R. & L. KREITZMAN. 2004. Rhythms of Life. Profile Books. London.
25. MONOD, J. & F. JACOB. 1961. Teleonomic mechanisms in cellular metabolism, growth and differentiation. Cold Spring Harbor Symposia Quantitative Biology **26:** 389–401.
26. JACOB, F. 1970. La Logique du vivant, une histoire de l'hérédité. Gallimard. Paris.
27. COEN, E. 1999. The Art of Genes. Oxford University Press. Oxford.
28. LAKOFF, G. & M. JOHNSON. 1980. Metaphors We Live By. University of Chicago Press. Chicago.
29. KÖVECSES, Z. 2002. Metaphor. A Practical Introduction. OUP. Oxford.
30. BRENNER, S. 2003. Lecture. Columbia Univ.
31. CRICK, F.H.C. 1994. The Astonishing Hypothesis: The Scientific Search for the Soul. Simon and Schuster. London.
32. RAMACHANRAN, V.S. 2004. The astonishing Francis Crick. Edge **147** (www.edge.org)
33. BENNETT, M.R. & P.M.S. HACKER. 2003. Philosophical Foundations of Neuroscience. Blackwell Publishing. Oxford.
34. STRAWSON, P.F. 1959. Individuals. Routledge. London.

Preface

Cardiac Control Pathways: Signaling and Transport Phenomena

SAMUEL SIDEMAN

Faculty of Biomedical Engineering, Technion, Israel Institute of Technology, Haifa, Israel

Signaling is part of a complex system of communication that governs basic cellular functions and coordinates cellular activity. Transfer of ions and signaling molecules and their interactions with appropriate receptors, transmembrane transport, and the consequent intracellular interactions and functional cellular response represent a complex system of interwoven phenomena of transport, signaling, conformational changes, chemical activation, and/or genetic expression. The well-being of the cell thus depends on a harmonic orchestration of all these events and the existence of control mechanisms that assure the normal behavior of the various parameters involved and their orderly expression. The ability of cells to sustain life by perceiving and responding correctly to their microenvironment is the basis for development, tissue repair, and immunity, as well as normal tissue homeostasis. Natural deviations, or human-induced interference in the signaling pathways and/or inter- and intracellular transport and information transfer, are responsible for the generation, modulation, and control of diseases. The present overview aims to highlight some major topics of the highly complex cellular information transfer processes and their control mechanisms. Our goal is to contribute to the understanding of the normal and pathophysiological phenomena associated with cardiac functions so that more efficient therapeutic modalities can be developed. Our objective in this volume is to identify and enhance the study of some basic passive and active physical and chemical transport phenomena, physiological signaling pathways, and their biological consequences.

Key words: environmental stimuli; signaling pathways; transport phenomena; control mechanisms; transmembrane transport

Introduction

The Puzzle of Life

The cell is the basic characteristic of all living entities, and all cells must differentiate and continuously exchange stimuli, nutrition, and waste with their environment in order to sustain and perpetuate life. These processes of differentiation and exchange of material and energy are closely governed and carefully monitored by molecular networks of activating and inhibitory signals. The signaling chemical molecules and ions transmit information to or from, between or within, the cells. The molecular networks and signaling pathways invoke various transport phenomena to maintain cellular integrity, sustain life, and secure the survival of the species. These molecules, which originate naturally in donor cells or are introduced artificially by

infusion or injection, migrate by convection or diffusion, either alone or by a carrier, toward the target cell(s) and penetrate or interact with protein receptors in the membrane of the target cells, triggering a single or chain response in the cell. The cell's functional behavior is thus dictated by these networks which translate chemical, electrical, or mechanical stimuli, either directly or (mostly) via receptor-binding interactions, into intracellular signals; these signals, in turn, affect transcriptional, metabolic, and/or cytoskeletal processes that modulate the cellular responses and the ensuing integrated organ functions. A pivotal theme in modern cell biology is deciphering the complexities of the transcription schemes, metabolic networks, and signaling pathways so as to unravel the puzzle of life. Cell biology, molecular biology, genetics, and system biology are relatively new chapters in the ever-going and ever-growing search for the secrets of life. The old alchemists' search for the nectar of life is now replaced by an army of molecular scientists and analytical explorers who seek to unravel the marvel of the transformation of "dead" molecules to living entities.

Address for correspondence: Professor Sam Sideman, DSc, DScHon, Faculty of Biomedical Engineering, Technion, IIT, Haifa, Israel 32000. Fax: +9724 829 4599.

Sam@bm.technion.ac.il

Ann. N.Y. Acad. Sci. 1123: xx–xli (2008). © 2008 New York Academy of Sciences.
doi: 10.1196/annals.1420.001

Goals and Means

Systems biology research aims to comprehend the underlying structure of cell signaling networks and the transmission pathways of matter and information. The goal is to understand how these signals influence the cell's response (i.e., to describe cellular function in terms of fundamental properties) so as to improve molecular and cell-based therapeutic technologies. This involves exploring and simulating receptor-mediated regulation of cellular functions, such as proliferation, adhesion, migration, differentiation, and death. Combining principles of engineering analysis with molecular biology techniques allows parametric studies of receptors or ligand properties, interaction parameters, and signaling network dynamics and may yield better understanding and improved health care by pharmaceutical and biotechnological technologies. The present manuscript highlights the major parameters which play significant roles in this complex drama of sustaining cellular function and keeping us alive and is particularly aimed at those of us of the engineering and exact sciences who have missed their biology classes.

This volume is a product of the 5th International Fairberg Cardiac Workshop and the 15th in the Goldberg/Fairberg Cardiac Workshops series.[1–14] It concentrates on natural and induced control and regulation of various interactive transport phenomena in the cardiac system and, as such, is part of The International Centre for Heat and Mass Transfer (ICHMT) series of seminars on Heat & Mass Transfer in Biological Systems. The main objectives of this workshop were:

1. to identify, understand, and analyze the control and regulation mechanisms of transport phenomena in the cardiac system so as to catalyze better therapeutic modalities for failing hearts;
2. to facilitate the interaction and enhance the cooperation between engineering disciplines and life sciences by bringing together leading scientists involved in research and clinical applications in the cardiac system.

The manuscripts herein highlight and explore various aspects of signaling mechanisms and transport phenomena; ions and metabolites, cellular membrane transport, and endocytosis; intracellular transport and energetics; blood–tissue exchange and inter-tissue transport (e.g., endothelial); and system biology, uni- and multiscale-transport models, and hierarchical analysis. Some related clinical aspects (e.g., cardiac protection, metabolic and pharmaceutical augmenta-

tion, and interferences) are discussed. Obviously, the interface between physics, mathematics, and cell biology is a fertile terrain for new discoveries. Effective dialogue between researchers of different backgrounds catalyzes creativity.

Transport Phenomena in Biological Systems

The term *transport phenomena* stems from the fundamental analytical similarity of fluid dynamics (momentum), heat transfer (energy), and mass transfer (chemical species). These phenomena can occur singularly or simultaneously in all active physical and biological systems.[15–23]

Cellular Transport and the Butterfly Effect

The cell is made of "dead" molecules and represents a biocomplex system wherein the dynamic interactions between the molecules turn them collectively into a vibrant living cell. The autonomous cell is the simplest representation of life, but life cannot be understood by monitoring the behavior of a cell from the outside. It requires insight into the biocomplexity of the actual intracellular molecular performance, the intercellular interactions, and the critical exchange between the cells and their dynamic environments. Quantitative experimental and theoretical procedures, including control models and system analyses, are needed to unravel this complexity.

Ideally, we seek the "butterfly effect": how do small changes in the intracellular components affect and control the behavior of the whole system (e.g., an organ). As these effects encompass different levels of biological organization, one faces a hierarchical control analysis. As the living body performance is affected and determined by the interactions between its various organs, we are faced with the horrendous task of analyzing and resolving the system characteristics of the whole system (i.e., deciphering the physiome).

General Modes of Transport

Biological tissues have a multiphase composition as about 55–70% of the volume of the human body consists of water. Accordingly, the transport of nutrients, such as oxygen, hormones, and cytokines, from their uptake location (intestine, lung) or production sites (glands, liver) to the cells of the various organs occurs in an aqueous media. Likewise, waste products are brought to the excreting organs (kidney, liver, lungs) in an aqueous phase. Also, intracellular transport is essentially fluid-based, dominated by or along carrier or directing molecules. Staying within the context of the cardiac system, we relate here mainly to cellular interactions and mass transfer phenomena, which play a major role in the formation, development, and maintenance of all living entities. However, we limit our discussion to nonequilibrium mass transfer, leaving out the important issues of concentrations and mixtures and scales of operation (molecular micro- and macrophases) states (e.g., equilibrium and biochemical reactions, such as order of reaction, temperature effects, Michaelis–Menters kinetics). Also essentially precluded are natural and forced convective mass transfer, either macroscopic (e.g., species conservation, flow across porous membranes, polarization, pharmakinetics) or microscopic (laminar versus turbulent, concentrations, boundary layers, chemical reactions, diffusion and tortuosity in porous media, effective diffusivity, steady versus transient states, multidimensionality) including methods of analyses (e.g., finite elements, compartmental analysis). In general terms, we seek only to highlight, at least nominally, the cause (signaling), the means (transport phenomena), and the effect (functional consequences) of some typical biocomplex pathways that keep us alive.

The basic driving mechanisms of mass transport in biological systems are:

- convection (driven by pressure or temperature gradients)
- diffusion (driven by concentration gradients)
- protein-mediated transport
a. passive transport by binding to specific protein transporters
b. active energy-dependant transport against concentration gradients
- vesicular transport (pinocytosis)
- transport by and of cells

With the exception of arterial-forced convection and muscular contractions, convection and diffusion are normally passive in nature and can usually be described systematically. In contrast, facilitated transport by carrier molecules and pinocytosis involves active processes, which are characterized by a great variety of mechanisms and usually defy a simple analytical description.

In general engineering practice, convection is considered rapid and effective while diffusion is slow and less efficient. This notion has to be cautiously viewed in biological systems where the transport distances covered by convection and diffusion differ greatly. For example, typical convection distances and times in the systemic circulation are of the order of centimeters to meters and seconds to minutes, respectively. The intravascular volume, typically 5 L blood, circulates every minute at rest (approximately 70 mL of blood × 70 beats per min). There is, however, substantial organ variability. The typical circulation time in highly perfused tissues (e.g., brain) is approximately 20 s, while it is up to 15 min in bone and joints. The average flow velocity in the aorta is about 20 cm/s at rest with a peak of about 150 cm/s. Under conditions of heavy exercise, these values are markedly increased.

Transport Phenomena in the Cardiac System

The inherent complexity and hierarchical structure of the cardiac system has inspired Jim Bassingthwaighte to champion the Cardiome concept.[24] The cardiome is a one-organ component of the physiome and represents the multiscale, multiparameter, and multidimensional complexity of the cardiac system and its numerous parameters that affect the cardiac function. The major elements of interest of the whole organ can be grouped[24] by *structure* (e.g., fibers, directionalities, physical properties, composition), *state* (e.g., contraction, relaxation, pathophysiology), *kinetics* (e.g., rates of deformations, ejection, outputs, excitation, pressures, volumes), and *functions* (e.g., activation, contraction, relaxation). The next level of interest brings us to the tissues, again defined by structure, state, kinetics, and function. Going a level deeper we face the cells, and then the organelles, and finally the molecular level, each defined by its structure, state, kinetics, and function. Clearly, an awesome array of interacting parameters!

The complexity increases significantly when we extend the analysis to include the development and maintenance of the cardiac system. The heart is the first organ to be formed and to function in the embryo, and all subsequent events in the life of the organism depend on the heart's ability to match its output to the organism's demands for oxygen and nutrients. A set of transcription factors initiates the program for cardiac gene expression and drives the morphogenic events involved in the formation of the multichambered heart. In contrast to the skeletal muscle in which a single transcription factor suffices to activate the complete program of muscle differentiation, the cardiac muscle differentiation depends on combinations of transcription factors. Analysis of regulatory DNA sequences responsible for cardiac transcription reveals an unexpected complexity of regulation in which individual genes are often controlled by multiple independent enhancers that direct expression in highly restricted patterns in the developing heart. In addition, cardiac transcription factors amplify and maintain their expression within mutually reinforcing transcriptional circuits involving positive feedback loops and protein–protein interactions.

As further amplified below, the performance of the cardiac system is regulated by an amazing network of ionic and molecular subsystems that sustains the normal cardiac performance and adjusts it in response to the internal and/or external environmental changes. For example, the circulation supplies cells with oxygen, nutrients, and removes carbon dioxide and other catabolic products. On the cellular scale, the variables involved in cardiovascular regulation, such as blood flow, blood pressure, oxygen blood concentration, and ATP concentration, are kept around their reference points by different feedback control mechanisms with different dynamic characteristics. Thus, the short-term control of blood flow and pressure on the cardiovascular system scale is performed by the autonomic nervous system through baroreceptor control loops. Relatively slower control mechanisms usually operate by hormonal regulation. In general, the extra-, inter-, and intracellular control mechanisms are based on ionic and molecular transport phenomena. Another example relates to the autonomic balance in the cardiac control of external stimulations. Briefly, stimulation of the sympathetic system produces tachycardia and changes in the left ventricular (LV) mechanical performance, whereas stimulation of the vegal nerves affects the heart rate and induces bradycardia. A balance between these two systems is thus required for a normal stable cardiac performance.

Following our earlier reports on the communicative cardiac cell[13] and the cellular communication maze,[25] we attempt here to highlight some of the basic cardiac control mechanisms (e.g., signaling pathways and the associated transport phenomena in the cellular and subcellular levels). This includes some aspects of, for example, transmembrane diffusion and permeability of nonelectrolytes and electrolytes, facilitated transport, active transport (Na^+–K^+ and ATP pumps), ion channels, receptor mediated transport, metabolism, signal transduction, and regulation of gene expression. Targeting the signaling pathway by selected or predesigned molecular messengers is a promising method of controlling undesired functions (e.g., cardiac pathologies, cancer cell growth, and others). A complementary review by Asrar B. Malik on the endothelial transport phenomena and controlling mechanisms in this barrier between the vascular blood and the surrounding tissue is presented in this volume.

Last, the versatility of the signaling and transport mechanisms in the cell is further emphasized by revis-

iting the important and extensively studied nucleocytoplasmic and mitochondrial transmembranes signaling and transport mechanisms, as well as the genetic expression phenomena.

Cellular Dynamic

Cellular Structure

Cells have a highly compartmental architecture and represent a unique collection of elements enclosed by a thin bilayered membrane. The main elements inside the membrane are the nucleus (the genetic household), the mitochondria (the chemical factory that produces the ATP), the sarcomere (the mechanical engine of muscle contraction and motion), the sarcoplasmic reticulum (SR; the dynamic Ca^{2+} store and supplier), the cytoskeleton (the fibrous network that maintains the shape of the cell), the endoplasmic reticulum (ER), and the Golgi complex (management of proteins and information transfer[25]). All these elements are engulfed in the cytoplasm, an aqueous gel-like fluid where proteins are formed and most enzymatic reactions occur.

Constant exchanges between cells, as well as between organelles within each cell, are vital to the maintenance of major biological functions. A variety of molecules are continuously exchanged via the cytoplasm between these intracellular compartments throughout the life of a cell. The traffic is quite complex. To communicate, the cell uses molecules that encode information. These molecules usually need specialized carriers, and the intracellular exchanges require setting up a genuine signaling network to structure and prioritize information.

Cells are attached together by different molecular "adhesives," including cadherins, the surface glycoproteins, which interact with themselves to maintain intercellular contracts. Cadherins play an essential part in tissue formation during embryonic development and in the cohesion of mature tissues. The cadherins form contacts that are both dynamic, thus allowing them to adapt quickly to their environment, and highly stable, thereby favoring lasting interactions.

Intracellular Mobility

Biopolymer gels, and in particular microtubules and filamentous actin (f-actin), make up the cytoskeleton and are rigid enough to distinguish between longitudinal and transverse directions at the level of the individual polymer. For example, the ER that pervades almost the entire cytoplasm is a dynamic network of interconnected microtubules. The microtubules and conventional kinesin (kinesin-1) are essential determinants in establishing and maintaining the structure of the ER by active membrane expansion. Furthermore, the microtubules and the motor families kinesin and dynein are largely responsible for spatial organization of the intracellular structures and for the transport of intracellular cargo.

Efficient internal transport systems maintain and manage trafficking between the different compartments. It is hypothesized that distinct types of microtubules are adapted for specific functions by post-translational modifications (acetylation, detyrosination, polyglycylation, and polyglutamylation), which affect the c-terminal tail domains of alpha- and beta-tubulin. The intracellular highways of microtubules and actin filaments and the motors that walk on these roads constitute the backbone of this transport phenomenon. These linear motor protein molecules (i.e., kinesins and dyneins) haul the molecular cargo to and from different locations inside cells by traveling along the microtubules network. The kinesins tow the tubes toward their destination. Several motors, which are bound to the membrane and are constantly renewed, combine dynamically to generate the force needed for the production of the mechanical work to pull the tubes and maintain the dynamic transport processes.

The cell can regulate its traffic by adjusting the concentration of kinesins or the vesicle membrane tension. This is the mechanism for almost all the long distance transport of materials inside a cell. One wonders how a cell manages such long-distance transport without loosing the cargo. And more importantly, how does it select the type of cargo for transport at a given instant of time? Many such questions are yet to be answered. A novel approach to control intracellular traffic by using preactivated nanocarriers has recently been reported.[26]

Intracellular Traffic Signals

Enzymes inside cells frequently add or remove different molecules from the surfaces of the microtubules. J. Gaertig and B. Edde[27] have recently identified a group of enzymes responsible for attaching glutamic acid tags to the sides of microtubules. Thus, polyglutamylase enzymes produce glutamic-acid chains of varying lengths and branching patterns, which appear to act like cellular traffic signals guiding the molecular motors as they travel along microtubules. The enzymes target different locations in the organism where they tag specific sites on nearby microtubules, and these tags act like stop signs for regulating dynein.

Rab proteins, a large family of the Guanosine triphosphate enzymes (GTPases), play a key role in the movement and targeting and docking of transport vesicles[28] with their acceptor compartments. GTPases Rab proteins are localized on the cytoplasmic surface of all organelles that are involved in the biosynthetic, secretory, and endocytic pathways. Rab6 is associated with membranes of the Golgi apparatus and the trans-Golgi network. Rab6A defines a novel retrograde pathway between Golgi and the ER. Resident Golgi enzymes and some lipids can continuously recycle through the ER by this Rab6-dependent pathway.

Cell Motion and Tissue Engineering

Animal cells can crawl along suitable substrates and appropriate artificial scaffolds. This phenomenon is responsible for the formation of organized tissues from individual elements and for some important processes and biomedical applications. The growth and mobility of anchorage-dependent mammalian cells are the cornerstone of the field of tissue engineering, which aims at cellular growth of three-dimensional (3-D) tissues. It is believed that the cell's generation of tension across its membrane is involved in the biochemical signaling that regulates the cycle of growth and replication.

Tissue engineering strategies involve implantation that stimulates angiogenesis, tissue integration, and/or tissue remodeling. The dynamic behavior of cell populations proliferating on 3-D scaffolds and the tissue growth rates determine the construct growth rate. This depends on various transport phenomena, including delivery of oxygen, nutrients, transporting factors, genes and cryoprotectants to cells, clearance of metabolic waste products, and intercellular communication via autocrine and paracrine factors. Transport and migration of cells through scaffolds and spatial organization of tissue structures is a challenging and exciting step in cardiac remolding. Intensive research efforts focus on tissue engineering as a potential tool for enhancing interstitial transport, curing, remodeling, or replacing failing organs, and numerous publications center on this exciting process of man-induced transport in the cardiac tissues.

Cellular Transport

The Membranes

The cell and most of its functional organelles are encased by phospholipid membranes that are essential for the integrity and function of the cell. To fulfill the function of the cell and its organelles, each membrane is embedded with specific functional proteins and lipid elements (e.g., receptors, ion pumps, caveolae) that enable it to perform its unique roles. The membranes are protective (e.g., the nucleous); regulate the transport of ions and molecules into and metabolic or waste products out of the cell or the subcellular domain; allow selective recognition receptivity and signal transduction by embedded transmembrane receptors that bind signaling molecules; provide anchoring sites for the cell to maintain its shape; allow cell or organelle mobility; provide a stable site for the binding and catalysis of enzymes; and regulate fusion of adjacent membranes in the cell and form specialized junctions (e.g., gap junctions). Redistribution of receptors within the cellular plasma membrane, as well as between the plasma membrane and membranes of various cell compartments, represents an important mode of regulating the cellular responsiveness to their cognate agonists.

The cell has an approximately 5-nm thick, bilayered, phospholipidic membrane. The nuclear envelope consists of two separate layers with a 20- to 40-nm wide space between them. The intermembrane entity is connected at points to the rough ER, a membrane enclosed organelle that pervades the entire cytoplasm and provides a means for designated material to reach all parts of the cell. The two nuclear envelope layers are connected at points by nuclear pores. The mitochondrion is also encased by two, inner and outer, phospholipidic layers with an intermembrane space between them. Typically, the outer membrane layer, like the cellular membranes, has a protein to phospholipids ratio of approximately 1:1 (by weight), whereas the ratio in the inner membrane is 3:1, indicating its higher functional activity. The outer membrane contains porin proteins, each with a 2–3 nm internal channel. The inner membrane is highly impermeable, requiring special transmembrane transporters, and its surface area is about five times larger than that of the outer membrane.

Transmembrane Receptors

With very few exceptions, cellular transmembrane receptors are phosphor-proteins made of peptide chains with seven loops that transverse the thickness of the bilayered plasma membrane of the cell. Signal transduction across the plasma membrane is possible through a shift in conformation. When the extracellular receptor recognizes the information-carrying molecule (e.g., a hormone), the whole receptor undergoes a structural shift that affects the intracellular receptor domain. The hormone itself does not pass through the plasma membrane into the cell.

The binding of the hormone by its specific receptor uses noncovalent mechanisms, such as hydrogen bonds, electrostatic forces, hydrophobic forces, and Van der Waals forces. The concentration of the circulating hormone determines the strength of the signal. The hormone-producing cells can store prehormones so as to quickly modify and release them if necessary. Also, the recipient cell can modify the sensitivity of the receptor, by phosphorylation for example. A variation of the number of receptors can change the total signal strength in the recipient cell.

There are generally two types of membrane receptors:

1. G-protein coupled receptors (GPCRs) representing the majority of membrane receptors; these receptors are proteins that bind the guanine nucleotides guanosine diphosphate (GDP) and guanosine triphosphate (GTP). When activated, GDP turns into GTP, which activates an effector, say the enzyme adenylyl cyclase in the plasma membrane, which catalyzes the conversion of ATP to form the second messenger cyclic adenosine monophosphate (cAMP). GTP is involved in energy transfer within the cell and is essential to signal transduction where it is converted to GDP through the action of GTPases. Thus, one GTP molecule is generated for every turn of the citric acid cycle. This is equivalent to the generation of one molecule of ATP since GTP is readily converted to ATP.

   The importance of the GPCRs is highlighted by the wide variety of GPCRs encoded in the human genome. The multiple mechanisms by which they are regulated have catalyzed great efforts to determine these mechanisms so as to enable control by pharmacological intervention of the GPCR function. Protein kinase-mediated desensitization of GPCR is a promising venue.

2. Other receptors, which transfer the signal directly to the intracellular target and demonstrate enzymatic activity or activate other enzymes inside the cell (e.g., phosphorylation of proteins).

The activated effector proteins usually stay close to the membrane or are anchored within the membrane by lipid anchors. Membrane-associated proteins can be activated in turn or come together to form a multiprotein complex that finally sends a signal via a soluble molecule into the cell.

Caveolae, Caveolin, and Lipid Rafts

The caveolae compartments are flax-shaped plasma membrane invaginations organized by caveolins. Caveolae consist of the protein caveolin-1 with a bilayer enriched in cholesterol and glycosphingolipids. When the caveolae lack caveolin expression they are sometimes referred to as lipid rafts. Caveolae and lipid rafts represent membrane compartments enriched in a large number of signaling molecules, many attached to Ca^{2+} signaling, whose structural integrity is essential for many signaling processes.[29] Uptake of extracellular molecules is believed to be specifically mediated via receptors in the caveolae.

Caveolin-1 is an essential structural component of cell surface caveolae, important for regulating trafficking and mobility of these vesicles. Caveolin-1 is found at many intracellular locations. Variations in the subcellular localization are paralleled by a plethora of ascribed functions for this protein in cellular signaling and the development of diseases (e.g., cancer).

Endocytosis

The cells constantly sample their environment by internalizing "samples" of the extracellular fluid and carrying the vacuoles to various intercellular targets. This mechanism is used by all cells because most substances of importance to them are polar and consist of big molecules, which cannot pass through the hydrophobic plasma membrane.

Phagocytosis is a process by which cells ingest large objects, such as cells that have undergone apoptosis or bacteria and viruses. The membrane folds around the object and the object is sealed off into a large vacuole known as a phagosome. Phagocytosis is *not* considered an endocytic process.

Pinocytosis or *endocytosis* is concerned with the uptake of solutes and single molecules, such as proteins. When the receptors for the extracellular molecules are part of this internalization process, we refer to it as receptor-mediated endocytosis. There are two basic types of endocytosis:

1. *Macropinocytosis* is the invagination of the cell membrane to form a pocket which then pinches off into the cell to form a vesicle filled with extracellular fluid and various molecules. The vesicle travels into the cytosol and fuses with other vesicles, such as endosomes and lysosomes. In this case, the cells bring in proteins and other types of ligands attached to the plasma membrane via the receptors.

2. *Clathrin-mediated endocytosis* is the specific uptake of large extracellular molecules, such as proteins, membrane-localized ligand–receptors complexes, and ion channels. These sucked receptors are associated with the formation of the cytosolic protein clathrin-coated pit vesicle by forming a crystalline coat on the inner surface of the cell's membrane. The coat promotes vesicular stability and aids the transport process. The coated pit vesicles on the cytoplasmic surface of the membrane appear as patches on the cell surface and normally coalesce to form an enclosed vesicle at one pole of the cell. The segregation of different cargo in the different clathrin-coated pits and the cargo-dependent control of the clathrin-coated pits dynamics suggest an early functional specialization in the endocytic pathway.[30] Once the vesicle has formed, the clathrin coat is lost by an energy-requiring process. The uncoated vesicles may join with other vesicles to form endosomes or receptosomes.

Adaptins, also named *Adaptor Protein-2*, are specific coat protein subunits of adaptor protein complexes involved in the formation of intracellular transport vesicles and in the selection of cargo for incorporation into the vesicles by trapping specific receptors that are destined for the clathrin-coated pits.[31] The adaptin receptor binding stimulates the attachment of clathrin that drives the formation of the coated pit vesicle. Receptors that do not have the right signal and other types of molecules are allowed to pass through. The adaptin insures that the receptors stay in the coated pit.

The receptors concentrated in the coated vesicles of mammalian cells include some hormone receptors, LDL receptors—which remove LDL from the blood circulation, as well as transfer receptors—which bring ferric ions into the cell. Receptor-mediated endocytosis plays a crucial role in turning off the response of a receptor to an activating signal. Thus, the growth factor receptor tyrosine kinase (RTK), which is a transmembrane protein that spans the plasma membrane just once, is stopped by being engulfed and the ligand receptor destroyed by the receptor-mediated endocytosis, thus preventing uncontrolled mitosis (i.e., cancer). Indeed, a number of the RTKs are proto-oncogenic.

Exocytosis

Exocytosis is performed by all cells and is the reverse of endocytosis. It plays an important role in cell signaling and regulatory functions and is the process by which a cell discharges large biomolecules through its membrane, including newly synthesized membrane proteins that are incorporated in the plasma membrane. Exocytosis is needed for the secretion of proteins (e.g., enzymes, peptide hormones, and antibodies from the cells), placement of integral membrane proteins, recycling of plasma membrane-bound receptors, and releasing neurotransmitters, either excitatory or inhibitory. It is also a mechanism for maintaining the steady turnover of the plasma membrane, restoring the normal membrane mass which is continuously depleted by endocytosis via the membrane-bound vesicles which fuse with the plasma membrane.

The molecules dissolved in the fluid contents of these fused vesicles are secreted into the extracellular fluid. Exocytic vesicles are created from endosomes traversing the cell or bud off from the ER and Golgi apparatus and migrate to the surface of the cell.

Nuclear–Cytoplasm Transport

Typical Nuclear Transport Mechanisms

The genetic system inside the nucleus provides codes for the production of proteins outside the nucleus. This nucleo–cytoplasmic exchange is an activity that requires the allocation of 100–200 different proteins. All nuclear proteins are imported from the cytoplasm, whereas transfer RNAs (tRNAs), messenger RNAs (mRNAs), and ribosomes (rRNs) are made in the nucleus and need to be exported to the cytoplasm. Ribosomal proteins are first imported into the nucleus, assembled in the nucleolus with rRN, and then exported as ribosomal subunits to the cytoplasm. The nuclear export of mRNA apparently relies largely on export mediators distinct from importin β-related factors.

Nuclear or cytoplasmic receptors are soluble proteins localized within the cytoplasm or the nucleus. The nuclear receptors are ligand-activated transcription activators. The typical ligands for nuclear receptors are lipophilic hormones, including steroid hormones (e.g., restosterone, progesterone, and cortisol) and derivatives of vitamin A and D. These hormones play a key role in the regulation of metabolism, organ function, developmental processes, and cell differentiation. The hormone-activated nuclear receptors attach to the DNA at receptor-specific hormone responsive elements. These DNA sequences are located in the promoter region of the genes. The signal strength is determined by the hormone concentration, which is regulated by a variety of mechanisms. These include feedback inhibition of the hormones to avoid overproduction, cell-stored hormones in the cystole for future use, and/or modification of the hormone in the target cell so as to delay or eliminate the triggering of the hormone receptor, thus effectively reducing the amount of available hormones.

Nuclear Pore Complex and Nucleoporins

The nuclear envelope is a double-layered membrane with a 20–40 nm space (the perinuclear space) between the layers. The perinuclear space is connected at points with the ER so as to provide the inroads for nuclear materials to reach all parts of the cell. With only a few exceptions, the molecular traffic across the nuclear boundary takes place through a special type of water-filled channel, the nuclear pore complex (NPC), which spans the dual membrane of the nuclear envelope and facilitates the process of translocating proteins through the hydrophobic membranes. This channel is actually a pump that separates molecules according to their identity. The membrane is perforated by some 3000–5000 NPCs in a proliferating human cell. The NPCs are estimated to be composed of approximately 50–100 different proteins that are often collectively called *nucleoporins*. The majority of the nucleoporins contain numerous short peptide repeats. They function as docking and interaction sites for nuclear transport factors. The NPCs allow diffusion of small molecules, such as metabolites, and can accommodate active transport of very large particles (e.g., export of ribosomal subunits [1.4 and 2.8×10^6 Da]). The passive NPC diffusion channel has an effective diameter of 9 nm. Passive diffusion becomes very inefficient as the size of the molecule approaches this 9 nm limit and is reasonably fast only for proteins of <20 to 30 kDa. Yet, proteins, such as histones or RNAs, that are smaller than 20 to 30 kDa normally cross the NPC in an active and carrier-mediated fashion. Transport of large protein assemblies or ribonucleoproteins through the NPCs is fundamentally different from protein import into the mitochondria, chloroplasts, or the rough ER where proteins cross the membrane singly and in a fully unfolded state. Understanding the NPC transport is crucial to many fields, including viral infection and gene therapy.

Adaptor Molecules: Importin and Exportin Receptors

As seen, a protein must be small enough to pass through the NPCs by diffusion or else be chaperoned by a transport element or carrier through the NPC.[32] The active import of proteins into the nucleus through the NPCs is largely mediated by nuclear transport receptors or carriers of the importin-β family that use direct RanGTP-binding to regulate the interaction with their cargoes. The carrier proteins are known as *karyopherins* (i.e., importins and exportins),[33] and shuttle between the two compartments. Members of the importin-β family can bind and transport cargo by themselves or form heterodimers with importin-α. As part of a heterodimer, importin-β mediates interactions with the NPC and the translocation from the cytoplasm to the nuclear side of the NPC. Binding with nuclear RanGTP releases the importin-β from the NPC. Importin-α acts as an adaptor–receptor protein in the cytoplasm and binds the nuclear localization signal (NLS) and importin-β to the cargo. Importin-α requires a specialized exportin (e.g., CAS) for its re-export.

The RanGTPase system is at the heart of the nuclear transport machinery to regulate the interactions with their cargoes. Ran is a Ras-related member of the small G-protein family that switches between GDP- and GTP-bound forms. RanGTPase helps to control the unidirectional transfer of cargo. Thus, the transport through the NPC is driven by a gradient of RanGTP. The cytoplasm contains primarily RanGDP and the nucleus contains mostly RanGTP. The RanGTP-binding proteins do the loading and unloading in the appropriate compartment and return to the cytoplasm as RanGTP complexes without their cargo. Complexes form between the target proteins, importin-α and importin-β in the cytoplasm, where Ran is in the GDP-bound form. Following transport through the NPC, RanGTP binds to importin-β, releasing importin-α and the target protein in the nucleus. The RanGTP–importin-β complex is then transported back to the cytoplasm where the Ran GTPase-activating protein (RanGAP) stimulates hydrolysis of the bound GTP to form RanGDP. This conversion of RanGTP to RanGDP is accompanied by the release of importin-β.

Exportins operate in the opposite direction and bind their substrates preferentially in the nucleus by forming a trimeric complex with RanGTP. The trimeric complex is transferred to the cytoplasm where it is disassembled. The substrate-free and Ran-free exportin can then re-enter the nucleus and bind and export the next cargo molecule. Examples include tRNA export by exportin-t.

A recent report has identified importin-α16, which recognizes inner nuclear membrane (INM) targeting signals. This finding provides the needed link for the transport of large INM proteins from the cytosolic face of the ER to the INM.[34]

Mitochondrial Transmembrane Transport

In addition to their well-known critical role in energy metabolism, mitochondria represent the chemical workshop where various catabolic and anabolic processes occur and various oxygen–nitrogen reactive species interact; where transport of calcium fluxes and other signal transduction pathways interact to maintain cell homeostasis and to mediate cellular responses to different stimuli.

Mitochondrial Import Transport

Structurally, all mitochondria are engulfed by two hydrophobic phospholipidic bilayer membranes with an intermembrane substance between them. The outer membrane delineates the organelle and is structurally similar to other cell membranes, being rich in cholesterol and permeable to ions. The inner membrane, which isolates the matrix, is virtually devoid of cholesterol, is rich in cardiolipin (which binds the proteins of the electron transport chain), and is impermeable to ions. This impermeability accounts for the generation of the electrochemical gradient that supplies the proton motive force for ATP generation.

Transport across the mitochondrial membranes involves the translocator outer membrane (TOM) and the translocator inner membranes (TIM) complexes.[35] These aqueous pathways are critical to the process of translocating proteins through the hydrophobic membranes. Both TOM and TIM channels are high conductance voltage-dependent channels that are slightly cation selective and are similarly affected by signal peptide sequences. The architecture of TOM recognizes the mitochondrial proteins and translocates them to the intermembrane space and the inner membrane. The architecture of the inner membrane complex allows electron flow and the functioning of ATPSynthase. The intermembrane sets up a membrane potential gradient that drives the entry of matrix and intermembrane proteins. The mitochondrial matrix contains DNA, RNA, and ribosomes, which encode and transfer the code for the desired key proteins. The architecture of the matrix allows the flow of solution to the enzyme complexes.

Proteins that go all the way to the mitochondrial matrix have an NH_2-cleavable signal, a peptide sequence, and are usually denoted as preproteins. Most other proteins must be uncoiled, unfolded, or stretched out to go through the translocators. This procedure involves ATP binding and is monitored and stabilized by chaperone proteins, including the heat stroke protein (HSP) 70.

Transport through the Outer Membrane

The outer mitochondrial membrane contains protein pores (porins) that form the aqueous channels through which proteins of up to 10,000 Da can pass and enter the intermembrane space. The TOM complex includes a number of import receptors (e.g., various TOMs) that recognize the signaling sequence (e.g., peptides). Different proteins use different TOM receptors. The receptors transfer the protein to a complex of proteins, known as the general import pore (GIP), made of TOM40 and others and facilitate the translocation of the presequence of the protein across the outer membrane. TOM40 appears to be the core element of the pore and traverses the membrane as 14-beta strands, which form a beta barrel. TOM40 interacts with the polypeptide chains passing through the pore. Other TOM components in the GIP are anchored to the outer membrane by helical transmembrane segments (hydrophobic anchors).

As with the nucleo–cytoplasmic exchange via the NPCs, TOM40 and other TOMs require cytosolic chaperones to become translocation competent. Specifically, it requires ATP and a partially folded protein state, which is mediated by the cytosolic chaperone (e.g., HSP70). Final insertion into the pre-existing TOM complexes requires an intact N terminus.

Transport through the Inner Membrane

Most proteins cannot get into the matrix unless they pass through the inner membrane, which contains the impermeable cardiolipin. Transport mechanisms across the inner membrane organize and regulate the translocation. Thus, the preproteins with amino terminal signals usually unfold by interacting with TOM20 and are then transferred to the GIP complex, where they interact with other TOMs that usher them on. In the inner membrane, the proteins interact with TIM23 and TIM17 and then enter the matrix via the pore complex. The entry depends on the membrane potential, which is set up by the electron transport complexes and helps pull the protein into the appropriate TIM channels. Pulling the protein into the matrix is ATP energy dependent.

Proteins that do not have a cleavable targeting signal but have an internal signal sequence (e.g., a positively charged amino terminal sequence) generally follow a similar route via the GIP complex and enter the special TIM pathway, requiring chaperons and ATP to reach the matrix. Other proteins interact with small TIMs of the intermembrane space and the inner membrane and do not usually require HSP70 chaperone or ATP for entry.

Electron transport complexes on the inner membrane pump H^+ protons across to the intermembrane space and maintain the matrix in a more negative state, thus creating a charge potential gradient that helps pull the protein into the TIM channels. The protein that enters the matrix has the cleavable preprotein clipped off by a protease and HSP60 (chaperonin) and folds into its tertiary sequence.

Permeability Transition Pore Complex

Mitochondrial membrane permeability transition (MPT) characterizes the permeability changes of the membranes due to the opening of the MPT pores. This increases the osmolar load of the organelles at certain pathological conditions, leading to mitochondrial swelling and cell death. The MPT pore forms when the mitochondria absorbs excess Ca^{2+} by, say, over activation of glutamate receptors or in response to oxidative stress, affecting depolarization and decreasing the voltage gradient. MPT is one of the major causes of cell death and ischemic damage.[36] MPT also allows Ca^{2+}, which can activate harmful calcium-dependent proteases (such as calpain), to leave the mitochondria. Reactive oxygen species (ROS) are also produced as a result of the opening of the MPT pores. MPT allows antioxidant molecules, such as glutathione, to exit the mitochondria, thus reducing the organelles' ability to neutralize ROS. The extent of MPT determines whether the cell may recover or undergo apoptosis or necrosis.

The voltage-dependent anion channel (VDAC) and the adenine nucleotide translocator (ANT) are major players in the MPT pore complex.[37] VDAC changes its structure either by voltage dependence or by interaction with ANT. The VDAC structure stimulates signals at the surface of protein-binding enzymes (e.g., hexokinase, glycerol kinase, and Bax). The VDAC–ANT complex changes conformation into an unspecific channel (uniporter), affecting MPT pore opening. Activity of bound hexokinase protects against Bax binding and employs the ANT as an antiporter. VDAC is sensitive to voltage at 30 mV and plays an important role in the coordination of communication in the management of transient formation of complexes with other proteins. VDAC may form a β-barrel channel composed of 16 β-strands. The pore diameter is 4 nm at a voltage smaller than 30 mV, and the pore is anion selective. Above 30 mV, the diameter of the pore is reduced to 2 nm and ion selectivity changes to cation selectivity. In general, most VDACs are in the low-conductance cation selective state.

It is noted that some specific proteins in the intermembrane space may also be involved in regulating the pore permeability.

Proton Conductance

Proton H^+ conductive pathways exist in the plasma membranes of a variety of cell types. The H^+ selective permeability of the conductive pathway supports large H^+ fluxes. The H^+ conductance is gated by depolarizing voltages and is promoted by modulating the intracellular and extracellular acidification. This facilitates a net H^+ efflux while precluding potentially deleterious H^+ uptake.

Ninety percent of the cellular O_2 utilization occurs in the inner mitochondrial membrane as the terminal electron acceptor in the electron transport chain. The produced electron flow creates a proton gradient that drives the ATP synthesis. During oxidative phosphorylation, most of the proton pumped out to the cytosol across the mitochondrial inner membrane returns to the matrix via the ATP synthase, which drives the ATP synthesis. However, some protons leak back to the matrix through a proton-conductance leakage pathway in the membrane. This is not a simple biophysical leak across the unmodified phospholipids bilayer, but the actual leakage mechanism is still unclear.

The basal proton conductance of mitochondria is an important contributor to metabolic rate changes due to the inner membrane. Changes in phospholipids fatty acyl (liver) composition and proton conductance per unit of mitochondrial protein may be secondary results of the changes in the surface area of the inner membrane and in the activity and concentration of ANT that are required for a higher capacity of mitochondrial ATP production.[37] In some cell types, activation of the conductance is controlled by physiological ligands and by second messengers.

The dependence of proton conductance of the mitochondria on ANT content may be due to a mechanism ensuring that the mitochondria with a higher ANT content and a high capacity for oxidative phosphorylation will also have a higher degree of mild uncoupling by the proton. Mg^{+2} is known to inhibit proton conductance in the mitochondria.[38]

Transmembrane ATPases and ATPsynthase

Transmembrane ATPases import many of the metabolites necessary for cell metabolism and export toxins, wastes, and solutes that can hinder cellular processes. An important example is the sodium–potassium exchanger (or Na^+/K^+ATPase), which establishes the ionic concentration balance that maintains the cell potential. Another example is the hydrogen potassium ATPase (H^+/K^+ATPase) or the gastric proton pump that acidifies the contents of the stomach.

Transmembrane ATPases harness the chemical potential energy of ATP to perform mechanical work by transporting molecules or ions against the concentration gradient. Mitochondrial carriers, such as ADP/ATP carriers, contain a mitochondrial import signal recognized by TIM10 and TIM12 in the intermembrane space. Transport is facilitated by binding to the TIM9.10.12 complex, which interacts with the precursors, and then inserting the carriers into the inner membrane by the TIM 22.54 complex and a potential gradient.

The ATPsynthase of the mitochondria is an anabolic inner membrane-bound enzyme in the form of a two sector (F1 and F0) rotary molecular motor. The F0 complex harnesses the flux of H^+ protons using the energy of a transmembrane proton concentration gradient for catalyzing (in the F1 core) the ATP synthesis by adding a phosphate ion to a molecule of ADP. The H^+ proton flux gives the F0 complex a spinning motion which energizes the creation of ATP. ATPsynthase can also function in reverse by using energy released by ATP hydrolysis to pump protons against their thermodynamic gradient. The catalytic function of the ATPsynthase can be affected by inhibitors.[39]

Mitochondrial Calcium Uniporter

Mitochondria accumulate significant amounts of Ca^{2+} from the cytosol during intracellular Ca^{2+} signaling. Ca^{2+} uptake is undertaken by the mitochondrial Ca^{2+} uniporter[40] located in the inner membrane; Ca^{2+} travels along the electrochemical gradient across this membrane without direct coupling to ATP hydrolysis or transport of other ions. This unique channel enables high Ca^{2+} selectivity. Ca^{2+} uptake controls the rate of energy production, shapes the amplitude and spatiotemporal patterns of intracellular Ca^{2+} signals, and is instrumental in cell death. It is noteworthy that recent studies[41] suggest that the mitochondria is significantly involved in the SR-related Ca^{2+}-induced Ca^{2+}-release (CICR) phenomenon and thus plays an important role in the mechanical energy production that controls cardiac contraction.

Signaling Mechanisms and Pathways

Cell functional behavior is governed by molecular networks that translate the environmental chemical, electrical, optical, or mechanical stimuli (usually via receptor-binding interactions) into intracellular signals. These signals, in turn, affect transcriptional, metabolic, and/or cytoskeletal processes that modulate cellular response and the ensuing integrated organ function.[25]

Cell Signaling

The signaling molecule is a chemical which transmits information between cells. These molecules, which originate from a donor cell, infusion, or injection, migrate by convection and/or diffusion (alone or assisted by a carrier) toward the target cell(s) and interact with the appropriate receptors in the target cells; this triggers a single or chain reaction in the cell. The cells respond to signaling molecules in the extracellular fluid (e.g., hormones from distant locations in multicellular organisms [endocrine signaling]), stimulation from nearby cells (paracrine cytokines), or stimulation from their own secretions (autocrine stimulation). The cells may also respond to molecules on the surface of adjacent cells (e.g., contact inhibition).

Signaling molecules may trigger a number of changes including:

1. a change in the metabolism of the cell (e.g., increased glycogenolysis and heart beat when the cell detects adrenaline);
2. a change in the electrical charge across the plasma membrane (e.g., the action potentials);
3. a change in gene expression (transcription)
a) by lipophylic steroids or nitric oxide (NO) diffusion into the nucleus and binding to internal receptors and/or
b) by the signal transducers and activators of transcription (STATs), which constitute a family of seven transcription factors regulating a multitude of cellular functions, such as growth, proliferation, differentiation, and apoptosis. In response to stimulation by extracellular signaling proteins, such as cytokines and growth factors, the STATs are transiently activated by phosphorylation through receptor-bound Janus kinases (JAKs) and accumulate rapidly in the nucleus where they switch on expression of their target genes.[42]

Cell signaling studies involve the spatial and temporal dynamics of the receptors and the components of signaling pathways that are activated by receptors in various cell types. It is interesting to determine how cells distinguish between adaptive and maladaptive signals when they appear to share the same molecular pathways. The ability to identify distinctive features of a pathophysiological response compared with a physiological response would allow for rational approaches to eliminate or diminish undesirable consequences of the former while preserving the beneficial effects of signals from the latter. These issues are especially pertinent in the heart and in understanding the development of heart failure.

Signaling Complexes

Cell signaling depends on the physical location of the proteins involved. Biophysical constraints imposed by macromolecular crowding and diffusion have a significant influence on the evolution of cell signaling pathways. Many of the protein kinases, protein phosphatases, transmembrane receptors, and others that carry and process messages inside a living cell are associated with signaling complexes[43] formed by compact clusters of molecules attached to cell membranes or the cytoskeleton. Proteins serve as adaptors, anchors, and scaffolds for signaling complexes. For example, the receptor complexes form when active and usually dissociate when inactive. Diffusion is important in controlling the numbers and locations of the signaling complexes. These clusters are important conceptually because they provide intermediate levels of organization and affect intracellular signal pathways.

Signal-mediated Transport

Active transport is a selective process triggered by specific transport signals (e.g., the NLS).[44] NLS is an amino acid sequence, which acts like a tag on the exposed surface of a protein. This sequence is used to convey the protein to the cell nucleus through the NPC and to direct a newly synthesized protein into the nucleus via its recognition by cytosolic nuclear transport receptors. Typically, this signal consists of a few short sequences of positively charged lysines or arginines. Different nuclear localized proteins may share the same NLS. Identified NLSs include the NLS-receptor importin-α, importin-β, Ran,[31] and nuclear transport factor 2. Importin-α accounts for NLS recognition and binds importin-β. Importin-β, in turn, mediates the interaction with the NPC of an entire superfamily of importin-β-related factors that mediate nuclear transport along pathways that are distinct from classical NLS import.

Signal Transduction

Most of the different molecules that buffer the cell's surface interact with specific receptors in the cell membrane and transduce the stimuli into the cytoplasm; some molecules, however, induce the opening of ion channels on the membrane. Strictly speaking, signal transduction refers only to the step that converts the *extra*cellular signal to an *intra*cellular one by binding the extracellular signaling molecules to receptors on the transmembrane proteins that face outward from the membrane. This binding consequently triggers intracellular events via the receptor effector site, which extends into the cytoplasm. These transductions take place via a change in shape or conformation of the receptor. Signal transduction thus relates to the process by which a receiving cell converts one type of signal or stimulus into another, which is then often followed by a sequence of biochemical reactions inside the cell that are carried out by enzymes and usually linked through second messengers. This chain of steps is referred to as a *signaling cascade* or a *second messenger pathway* whereby a small stimulus elicits a large response.

Complex multicomponent signal transduction pathways inside the cell provide opportunities for feedback, signal amplification, and interactions between multiple signals and signaling pathways. Some signaling transduction pathway responses depend on the amount of signaling molecules received by the cell. For example, the Hedgehog cell signaling[45] activates different genes depending on the amount of Hedgehog protein present.

Extracellular Signaling

Most extracellular chemical signals are carried in the blood as specific individual molecules or as carrier-bound molecules. The signaling molecules must be recognized by the extracellular end of the membrane-bound receptors on the target cell. This selective matching ensures that crucial activities occur in the right cells at the right time and in synchrony with the other cells.

Lipophylic steroids and NO represent examples of extracellular signals that can cross the cell membrane and permeate the cell. However, most extracellular chemical signals are hydrophilic and are unable to penetrate the lipid membrane that surrounds the cell. Common extracellular signals are nutrients providing energy or structural molecules that cannot be synthesized in the cell. In complex organisms this includes ligands responsible for sensations of smell and taste.

Another mode of signaling involves the extracellular bead-like vesicles composed of proteins and lipids that can act either on close-by receptors or, like endocrine hormones signaling, at a distance from their origin.

Intercellular Signaling

Understanding cell–cell communication, cell polarity, and cellular compartmentalization is one of the central problems in modern cell biology. Intercellular signaling complements extracellular signaling and is commonly subdivided into four types:

- endocrine signals, usually hormones, which are produced by the endocrine cells and travel through the blood to reach all parts of the body;
- paracrine signals, usually by cytokines, which target only cells in the vicinity of the emitting cell (e.g., neurotransmitters). Gap junctions between adjacent cells in cardiac muscle allow for action-potential propagation from the cardiac pacemaker region of the heart to spread and coordinate contraction of the heart;
- autocrine signal stimulation, which affects only cells that are of the same cell type as the emitting cell;

- juxtacrine signals, which are transmitted along cell membranes via protein or lipid components and are capable of affecting either the emitting cell or cells immediately adjacent.

Some signaling molecules can function as hormones as well as neurotransmitters. For example, epinephrine and norepinephrine can function as hormones when released from the adrenal gland and transported to the heart by way of the blood stream. Norepinephrine can also be produced by neurons and function as a neurotransmitter in the brain.

Intracellular Signaling and Transport

The intracellular events triggered by the external signals are sometimes distinct from a transduction event (i.e., the conversion of the extracellular signal to an intracellular one). However, intracellular signaling molecules, often denoted second messengers, include activation of heterotrimeric G-proteins, small GTPases, cyclic nucleotides (such as cAMP and cyclic guanosine monophosphate [cGMP]), calcium ions, some organic phosphate derivatives, diacylglycerol and inositol-triphosphate, phosphatases, and various protein kinases that catalyze phosphorylation. These modify the activity of some proteins and affect binding to other proteins. A distinct signaling system may be activated or inhibited via the ATP-dependent ubiquitin–proteasome pathway. It is noted that the phosphates are important in signal transduction since they regulate the proteins they are attached to. Removing the phosphate by hydrolysis or protein phosphatases can reverse their regulatory effect.

The existence of a large number of receptors coupled to the heterotrimeric guanine nucleotide-binding proteins (G-proteins) raises the question of how a particular receptor selectively regulates specific targets. Electrophysiological recordings demonstrate highly localized signal transduction from the receptor to the channel. This signaling complex provides a mechanism that ensures specific and rapid signaling by a GPCR.

Sorting Signals

Edward B. Lewis, Christiane Nüsslein-Volhard, and Eric F. Wieschaus were awarded the 1995 Nobel prize for identifying genetic mutations of genes involved in the development of body segmentation. The sorting signals are mostly simple peptides with complex functions. They usually consist of short linear sequences of amino acid that are attached to transported proteins and selectively guide the distribution of the proteins to specific cellular compartments. For example, sorting of transmembrane proteins to endosomes and lysosomes is mediated by signals present within the cytosolic domains of the proteins.

In addition to peptide motifs, ubiquitination of cytosolic lysine residues also serves as a signal for sorting at various stages of the endosomal–lysosomal system.[46] Regarding connexin 43 (Cx43) gap channel degradation, it is suggested that a tyrosine-based sorting signal (YKLV) in the C terminus of Cx43 controls gap junction turnover by affecting internalization and targeting Cx43 for degradation in the endosomal–lysosomal compartment. This complex array of signals and recognition proteins ensures the dynamic but accurate distribution of transmembrane proteins to different compartments of the endosomal–lysosomal system. Phosphorylation events also regulate signal recognition.[46] A large number of cell sorting signal servers are commercially available for predicting protein localization sites in cells based on amino acid sequences and origins.

Hedgehog Signaling

The Hedgehog signaling pathway[45] is one of the key regulators of animal development. It takes its name from its polypeptide ligand (an intercellular signaling molecule), which is one of the segment polarity gene products. It exists in three homologs in mammals, the most studied being the Sonic Hedgehog whose pathway is important during embryonic development. The absence of the Sonic Hedgehog signaling pathway causes faulty development and has also been implicated in some cancer developments. Recent studies point to the role of Hedgehog signaling in regulating adult stem cells involved in maintenance and regulation of adult tissue.

Steroid Signaling

Steroids are small hydrophobic molecules that can freely diffuse across the plasma membrane, through the cytosol, and into the nucleus. Quabain, a steroid ligand of the Na/K–ATPase plasma protein, binds specifically to the integral plasma membrane protein that establishes the electrochemical gradient across the plasma membrane and activates a Ca^{2+} oscillatory signaling pathway. Cardiac steroids are specific inhibitors of Na^+/K–ATPase activity and are a major determinant of the Na^+ and K^+ electrochemical gradient. Cardiac steroids are thus important in regulating cell volume, cytoplasmic pH, and plasma membrane electric potential and are physiological regulators of recycling endocytosed membrane proteins and cargo.[47]

Natural and synthetic glucocorticoids (GC) are widely used in the therapy of GC-responsive diseases. The effects of GC are mediated by specific nuclear receptors. Steroids that regulate gene expression include: GCs (e.g., cortisol), mineralo-corticoids (e.g., aldosterone), and sex hormones (e.g., estradiol, progesterone, testosterone). Steroid receptors reside within the nucleus, except for the GC receptor (GR), which resides in the cytosol. Recent findings[48] show the existence of additional plasma membrane-located steroid receptors, which are termed *steroid hormone recognition and effector complex* (SHREC). SHREC are thought to be responsible for rapid nongenomic responses and play a pivotal role in the complex network of plasma-related nongenomic responses that lead to GR-mediated genomic effects.

Nitric Oxide Signaling

NO is a free radical which diffuses through the plasma membrane and affects nearby cells.[49] NO binds to protein receptors with a metal ion or cysteine in the cell and is quickly consumed. NO is made from arginine and oxygen by the enzyme NOsynthase, with citrulline as a byproduct. NO, an endogenous signaling molecule, works mainly through activation of its target receptor, the enzyme-soluble guanylate cyclase, which when activated produces the second messenger cGMP. cGMP-elevating drugs, such as glyceryl trinitrate, have been used for more than 100 years to treat cardiovascular diseases! Evidence suggests that cGMP-dependent protein kinases (cGKs) are major mediators of cGMP signaling in the cardiovascular system, particularly in regulating vascular remodeling and thrombosis. Consequently, cGKs are novel drug targets for the treatment of human cardiovascular disorders. NO can also act through covalent modification of proteins or their metal cofactors. NO is toxic in high concentrations and is thought to be responsible for some damage after a stroke. Signaling cascades initiated by NO and natriuretic peptides play an important role in the maintenance of cardiovascular homeostasis. NO serves multiple functions, including relaxation of blood vessels, regulation of exocytosis of neurotransmitters, modulation of the hair cycle, cellular immune response, and production, as well as maintenance of penile erections.

Signaling Hormones

Hormones represent a large body of molecules that promote signaling between cells. Once recognized by the appropriate receptor, they lead to conformational changes that can lead to downregulation. Hormones and other signaling molecules may exit the sending cell by exocytosis or other means of membrane transport. As already noted, there are two classes of hormone receptors: membrane-associated receptors and intracellular or cytoplasmic receptors, the latter responding to hormones that can cross the membrane intact.

Hormone signaling is elaborate and hard to dissect. The reason for this complicated phenomenon is that while the donor cell is typically of a specialized type, its recipients may be of one or more types. Thus, a cell can have several different receptors that recognize the same hormone but activate different signal transduction pathways; or different hormones and their receptors can invoke the same biochemical pathway. Furthermore, different tissue types can answer differently to the same hormone stimulus. Also, a hormone-activated receptor can activate many downstream effector proteins. This signal amplification can be deactivated by deactivating the activated receptor, either by phosphorylation or by internalization via ubiquitin processing.

Ca²⁺ Signaling

Calcium is the most universal carrier of signals to cells and has a unique flexibility as a ligand, from fertilization to apoptotic suicide at the end of life cycle. It is unique in its ability to perform both as a first messenger and as a second messenger function and plays a crucial role in autoregulatory mechanisms (e.g., the crossbridge recruitment in the sarcomere's contraction and its response to changes in the afterload).[50] Thus, Ca^{2+} appears in a multitude of processes, among them muscle contraction, release of neurotransmitter, vision in retina cells, proliferation, secretion, cytoskeleton management, cell migration, gene expression, and metabolism. The three main pathways that lead to Ca^{2+} activation are: 1) G-protein-regulated pathways; 2) pathways regulated by RTKs; 3) ligand- or current-regulated ion channels.

Ca^{2+} can regulate proteins by the direct recognition of Ca^{2+} by the protein or by Ca^{2+} binding in the active site of the enzyme. One of the best studied interactions is the regulation of calmodulin protein by Ca^{2+}. The Ca^{2+}/calmodulin complex[51] plays an important role in proliferation, mitosis, and neural signal transduction. L-type Ca^{2+} channels play a critical role in regulating Ca^{2+}-dependent signaling in cardiac myocytes, including excitation–contraction coupling. A subpopulation of L-type Ca^{2+} channels is localized in caveolae in ventricular myocytes as part of a macromolecular signaling complex necessary for β_2 adrenergic receptor regulation of I_{caLs}, the current density of the L-type Ca^{2+} channels.

Signals Crosstalk

The signaling elements discussed here represent only part of the major players in the cardiac system,

and even the description of these is greatly oversimplified because many of these signaling pathways interconnect with one another, creating "crosstalk." This crosstalk is probably essential for precisely modulating the genetic response to a variety of ligands reaching the cell at the same time and in varying intensities.

Receptors

Adrenergic Receptors

Major factors contributing to worsening heart failure include a number of compensatory neurohormonal signals intended to counteract decreased cardiac output.

Adrenegic signaling in the myocardium contributes to the control of heart rate (chronotropy), strength of contraction (inotropy), and rate of relaxation (lusitropy) by changing the levels of intracellular Ca^{2+} or by altering the sensitivity of critical regulatory proteins to Ca^{2+}. The β-adrenergic signaling cascade and its clinical implications in arrythmias and sudden cardiac death were recently reviewed in detail by Rudy and Silvia.[52] Signaling is mediated predominantly by two distinct β-adrenergic receptors, β_1 and β_2, which differ in their abundance, distribution, and downstream signal transducers. Approximately 75% of the cardiac β-adrenergic receptors are β_1, which appear to be distributed globally throughout the sarcolemma. β_1 receptors couple to the G_s heterotrimeric G-protein. The less abundant β_2 receptors reside predominantly in the caveolae.[53] Besides their distinct homes, β_2 receptors also differ from β_1 receptors in that they couple to both G_s and G_i. Nevertheless, stimulation of either β_1 or β_2 activates adenylyl cyclase to increase intracellular cAMP. In turn, cAMP activates protein kinase A (PKA) resulting in the phosphorylation of key elements of the contractile apparatus and of proteins that control internal Ca^{2+} levels.

Ryanodine Receptors

Prominent among the PKA targets are the L-type voltage-gated Ca^{2+} channels ($Ca_V 1.2$), which open upon membrane depolarization and allow Ca^{2+} to enter the cell. PKA phosphorylation of L-type Ca^{2+} channels potentiates inward Ca^{2+} current and thereby augments contraction.[54] The receptors for this Ca^{2+} signal are ryanodine receptors (RyR), which form a class of calcium channels in various forms of muscle and other excitable animal tissue (RyR1 is expressed in skeletal muscle, and RyR3 is expressed especially in the brain). RyR2, which are the Ca^{2+} release channels on the SR in the myocardium, flood the cytoplasm with additional Ca^{2+}. Thus, the (RyR2) receptors are stimulated to transport Ca^{2+} into the cytosol by recognizing Ca^{2+} on its cytosolic side and establishing a positive feedback mechanism; a small amount of Ca^{2+} in the cytosol near the receptor will cause it to release even more Ca^{2+} that then initiates cardiac contraction. Ca^{2+} is the major cellular mediator of CICR in animal cells.

Depletion of the channel-stabilizing protein calstabin2 (FKBP12.6) from the RyR2 complex causes an intracellular Ca^{2+} leak that can trigger fatal cardiac arrhythmias. A derivative of 1,4-benzothiazepine (JTV519) increases the affinity of calstabin2 for RyR2, which stabilizes the closed state of RyR2 and prevents the Ca^{2+} leak that triggers arrhythmias. Thus, enhancing the binding of calstabin2 to RyR2 may be a therapeutic strategy for common ventricular arrhythmias.[54]

Cytokine Receptors

Cytokines are a group of signaling compounds which represent a diverse class of small water-soluble proteins and glycoproteins. Cytokines are produced by a wide variety of cell types and can have effects on nearby cells or effects throughout the organism; like hormones and neurotransmitters, they are used extensively in intercellular communication. Each cytokine binds to a specific cell surface receptor, which then signals the cell via second messengers (often tyrosine kinases) to alter its function: upregulation and/or downregulation of several genes and their transcription factors, in turn resulting in the production of other cytokines, an increasing number of surface receptors for other molecules, or suppressing their own effect by feedback inhibition.

Cytokines are critical to the functioning of both innate and adaptive immune responses. They play a major role in a variety of immunological, inflammatory, and infectious diseases; a deficiency of cytokine receptors is directly linked to certain debilitating immunodeficiency states. Cytokines are also involved in several developmental processes during embryogenesis and myocardial dysfunction in heart failure development, and LV remodeling. Many cytokines can share similar functions. There are dozens of cytokine receptors, and most fall into one of two major families:

1) RTKs[55]—the human genome encodes 90 different tyrosine kinases. Some of the cytosolic tyrosine kinases act directly on gene transcription by entering the nucleus and transferring their phosphate to transcription factors. Transmembrane protein receptors span the plasma membrane just once and are triggered by ligands, such as mitogen, insulin, vascular endothelial or platelet derived growth factors, and others. Others act indirectly through the production of second messengers. This is achieved by binding of the ligand to two adjacent receptors to form an active dimmer, a tyrosine kinase enzyme that attaches phosphate groups to certain tyrosine residues; a cascade of expanding phosphorylations then occurs within the cytosol. Turning RTKs off is done by quickly engulfing and destroying the ligand–receptor complex by receptor-mediated endocytosis. It is noteworthy that the mitogen-activated protein kinase pathway is one of the most important and intensively studied signaling pathways, being part of a network that governs the growth, proliferation, differentiation, and survival of most cell types. It is deregulated in a wide range of diseases and thus represents an important experimental and theoretical drag target.[55]

2) *JAK-STAT generating receptors*[56]—these receptors consist of two identical single-pass transmembrane proteins (i.e., homodimers) embedded in the plasma membrane. Each of their cytoplasmic ends binds a molecule of a JAK, a member of the kinase family. Ligands that trigger JAK-STAT pathways are interleukins, growth hormone, erythropoietin, and others. Binding of the ligand activates the JAK molecules, which phosphorylate certain tyrosine residues as well as STAT proteins. These, in turn, form dimers, which enter the nucleus and bind to specific DNA sequences in the promoters of genes that begin transcription. As the JAK-STAT pathways are shorter than the RTKs, the effect of these ligands is more rapid.

Ion Channel Receptors

The normal electrophysiologic behavior of the heart is determined by ordered propagation of excitatory stimuli that result in rapid and slow repolarization, thereby generating action potentials in individual myocytes. An ion channel can open when the receptor is activated by a change in the cell's transmembrane potential gradient. When such a membrane potential change occurs, the ligand-activated ion channel opens and allows ions to pass through the voltage-gated ion channels. In neurons, this mechanism underlies the action potential impulses along nerves. Abnormalities of impulse generation, propagation, or the duration and configuration of individual cardiac action potentials form the basis of disorders of cardiac rhythm. An extensive analytical review of the cardiac ion channels and cell electrophysiology was recently reviewed by Rudy and Silva.[52]

Cardiac M(3) Receptors

The M(3) subtypes of the muscarinic acetylcholine receptors M(3)-mAChR[57] are proteins that regulate physiological processes in a variety of cell types and play important roles in regulating and maintenance of cardiac functions, such as heart rate and repolarization, and in generation and progression of heart diseases (e.g., atrial fibrillation, contraction of smooth muscle, and adhesion of epithelial cells). Abnormal signaling of these proteins may contribute to asthma, hypertension, cardiomyopathy, and cancer. Signal transduction mechanisms underlying these pathophysiological functions suggest that M(3)-mAChR activates a delayed rectifying K^+ current to participate in cardiac repolarization and negative chronotropic actions; M(3)-mAChR also suppresses ischemic dysrhythmias as well as facilitates atrial fibrillation actions. M(3)-mAChR interacts with gap-junctional channel Cx43 to maintain cell–cell communication and excitation propagation, regulates intracellular phosphoinositide hydrolysis to improve cardiac contraction and hemodynamic function, activates antiapoptotic signaling molecules, enhances endogenous antioxidant capacity, and diminishes intracellular Ca^{2+} overload, all of which contribute to protecting the heart against ischemic injuries.

Regulation of Gene Expression

Genetic Expression and Protein Synthesis

The genetic code is not the only player in the cell. Although the rest of the cell inheritance machinery is not coded, the germline cells are just as immortal as the DNA because of the endless process of cell divisions. The inherited cellular elements include the uncoded phosphor-lipid membranes of the cell and the cell's organelles as well as other chemicals specific to the cellular compartments. There will be no cell without them. The actual network that must be inherited to maintain the species is thus a gene–protein–lipid–cell network, not just the gene network.[58] The large number of different cell types that must maintain a harmonious interaction in the organism indicates the enormous complexity of our inheritance. As stated by Denis Noble,[58] it took 2 billion years of experimentations to achieve this harmony, so even if it is not perfect, it works. We limit our discussion here to what we can study best (i.e., the genetic code).

Genes are expressed as proteins, which are involved in all processes of life as enzymes, transcription factors, or other regulators of metabolic activity. The fundamental principles for the regulation of gene expression were identified more than 40 years ago by the Nobel Laureates François Jacob and Jacques Monod: The genome operates by sending instructions, contained in the DNA in the nucleus of the cell, by mRNA for the manufacture of proteins by the protein synthesizing machinery in the cytoplasm. The flow of genetic information from DNA via mRNA to protein has long been the central dogma of molecular biology.

The human genome consists of approximately 30,000 genes. The selection of genes to be expressed (i.e., govern the synthesis of new proteins) is controlled by molecular signals that determine when and how often a given gene is transcribed (i.e., activate the transcription machinery that copies DNA to mRNA by cell-type specific components of the transcription machinery). As noted below, this process of gene expression can be modulated by various factors which can act at different steps in the gene expression pathway.

An initial extracellular stimulus can trigger, via signal transduction, the expression of an entire suite of genes and a myriad of physiological events. The protein product encoded by one gene often regulates expression of other genes. The time delay to reach an effective level to control the next promoter after activation of the first promoter depends on the rate of protein accumulation. Furthermore, each of the very many different cell types in the higher eukaryote species—some 200+ in humans—synthesizes a different protein. In contrast, and with exceptions, the genes encoding these different proteins are identical in all the different cell types. Consequently, gene expression must be regulated so that different genes within the DNA are active in producing the proteins in the different cell types.[59] Furthermore, complex control mechanisms orchestrate these processes. For example, control of gene expression in eukaryotic cells is known to occur at the levels of chromatin structure, transcriptional initiation, processing and stability, translational initiation, post-translational modification, protein stability, and others. A recently identified gene expression control mode involves microRNAs, and their activity in relation to cardiogenesis and cardiovascular pathophysiology is reported here by Catalucci, Latronico, and Condorelli.[60] Some other interesting control mechanisms involving the interaction of signaling and transport phenomena in the regulation of the genetic expression are noted below.

Gene Control by Selective Protein Degradation: Ubiquitin

Selective degradation of proteins, which is primarily conducted by the ubiquitin–proteosome system, was highlighted in 2004 by the Nobel prize awarded to A. Chechanover, A. Hershko, and I. Rose. Accordingly, a protein marker, the ubiquitin, attaches to the protein doomed to be destroyed. The marked complex then moves to the proteasomes, the molecular disassembly complex, where the ubiquitin is released and the doomed protein is broken down, allowing its subunits to be reused. This process is well known in cell division DNA repair and in parts of the immune system, with cystic fibrosis and cancer as the undesired diseases.

RNA Interference

A. Fire and G. Mellow's 1998 report on RNA interference and gene silencing brought them the 2006 Nobel prize. The double-stranded pairs of RNA activate biochemical machinery, which degrades the mRNA molecules that carry a genetic code identical to that of the double-stranded RNA. Hundreds of genes in our genome encode small RNA molecules. These microRNA molecules can form a double-stranded structure and activate the RNA interference (iRNA) machinery to block protein synthesis.[60] Consequently, the genetic signaling and marking is culminated by silencing a selected desired–undesired gene. iRNA is a natural phenomenon, which is important for the natural regulation of gene expression in the development of the organism and the control of cellular health care including the defense against viral infections and keeping unruly genes under control. Commercial or homemade libraries of double-stranded RNAs are now used to silence the expression of any gene they choose. This greatly extends the ability to determine the function of individual genes and analyze the complex pathways. Plasmid-based iRNA is also commercially available for experimental studies in human, mouse, and rat genes; executed properly, it may lead to novel future therapies.

Controlling Gene Expression: Wnt Signaling

Wnt constitutes a large family of secreted and hydrophobic glycoproteins that control a variety of embryonic developmental and adult regulatory processes in all metazoan organisms. Distinct endocytic pathways correlate with the specificity of Wnt signaling events.[61] Wnt activates multiple intracellular cascades affecting the regulation of cellular proliferation, migration, and polarity. Yet, it is not clear how Wnt activates these pathways after it binds to the Frizzled receptors. These receptors, like the GPCRs, are transmembrane proteins that wind seven times back and forth through the plasma membrane and are activated by Wnt proteins. The signal is subsequently transduced

through several cytoplasmic components to β-catenin, which enters the nucleus and activates the transcription of several genes important in development. β-catenin molecules connect actin filaments to the cadherins that make up adherens junctions that bind cells together. Any excess β-catenin is quickly destroyed by a multiprotein degradation complex that phosphorylates β-catenin so that ubiquitin molecules can attach to it to prepare it for destruction in proteasomes.

Another pathway, initiated by the binding of a Wnt ligand to the Frizzled receptor, inhibits the β-catenin degradation complex so that β-catenin escapes destruction by proteasomes and is free to enter the nucleus where it binds to the promoters of its target genes.

Gene Therapy

The ultimate goal of the genome project is its application in gene therapy (i.e., genetic manipulations that will assure that desired genes replace pathology affecting ones, thus enabling healing of some ills that affect the cardiac system and result in early mortality). Genetic engineering is the art and science of modulating a particular gene, or a group of genes, so as to affect new characteristics in the cell. For example, new transport technologies, such as positional cloning, have been developed to identify the specific gene mutations that cause familial aortic aneurysm disease, which represents some 20% of all aortic aneurysm cases. Another target is the genetic form of dilated cardiomyopathy, which accounts for over 25% of idiopathic human cases. Transvascular gene transfer by viral vectors, such as adeno-associated virus, is a promising approach to improve cardiac function. Another promising approach involves synthesizing messenger RNA to be capable of manipulating genetic abnormalities and combating infectious agents and genes responsible for specific pathologies.

A clearer insight into the genetic maze is needed to modulate the complex genome system. The technological challenges are amplified by the inherent challenges imposed by the genome. While only some 5000 out of 30,000 human genes are functionally defined, we face the added complexity that the same genes can code several proteins while a number of genes can contribute to the expression of the same functional characteristic. Our goal, to treat the cause rather than the effect, is still an ideal to be fulfilled.

Immunosuppressive Gene Therapy

Allograft-targeted immunosuppressive gene therapy may inhibit recipient immune activation and provide an alternative to systemic immunosuppression. The therapy involves the inhibition of immune reactions by shifting the cytokine balance using cytokine inhibitors or inhibitory cytokines that dampen Th1 proinflammatory reactions. In addition, researchers have used proapoptotic molecules, such as FAS-L, an antibody that initiates cell death, or inhibitors of accessory molecules, such as CTLA4 Ig. The optimal technique and efficacy of intracoronary gene transfer involves viral interleukin-10 and human transforming growth factor-beta1 in (rabbit) heterotopic heart transplantation.[62] The emerging disciplines of tissue engineering (e.g., hepatocytes secreting insulin) and immunosuppressive gene therapy are of great potential in inflammatory diseases.

Cardioprotection

Cardioprotection (CP) against ischemic reperfusion has been enhanced by studies of preconditioning, whereby brief periods of ischemia protect the heart against subsequent prolonged periods without oxygen. Molecular mechanisms of ischemic preconditioning CP (ICP) act to reduce the infarct size resulting from sustained ischemia–reperfusion by preceding brief episodes of ischemia and reperfusion.

Redundant mechanisms are common in biological systems, and a number of factors are believed to confer protection against ischemia–reperfusion injury. ICP is a receptor-mediated process triggered by cell signaling and signaling pathways, either parallel or in series that have been shown to mediate CP. An attempt to map the preconditioning signaling pathways has been reported by Downey *et al.*[63] As noted by them, the temporal sequence of events is critical to the ICP mechanism and different components act differently at different times.

Improved CP after exercise was associated with a significant increase in HSP72. Proteins, including mitochondrial Cx43, are imported into the mitochondria via the TOM receptor complex (e.g., TOM20) by HSPs, such as HSP90.[64,65] TIM complexes direct the proteins to their final destination in the inner membrane or the mitochondrial matrix. The localization of Cx43 at the inner mitochondrial membrane seems central for ischemic and pharmacological preconditioning CP. A normal protein level of Cx43 is important for CP, and Cx43 has been suggested to act as an end effector in the signaling cascade of ICP. Cx43 may be involved in the gating of the K_{ATP} channel, in regulating the formation of ROS in response to diazoxide and in modulating the amount

of molecules triggering ICP. A genetic reduction of Cx43 abolishes both pharmacological and ICP. It is strongly suggested the ICP prevents the formation of mitochondrial permeability transition (MPT) pores. While ICP is effective in healthy hearts of most animal species tested to date, there is evidence that it may not be operative in post-myocardial infarction or failing hearts associated with a reduction in myocardial Cx43 expression.

Corticosteroids are cardioprotective, protecting cardiac tissue from damage following a heart attack. This protection occurs possibly through binding to the glucocorticoid receptor (a steroid hormone receptor), which acts as a ligand-dependent transcription factor. Some of the consequences include reduced expression of cell adhesion molecules, decreased neutrophil infiltration, and increased NO, which diffuses into surrounding tissues to prevent clotting and to cause vasodilatation. Research efforts aim to retain the cardioprotective activities of glucocorticoid responses without inhibiting wound healing

Reoxygenation after prolonged hypoxia reduces the ROS threshold for the MPT in the cardiomyocytes; cell survival is inversely correlated with the fraction of depolarized mitochondria. Cell protection by preconditioning results from triggered mitochondrial swelling that causes enhanced substrate oxidation and ROS production, leading to redox activation of protein kinase C, which inhibits glycogen synthase kinase-3β (GSK-3β). The inhibition of GSK-3β on the end effector, the permeability transition pore complex, to limit the MPT induction is suggested as the general mechanism of cardiomyocyte protection.

Many cardioprotector drugs concentrate on the inhibition of the enzyme GSK-3β. Early activation of mitochondria-regulated cell death pathway can reduce ischemia–reperfusion cell death. GSK-3β is involved in the control of glycogen metabolism. Sollott *et al.*[66] have shown that CP agents affect phosphorylation and inhibition of mitochondrial GSK-3β, which in turn affect inhibition of apoptosis and the MPT pore, the end effector of the protection signaling.

Modeling and Control

The Multi-scale Control of the Cardiovascular System

The performance of the cardiac system is regulated by an amazing network of interactions between the systems that adjust the cardiac ventricles to the continuous changes in the internal and/or external environmental stresses and the consequent varying needs of the body. For example,[67] the circulation system supplies the cells with oxygen and nutrients and removes carbon dioxide and other catabolites. Going from the cardiovascular system scale to the cellular scale, the variables involved in cardiovascular regulation, such as blood flow, blood pressure, oxygen blood concentration, and ATP concentration, are kept around their reference points by different feedback control mechanisms with different dynamics in each of these scales. Thus, the short-term control of blood flow and pressure on the cardiovascular system scale is performed by the autonomic nervous system through baroreceptor control loops. Longer horizons with slower control mechanisms usually operate by hormonal regulation. The inter- and intracellular controls are, in general terms, based on ionic transfer rates.

Model-based Signal Processing

The model-based signal processing approach is used to estimate the classical discrete-time feedback loop sensitivities of practical interest. The objective is to assess the short-term control of the cardiovascular system. The control viewpoint is useful in accounting for macroscopic properties (such as the Starling law or Hill force–velocity relation) on lower scales and defining performance indexes of the electro–mechanical coupling on each scale.

Mechano–energetics models on the cardiovascular system scale provide constitutive laws for the cardiac tissue, using 3-D models image processing for computing stress, strain, and action potential fields in the heart. Clearly, the appropriate model must relate to the fact that the cardiac muscle adapts its function and energy consumption to changes in the afterload and the preload and must link the ventricular function to the collective behavior of actin–myosin molecular motors converting chemical energy into mechanical energy. The control mechanisms in the different scales must be synchronized to assert life. Models of the electro–mechanical activity of the cardiac muscle provide tools for control analysis and signal processing applications.

Analytical and Computational Approach

Molecular networks are integral elements within individual cells and organisms. A pivotal theme in modern cell biology is deciphering complex systems, such as transcription schemes, metabolic networks, and signaling pathways. An excellent insight into the analytical complexities involved is presented by J.B. Bassingthwaighte and K.C. Vinnakota.[67] An extensive computational study of the cardiac ion channels cell electrophysiology is given by Y. Rudy and J.R. Silva.[52]

Last century brought understanding of the molecular basis of life. Given the limitations of test tube approaches with regards to a "random chemistry" scenario, and as computational tools for the reconstruction of molecular interactions improve rapidly, it may eventually be possible to perform appropriate computer-based simulations of prebiotic evolution. An interesting computational attempt to unveil the origin of life by computer simulations of realistic complex prebiotic chemical networks[68] is based on a novel algorithmic approach constituting a hybrid of molecular dynamics and stochastic chemistry (http://ool,weizmann.ac.il/CORE).

Summary

The present overview aims to present the physical–physiological framework of the major parameters involved in cellular signaling pathways and their associated transport phenomena by elucidating the roles of the ligands, receptors, and some major signals generation and propagation pathways. Obvious limitations of space and scope make this presentation an incomplete picture viewed through a spotted glass window. Alas, more often than not, the window reflects our ignorance of the marvels of creation and fills us with humility when facing these outstanding, almost unreal, phenomena used by nature to sustain life. It is hoped that this humble attempt of highlighting some of these phenomena may eventually help to catalyze better understanding and insight into the secrets of life.

Acknowledgments

We gratefully thank all those who helped transform this workshop from vision to reality. First and foremost, we salute the workshop participants whose excellence made the meeting an exciting meaningful affair. Next, we heartily thank the International Program Committee and the Members of the Organizing Committee for endless suggestions and continuous vigilance. Special thanks go to Professor Faruk Arinc, General Secretary of the International Centre for Heat & Mass Transfer (ICHMT), who helped organize the meeting in beautiful Antalya, Turkey, and to Dr. Daniele Schwarz and Professor Gianluigi Condorelli of the MultiMedica Hospital Research Center, Milan, Italy, who contributed significantly in soul and matter to make this meeting a successful and most rewarding experience. Thanks are also due to our secretary, Mrs. Betty Kazin, for devotion and care, and Ms. Edita Leonardo of the American Technion Society, NY, for her efficient help in organizing this workshop. Finally, we thank our sponsors, The Technion, Israel Institute of Technology, the ICHMT, the MultiMedica Hospital in Milan, the Middle East Technical University, and the Scientific and Technological Research Council of Turkey (TUBITAK), Ankara, Turkey. Most sincerely, we gratefully remember our departed friends, Henry Goldberg, whose generosity and curiosity helped to start this series of cardiac workshops, and Horti and Larry Fairberg of the American Technion Society, New York, USA, whose endowment facilitated this important international event.

Conflicts of Interest

The author declares no conflicts of interest.

References

1. SIDEMAN S. & R. BEYAR (Eds). 1985. Simulation and Imaging of the Cardiac System – State of the Heart. Proc 1st Henry Goldberg Workshop, Technion, Haifa, 1984, Martinus Nijhoff Publ. Dordrecht/Boston.

2. SIDEMAN S. & R. BEYAR (Eds). 1987. Simulation and Control of the Cardiac System. Parts I, II, III, Proc 2nd Henry Goldberg Workshop, Haifa. 1985. CRC Press. Florida.

3. SIDEMAN S. & R. BEYAR (Eds). 1987. Activation, Metabolism and Perfusion of the Heart. Simulation and Experimental Models. Proc 3rd Henry Goldberg Workshop, Rutgers Univ, USA, 1986. Martinus Nijhoff Publ. Dordrecht/Boston.

4. SIDEMAN S. & R. BEYAR (Eds). 1989. Analysis and Simulation of the Cardiac System – Ischemia. Proc 4th Henry Goldberg Workshop, Tiberias, May 1987. CRC Press. Florida (3 vols).

5. SIDEMAN S. & R. BEYAR (Eds). 1990. Imaging, Analysis and Simulation of the Cardiac System. Proc 5th Henry Goldberg Workshop, Cambridge, UK, 1988. Freund Publ. London.

6. SIDEMAN S. & R. BEYAR (Eds). 1991. Imaging, Measurements and Analysis of the Heart. Proc 6th Henry Goldberg Workshop, Eilat, Dec. 1989. Hemisphere Publ. NY.

7. SIDEMAN S., R. BEYAR & A. KLEBER (Eds). 1991. Cardiac Electrophysiology, Circulation and Transport. Proc 7th Henry Goldberg Workshop, Gwatt, Switzerland, 1991. Klüwer Publ. New York.

8. SIDEMAN S. & R. BEYAR (Eds). 1993. Interactive Phenomena in the Cardiac System. Proc of the 8th Henry Goldberg Workshop, Bethesda, MD, 1992. Adv. Exp. Med. Biol. **346**. Plenum Press. New York.

9. SIDEMAN S. & R. BEYAR (Eds). 1995. Molecular and Subcellular Cardiology: Effects on Structure and Function. Proc. of the 9th Henry Goldberg Workshop, Haifa, Israel. 1994, Adv. Exp. Med. Biol. **382**. Plenum Press. New York.

10. SIDEMAN S. & R. BEYAR (Eds). 1997. Analytical and Quantitative Cardiology: From Genetics to Function. Proc. of the 10th Henry Goldberg Workshop, Haifa, Israel. Adv. Exp. Med. Biol. **430**. Plenum Press. New York.

11. SIDEMAN S. & A. LANDESBERG (Eds). 2002. Visualization and Imaging in Transport Phenomena. Proc of the 1st Larry and Horti Fairberg Workshop, Antalya, Turkey. Ann. N.Y. Acad. Sci. Vol **972**.

12. SIDEMAN S. & R. BEYAR (Eds). 2004. Cardiac Engineering: From Genes and Cells to Structure and Function. Proc of the 2nd Larry and Horti Fairberg Workshop, Erice, Sicily. Ann. N.Y. Acad. Sci. Vol **1015**.

13. SIDEMAN, S., R. BEYAR & A. LANDESBERG. 2005. The Communicative Cardiac Cell. Proc of 3rd Larry & Horti Fairberg Workshop, Sintra, Portugal. Ann. N.Y. Acad. Sci. Vol. **1047**.

14. SIDEMAN, S., R. BEYAR & A. LANDESBERG (Eds). 2006. Interactive and integrative cardiology. Proc. 4th Larry & Horti Fairberg Workshop, Charleston, SC, USA. Ann. N.Y. Acad. Sci. Vol **1080**.

15. BIRD, R.B. 1960. Transport Phenomena. John Wiley & Sons. New York.

16. LIGHTFOOT, E.N. 1974. Transport Phenomena and Living Systems, Biomedical Aspects of Momentum and Mass Transport. J. Wiley & Sons. New York.

17. BIRD, R.B., W.E. STEWART & E.N. LIGHTFOOT. 2001. Transport Phenomena, 2nd ed. John Wiley & Sons. New York.

18. SCHIESSER W.E. & C.A. SILEBI. 1997. Computational Transport Phenomena. Cambridge Univ Press. Cambridge, UK.

19. DEEN, W.M. 1998. Analysis of Transport Phenomena. Oxford Univ Press. Chem. & Biochem, Oxford, UK.

20. SAATDJIAN, E. 2000. Transport Phenomena – Equations of Numerical Solutions. John Wiley & Sons., Inc. New York.

21. TRUSKEY, G.A., Y. FAN & D.F. KATZ. 2003. Transport Phenomena in Biological Systems. Prentice Hall. New York.

22. PETERS, M.H. 2005. Molecular Thermodynamics and Transport Phenomena. McGraw Hill Publ. New York.

23. FOURNIER R. 2006. Basic Transport Phenomena in Biom Eng, 2nd ed. Taylor & Francis Publ. New York.

24. BASSINGTHWAIGHTE, J.B. 1995. Toward modeling the human physiome. Adv. Exp. Med. Biol. **382**: 331–339.

25. SIDEMAN, S. 2005. Preface: The Cellular Communications Maze. Ann. N.Y. Acad. Sci. N.Y. **1047**: xi–xxiv.

26. MURO, M., C., MATEESCU, M. GAJEWSKI, *et al.* 2006. Control of intracellular trafficking of ICAM-1-targeted nanocarriers by endothelial Na+/H+ exchanger proteins. Am. J. Physiol. Lung Cell Mol. Physiol. **290**: L809–L817.

27. GAERTIG, J. & B. EDDE. 2005. Proteins that direct intracellular transport and locomotion. www.Physorg.com/news4101.html.

28. PFEFFER, S.R. 1999. Transport vesicle targeting: tethers before snares. Nature Cell. Biol. **1**: E17–E22.

29. INSEL, P.A. & B.P. HEAD, *et al.* 2005. Caveolae and lipid rafts: G-protein-coupled receptor signaling microdomains in cardiac myocytes. Ann. N.Y. Acad. Sci. **1047**: 166–172

30. PATHENVEEDU, M.A. & M. VON ZASTROW. 2006. Cargo regulates clathrin-coated pit dynamics. Cell **127**: 113–124.

31. BOEHM, M. & J.S. BONIFACINO. 2001. Adaptins, the final recount. Mol. Biol. **12**: 2907–2920.

32. LUSK, C.P., G. BLOBEL & M.C. KING. 2007. Highway to the inner nuclear membrane: rules for the road. Nat. Rev. Mol. Cell. Biol. **8**: 414–420.

33. ULLMAN K.S., M.A. POWERS & D.J. FORBES. 1997. Nuclear export receptors: from importin to exportin. Cell **90:** 967–970.

34. REXACH, M.F. 2006. A. sorting importin on Sec61. Nat. Structur. Mol. Biol. **13:** 476–478.

35. MURO, C., S.M. GRIGORIEV, D. PIETKIEWICZ, K.W. KINNALLY, M.L. CAMPO. 2003. Comparison of the TIM and TOM channel activities of the mitochondrial protein import complexes. Biophys. J. **84:** 2981–2989.

36. BERNARDI, P., A. KRAUSKOPF, E. BASSO, *et al.* 2006. The mitochondrial permeability transition from in vitro artifact to disease target. FEBS J. **273:** 2077–2099.

37. VYSSOKIKH, M.Y. & D. BRDICZKA. 2003. The function of complexes between the outer mitochondrial membrane pore (VDAC) and the adenine nucleotide translocase (ANT) in regulation of energy metabolism and apoptosis. Acta. Biochimica Polonica **50:** 389–404.

38. CADENAS, S. & M.D. BRAND. 2000. Effects of magnesium and nucleotides on the proton conductance of rat skeletal-muscle mitochondria. Biochem. J. **348**(Pt 1): 209–213.

39. GLEDHILL, J.R. & J.E. WALKER. 2006. Inhibitors of the catalytic domain of mitochondrial ATPsynthase. Biochem. Soc. Trans. **34:** 993–996.

40. KIRICHOK, Y., G. KRAPIVINSKY, D.E. CLAPHAM. 2004. The mitochondrial calcium uniporter is a highly selective ion channel. Nature **427:** 360–364.

41. BELMONTE, S. & M. MORAD. 2008. Shear fluid-induced Ca^{2+} release and the role of mitochondria in rat cardiac myocytes. Ann. N.Y. Acad. Sci. **1123:** 58–63. This volume.

42. EL-ADAWI H., L. DENG, *et al.* 2003. The functional role of the JAK–STAT pathway in post-infarction remodeling. Cardiovas Res. **57:** 129–138.

43. BRAY, D. 1998. Signaling complexes: biophysical constraints on intracellular communication. Ann. Rev. Biophys. Biomol. Struc. **27:** 59–75.

44. LANGE, A. & R.E. MILLS, *et al.* 2007. Classical nuclear localization signals: definition, function, and interaction with Importin α. J. Biol. Chem **282:** 5101–5105.

45. VARJOSALO, M. & J. TAIPALE. 2007. Hedgehog signaling. J. Cell Sci. **120:** 3–6.

46. BONIFACINO, J.S & L.M. TRAUB. 2003. Signals for sorting of transmembrane proteins to endosomes and lysosomes. Annu. Rev. Biochem. **72:** 395–447.

47. ROSEN H , V. GLUCKHMAN, *et al.* 2004. Cardiac steroids induce changes in recycling of the plasma membrane in human NT2 cells. Mol. Biol. Cell **15:** 1054–1044.

48. DAUFELDT S, R. KLEIN, L. WILDT & A. ALLERA. 2006. Membrane initiated steroid signaling (MISS): computational in vitro and in vivo evidence for a plasma membrane protein initially involved in genomic steroid hormone effects Membrane initiated steroid signaling (MISS). Mol. Cell Endocrinol. **246:** 42–52.

49. BELGO C, B. MASSION, M. PELAT & J.L. BALLINGAND. 2005. Nitricoxid and the heart: update on new paradigms. Ann. N.Y. Acad. Sci. **1047:** 173–182.

50. LANDESBERG, A., C. LEVY, Y. YANIV & S. SIDEMAN. 2004. The adaptive intracellular control of cardiac muscle function. Ann. N.Y. Acad. Sci. **1015:** 71–83.

51. SAIMI, Y. & C. KING. 2002. Camodulin as an ion-channel subunit. Ann. Rev. Physiol. **64:** 289–311.

52. RUDY, Y. & J.R. SILVA. 2006. Computational biology in the study of cardiac ion channels and cell electrophysiology. Quart. Rev. Biophys. **39:** 57–116.

53. BALIJEPALLI, R.C., J.D. FOEL, *et al.* 2006. Localization of cardiac L-type Ca2+ channels to a caveolar macromolecular signaling complex is required for β2-adrenergic regulation. Proc. Natl. Acad. Sci. U.S.A. **103:** 7500–7505.

54. WEHRENS X.H., S.E. LEHNART. 2004. Protection from cardiac arrhythmia through ryanodine receptor-stabilizing protein calstabin2. Science **5668:** 292–296.

55. ORTON R.J., O.E. STURM, *et al.* 2005. Computational modeling of the receptor-tyrosine- kinase-activated MAPK pathway. Biochem. J. **392:** 249–261.

56. SCHNEIDER, C. 1999. Cytokines and JAK-STAT signaling. Exp. Cell Res. **253:** 7–14.

57. WANG, H., Y. LU, Z., WANG. 2007. Function of cardiac M(3) receptors. Auton. Autacoid Pharmacol. **27:** 1–11.

58. NOBLE, D. 2008. Mind over molecule: Activating biological demons. Ann. N.Y. Acad. Sci. Control and Regulation of Transport Phenomena in the Cardiac System. In Press.

59. LATCHAM, D.S. 1997. Gene organization and regulation in colony stimulating factors. *In* Molecular & Cellular Biology, 2nd ed. J.M. Garland, P.J. Quesenberry & D.J. Hilton, Eds.: 97–106. Marcel Dekker Inc. New York.

60. CATALUCCI, D., M.V.G. LATRONICO & G. CONDORELLI. 2008. MicroRNAs control gene expression: importance for heart development and pathophysiology. Ann. N.Y. Acad. Sci. **1123:** 20–29. This volume.

61. KIKUCHI, A. & H. YAMAMOTO. 2007. Regulation of Wnt signaling receptor mediated endocytosis. J. Biochem. (Japan) PubM<ed ID: A317692 (Epub ahead of print).

62. HITT, M.M. & F.F. GRAHAM. 2000. Adenovirus vectors for human gene therapy. Adv. Virus Res. **55:** 479–505.

63. DOWNEY, J.M., T. KRIEG & M.V. COHEN. 2008. Mapping preconditioning's signaling, Pathways: An engineering approach. Ann. N.Y. Acad. Sci. Control and Regulation of Transport Phenomena in the Cardiac System. In Press.

64. YUE, P., Y. ZHANG, *et al.* 2006. Ischemia impairs the association between Connexin 43 and M3 subtype of acetyicholine muscarinic receptor (M3-mAChR) in ventricular myocytes. Cell Physiol. Biochem. **17:** 129–136.

65. RODRIGUEZ-SINOVAS, A., K. BOENGLER, *et al.* 2006. Translocation of connexin43 to the inner mitochondrial membrane of cardiomyocytes through the HSP90-dependent TOM pathway and its importance for cardioprotection. Circ. Res. **99:** 93–101.

66. JUHASZOVA, M. *et al.* 2004. Glycogen synthase kinase-3β mediates convergence of protection signaling to inhibit the mitochondrial permeability transition pore. J. Clin. Invest. **113:** 1535–1549.

67. BASSINGTHWAIGHTE, J.B. & K.C. VINNAKOTA. 2004. The computational integrated myocyte. A view into the virtual heart. Ann. N.Y. Acad. Sci. **1015:** 391–405.

68. SHENHAV, B. & D. LANCET. 2004. Prospects of a computational origin of life endeavor. Orig. Life Evol. Biosph. **34:** 181–194.

The Death of Transcriptional Chauvinism in the Control and Regulation of Cardiac Contractility

SAKTHIVEL SADAYAPPAN AND JEFFREY ROBBINS

Cincinnati Children's Hospital, University of Cincinnati, Cincinnati, Ohio 45229-3039, USA

In the last 25 years we have witnessed the triumph of the genome. There are now well over 200 complete genome sequences. The application of modern solid state technologies to genomic sequencing promises affordable personalized sequences for the individual in the very near future. With this explosion in DNA sequence data, the focus in the immediate past has been on the primary DNA sequence, the cis-trans interactions that underlie controlled transcription, cataloging the transcriptome, and applying rudimentary systems analysis to those data sets in an attempt to assign molecular signatures to normal and abnormal physiological states. However, it is becoming clear that the post-transcriptional processes, which operate at the levels of RNA stability and selection for translational initiation, as well as the post-translational processes of protein stability, trafficking, and secondary modifications, such as phosphorylation, all play key roles in the homeostasis of the contractile apparatus and its overall function. Defining the interplay of these processes, in concert with the signaling pathways that allow transcription, translation, and post-translational processes to be quickly modified in response to events outside of the cardiomyocyte are leading to an understanding of the spatial and temporal requirements for each of these processes in controlling cardiac output. In order to confirm the importance of post-translational modification in controlling cardiac contractility *in vivo*, we examined the role that post-translational modification of an important component of the cardiac contractile apparatus, myosin binding protein C (MyBP-C), plays in the normal and diseased heart by creating transgenic mice in which the effects of chronic cardiac MyBP-C phosphorylation and dephosphorylation could be determined.

Key words: cardiac myosin binding protein C; phosphorylation; hypertrophy; signaling

Introduction

Cardiac myosin binding protein C (cMyBP-C) is a large, 140-kDa thick, filament-associated protein that is localized to the A band of the sarcomere (FIG. 1). cMyBP-C is localized in the C region of the A band and has a unique organization with seven to nine axial bands situated on either side of the central M-line in each half sarcomere. cMyBP-C binds myosin at the S2 region and light meromyosin where it modulates myosin assembly, actin–myosin interaction in sarcomeres, and stabilizes thick filaments; this suggests that cMyBP-C is required for maintaining sarcomere integrity. It also binds titin, via domains C8–C10, and actin in the pro-ala-rich sequences between the C0–C1 domains.[1] Discovered over 27 years ago,[2] as is the case for many of the contractile proteins, cMyBP-C is represented in the different muscle types by multiple isoforms. Interest in the protein's role(s) intensified after multiple mutations in the polypeptide were linked to familial hypertrophic cardiomyopathy (FHC).[3] A significant amount of effort has been directed at the transcriptional and structural levels of the polypeptide, understanding the extent and diversity of these mutations, and studying the resultant human pathologies.[4] Although mutations in many different sarcomeric genes can lead to FHC, it is estimated that the spectrum of single base pair changes leading to missense and nonsense mutations in cMyBP-C can account for as much as 40% of all FHC associated with genetic defects.[5]

cMyBP-C belongs to the intracellular immunoglobulin (Ig) superfamily and is composed of repeated Ig and fibronectin domains.[6] *In vitro* modeling and reconstitution experiments, as well as experiments carried out using cell transfections, indicate that the protein probably plays an important role in assembling and maintaining the overall architecture of the

Address for correspondence: Professor Jeffrey Robbins, PhD, Molecular Cardiovascular Biology, MLC 7020, Cincinnati Children's Hospital, 3333 Burnet Avenue, Cincinnati, OH 45229-3039. Voice: +1-513-636-8098; fax: +1-513-636-59578.

jeff.robbins@cchmc.org

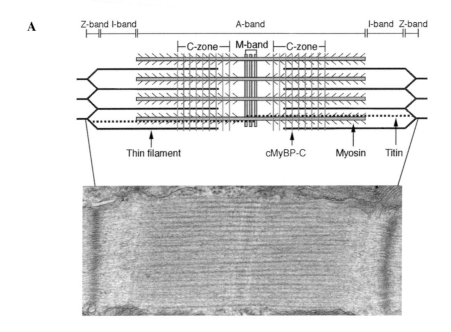

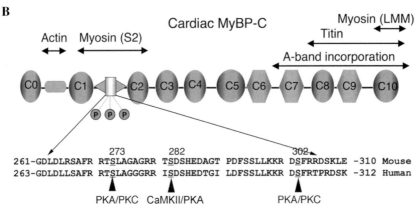

FIGURE 1. cMyBP-C location and functional domains. **(A)** Shown is a schematic diagram of a sarcomere that highlights the three filament systems: the thick filament, which contains myosin, the thin filament containing actin, and the third filament, titin. cMyBP-C is located on each side of the M-band and is thought to associate with myosin in a collar-like structure consisting of a trimer.[30] **(B)** General domain structure of mouse cMyBP-C. There are eight Ig-type (*ovals*) and three fibronectin-type (*octagon*) domains. The interacting sites with actin, myosin, and titin are shown (*double arrows*). The cardiac-specific phosphorylation motif, which is highly conserved between species, is located between domains C1 and C2 and the sequence shown. The cardiac domain contains three sites that can be phosphorylated by PKA, CaMKII, and PKC. LMM = light meromyosin; S2 = subfragment 2 of myosin, which contains the ATPase-containing head region. (In color in *Annals* online.)

sarcomere.[7,8] However, the cardiac-specific isoform has a number of unique characteristics indicating that it may also play an important role in the heart's response to β-adrenergic stimulation. Compared to the skeletal isoforms, which have a single site that can be phosphorylated via the cyclic adenosine monophos-

phate (cAMP)-dependent protein kinase (PKA) and/or fiber-associated Ca^{2+}/calmodulin-dependent (kinase CaMKII), the human cardiac isoform has four potential sites that can serve as differential substrates for PKA, protein kinase C (PKC), and CaMKII (FIG. 1). cMyBP-C phosphorylation can change both filament orientation and contractile mechanics.[8-10] *In vitro*-based reconstitution studies showed that PKA-mediated phosphorylation of cMyBP-C extends the myosin–actin crossbridges from the backbone of the thick filament, changes their orientation, increases their degree of order, and decreases crossbridge flexibility.[11] The phosphorylation sites of cMyBP-C are not equivalent in terms of their substrate specificity for the different kinases. *In vitro* studies showed that PKA phosphorylates Ser-273, Ser-282, and Ser-302 whereas PKC only phosphorylates Ser-273 and Ser-302. One site, Ser-282, located at the cardiac isoform-specific insertion that forms a surface loop,[9] is particularly interesting. When deleted or mutated, overall cMyBP-C phosphorylation is markedly decreased. Thus, phosphorylation at this site may function as a conformational switch, rendering other sites on the protein more accessible to the relevant kinase(s) and defining a hierarchy for this post-translational process.[9]

Three of the phosphorylation sites are located at or near a region of the protein that interacts with the myosin heavy chain's (MHC) head (FIG. 1). Biochemical studies showed that phosphorylation of these residues by PKA could abolish the cMyBP-C–myosin interaction. In the absence of cMyBP-C phosphorylation, the C1–C2 domain of cMyBP-C is bound to myosin in the S2 region,[10,12] but when phosphorylated, it releases its interaction with myosin, binding to actin.[13] Therefore, cMyBP-C probably helps regulate force generation by modulating thick–thin filament interactions.

cMyBP-C is phosphorylated in a dynamic manner in response to either β-adrenergic agonists or tissue plasminogen activator (TPA). β-Adrenergic stimulation is accompanied by increases in systolic tension development while stimulation of cMyBP-C phosphorylation by TPA parallels the inhibition of the myofibrillar Ca^{2+}-dependent Mg^{2+}-ATPase.[14,15] Studies concerning the structure–function relationships of these sites have been carried out *in vitro* using either transfection of cell culture with complementary DNAs (cDNAs) or biochemically defined filament reconstitutions with fragments of the protein.[6,10] As noted above, X-ray diffraction analysis on thick filaments showed that the phosphorylation state of cMyBP-C affects crossbridge extension and increases the overall order of the thick filament,[8,11] and the authors suggested that phosphorylation of residues within the regulatory motif of cMyBP-C changes the end-to-end interactions of adjacent molecules within the thick filament. This, in turn, could change the orientation or flexibility of the crossbridges, altering the rate constants of attachment and detachment.[8] Using a variety of biochemical and *in vitro* approaches, Gautel and colleagues demonstrated the potential importance of phosphorylation of this domain in regulating filament mechanics. They noted that five of the six parameters measured were modified when exchanging the phosphorylated and non-phosphorylated fragments,[10] with most of the data supporting a positive inotropic role for cMyBP-C phosphorylation.

Direct manipulation of the contractile endpoints, bypassing the transcriptional apparatus and receptor–ligand interactions, is now a focus of drug development. The PKC and PKA pathways mediate a major mechanism by which the myofilament modulates changes in contraction in response to alterations in both the immediate and larger environments. Although these alterations have been documented for almost 30 years, there is no agreement or comprehensive understanding as to the roles that they actually play *in vivo*, as most of the studies have been done *in vitro* or *in situ*. cMyBP-C represents a key target for this critical post-translational process. Biochemical studies have established in highly artificial systems that cMyBP-C phosphorylation can positively affect the manner and magnitude of tension and force development, but whether and how this translates into whole organ function and animal cardiac physiology is not known.

We wished to determine the physiological importance of chronic dephosphorylation and phosphorylation of cMyBP-C in terms of cardiac structure and function. To this end, we created transgenic mice in which endogenous cMyBP-C was replaced with the transgenically encoded protein. One protein, termed *AllP−*, contained alanine residues substituted for the phosphorylatable serines, while another protein, termed *AllP+*, substituted the charged amino acid aspartate for the three phosphorylatable serines in order to mimic chronic phosphorylation of the protein.

Materials and Methods

Transgenic and Gene-targeted Mice

The cDNA encoding normal wild-type mouse cMyBP-C ($cMyBP-C^{WT}$) was used to convert the known cMyBP-C phosphorylation sites (Ser-273, Ser-282, and Ser-302) to either alanine ($cMyBP-C^{AllP-}$) or to aspartic acid ($cMyBP-C^{AllP+}$). A sequence encoding the myc epitope was also incorporated so that expression and correct trafficking of the mature protein could be independently verified. The constructs were then used to generate multiple transgenic (TG) mouse lines with high levels of cardiomyocyte-specific expression as described.[16,17] To ensure the absence of endogenous dephosphorylated cMyBP-C, the TG were bred with cMyBP-C null mice that expressed less than 10% of normal cMyBP-C levels (cMyBP-C$^{t/t}$).[18] cMyBP-C^{WT}, an animal that expressed the normal cardiac isoform at equivalent levels to cMyBP-C^{AllP-} or cMyBP-C^{AllP+}, was also bred to the nulls to serve as a control.[16]

Measurement of cMyBP-C Phosphorylation in Heart Failure Models

One-dimensional isoelectric focusing (IEF) was used to detect de-, mono-, bi-, and triphosphorylated cMyBP-C. IEF slab gels were prepared using empty criterion cassettes (Bio-Rad, Hercules, CA) consisting of 5% Duracryl (Genomic Solutions, Ann Arbor, MI), 5% glycerol, and 1× of both pH 5.0–7.0 and pH 3.0–10.0 ampholites (Bio-Rad), which were degassed for longer than 1 h. The three catalysts (riboflavin-5′-phosphate, ammonium persulfate, and N,N,N,N,-tetramethylethylenediamine) were added separately. Total myofibrillar protein was loaded in buffer containing 10 mmol/L urea, 40% glycerol, 1× of both pH 5.0–7.0 and pH 3.0–10.0 ampholites, and 2% Triton X-100 with 0.025% bromophenol blue. The cathode buffer consisted of 30 mmol/L NaOH (pH 10.1), and the anode buffer consisted of 15 mmol/L phosphoric acid (pH 2.4; Bio-Rad). Standard IEF markers (Bio-Rad) were used for calculating isoelectric points. The running conditions were 1 W for 2 h, 2 W for 30 min, and 500 V for 2 h per gel. The transfer took place in 0.7% acetic acid (pH 3.0), placing the nitrocellulose membrane toward the anode at 200 mA for 10–12 h at 4 °C. After the transfer, an immunoblotting procedure was performed using a polyclonal anti-cMyBP-C antibody that reacted to the C0–C1 domains.

Echocardiography Measurements

For two-dimensional M-mode echocardiography, mice were anesthetized with 2% isoflurane and hearts visualized with a Hewlett Packard (Philips, Bothell, WA) Sonos 5500 instrument and a 15 MHz transducer as described.[19] Measurements were taken three times per animal from different areas and then averaged for left ventricular (LV) diastolic and systolic dimensions (LVEDD and LVESD, respectively) and septal and posterior wall thickness (SWT and PWT, respectively) from which fractional shortening (FS) and LV mass were derived. Pulsed-wave Doppler was used to measure aortic ejection time (E-TIME) and calculate the velocity of circumferential shortening, V_{cf} (FS/E-TIME). Diastolic transmitral inflow Doppler indices included the following: peak E-wave velocity, peak A-wave velocity, E-wave deceleration time, and isovolumic relaxation time.

Histochemistry

Heart weight to body weight was measured to determine if hypertrophy had occurred. For histopathological examinations, the hearts were removed while still beating from deeply anesthetized mice, drained of blood, and fixed in 10% formalin. The hearts were bisected longitudinally, dehydrated through a graded series of alcohols, and laid open before being paraffin embedded. Step-serial sections (5 μm) were taken from two to three hearts per group. Sections were stained with hematoxylin and eosin stain or Masson's trichrome stain and images evaluated with the SPOT software (Diagnostic Instruments, Inc., Sterling Heights, MI) using an Olympus BX60 microscope. The presence of necrosis, fibrosis, myocyte disarray, and calcification were evaluated by an expert who was blinded to genotype.

Statistical Analysis

All results were presented as mean ± SE. For comparisons of data from two groups, Student's *t*-test was used. For comparisons of multiple groups, one-way analysis of variance (ANOVA) or ANOVA for repeated measurements followed by the Tukey–Kramer multiple comparisons test was used (SigmaStat V3.0; Systat Software, Inc., San Jose, CA). A value of $P < 0.05$ was considered significant.

Results

cMyBP-C Phosphorylation In Vivo

Although *de novo* transcription of the contractile proteins clearly plays a critical role in the controlled contractile function of the heart, post-translational processes are largely responsible for the minute-to-minute and second-to-second tuning of cardiac systole and diastole. A number of sarcomeric proteins can be post-translationally modified, and reversible phosphorylation of the myosin light chains, troponin I, troponin T, and cMyBP-C plays a role in tuning the contractile parameters.[20] To establish a basis formulating the phosphorylation status of cMyBP-C in the mouse, we established the phosphorylation states of the endogenous protein under baseline and stressed conditions. We quantitated cMyBP-C phosphorylation using IEF followed by Western blot using a cMyBP-C C0–C1 motif-specific antibody.[21] Myofibrillar proteins were extracted from unstressed nontransgenic (NTG) adults as well as animals that had undergone transverse aortic constriction to produce pressure-overload hypertrophy. A genetically induced heart failure model,

A

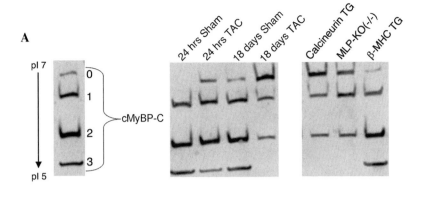

B

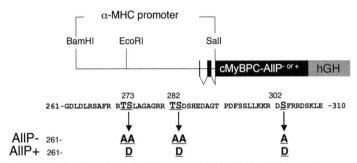

FIGURE 2. cMyBP-C phosphorylation levels in normal and pathological states. **(A)** One-dimensional IEF (pH 5.0–7.0) of total myofilament proteins followed by Western blot analysis using cMyBP-C antibody. The first lane shows the distribution of phosphorylated forms in a NTG mouse heart under unstressed conditions. Note the low amount of the completely non-phosphorylated (0) species. The phosphorylated forms of cMyBP-C were obtained from the following samples: hearts that had undergone a sham operation for transverse aortic constriction (TAC), 24 h; 24 h post-TAC; 18 days sham; 18 days post-TAC; 8-week-old calcineurin TG mouse; 7- to 8-month-old *mlp*-deficient mouse; a β-MHC TG mouse. The four forms of cMyBP-C based on its phosphorylation status are shown as 0, 1, 2, and 3, which correspond to the de-, mono-, bi-, and triphosphorylated forms, respectively, with isoelectric points of 6.1, 5.9, 5.7, and 5.5. **(B)** The TG constructs used to effect replacement of endogenous cMyBP-C with either non-phosphorylatable or phosphomimetic cMyBP-C. Cardiomyocyte-specific expression was driven by the α-MHC promoter. A human growth hormone polyadenylation signal sequence (hGH) was placed 3′ to the cMyBP-C cDNAs. Sites indicating the amino acid substitutions are shown (*arrowheads*).

produced by cardiomyocyte-specific overexpression of calcineurin,[22] was tested. Another genetically based heart failure model in which the striated muscle-specific LIM-only protein *MLP* (muscle LIM protein) gene was ablated (supplied by Dr. Pico Caroni, Friedrich Miescher Institut Basel, Switzerland) was also used: the adult homozygous nulls develop dilated cardiomyopathy and heart failure.[23] Additionally, because the degree to which cMyBP-C and its phosphorylated forms regulate cardiac function may depend, in part, upon the MHC isoform that is present,[8] we also quantitated cMyBP-C phosphorylation levels in a mouse model in which 70% of the normal α-MHC was replaced with β-MHC.[24,25]

In the unstressed NTG animals, the mono-, bi-, and triphosphorylated species of cMyBP-C made up >90% of the cMyBP-C population (FIG. 2, left panel, top). A β-MHC shift had no effect on the normal phosphorylation pattern (FIG. 2, right panel, top). In the stressed hearts produced either by surgically induced pressure overload or genetic manipulation (*mlp* knockout) in which cardiac failure was imminent, total cMyBP-C phosphorylation, particularly the triphosphorylated species, was strikingly decreased. Other heart failure models were tested as well,[32] and all data showed that decreased phosphorylation of cMyBP-C is associated with decreased contractility and pending heart failure.

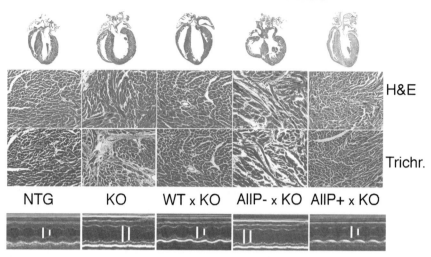

FIGURE 3. Rescue and nonrescue of the cMyBP-C$^{t/t}$ phenotype by transgenic overexpression. Longitudinal sections of the entire heart were stained with hematoxylin and eosin (H&E) stain, and Masson's trichrome (Trichr.) stain to assess fibrosis. Shown are representative paraffin-embedded sections prepared from 10% buffered formalin perfusion-fixed hearts from 12-week-old mice. Interstitial fibrosis, calcification, and myocardial disarray are seen in the cMyBP-C$^{t/t}$ (t/t) and cMyBP-C$^{AllP-:t/t}$ (AllP−) sections. In contrast, either normal cMyBP-C TG expression (WT) or expression of the phosphomimetic (AllP+) rescues the cMyBP-C$^{t/t}$ phenotype. M-mode echocardiographic tracings show the left ventricular chamber motion of the various mice. Again, normal cMyBP-C TG expression or expression of the phosphomimetic (AllP+) rescues the cMyBP-C$^{t/t}$ phenotype. (In color in *Annals* online.)

Transgenic Replacement of Endogenous cMyBP-C

The above data provided a firm rationale for manipulating the phosphorylation state cMyBP-C *in vivo*. To that end, the cDNA encoding the endogenous form of mouse cMyBP-C (cMyBP-C^{WT}) was obtained by RT-PCR using total RNA isolated from the mouse cardiac ventricle. Full-length cMyBP-C^{WT} (approximately 3.8 kb) was subcloned, completely sequenced in both directions, and compared with Genebank cDNA database (Accession number: NM_008653). The known phosphorylation sites (Ser-273, Ser-282, and Ser-302) and two neighboring potentially alternative phosphorylation sites (Thr-272 and Thr-281) were converted either to alanines (cMyBP-C^{AllP-}) or aspartates (cMyBP-C^{AllP+}) using standard PCR-based methods and subcloned into the ApoI and SphI restriction sites. The cDNAs were subcloned into the mouse α-MHC promoter to drive cardiomyocyte-restricted expression, the constructs purified from the plasmid backbone after NotI digestion, and the DNAs used to generate multiple lines of FVB/N TG mice (FIG. 2B).[26] Multiple founders were obtained for each construct and a line chosen for each that showed roughly equivalent levels of TG expression. Each line was subsequently bred into the cMyBP-C knockout mice (cMyBP-C$^{t/t}$),[27] which produce only trace

amounts of a truncated cMyBP-C and phenocopies of a cMyBP-C knockout. cMyBP-C^{AllP-}, cMyBP-C^{AllP+}, and cMyBP-C^{WT} mice were crossed with homozygous cMyBP-C$^{t/t}$ mice to ensure the absence of endogenous cMyBP-C. Five to seven mice that were 12- to 15-weeks old of mixed gender were used in each experiment after preliminary experiments showed no gender differences.

The cMyBP-C$^{t/t}$ mice have been characterized previously. The homozygotes are viable but soon after birth display a progressive dilated cardiomyopathy. Myocyte hypertrophy, disarray, fibrosis, and calcification all progress as the animals mature. The compromised gross structure of the heart and chamber dilation are apparent (FIG. 3), and histological analyses reveal the characteristic myocyte disarray and fibrosis typical of hypertrophic cardiomyopathy. Strikingly, cMyBP-C^{WT} expression was completely effective in preventing the hypertrophic phenotype from presenting at the anatomical level. However, the phosphoincompetent cMyBP-C^{AllP-} was unable to rescue the phenotype even though cMyBP-C^{AllP-} protein was present at normal levels.[32] In contrast to the inability of cMyBP-C^{AllP-} to rescue the null phenotype, expression of cMyBP-C^{AllP+} was as effective as cMyBP-C^{WT} at rescuing the cMyBP-C$^{t/t}$ mice (FIG. 3).

These results were recapitulated when cardiac function of the different lines was determined at 12 weeks. The inability of cMyBP-C^{AllP-} to rescue the cMyBP-C$^{t/t}$ phenotype was confirmed at the functional level. M-mode echocardiography showed the cMyBP-C$^{AllP-:t/t}$ and cMyBP-C$^{t/t}$ mice had increased LV end diastolic and systolic end dimensions as well as reduced fractional shortening, while normal shortening fractions were observed in the cMyBP-C$^{WT:t/t}$ and cMyBP-C$^{AllP+:t/t}$ hearts (FIG. 3 and Ref. 34).

Discussion

Heart disease and cardiac failure are often associated with diminished β-adrenergic responsiveness, loss of cardiac contractility, abnormalities in Ca^{2+} flux, and altered contractile protein phosphorylation.[28] Although cMyBP-C is extensively phosphorylated under basal conditions, it becomes partially dephosphorylated during the development of heart failure or pathological hypertrophy, with the triphosphorylated form largely or completely absent in the advanced stages of heart failure (FIG. 2). This phenomenon appears to be relatively independent of the type of cardiac stress as pressure overload or genetic alterations in the cardiac machinery both resulted in significantly decreased phosphorylation. Indeed, we have determined that the triphosphorylated form is significantly diminished or lacking in hearts that have been subjected to ischemia–reperfusion injury as well. Basal levels of cMyBP-C phosphorylation may be necessary for maintaining thick filament orientation, dynamic regulation, and contractile mechanics,[9,10] and altered phosphorylation in this protein clearly impacts on its ability to function as the phospho-incompetent cMyBP-C^{AllP-} is unable to rescue the phenotype resulting from the cognate gene's ablation (FIG. 3). The inability of cMyBP-C^{AllP-} to rescue the cMyBP-C$^{t/t}$ phenotype is consistent with the hypothesis that cMyBP-C phosphorylation is essential for normal cardiac function. Strikingly, equivalent TG expression of cMyBP-C^{WT} effectively rescued the null mice, resulting in restoration of normal morphology and function and proving that we are not looking at epiphenomena; instead, the results are directly due to changes in the cMyBP-C complement. Protein phosphatase activity increases in heart failure,[29] and we hypothesize that this increased activity, along with decreased β-adrenergic responsiveness, results in decreased cMyBP-C phosphorylation.

In addition to the structural roles of cMyBP-C[30] and its ability to bind to titin,[31] phosphorylation of cMyBP-C apparently plays a critical role in the protein's ability to normally function in the sarcomere. cMyBP-C is the only thick filament protein that is differentially phosphorylated at multiple sites by the three enzymes PKA, PKC, and CaMKII,[9] potentially providing a sensitive nodal control point for tuning the activity of the contractile apparatus in response to different environmental stimuli. While our data do not address crossbridge mechanics directly, we have confirmed that cMyBP-C^{AllP-} is present in the sarcomere in a pattern that is indistinguishable from normal protein.[32] However, it appears to lack some critical function necessary for maintaining the overall sarcomere architecture as manifested by alterations observed in the sarcomere–mitochondrial spatial relationships and the sarcomere itself.[32] A number of investigators have modeled the interactions of cMyBP-C with the other contractile proteins, and it has been hypothesized that a trimer of cMyBP-C serves as a "collar" for the thick filament and can affect the position of the head of the myosin with respect to both the rest of the thick filament and with the thin filament.[30] Indeed, it is known that flexibility of the myosin head is required for efficient force generation.[10,33] In the absence of cMyBP-C phosphorylation, the C1-C2 domain of cMyBP-C is bound to myosin in the S2 region, but, when phosphorylated, it releases its interaction with myosin,[34] binding to actin.[35] Therefore, cMyBP-C not only can participate in maintenance of the structural integrity of the sarcomere but can also actively help regulate force generation by modulating thick–thin filament interactions.

Summary

We examined the effects of total chronic phosphorylation of cMyBP-C by generating TG mice in which the known cMyBP-C phosphorylation sites are either converted to aspartic acid (cMyBP-C^{AllP+}) in order to mimic a constant state of constitutive phosphorylation or changed to alanine (cMyBP-C^{AllP-}), rendering the protein phospho-incompetent. The mice were then bred into a homozygous cMyBP-C null background to obtain complete replacement of endogenous cMyBP-C with cMyBP-C^{AllP-} or cMyBP-C^{AllP+}. The phosphorylation mimetic effectively rescued the null mice while cMyBP-C^{AllP-} could not, implying a critical role for cMyBP-C phosphorylation for the normal function of the protein.

Acknowledgments

This research was supported by grants from the National Institutes of Health (HL69799, HL60546, HL52318, HL60546, HL56370; J.R.) and by the

American Heart Association, Ohio Valley Affiliate (S.S.). The authors would like to thank Dr. Robert S. Decker for sharing his technical expertise in isolating and analyzing the phosphorylated forms of cMyBP-C.

Conflicts of Interest

The authors declare no conflicts of interest.

References

1. FREIBURG, A. & M. GAUTEL. 1996. A molecular map of the interactions between titin and myosin-binding protein C. Implications for sarcomeric assembly in familial hypertrophic cardiomyopathy. Eur. J. Biochem. **235:** 317–323.

2. OFFER, G., C. MOOS & R. STARR. 1973. A new protein of the thick filaments of vertebrate skeletal myofibrils. Extractions, purification and characterization. J. Mol. Biol. **74:** 653–676.

3. WATKINS, H., D. CONNER, L. THIERFELDER, *et al.* 1995. Mutations in the cardiac myosin binding protein-C gene on chromosome 11 cause familial hypertrophic cardiomyopathy. Nat. Genet. **11:** 434–437.

4. ASHRAFIAN, H. & H. WATKINS. 2007. Reviews of translational medicine and genomics in cardiovascular disease: new disease taxonomy and therapeutic implications cardiomyopathies: therapeutics based on molecular phenotype. J. Am. Coll. Cardiol. **49:** 1251–1264.

5. AHMAD, F., J.G. SEIDMAN & C.E. SEIDMAN. 2005. The genetic basis for cardiac remodeling. Annu. Rev. Genomics Hum. Genet. **6:** 185–216.

6. EINHEBER, S. & D.A. FISCHMAN. 1990. Isolation and characterization of a cDNA clone encoding avian skeletal muscle C-protein: an intracellular member of the immunoglobulin superfamily. Proc. Natl. Acad. Sci. USA **87:** 2157–2161.

7. SEILER, S.H., D.A. FISCHMAN & L.A. LEINWAND. 1996. Modulation of myosin filament organization by c-protein family members. Mol. Biol. Cell. **7:** 113–127.

8. WINEGRAD, S. 2000. Myosin binding protein C, a potential regulator of cardiac contractility. Circ. Res. **86:** 6–7.

9. GAUTEL, M., O. ZUFFARDI, A. FREIBURG, *et al.* 1995. Phosphorylation switches specific for the cardiac isoform of myosin binding protein-C: a modulator of cardiac contraction? EMBO J. **14:** 1952–1960.

10. KUNST, G., K.R. KRESS, M. GRUEN, *et al.* 2000. Myosin binding protein C, a phosphorylation-dependent force regulator in muscle that controls the attachment of myosin heads by its interaction with myosin S2. Circ. Res. **86:** 51–58.

11. WEISBERG, A. & S. WINEGRAD. 1996. Alteration of myosin cross-bridges by phosphorylation of myosin-binding protein C in cardiac muscle. Proc. Natl. Acad. Sci. USA **93:** 8999–9003.

12. LEVINE, R., A. WEISBERG, I. KULIKOVSKAYA, *et al.* 2001. Multiple structures of thick filaments in resting cardiac muscle and their influence on cross-bridge interactions. Biophys. J. **81:** 1070–1082.

13. KULIKOVSKAYA, I., G. MCCLELLAN, J. FLAVIGNY, *et al.* 2003. Effect of MyBP-C Binding to actin on contractility in heart muscle. J. Gen. Physiol. **122:** 761–774.

14. GARVEY, J.L., E.G. KRANIAS & R.J. SOLARO. 1988. Phosphorylation of C-protein, troponin I and phospholamban in isolated rabbit hearts. Biochem. J. **249:** 709–714.

15. VENEMA, R.C. & J.F. KUO. 1993. Protein kinase C-mediated phosphorylation of troponin I and C-protein in isolated myocardial cells is associated with inhibition of myofibrillar actomyosin MgATPase. J. Biol. Chem. **268:** 2705–2711.

16. YANG, Q., A. SANBE, H. OSINSKA, *et al.* 1998. A mouse model of myosin binding protein C human familial hypertrophic cardiomyopathy. J. Clin. Invest. **102:** 1292–1300.

17. YANG, Q., A. SANBE, H. OSINSKA, *et al.* 1999. In vivo modeling of myosin binding protein C familial hypertrophic cardiomyopathy. Circ. Res. **85:** 841–847.

18. MCCONNELL, B.K., K.A. JONES, D. FATKIN, *et al.* 1999. Dilated cardiomyopathy in homozygous myosin-binding protein-C mutant mice. J. Clin. Invest. **104:** 1235–1244.

19. HAHN, H.S., Y. MARREEZ, A. ODLEY, *et al.* 2003. Protein kinase Calpha negatively regulates systolic and diastolic function in pathological hypertrophy. Circ. Res. **93:** 1111–1119.

20. SUMANDEA, M.P., E.M. BURKART, T. KOBAYASHI, *et al.* 2004. Molecular and integrated biology of thin filament protein phosphorylation in heart muscle. Ann. N.Y. Acad. Sci. **1015:** 39–52.

21. DECKER, R.S., M.L. DECKER, I. KULIKOVSKAYA, *et al.* 2005. Myosin-binding protein C phosphorylation, myofibril structure, and contractile function during low-flow ischemia. Circulation **111:** 906–912.

22. MOLKENTIN, J.D., J.R. LU, C.L. ANTOS, *et al.* 1998. A calcineurin-dependent transcriptional pathway for cardiac hypertrophy. Cell **93:** 215–228.

23. ARBER, S., J.J. HUNTER, J. ROSS, JR., *et al.* 1997. MLP-deficient mice exhibit a disruption of cardiac cytoarchitectural organization, dilated cardiomyopathy, and heart failure. Cell **88:** 393–403.

24. KRENZ, M. & J. ROBBINS. 2004. Impact of beta-myosin heavy chain expression on cardiac function during stress. J. Am. Coll. Cardiol. **44:** 2390–2397.

25. KRENZ, M., A. SANBE, F. BOUYER-DALLOZ, *et al.* 2003. Analysis of myosin heavy chain functionality in the heart. J. Biol. Chem. **278:** 17466–17474.

26. PALERMO, J., J. GULICK, M. COLBERT, *et al.* 1996. Transgenic remodeling of the contractile apparatus in the mammalian heart. Circ. Res. **78:** 504–509.

27. MCCLELLAN, G., I. KULIKOVSKAYA & S. WINEGRAD. 2001. Changes in cardiac contractility related to calcium-mediated changes in phosphorylation of myosin-binding protein C. Biophys. J. **81:** 1083–1092.

28. VAN DER VELDEN, J., Z. PAPP, R. ZAREMBA, *et al.* 2003. Increased Ca2+-sensitivity of the contractile apparatus in end-stage human heart failure results from altered phosphorylation of contractile proteins. Cardiovasc. Res. **57:** 37–47.

29. NEUMANN, J. 2002. Altered phosphatase activity in heart failure, influence on Ca2+ movement. Basic Res Cardiol. **97**(Suppl 1): I91–95.

30. FLASHMAN, E., C. REDWOOD, J. MOOLMAN-SMOOK, *et al.* 2004. Cardiac myosin binding protein C: its role in physiology and disease. Circ. Res. **94:** 1279–1289.

31. SQUIRE, J.M., P.K. LUTHER & C. KNUPP. 2003. Structural evidence for the interaction of C-protein (MyBP-C) with actin and sequence identification of a possible actin-binding domain. J. Mol. Biol. **331:** 713–724.

32. SADAYAPPAN, S., J. GULICK, H. OSINSKA, *et al.* 2005. Cardiac myosin-binding protein-C phosphorylation and cardiac function. Circ. Res. **97:** 1156–1163.

33. WEISBERG, A. & S. WINEGRAD. 1998. Relation between crossbridge structure and actomyosin ATPase activity in rat heart. Circ. Res. **83:** 60–72.

34. SADAYAPPAN, S., H. OSINSKA, R. KLEVITSKY, *et al.* 2006. Cardiac myosin binding protein C phosphorylation is cardioprotective. Proc. Natl. Acad. Sci. USA **103:** 16918–16923.

35. KULIKOVSKAYA, I., G. MCCLELLAN, R. LEVINE, *et al.* 2003. Effect of extraction of myosin binding protein C on contractility of rat heart. Am. J. Physiol. Heart Circ. Physiol. **285:** H857–865.

Genetic Mechanisms Controlling Cardiovascular Development

JAMIE BENTHAM AND SHOUMO BHATTACHARYA

Department of Cardiovascular Medicine and Wellcome Trust Centre for Human Genetics, University of Oxford, Oxford, United Kingdom

Congenital heart disease (CHD) is a major cause of childhood morbidity and death in the West; the incidence is approximately 1 in 145 live births. Mendelian and chromosomal syndromes account for approximately 20% of CHD. The genetic mechanisms underlying non-chromosomal or non-Mendelian "sporadic" CHD, which account for the remaining 80%, are poorly understood. The genetic architecture of sporadic CHD likely includes accumulation of rare nonsynonymous variants in cardiac developmental genes leading to mutational loading of cardiac developmental networks, copy number variation in cardiac developmental genes, and common variants that may not be obviously linked to cardiac development but may alter genetic buffering pathways (e.g., folate metabolism). The rare mutations typically associated with sporadic CHD likely arise from the severe decrease in reproductive fitness selecting against any CHD-causing gene variant. The resulting allelic heterogeneity reduces the power of genome-wide association studies for CHD. A complementary approach to the genetic analysis of CHD is to resequence candidate genes that have been shown to be necessary for mouse heart development. The number of such genes likely exceeds 1700. To identify these genes, we have developed an enabling technology (high-throughput magnetic resonance imaging of mouse embryos), which is used in combination with N-ethyl-N-nitrosourea/transposon mutagenesis and knockout techniques. Key future challenges now involve translating discoveries made in mouse models to human CHD genetics and understanding the mechanisms that create and disrupt genetic buffering. A long-term goal in CHD is to manipulate these pathways to enhance buffering and prevent disease in a manner analogous to the use of folate in preventing neural tube defects.

Key words: genetics; congenital heart disease; MRI; mutagenesis; pleiotropy

Introduction

Congenital heart disease (CHD), the most common birth defect, is a major cause of childhood morbidity and death in the West. The incidence of CHD in live-born infants ranges from 0.4 to 1.2%.[1,2] In the UK, approximately 4600 babies are born with CHD each year. In addition, there are approximately 150,000 people over the age of 16 living with CHD, of whom about 11,500 have complex disease requiring lifelong care.[3] The total adult population in the USA with CHD is estimated at greater than one million and is increasing at a rate of 5% per year.[4] Although surgical advances have made a tremendous impact on mortality, in our opinion prevention of CHD should be a major translational goal. The paradigm here is the prophylaxis of

neural tube defects using folate,[5] an intervention that has led to the prevention of 50–75% of cases.[6] Identification of pathways that could be manipulated—for example, by micronutrients—to prevent CHD is a major aim of genetic studies. A secondary aim is to elucidate the genetic architecture of CHD in order to further understand the underlying biological mechanisms of this disease and to allow improved risk stratification.

Genetic Architecture of CHD

Chromosomal and Mendelian Syndromes

Human CHD is typically characterized by lesions (including atrial septal defects [ASD], ventricular septal defects [VSD], and atrioventricular septal defects [AVSD]), outflow tract malformations (e.g., Tetralogy of Fallot [TOF], common arterial trunk [CAT], and transposition of great arteries [TGA]), and aortic arch malformations (e.g., patent ductus arteriosus [PDA] and aortic coarctation).[7] Chromosomal and Mendelian syndromes account for approximately 20%

Address for correspondence: Professor Shoumo Bhattacharya, MD, FRCP, Department of Cardiovascular Medicine, University of Oxford, Wellcome Trust Centre for Human Genetics, Roosevelt Drive, Oxford OX37BN, UK. Voice: 44-1865-287771; Fax: 44-1865-287742.
shoumo.bhattacharya@well.ox.ac.uk

Ann. N.Y. Acad. Sci. 1123: 10–19 (2008). © 2008 New York Academy of Sciences.
doi: 10.1196/annals.1420.003

(11.9% and 7.4%, respectively) of CHD[8,9] and have been recently reviewed.[10] In addition to these eponymous syndromes, studies in nonsyndromic large families with Mendelian inheritance patterns have established the role of genes, such as *ZIC3*[11,12] (heterotaxy), *NOTCH1*[13] (aortic stenosis and bicuspid aortic valve), *NKX2.5*[14] (ASD), *NKX2.6*[15] (CAT), *MYH6*[16] (ASD), *MYH11*[17] (PDA), *JAG1*[18] (TOF), *ACTC1* (ASD),[18a] and *GATA4*[19] (ASD), in the etiology of human CHD. The identification of the genes involved in these "experiments of nature" through family-based linkage or cytogenetic studies has contributed greatly to our understanding of genetic mechanisms in cardiac development and CHD.[10]

Non-Mendelian/Non-chromosomal CHD

The genetic mechanisms underlying non-Mendelian/non-chromosomal ("sporadic") CHD, which account for the remaining 80%, are, however, poorly understood. Even for these seemingly sporadic cases, epidemiological studies have demonstrated an increased risk of CHD recurrence of 2–5% in siblings and offspring, indicating the potential role of shared genes and/or environment.[2] The recurrence risk varies with the index case lesion. Aortic valve anomalies,[20–22] ASD,[22–24] and AVSD[25–28] have recurrence risks much greater than TOF[25,28–31] (i.e., 11.5%, 11.9%, 9.9% versus 2.4%, respectively).

Role of Variations in Genes Controlling Embryonic Cardiac Development

Supporting the idea that genetic variants can cause non-Mendelian/non-chromosomal CHD, non-synonymous disease-associated mutations in genes controlling cardiac development have been found in case–control studies.[32–41b] These genes include *CFC1*, *CITED2*, *CRELD1*, *FOG2*, *GATA4*, *GDF-1*, *LEFTA*, *NKX2-5*, *NOTCH1*, *PROSIT240*, *TBX1*, *TBX20*, and *ZIC3*. For unselected CHD, the frequency of these gene variants ranges from 0.77% for *CITED2*[33] to 2% for *NKX2-5*.[39,42] An important feature of non-Mendelian/non-chromosomal CHD is allelic heterogeneity. Examination of 51, disease-associated, nonsynonymous mutations identified in 3108 patients indicates that these mutations are typically individually unique, with the sole exception of the *CFC1*-G174del1 and *GATA4*-E261D mutations (see TABLE 1 for studies reporting variants and[43–46] for studies reporting negative results). These variants are typically individually unique, resulting in allelic heterogeneity. This is not surprising as untreated CHD has a 60% infant mortality[47]; such a decrease in reproductive fitness would be expected to eliminate highly penetrant CHD-causing alleles from the population.[48] Importantly, this will reduce the power of genome-wide association approaches to detect such alleles, making large-scale studies desirable.[48,49]

Role of Copy Number Variation

Large-scale structural variation has a clear role in many classic chromosomal syndromes that have CHD as a major component, such as Alagille, Down, Di-George, and Trisomy 18,[50] and also in CHD associated with learning disabilities or dysmorphic features.[51,51a] The role of intermediate-scale copy number variation (CNV; 1 kb–3 Mb) in non-Mendelian/non-chromosomal CHD is unknown, but, given that many developmentally active genes have such CNVs, it is likely to play a role.[52]

Role of Genetic Buffering and Mutational Load

Where reported, mutations in non-Mendelian/non-chromosomal CHD are invariably heterozygous and are transmitted from unaffected parents, indicating that these mutations are partially penetrant.[38,39] Despite this, many of the nonsynonymous mutations identified in case–control studies have functional effects in biological assays. For instance, functional mutations have been identified in *CFC1*,[32] *CITED2*,[33] and *ZIC3*[41] amongst others. One explanation for the reduced penetrance observed in these studies of sporadic CHD is buffering.[53,54] This can result from compensation by a normally functioning second allele or a duplicated gene or a pathway that maintains residual function, from negative feedback mechanisms that regulate flux, and also from epigenetic and environmental mechanisms.[53–59] Genetic evidence in the mouse indicates that increased mutation loading of a genetic network can result in cardiac malformation, presumably through loss of the buffering properties of the network. For instance, synthetic phenotypes are created by compound mutations of members of networks required for cardiac development. Thus, in the left–right patterning network, interactions can be detected between *iv* and *Actr2b*, *Nodal* and *Actr2b*[60]; *Nodal* and *Zic3*[61]; *Nodal* and *Smad2*[62]; and *Nodal* & *Foxa2*.[63] Similarly, interactions can be detected in the development of the secondary heart field-derived structures (e.g., between *Pitx2* and *Tbx1*),[64] in the development of the atrial septum (*Tbx20* and *Nkx2-5*),[65] or in the development of the cardiac chambers (*Nkx2-5* and *Hand2*,[66] *Gata4* and *Gata6*[67]). Extrapolating this to human disease we may expect that patients with CHD have high mutational loads in left–right patterning or other networks that are necessary for cardiac development.

TABLE 1. Case–control studies in non-Mendelian, non-chromosomal CHD^a

Gene	Phenotype	Cases	Control	NS/NP	NS/NP frequency	Variants identified in cases	Parental transmission	Polymorphisms present in controls	Reference
CFC1	Heterotaxy	144	200	3	2.08%	R112C* (1); G174del1* (2); R189C; R78W* (5)	G174del1	R78W (1) – in African Americans	32
CFC1	d-TGA (58) or DORV (22) or L-TGA (6)	86	200	2	2.33%	G174del1* (1); splice dup (1); R78W* (1)	G174del1	R78W (1)	74
CITED2	CHD	392	384	3	0.77%	S170-G178del* (1); G178-S179ins9* (1); S198G199del* (1); H39del (2); H160L (1)		H39del(1); H160L (1)	33
CRELD1	AVSD	50	100	3	6.00%	T311I (1); R329C (1); R107H (1)	R329C		34
CRELD1	AVSD	49	100	1	2.04%	P162A (1)	P162A		35
FOG2	TOF	47	240	2	4.26%	E30G (1), S657G (1)	E30G, S657G		36
GATA4	TOF	26	446	2	7.69%	E216D* (2)	E216D*		37
GATA4	CHD	31	423	2	6.45%	V267M (1), V380M (1)	Both de novo		115
GDF-1	CHD	375	423	8	2.13%	R68H (1); G162D (1); C227X (1); G262S (1); C267Y (1); S309P (1); P312T (1); A318T (1)			41a
LEFTYA	Heterotaxy	112	200	2	1.79%	R314X (1), S342L (1)	R314X & S342L		38
NKX2-5	ASD	84	100	4	4.76%	K151 (1); A127E (1); D235AFSter (1); Y250X (1)	K151; D235AFSter		39, 42
NKX2-5	TOF	201	100	5	2.49%	R25C (4); E21Q (1); Q22P (1); R216C (1); A219V (1); A323T (1)	E21Q; R25C	R25C present in 2/43 African Americans	39
NKX2-5	DORV	31	100	1	3.23%	del291N (1)	del291N		39
NKX2-5	Int AoA	23	100	0	0.00%	R25C (1)		R25C	39
NKX2-5	CAT	22	100	0	0.00%	R25C (1)		R25C	39
NKX2-5	L-TGA	7	100	1	14.29%	A63V (1)			39
NKX2-5	D-TGA	86	100	0	0.00%				39
NKX2-5	Ebstein's	7	100	0	0.00%				39
NKX2-5	HLHS	80	100	0	0.00%	R25C (1)		R25C	39
NKX2-5	Coarct	59	100	1	1.69%	P275T (1)			39
NKX2-5	AS	21	100	0	0.00%				39
NOTCH1	BCAV, AS	48	654	2	4.20%	T596M (1); P1797H (1)			84
PROSIT240	D-TGA	97	400	3	3.09%	E251G (1); R1872H (1); D2023G (1)	E251G		40
TBX1	AoA abnormalities	35	100	2	5.71%	G350D (1); P396L (1)			116
TBX1	Int AoA	11	100	1	9.09%	466_476dup10A (1)			116
TBX20	CHD	352	200	2	0.57%	Q195X (1); 1152M (1)			41b
ZIC3	Heterotaxy	145	145	1	0.69%	K405E (1)			41
ZIC3	Non-heterotaxy CHD	29	145	1	3.45%	P217A* (1)			41

^{a}Case–control studies where a disease-associated mutation was published were identified by searching PubMed, the Human Gene Mutation Database; http://www.hgmd.cf.ac.uk/ac/all.php) and the Online Mendelian Inheritance in Man database. Numbers in parentheses following a mutant allele indicate the number of times it was observed. Abbreviations: AoA = aortic arch; BCAV = bicuspid aortic valve; Coarct = coarctation of aorta; D-TGA = simple transposition great arteries; HLHS = hypoplastic left heart syndrome; IntAoA = interrupted aortic arch; L-TGA = congenitally corrected TGA; NP = non-polymorphic (i.e., not present in controls); NS = non-synonymous; see text for other abbreviations.

* Mutations where an effect on biological function has been demonstrated.

Gene–Environment Interactions in Metabolic Pathways

The maternal methylene tetrahydrofolate reductase variant 677CT and 677TT genotypes are associated with threefold and sixfold increases, respectively, in CHD risk to their children.[68] In contrast, no association was seen with the fetal 677T allele. In a randomized controlled trial, periconceptional folate significantly reduced the risk of CHD (OR = 0.6).[69] These studies show how genetic and environmental influences can interact in the multifactorial pathogenesis of CHD. Importantly, environmental factors, such as maternal diabetes and obesity, may disrupt buffering mechanisms to result in CHD.[70–73]

Pleiotropy of CHD Phenotypes

Mutation in certain genes (e.g., *CFC1*,[32,74] *CITED2*,[33] *CREBBP/EP300*,[75–77] *CRELD1*,[34] *CHD7*,[78–80] *GATA4*,[19,81] *NKX2-5*,[39,82,83] *NOTCH1*,[13,84,85] and *ZIC3*[11,12,41]) results in pleiotropic and variable cardiovascular malformations in humans. Importantly, in studies of human disease recurrence, the recurrent lesions also frequently do not resemble the index lesion. For instance, in studies of 21 recurrences where the parent had TOF, only two had TOF and the others had pulmonary stenosis

(PS), aortic stenosis (AS), ASD, VSD, PDA, and CAT.[25,29–31,86] Concordance with parental diagnosis is higher for lesions, such as ASD (68%)[22,86] and VSD (45%).[22,86,87] Cardiac pleiotropy resulting from a single gene mutation is also seen in the mouse. Examination of the Mouse Genome Informatics database reveals that of 267 gene mutations with cardiac phenotypes resembling CHD, 122 have pleiotropic malformations that affect more than one cardiac structure (i.e., atria, ventricles, valves, outflow tract, or aortic arches; TABLE 2). The variability in phenotype resulting from a single gene mutation can be explained, in part, by variation in genetic background (e.g., mice lacking *Cited2*)[88,89] and also likely by environmental or epigenetic mechanisms.

Genetic Analysis of CHD

Future investigations of the genetic architecture of CHD will require, in addition to genome-wide association studies to investigate the impact of common variants, candidate gene resequencing approaches to study relatively rare higher penetrance mutations. Here, disease association can be detected by measuring the difference in a gene's "nonsynonymous

---

TABLE 2. Mouse genes with mutant phenotypes resembling human congenital heart disease[a]

Pleiotropy score = 1: Adrbk1, Agtr1b, Apc, Apoe, Axin1, Bmp1, Btc, Calr, Cav1, Cav3, Ccnd1, Ccnd2, Ccnd3, Cdh2, Chd7, Chmp5, Col2a1, Cst3, Ctbp2, Cxcl12, Cxcr4, Cyp26a1, Disp1, Dll1, Dll4, Dnahc11, Dnmt3b, Dvl1, Dync2h1, Dync2li1, Efnb2, Egfr, Egln1, Ep300, Ephb4, Erbb3, Erbb4, Evi1, F2r, Fgf10, Fgf2, Fgf9, Fgfr1, Fkbp1a, Fkbp1b, Foxp4, Frem2, Furin, Fxn, Gab1, Gas1, Gna11, Gnaq, Gys1, Hbegf, Hdac5, Hdac7a, Hdac9, Hgs, Htr1b, Ift172, Ift57, Ift88, Jarid2, Jmjd6, Jup, Kdr, Kif3a, Kif3b, Kl, Krit1, Krt1, Lats2, Lox, Ly6e, Map3k7, Map3k7ip1, Mapk7, Men1, Mgat1, Mib1, Mixl1, Mkl1, Mospd3, Mycn, Myl2, Myst3, Nckap1, Ncoa6, Nkx2-6, Nog, Notch1, Nr2f2, Nrg1, Ofd1, Ovol2, Pbrm1, Pdgfa, Pdgfc, Pdpk1, Plce1, Pnpla2, Por, Postn, Pparbp, Pparg, Prkar1a, Prrx1, Prrx2, Psen1, Psen2, Pten, Rbl2, Rbp4, Rere, Rfx3, Ror1, Ror2, Rttn, Sall1, Sgsh, Shc1, Shh, Slc6a4, Slc8a1, Smad1, Smad2, Smad4, Smad5, Smyd1, Sufu, T, Tall1, Tceb3, Tgfbr3, Th, Thbs1, Thrap4, Tll1, Txnrd2, Vcam1, Vcan, Vcl, Vezf1, Wasf2, Wrn

Pleiotropy score = 2: Adam12, Adam17, Adam9, Atp2a2, Bmp10, Bmpr1a, Bmpr2, Cdk2, Cdk4, Dand5, Dvl2, Ece2, Edn1, Eng, Erbb2, Fbn1, Foxj1, Foxm1, Foxo1, Gbx2, Gdf1, Gja7, Hey1, Hspg2, Jag1, Map2k5, Mgrn1, Nfatc1, Nodal, Nos3, Notch2, Nrp2, Osr1, Pcsk6, Pdlim3, Pkd2, Plxnd1, Smad6, Sox11, Sox4, Srf, Tbx20, Tcfap2a, Tmod1, Ubr1, Ubr2

Pleiotropy score = 3: Acvr2b, Acvrl1, Ate1, Ccne1, Ccne2, Chrd, Crebbp, Crkl, Cxadr, Ednra, Fgfr2, Foxc1, Foxg1, Foxp1, Gata6, Hand1, Has2, Hey2, Hhex, Hif1a, Hoxa3, Invs, Jun, Mef2c, Mkl2, Myh10, Myl7, Nrp1, Pax3, Pdgfra, Pkd1, Rarg, Sall4, Sirt1, Sox9, Ssr1, Tbx5, Tgfbr2, Vegfa, Zfpm1, Zic3,

Pleiotropy score = 4: Acvr1, Adam19, Bmp2, Cfc1, Ece1, Fgf15, Flna, Foxc2, Foxh1, Gja1, Gja5, Hand2, Isl1, Nf1, Ntf3, Ntrk3, Ptpn11, Rara, Rarb, Rxra, Sema3c, Smo, Tbx2, Zfpm2

Pleiotropy score = 5: Aldh1a2, Bmp4, Cited2, Fgf8, Gata4, Mesp1, Nkx2-5, Pitx2, Tbx1, Tgfb2

---

[a]Data are based on searches of the Mouse Genome Informatics database for the following keywords: MP:0000267, MP:0004055, MP:0006356, MP:0006354, MP:0006355, MP:0002672, MP:0004113, MP:0002977, MP:0000282, MP:0004225, MP:0006107, MP:0004187, MP:0000644, MP:0000298, MP:0003808, MP:0000299, MP:0000297, MP:0000301, MP:0000300, MP:0002740, MP:0000270 MP:0003923, MP0004252, MP0004251, MP0000269 MP0006126 MP0005674 MP0002633, MP0006061 MP:0003922, MP0003228, MP0004110, MP0006127, MP:0006115, MP:0002747, MP:0006047, MP:0006117, MP:0002745 MP:0003958, MP:0000287, MP:0006119, MP:0000285, MP:0006045, MP:0006122, MP:0000286, MP:0006130, MP:0006048, MP:0006128, MP:0002748, MP:0002746, MP:0006049, MP:0006123, MP:0006044, MP:0006124, MP:0000279, and MP:0000281. For each gene, aortic arch, outflow tract, valve/endocardial cushion, ventricle/heart tube and atrial abnormalities were scored as 0 or 1 if absent or present, respectively. These scores were summed to create a pleiotropy score for each gene. Of 267 genes identified, 122 had a pleiotropy score >1 (i.e., mutation in these genes affected more than one cardiac structure).

mutational load"—a powerful technique used to determine the role of a candidate gene in case–control approaches.[90–93] Because of the pleiotropic nature of CHD, it may not be advantageous to subclassify CHD into discrete anatomical lesions.

Identification of Candidate Genes

One approach in identifying candidate genes for CHD is to identify those that have a major role in mouse heart development. Mouse models are powerful experimental tools for understanding human CHD. Similar to humans, the mouse has a four-chambered heart with a septated outflow tract, left-sided great arteries, and parallel pulmonary and systemic circulations.[94] Thus, the mouse is a good anatomical model for common cardiac malformations (e.g., ASD, VSD, TOF, TGA, CAT, PDA) that cannot be identified in other genetically tractable model organisms, such as the fruitfly or zebrafish. The availability of complete mouse and human genomic sequence, extensive chromosomal synteny, and phylogenetic closeness, makes the identification of homologous human genes considerably easier. The usefulness of the mouse as a model for CHD is borne out by genetic evidence. Several mouse mutations (e.g., in *Cbp*,[76,95] *Tfap2*,[96,97] *Tbx5*,[98–100] *Tbx1*,[101–103] and *Zic3*[11,104]), generated using transgenic technology, recapitulate the cardiac malformations observed in patients with mutations in these genes. The identification of several human CHD genes (e.g., *NKX2.5*,[14] *GATA4*,[19] *CFC1*,[32] *LEFTY2*,[38] *ACVR2B*,[105] *FOG2*,[36] and *CITED2*[33]) has resulted directly from an initial understanding of their function in mice. A key question is the number of genes that control cardiovascular development. Examination of the Mouse Genome Informatics database reveals that of 4373 genes that have been knocked out, there are 267 with abnormal cardiac morphology or development resembling CHD. If extrapolated to 28,000 genes (the estimated number of genes in the mouse genome[106]), we expect that there will be over 1700 genes that are necessary for heart development. These genes can be efficiently identified using mouse genetic screens that use such techniques as knockouts, *N*-ethyl-*N*-nitrosourea (ENU), and transposon mutagenesis.

High-throughput Magnetic Resonance Imaging

A key limitation to the use of the mouse as a model is that the embryo is opaque at a time that common congenital malformations can be identified. Histology, the traditional approach used at late-developmental stages, is labor intensive and loses 3-D information necessary for interpretation of complex cardiac malformations. Traditional approaches to 3-D analysis (essential, for instance, for aortic arch morphology) use corrosion casting of plastic dye injections, a labor-intensive and highly demanding technique that is not suitable for high throughput. To further mouse genetic approaches, we have developed a key enabling technology—high-throughput magnetic resonance imaging (MRI)—that allows us to rapidly and accurately image cardiac and other embryonic malformations in mouse embryos at a reasonable cost.[89,107–112] We can image approximately 32 mouse embryos at 15.5 days post-coitum simultaneously in single overnight runs [108] to obtain a final image resolution of $25.4 \times 25.4 \times 24.4\,\mu m$ per voxel. All common congenital malformations can be identified with a high degree of accuracy. These include ASD, AVSD, VSD, double outlet right ventricle (DORV), CAT, TGA, interrupted and aberrant aortic arches, and more complex malformations, such as isomerism and abnormal ventricular topology. This approach has a potential throughput of over 10,000 embryos per year. However, screening these images visually is a key limitation of embryo MRI: it is labor intensive, is highly dependent on observer expertise and training, and is now the major limitation to throughput. The time taken to perform quantitative analysis (e.g., to measure the volume of individual organs) means that this important informational component is not routinely extracted. Thus, there is a great need to develop automated analytical procedures (e.g., using mechanical deformation approaches) for rapidly extracting information from the image data sets.

MRI and High-throughput Mouse Genetics

We are now using MRI in combination with ENU and transposon mutagenesis and will begin high-throughput knockout screens in the near future. In the last 2 years our laboratory has analyzed over 4000 embryos using this technique and has characterized malformations in mice lacking *Cited2, Pitx2c, Cyclin E1/E2, Ptdsr, Flna,* and *Lims1*[89,107,113,114] in addition to identifying new mouse mutant lines with cardiac malformation. The combination of MRI and high-throughput mouse genetics will lead to substantial advances in our understanding of the genetic mechanisms in heart development and the identification of new candidate genes for CHD. In the recently announced European and North American Conditional Mouse Mutagenesis programs (EUCOMM and NorCOMM, respectively; http://www.prime-eu.org/newsletters.html), 20,000 mouse genes will be systematically knocked out over the next 5 years. Phenotyping of these knockouts by high-throughput MRI in the long term will lead to the identification of many mouse genes controlling heart development. We envisage that MRI will also allow us to begin testing genetic interactions in high throughput. This will allow the identification of synthetic phenotypes created by mutation in more than one gene—and subsequent analysis of these genetic combinations in human CHD will be facilitated. In addition to identifying the genetic mechanisms, embryo MRI allows us to investigate systematically the buffering effects of genetic background and of the disruption of buffering by environmental variation.

Summary

Sporadic CHD is likely to arise from an accumulation of rare nonsynonymous variants in cardiac developmental genes, from CNV in cardiac developmental genes and from common variants. Common variants that may alter genetic buffering pathways (e.g., folate) and may not be obviously linked to cardiac development. This accumulation will, in turn, lead to mutational loading of cardiac developmental networks. The problems associated with allelic heterogeneity in understanding the genetic mechanisms underlying common CHD can be overcome by using a candidate gene resequencing approach. Candidate genes can be efficiently isolated using mouse genetic techniques combined with high-throughput MRI approaches and applied to human disease. Key future challenges now involve understanding the mechanisms that create and disrupt genetic buffering. A long-term goal in CHD is to manipulate these pathways to enhance buffering and prevent disease.

Conflicts of Interest

The authors declare no conflicts of interest.

References

1. HOFFMAN, J.I.E. 2002. Incidence, mortality and natural history. *In* Paediatric Cardiology. R.H. Anderson, E.J. Baker, F.J. Macartney, *et al.*, Eds.: 111–139. Churchill Livingstone. London.

2. BURN, J., J. GOODSHIP. 2002. Congenital heart disease. *In* Principles and Practice of Medical Genetics. D.L. Rimoin, J.M. Connor, R.E. Pyeritz, *et al.*, Eds. Churchill Livingstone. London.

3. PETERSEN, S., V. PETO & M. RAYNER. 2003. Congenital heart disease statistics 2003. British Heart Foundation Statistics Website www.heartstats.org.

4. BRICKNER, M.E., L.D. HILLIS & R.A. LANGE. 2000. Congenital heart disease in adults. First of two parts. N. Engl. J. Med. **342:** 256–263.

5. MRC VITAMIN STUDY RESEARCH GROUP. 1991. Prevention of neural tube defects: results of the Medical Research Council Vitamin Study. Lancet **338:** 131–137.

6. BLOM, H.J., G.M. SHAW, M. DEN HEIJER, *et al.* 2006. Neural tube defects and folate: case far from closed. Nat. Rev. Neurosci. **7:** 724–731.

7. CLARK, E.B. 2001. Etiology of congenital cardiac malformations: epidemiology and genetics. *In* H.D. Allen, H.P. Gutgessell, E.B. Clark, *et al.*, Eds. Moss and Adams' Heart Disease in Infants, Children, and Adolescents, 6th ed. Lipincott Williams & Wilkins. Philadelphia.

8. FERENCZ, C., J.A. BOUGHMAN, C.A. NEILL, *et al.* 1989. Congenital cardiovascular malformations: questions on inheritance. Baltimore-Washington Infant Study Group. J. Am. Coll. Cardiol. **14:** 756–763.

9. FERENCZ, C. & J.A. BOUGHMAN. 1993. Congenital heart disease in adolescents and adults. Teratology, genetics, and recurrence risks. Cardiol. Clin. **11:** 557–567.

10. PIERPONT, M.E., C.T. BASSON, D.W. BENSON, JR., *et al.* 2007. Genetic basis for congenital heart defects: current knowledge: a scientific statement from the American Heart Association Congenital Cardiac Defects Committee, Council on Cardiovascular Disease in the Young: endorsed by the American Academy of Pediatrics. Circulation **115:** 3015–3038.

11. GEBBIA, M., G.B. FERRERO, G. PILIA, *et al.* 1997. X-linked situs abnormalities result from mutations in ZIC3. Nat. Genet. **17:** 305–308.

12. MEGARBANE, A., N. SALEM, E. STEPHAN, *et al.* 2000. X-linked transposition of the great arteries and incomplete penetrance among males with a nonsense mutation in ZIC3. Eur. J. Hum. Genet. **8:** 704–708.

13. GARG, V., A.N. MUTH, J.F. RANSOM, *et al.* 2005. Mutations in NOTCH1 cause aortic valve disease. Nature **437:** 270–274.

14. SCHOTT, J.J., D.W. BENSON, C.T. BASSON, *et al.* 1998. Congenital heart disease caused by mutations in the transcription factor NKX2–5. Science **281:** 108–111.

15. HEATHCOTE, K., C. BRAYBROOK, L. ABUSHABAN, *et al.* 2005. Common arterial trunk associated with a homeodomain mutation of NKX2.6. Hum. Mol. Genet. **14:** 585–593.

16. CHING, Y.H., T.K. GHOSH, S.J. CROSS, *et al.* 2005. Mutation in myosin heavy chain 6 causes atrial septal defect. Nat. Genet. **37:** 423–428.

17. ZHU, L., R. VRANCKX, P. KHAU VAN KIEN, *et al.* 2006. Mutations in myosin heavy chain 11 cause a syndrome associating thoracic aortic aneurysm/aortic dissection and patent ductus arteriosus. Nat. Genet. **38:** 343–349.

18. ELDADAH, Z.A., A. HAMOSH, N.J. BIERY, *et al.* 2001. Familial Tetralogy of Fallot caused by mutation in the jagged1 gene. Hum. Mol. Genet. **10:** 163–169.

18a. MATSSON, H., J. EASON, C.S. BOOKWALTER, *et al.* 2007. Alpha cardiac actin mutations produce atrial septal defects. Hum. Mol. Genet. Oct 18 [Epub ahead of print].

19. GARG, V., I.S. KATHIRIYA, R. BARNES, *et al.* 2003. GATA4 mutations cause human congenital heart defects and

reveal an interaction with TBX5. Nature **424:** 443–447.

20. LEWIN, M.B., K.L. McBRIDE, R. PIGNATELLI, *et al.* 2004. Echocardiographic evaluation of asymptomatic parental and sibling cardiovascular anomalies associated with congenital left ventricular outflow tract lesions. Pediatrics **114:** 691–696.

21. HUNTINGTON, K., A.G. HUNTER & K.L. CHAN. 1997. A prospective study to assess the frequency of familial clustering of congenital bicuspid aortic valve. J. Am. Coll. Cardiol. **30:** 1809–1812.

22. ROSE, V., R.J. GOLD, G. LINDSAY, *et al.* 1985. A possible increase in the incidence of congenital heart defects among the offspring of affected parents. J. Am. Coll. Cardiol. **6:** 376–382.

23. CAPUTO, S., G. CAPOZZI, M.G. RUSSO, *et al.* 2005. Familial recurrence of congenital heart disease in patients with ostium secundum atrial septal defect. Eur. Heart J. **26:** 2179–2184.

24. CZEIZEL, A., A. PORNOI, E. PETERFFY, *et al.* 1982. Study of children of parents operated on for congenital cardiovascular malformations. Br. Heart J. **47:** 290–293.

25. BURN, J., P. BRENNAN, J. LITTLE, *et al.* 1998. Recurrence risks in offspring of adults with major heart defects: results from first cohort of British collaborative study. Lancet **351:** 311–316.

26. DRENTHEN, W., P.G. PIEPER, K. VAN DER TUUK, *et al.* 2005. Cardiac complications relating to pregnancy and recurrence of disease in the offspring of women with atrioventricular septal defects. Eur. Heart J. **26:** 2581–2587.

27. SANCHEZ-CASCOS, A. 1978. The recurrence risk in congenital heart disease. Eur. J. Cardiol. **7:** 197–210.

28. EMANUEL, R., J. SOMERVILLE, A. INNS, *et al.* 1983. Evidence of congenital heart disease in the offspring of parents with atrioventricular defects. Br. Heart J. **49:** 144–147.

29. MEIJER, J.M., P.G. PIEPER, W. DRENTHEN, *et al.* 2005. Pregnancy, fertility, and recurrence risk in corrected tetralogy of Fallot. Heart **91:** 801–805.

30. ZELLERS, T.M., D.J. DRISCOLL & V.V. MICHELS. 1990. Prevalence of significant congenital heart defects in children of parents with Fallot's tetralogy. Am. J. Cardiol. **65:** 523–526.

31. DIGILIO, M.C., B. MARINO, A. GIANNOTTI, *et al.* 1997. Recurrence risk figures for isolated tetralogy of Fallot after screening for 22q11 microdeletion. J. Med. Genet. **34:** 188–190.

32. BAMFORD, R.N., E. ROESSLER, R.D. BURDINE, *et al.* 2000. Loss-of-function mutations in the EGF-CFC gene CFC1 are associated with human left-right laterality defects. Nat. Genet. **26:** 365–369.

33. SPERLING, S., C.H. GRIMM, I. DUNKEL, *et al.* 2005. Identification and functional analysis of CITED2 mutations in patients with congenital heart defects. Hum. Mutat. **26:** 575–582.

34. ROBINSON, S.W., C.D. MORRIS, E. GOLDMUNTZ, *et al.* 2003. Missense mutations in CRELD1 are associated with cardiac atrioventricular septal defects. Am. J. Hum. Genet. **72:** 1047–1052.

35. ZATYKA, M., M. PRIESTLEY, E.J. LADUSANS, *et al.* 2005. Analysis of CRELD1 as a candidate 3p25 atrioventicu-lar septal defect locus (AVSD2). Clin. Genet. **67:** 526–528.

36. PIZZUTI, A., A. SARKOZY, A.L. NEWTON, *et al.* 2003. Mutations of ZFPM2/FOG2 gene in sporadic cases of tetralogy of Fallot. Hum. Mutat. **22:** 372–377.

37. NEMER, G., F. FADLALAH, J. USTA, *et al.* 2006. A novel mutation in the GATA4 gene in patients with Tetralogy of Fallot. Hum. Mutat. **27:** 293–294.

38. KOSAKI, K., M.T. BASSI, R. KOSAKI, *et al.* 1999. Characterization and mutation analysis of human LEFTY A and LEFTY B, homologues of murine genes implicated in left-right axis development. Am. J. Hum. Genet. **64:** 712–721.

39. McELHINNEY, D.B., E. GEIGER, J. BLINDER, *et al.* 2003. NKX2.5 mutations in patients with congenital heart disease. J. Am. Coll. Cardiol. **42:** 1650–1655.

40. MUNCKE, N., C. JUNG, H. RUDIGER, *et al.* 2003. Missense mutations and gene interruption in PROSIT240, a novel TRAP240-like gene, in patients with congenital heart defect (transposition of the great arteries). Circulation **108:** 2843–2850.

41. WARE, S.M., J. PENG, L. ZHU, *et al.* 2004. Identification and functional analysis of ZIC3 mutations in heterotaxy and related congenital heart defects. Am. J. Hum. Genet. **74:** 93–105.

41a. KARKERA, J.D., J.S. LEE, S. ROESSLER, *et al.* 2007. Loss-of-function mutations in growth differentiation factor-1 (GDF-1) are associated with congenital heart defects in humans. Am. J. Hum. Genet. **81**: 987–994.

41b. KIRK, E.P., M. SUNDE, M.W. COSTA, *et al.* 2007. Mutations in cardiac T-box factor gene TBXZO are associated with diverse cardiac pathologies including defects of septation and valvologenesis and cardiomyopathy. Am. J. Hum. Genet. **81:** 280–291.

42. SARKOZY, A., E. CONTI, C. NERI, *et al.* 2005. Spectrum of atrial septal defects associated with mutations of NKX2.5 and GATA4 transcription factors. J. Med. Genet. **42:** e16.

43. SCHLUTERMAN, M.K., A.E. KRYSIAK, I.S. KATHIRIYA, *et al.* 2007. Screening and biochemical analysis of GATA4 sequence variations identified in patients with congenital heart disease. Am. J. Med. Genet. A. **143:** 817–823.

44. ZHANG, L., Z. TUMER, J.R. JACOBSEN, *et al.* 2006. Screening of 99 Danish patients with congenital heart disease for GATA4 mutations. Genet. Test **10:** 277–280.

45. SARKOZY, A., E. CONTI, & R. D'AGOSTINO, *et al.* 2005. ZFPM2/FOG2 and HEY2 genes analysis in nonsyndromic tricuspid atresia. Am. J. Med. Genet. A **133:** 68–70.

46. OZCELIK, C., N. BIT-AVRAGIM, A. PANEK, *et al.* 2006. Mutations in the EGF-CFC gene cryptic are an infrequent cause of congenital heart disease. Pediatr. Cardiol. **27:** 695–698.

47. MACMAHON, B., T. McKEOWN & R.G. RECORD. 1953. The incidence and life expectation of children with congenital heart disease. Br. Heart. J. **15:** 121–129.

48. PRITCHARD, J.K. 2001. Are rare variants responsible for susceptibility to complex diseases? Am. J. Hum. Genet. **69:** 124–137.

49. SLAGER, S.L., J. HUANG & V.J. VIELAND. 2000. Effect of allelic heterogeneity on the power of the transmission disequilibrium test. Genet. Epidemiol. **18**: 143–156.

50. POLLEX, R.L. & R.A. HEGELE. 2007. Copy number variation in the human genome and its implications for cardiovascular disease. Circulation **115**: 3130–3138.

51. MENTEN, B., N. MAAS, B. THIENPONT, *et al.* 2006. Emerging patterns of cryptic chromosomal imbalance in patients with idiopathic mental retardation and multiple congenital anomalies: a new series of 140 patients and review of published reports. J. Med. Genet. **43**: 625–633.

51a. THIENPONT, B., L. MERTENS, T. DE RAVEL, *et al*, 2007. Submicroscopic chromosomal imbalances detected by array-CGH are a frequent cause of congenital heart defects in selected patients. Eur. Heart J. **28**: 2778–2784.

52. REDON, R., S. ISHIKAWA, K.R. FITCH, *et al.* 2006. Global variation in copy number in the human genome. Nature **444**: 444–454.

53. HARTMAN, J.Lt., B. GARVIK & L. HARTWELL. 2001. Principles for the buffering of genetic variation. Science **291**: 1001–1004.

54. RUTHERFORD, S.L. & S. HENIKOFF. 2003. Quantitative epigenetics. Nat. Genet. **33**: 6–8.

55. SOLLARS, V., X. LU, L. XIAO, *et al.* 2003. Evidence for an epigenetic mechanism by which Hsp90 acts as a capacitor for morphological evolution. Nat. Genet. **33**: 70–74.

56. COWEN, L.E. & S. LINDQUIST. 2005. Hsp90 potentiates the rapid evolution of new traits: drug resistance in diverse fungi. Science **309**: 2185–2189.

57. QUEITSCH, C., T.A. SANGSTER & S. LINDQUIST. 2002. Hsp90 as a capacitor of phenotypic variation. Nature **417**: 618–624.

58. TRUE, H.L., I. BERLIN & S.L. LINDQUIST. 2004. Epigenetic regulation of translation reveals hidden genetic variation to produce complex traits. Nature **431**: 184–187.

59. LEHNER, B., C. CROMBIE, J. TISCHLER, *et al.* 2006. Systematic mapping of genetic interactions in Caenorhabditis elegans identifies common modifiers of diverse signaling pathways. Nat. Genet. **38**: 896–903.

60. OH, S.P. & E. LI. 2002. Gene-dosage-sensitive genetic interactions between inversus viscerum (iv), nodal, and activin type IIB receptor (ActRIIB) genes in asymmetrical patterning of the visceral organs along the left-right axis. Dev. Dyn. **224**: 279–290.

61. WARE, S.M., K.G. HARUTYUNYAN & J.W. BELMONT. 2006. Heart defects in X-linked heterotaxy: evidence for a genetic interaction of Zic3 with the nodal signaling pathway. Dev. Dyn. **235**: 1631–1637.

62. NOMURA, M., & E. LI. 1998. Smad2 role in mesoderm formation, left-right patterning and craniofacial development. Nature **393**: 786–790.

63. COLLIGNON, J., I. VARLET & E.J. ROBERTSON. 1996. Relationship between asymmetric nodal expression and the direction of embryonic turning. Nature **381**: 155–158.

64. NOWOTSCHIN, S., J. LIAO, P.J. GAGE, *et al.* 2006. Tbx1 affects asymmetric cardiac morphogenesis by regulating Pitx2 in the secondary heart field. Development **133**: 1565–1573.

65. STENNARD, F.A., M.W. COSTA, D. LAI, *et al.* 2005. Murine T-box transcription factor Tbx20 acts as a repressor during heart development, and is essential for adult heart integrity, function and adaptation. Development **132**: 2451–2462.

66. YAMAGISHI, H., C. YAMAGISHI, O. NAKAGAWA, *et al.* 2001. The combinatorial activities of Nkx2.5 and dHAND are essential for cardiac ventricle formation. Dev. Biol. **239**: 190–203.

67. XIN, M., C.A. DAVIS, J.D. MOLKENTIN, *et al.* 2006. A threshold of GATA4 and GATA6 expression is required for cardiovascular development. Proc. Natl. Acad. Sci. USA **103**: 11189–11194.

68. VAN BEYNUM, I.M., L. KAPUSTA, M. DEN HEIJER, *et al.* 2006. Maternal MTHFR 677C>T is a risk factor for congenital heart defects: effect modification by periconceptional folate supplementation. Eur. Heart. J. **27**: 981–987.

69. CZEIZEL, A.E., M. DOBO & P. VARGHA. 2004. Hungarian cohort-controlled trial of periconceptional multivitamin supplementation shows a reduction in certain congenital abnormalities. Birth Defects Res. A Clin. Mol. Teratol. **70**: 853–861.

70. LOFFREDO, C.A., P.D. WILSON & C. FERENCZ. 2001. Maternal diabetes: an independent risk factor for major cardiovascular malformations with increased mortality of affected infants. Teratology **64**: 98–106.

71. LOFFREDO, C.A., J. HIRATA, P.D. WILSON, *et al.* 2001. Atrioventricular septal defects: possible etiologic differences between complete and partial defects. Teratology **63**: 87–93.

72. KUEHL, K.S. & C. LOFFREDO. 2002. Risk factors for heart disease associated with abnormal sidedness. Teratology **66**: 242–248.

73. WATKINS, M.L., S.A. RASMUSSEN, M.A. HONEIN, *et al.* 2003. Maternal obesity and risk for birth defects. Pediatrics **111**(5 Part 2): 1152–1158.

74. GOLDMUNTZ, E., R. BAMFORD, J.D. KARKERA, *et al.* 2002. CFC1 mutations in patients with transposition of the great arteries and double-outlet right ventricle. Am. J. Hum. Genet. **70**: 776–780.

75. STEVENS, C.A. & M.G. BHAKTA. 1995. Cardiac abnormalities in the Rubinstein-Taybi syndrome. Am. J. Med. Genet. **59**: 346–348.

76. PETRIJ, F., R.H. GILES, H.G. DAUWERSE, *et al.* 1995. Rubinstein-Taybi syndrome caused by mutations in the transcriptional co-activator CBP. Nature **376**: 348–351.

77. ROELFSEMA, J.H., S.J. WHITE, Y. ARIYUREK, *et al.* 2005. Genetic heterogeneity in Rubinstein-Taybi syndrome: mutations in both the CBP and EP300 genes cause disease. Am. J. Hum. Genet. **76**: 572–580.

78. VISSERS, L.E., C.M. VAN RAVENSWAAIJ, R. ADMIRAAL, *et al.* 2004 Mutations in a new member of the chromodomain gene family cause CHARGE syndrome. Nat. Genet. **36**: 955–957.

79. BOSMAN, E.A., A.C. PENN, J.C. AMBROSE, *et al.* 2005. Multiple mutations in mouse Chd7 provide models for CHARGE syndrome. Hum. Mol. Genet. **14**: 3463–3476.

80. JONGMANS, M.C., R.J. ADMIRAAL, K.P. VAN DER DONK, *et al.* 2006. CHARGE syndrome: the phenotypic spectrum

of mutations in the CHD7 gene. J. Med. Genet. **43:** 306–314.

81. HIRAYAMA-YAMADA, K., M. KAMISAGO, K. AKIMOTO, *et al.* 2005. Phenotypes with GATA4 or NKX2.5 mutations in familial atrial septal defect. Am. J. Med. Genet. A **135:** 47–52.

82. GOLDMUNTZ, E., E. GEIGER & D.W. BENSON. 2001. NKX2.5 mutations in patients with tetralogy of fallot. Circulation **104:** 2565–2568.

83. BENSON, D.W., G.M. SILBERBACH, A. KAVANAUGH-MCHUGH, *et al.* 1999. Mutations in the cardiac transcription factor NKX2.5 affect diverse cardiac developmental pathways. J. Clin. Invest. **104:** 1567–1573.

84. MOHAMED, S.A., Z. AHERRAHROU, H. LIPTAU, *et al.* 2006. Novel missense mutations (p.T596M and p.P1797H) in NOTCH1 in patients with bicuspid aortic valve. Biochem. Biophys. Res. Commun. **345:** 1460–1465.

85. KREBS, L.T., Y. XUE, C.R. NORTON, *et al.* 2000. Notch signaling is essential for vascular morphogenesis in mice. Genes. Dev. **14:** 1343–1352.

86. WHITTEMORE, R., J.C. HOBBINS & M.A. ENGLE. 1982. Pregnancy and its outcome in women with and without surgical treatment of congenital heart disease. Am. J. Cardiol. **50:** 641–651.

87. DENNIS, N.R. & J. WARREN. 1981. Risks to the offspring of patients with some common congenital heart defects. J. Med. Genet. **18:** 8–16.

88. BAMFORTH, S.D., J. BRAGANCA, J.J. ELORANTA, *et al.* 2001. Cardiac malformations, adrenal agenesis, neural crest defects and exencephaly in mice lacking Cited2, a new Tfap2 co-activator. Nat. Genet. **29:** 469–474.

89. BAMFORTH, S.D., J. BRAGANCA, C.R. FARTHING, *et al.* 2004. Cited2 controls left-right patterning and heart development through a Nodal-Pitx2c pathway. Nat. Genet. **36:** 1189–1196.

90. COHEN, J.C., R.S. KISS, A. PERTSEMLIDIS, *et al.* 2004. Multiple rare alleles contribute to low plasma levels of HDL cholesterol. Science **305:** 869–872.

91. FRIKKE-SCHMIDT, R., B.G. NORDESTGAARD, G.B. JENSEN, *et al.* 2004. Genetic variation in ABC transporter A1 contributes to HDL cholesterol in the general population. J. Clin. Invest. **114:** 1343–1353.

92. COHEN, J., A. PERTSEMLIDIS, I.K. KOTOWSKI, *et al.* 2005. Low LDL cholesterol in individuals of African descent resulting from frequent nonsense mutations in PCSK9. Nat. Genet. **37:** 161–165.

93. COHEN, J.C., A. PERTSEMLIDIS, S. FAHMI, *et al.* 2006. Multiple rare variants in NPC1L1 associated with reduced sterol absorption and plasma low-density lipoprotein levels. Proc. Natl. Acad. Sci .USA **103:** 1810–1815.

94. KENT, G.C. & R.K. CARR. 2001. Comparative Anatomy of the Vertebrates, 9 ed. McGraw Hill. Boston.

95. OIKE, Y., A. HATA, T. MAMIYA, *et al.* 1999. Truncated CBP protein leads to classical Rubinstein-Taybi syndrome phenotypes in mice: implications for a dominant-negative mechanism. Hum. Mol. Genet. **8:** 387–396.

96. SATODA, M., F. ZHAO, G.A. DIAZ, *et al.* 2000. Mutations in TFAP2B cause Char syndrome, a familial form of patent ductus arteriosus. Nat. Genet. **25:** 42–46.

97. BREWER, S., X. JIANG, S. DONALDSON, *et al.* 2002. Requirement for AP-2alpha in cardiac outflow tract morphogenesis. Mech. Dev. **110:** 139–149.

98. BASSON, C.T., D.R. BACHINSKY, R.C. LIN, *et al.* 1997. Mutations in human TBX5 cause limb and cardiac malformation in Holt-Oram syndrome. Nat. Genet. **15:** 30–35.

99. LI, Q.Y., R.A. NEWBURY-ECOB, J.A. TERRETT, *et al.* 1997. Holt-Oram syndrome is caused by mutations in TBX5, a member of the Brachyury (T) gene family. Nat. Genet. **15:** 21–29.

100. BRUNEAU, B.G., G. NEMER, J.P. SCHMITT, *et al.* 2001. A murine model of Holt-Oram syndrome defines roles of the T-box transcription factor Tbx5 in cardiogenesis and disease. Cell **106:** 709–721.

101. LINDSAY, E.A., F. VITELLI, H. SU, *et al.* 2001. Tbx1 haploinsufficieny in the DiGeorge syndrome region causes aortic arch defects in mice. Nature **410:** 97–101.

102. MERSCHER, S., B. FUNKE, J.A. EPSTEIN, *et al.* 2001. TBX1 is responsible for cardiovascular defects in velo-cardio-facial/DiGeorge syndrome. Cell **104:** 619–629.

103. JEROME, L.A. & V.E. PAPAIOANNOU. 2001. DiGeorge syndrome phenotype in mice mutant for the T-box gene, Tbx1. Nat. Genet. **27:** 286–291.

104. PURANDARE, S.M., S.M. WARE, K.M. KWAN, *et al.* 2002. A complex syndrome of left-right axis, central nervous system and axial skeleton defects in Zic3 mutant mice. Development **129:** 2293–2302.

105. KOSAKI, R., M. GEBBIA, K. KOSAKI, *et al.* 1999. Left-right axis malformations associated with mutations in ACVR2B, the gene for human activin receptor type IIB. Am. J. Med. Genet. **82:** 70–76.

106. WATERSTON, R.H., K. LINDBLAD-TOH, E. BIRNEY, *et al.* 2002. Initial sequencing and comparative analysis of the mouse genome. Nature **420:** 520–562.

107. BOGANI, D., C. WILLOUGHBY, J. DAVIES, *et al.* 2005. Dissecting the genetic complexity of human 6p deletion syndromes by using a region-specific, phenotype-driven mouse screen. Proc. Natl. Acad. Sci. U. S. A. **102:** 12477–12482.

108. SCHNEIDER, J.E., J. BOSE, S.D. BAMFORTH, *et al.* 2004. Identification of cardiac malformations in mice lacking Ptdsr using a novel high-throughput magnetic resonance imaging technique. BMC Dev. Biol. **4:** 16.

109. SCHNEIDER, J.E. & S. BHATTACHARYA. 2004. Making the mouse embryo transparent: identifying developmental malformations using magnetic resonance imaging. Birth Defects Res. C Embryo. Today **72:** 241–249.

110. SCHNEIDER, J.E., S.D. BAMFORTH, S.M. GRIEVE, *et al.* 2003. High-resolution, high-throughput magnetic resonance imaging of mouse embryonic anatomy using a fast gradient-echo sequence. Magma **16:** 43–51.

111. SCHNEIDER, J.E., S.D. BAMFORTH, C.R. FARTHING, *et al.* 2003. Rapid identification and 3D reconstruction of complex cardiac malformations in transgenic mouse embryos using fast gradient echo sequence magnetic resonance imaging. J. Mol. Cell. Cardiol. **35:** 217–222.

112. SCHNEIDER, J.E., S.D. BAMFORTH, C.R. FARTHING, *et al.* 2003. High-resolution imaging of normal anatomy, and

neural and adrenal malformations in mouse embryos using magnetic resonance microscopy. J. Anat. **202:** 239–247.

113. GENG, Y., Q. YU, E. SICINSKA, *et al.* 2003. Cyclin E ablation in the Mouse. Cell **114:** 431–443.

114. HART, A.W., J.E. MORGAN, J. SCHNEIDER, *et al.* 2006. Cardiac malformations and midline skeletal defects in mice lacking filamin A. Hum. Mol. Genet. **15:** 2457–2467.

115. TANG, Z.H., L. XIA, W. CHANG, *et al.* 2006. Two novel missense mutations of GATA4 gene in Chinese patients with sporadic congenital heart defects. Zhonghua Yi Xue Yi Chuan Xue Za Zhi **23:** 134–137.

116. GONG, W., S. GOTTLIEB, J. COLLINS, *et al.* 2001. Mutation analysis of TBX1 in non-deleted patients with features of DGS/VCFS or isolated cardiovascular defects. J. Med. Genet. **38:** E45.

MicroRNAs Control Gene Expression

Importance for Cardiac Development and Pathophysiology

Daniele Catalucci,[a,b] Michael V. G. Latronico,[b] and Gianluigi Condorelli[a,b]

[a]Department of Medicine, Division of Cardiology, University of California San Diego,
La Jolla, California, 92093-0613, USA

[b]Science and Technology Pole, I.R.C.C.S. MultiMedica, Milan 20100, Italy

Growing evidence indicates that microRNAs (miRNAs) are involved in a variety of basic biological processes, including cell proliferation, apoptosis, stress response, hematopoesis, and oncogenesis. In fact, bioinformatic analysis predicts that each miRNA may regulate hundreds of targets, suggesting that miRNAs may play roles in almost every biological pathway. Information from recent studies indicate that miRNAs are involved in the regulation of cardiac development and pathophysiology. Notably, knockout of miRNA-1 was associated with cardiac defects, including regulation of cardiac morphogenesis, electrical conduction, and cell cycle control. Our group has identified a critical role of miRNA-1 and miRNA-133 in determining cardiac hypertrophy and has shown an inverse correlation of expression with cardiac hypertrophy, *in vitro,* in murine models and in human disease states associated with cardiac hypertrophy. Remarkably, *in vivo* experiments with a single infusion of antagomir-133 oligonucleotide, a small cholesterol-conjugated RNA sequence suppressing endogenous miRNA, induced marked and sustained cardiac hypertrophy. Shedding light on the role of this new class of RNA molecules in heart physiology and pathology may reveal possible future therapeutic applications for the treatment of heart diseases.

Key words: microRNA; cardiac hypertrophy; heart disease; antagomir

Introduction

Gene expression is the process by which the DNA sequence of a gene is converted into the final product (i.e., proteins or, sometimes, RNA). Each cell of a multicellular organism contains the same set of genes, yet, each with a distinctive pattern of gene expression. This permits the array of different cellular functions needed to sustain life. In the functioning of a particular process, cells constantly adjust the type and/or the amount of protein made by changing the set of expressed genes. This fact is fundamental for development (e.g., the formation of the heart chambers), growth (e.g., physiological hypertrophy of the heart with growth of the child), and adjustment to the environment (e.g., hypertrophy of the heart with exercise). However, when the accurate control of gene expression breaks down, the consequences are varied. Birth defects, such as septal abnormalities, can result when the regulation of one or more genes important for development is modified. When alterations arise in the postnatal organism, a myriad of pathological conditions—stretching from benign alterations of conduction to potentially life-threatening cardiomyopathies—arise. Moreover, the considerable progress in the management of once life-incompatible congenital disorders has unmasked more subtle alterations that now often need treatment later in life.

Complex and, as yet, only partially understood control mechanisms orchestrate these processes. To date, control of gene expression in eukaryotic cells is known to occur at several levels, including chromatin structure, transcriptional initiation, transcript processing, transcript stability, translational initiation, post-translational modification, and protein stability (TABLE 1). At each of these levels, the many operant mechanisms are mediated by histone deacetylases (HDACs), transcription factors, small interfering (si) RNAs, kinases, tissue-specific proteins, and others. Some of these mechanisms are discussed with regard to the cardiac system in other articles in this Special Issue. However, the recent discovery of the existence of microRNAs (miRNAs) has introduced an additional level of control into that already complex phenomenon that

Address for correspondence: Professor Gianluigi Condorelli, MD, PhD, Division of Cardiology, Department of Medicine, University of California San Diego, 9500 Gilman Drive, La Jolla, CA 92093-0613. Fax: +1 858 822 3027.

gcondorelli@ucsd.edu

Ann. N.Y. Acad. Sci. 1123: 20–29 (2008). © 2008 New York Academy of Sciences.
doi: 10.1196/annals.1420.004

TABLE 1. Overview of conventional gene expression control mechanisms

Regulatory level	Description
Chromatin structure	DNA can be more (heterochromatin) or less (euchromatin) compacted. The structuring of DNA can, thus, affect the ability of transcription factors and RNA polymerases to access genes and initiate transcription. Moreover, whether euchromatin is under active transcription or not depends on the state of the histone proteins (methylation, phosphorylation, acetylation) to which the DNA is attached.
Transcriptional initiation	Promoter strength, the presence of enhancer sequences, and the interaction between multiple activator or inhibitor proteins constitutes the most important level of control of gene expression.
Transcript processing	mRNA must be capped and polyadenylated. Moreover, introns must be removed, but alternative splicing allows different proteins to be produced from the same gene.
mRNA translocation	Translation takes place at the endoplasmic reticulum, so mRNA is exported from the nucleus for this to occur.
Transcript stability	Stability of mRNAs can vary depending on sequences found on the 3′ UTR.
Translational initiation	Ability of ribosomes to recognize the start codon can affect translation efficiency.
Post-translational modification	Processes, such as glycosylation, acetylation, and acylation, can affect the location and stability of a protein once it is translated.

is gene expression. This will be reviewed here with emphasis on the relationship with cardiogenesis and cardiovascular pathophysiology. The new understandings being gained are already beginning to modify conventional views of cardiac biology.

Characteristics of MicroRNAs

MicroRNAs are an abundant class of small, roughly 22-nucleotide (nt)-long, endogenous, noncoding RNAs. They were stumbled upon more than 20 years ago when a mutation of a heterochronic gene of *Caenorhabditis elegans* was found to be responsible for the alteration of normal development in this nematode worm.[1] Subsequent studies then demonstrated that the mechanism responsible for the alterations operated not through a canonical protein product but rather via a newly defined RNA–RNA interaction.[2,3]

Biogenesis of MicroRNAs and Assembly into a Ribonucleoprotein Complex

In mammals, the majority of miRNAs are located within introns of either protein-coding or noncoding host genes, while others, depending on the occurrence of alternative splicing, are present either in an exon or an intron.[4] A significant number of miRNAs are also assembled in clusters in which two or three miRNAs are generated from a common parent mRNA. Many of the discovered miRNAs are specifically expressed as tissue-stage and/or developmental stage miRNAs, and this can be attributed to regulatory sequences present in their promoters.[5–8] A specific expression pattern can be imposed by host genes when miRNAs are located in their respective introns.[4,9]

RNA polymerase II is responsible for the transcription of miRNAs to form a primary transcript, referred to as the *primary miRNA precursor* (pri-miRNA) (FIG. 1, left box). This is several hundred or thousands of nts long and has a characteristic stem-loop configuration with a 5′-end cap structure and a polyadenylated 3′-tail sequence. While still within the nucleus, pri-miRNA is cleaved into a shorter transcript, termed *pre-miRNA*, by the nuclear ribonuclease III Drosha.[10] The pre-miRNA is 60–70-nt long and is characterized by the usual single-stranded, 2–3-nt, overhanging 3′-end and a phosphate group at the 5′-end, distinctive of RNase III cleavage of double-stranded (ds) RNA. Pre-miRNA is then exported from the nucleus by exportin-5. Within the cytoplasm, pre-miRNA is processed into an 18–22-nt miRNA:miRNA* duplex by another RNase III, Dicer. Notably, further differential processing may occur at this step, intervening as an additional control of miRNA expression patterning.[11,12] Very recently, intronic miRNA precursors (called by the authors *mirtrons*) have been shown to enter the processing pathway without the Drosha-mediated cleavage, thus introducing a further complexity in miRNA biogenesis.[13] The miRNA is then assembled into a ribonucleoprotein particle (RNP) called the *miRNA-induced silencing complex* (miRISC).[14] As it is loaded into the RNP, the duplex is unwound by a helicase; miRNA*, the *passenger strand*, is degraded leaving the *guide strand* within the miRISC (for excellent reviews, see Refs. 9, 15). miRNAs are then ready to be functional.

More than Just One Function?

The effects of miRNAs are produced mainly within the cytoplasm of cells through their base pairing with complementary sequences present at the

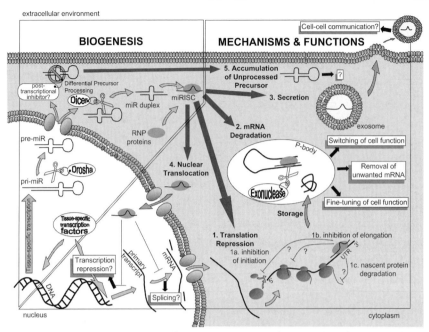

FIGURE 1. Schematic representation of miRNA biogenesis (left box) and mechanisms (right box). Functions are given in the white boxes; see text for explanation. (In color in *Annals* online.)

3′-untranslated region (UTR) of target mRNAs (FIG. 1, right box). The binding specificity between miRNAs and these mRNAs is dictated by only 6–7 out of the approximately 22 nt that compose an miRNA. This sequence is called the *seed sequence* and is located at the 5′-end of the miRNA molecule, often in multiple copies.[14] The rest of the molecule usually binds with only partial complementarity, producing characteristic mismatch bulges, especially in the central region and to a lesser extent at the 3′-end. Occasionally, the pairing of the miRNA seed can be marginally suboptimal, but miRNA–mRNA annealing can be stabilized by a higher degree of complementarity at the 3′-end. Of relevance are the thermodynamic properties of UTR target sites.[7] In fact, while an unstructured secondary configuration located in an accessible region may facilitate miRNA pairing, a more stable and complex structure may interfere with the binding of miRNAs even with high sequence complementarity. However, in particular cellular conditions, an unfolding of these stable secondary structures might be promoted, thus rendering the same site accessible. RNA-binding proteins or miRNAs could, therefore, function cooperatively to alter the complexity of regions by binding specific neighboring binding sites, thus promoting or inhibiting the binding of other miRNAs. This might introduce another level of miRNA target selection.

Dfferences in pairing seem important also for the type of post-translational control that would be produced. When binding is partial, miRNAs are responsible for the reduced translation of targeted mRNAs (FIG. 1, 1. Translation Repression). The exact mechanism of action is still not clear since steps both before and after translation initiation have been reported as the point of repression. In fact, inhibition of initiation factor (IF)4E-dependent initiation,[16,17] elongation,[2,3,18] and degradation of nascent protein,[19,20] among others, have been reported (FIG. 1, 1a–c). When miRNAs bind with precise complementarity to target mRNAs, they behave similarly to siRNAs and signal for mRNA degradation.[7,21–25] In contrast, because miRNAs are transported into the cytoplasm, miRNA-mediated mRNA degradation occurs not via an siRNA-like mechanism of endonucleolytic cleavage but rather through the normal pathway of deadenylation followed by decapping and subsequent degradation by exonuclease activity (FIG. 1, 2. mRNA Degradation). It has also been shown that miRISC components localize to structures called *processing bodies* (P-bodies).[26] These are cytoplasmic foci containing enzymes important in the normal pathway of mRNA degradation. Within these P-bodies, translationally repressed mRNA is either sequestered in storage structures or processed for degradation (FIG. 1).

The first insights into the function of miRNAs were seen in *C. elegans* where they have an important "switch" effect during development.[2,3] Later observations pointed out the importance of miRNAs also for the normal development of mammals. Mouse and human embryonic stem (ES) cells, for example, were found to specifically express a set of miRNAs that, upon differentiation, are downregulated.[27,28] Moreover, ES cells deficient for *dicer*, even if viable, were not able to differentiate either *in vitro* or *in vivo*,[29] and *dicer* mutant mouse embryos died during gastrulation and lost ES cells.[30] In addition, it was proposed,[31] that tissue-specific miRNAs might confer robustness to tissue-specific gene expression by blocking potentially large sets of mRNAs that are expressed inappropriately in tissues in which their presence would be otherwise detrimental. Analysis of *Drosophila melanogaster* revealed that mRNAs with miRNA-1 target sites are expressed largely where miRNA-1 target sites are not expressed, that is in nonmuscle tissues.[31] In mice, mRNAs were found coexpressed with miRNAs that had evolved to be devoid of the target sequences for these miRNAs.[32] Therefore, miRNAs also have important functions after the completion of development, e.g. to silence mRNAs that are unwanted in specific cell lineages.[33]

MicroRNAs have been recently implicated in a cell to cell communication mechanism whereby cells are involved in an exchange of genetic material.[34] Many types of cells are known to release proteins into the extracellular environment via exosomes. These structures have now been demonstrated to contain molecules of RNA, including miRNAs (FIG. 1, 3. Secretion). This finding has raised the exciting possibility that cells can modify gene expression not only of other nearby cells but also, if released into the circulation, of cells at distant sites, with miRNAs acting akin to hormones.

All the above described functions depend on post-translational repression (either via reduced translation efficiency or degradation of the targeted mRNA), thought to be the sole mechanism through which miRNAs act. However, this view has been recently challenged. Although relatively small in size, miRNAs can still contain specific sequences at the 3′-end that are responsible for controlling their post-transcriptional behavior. In fact, miRNAs, such as miRNA-29b, may contain a distinctive 3′-hexanucleotide terminal motif responsible for its relocation back into the nucleus during the cycling phase of cells (FIG. 1, 4. Nuclear Translocation).[35] The exact function of nuclear relocated miRNAs, however, is not yet understood but has been speculated to involve either transcriptional control or splicing regulation. Furthermore, differential precursor processing has been documented for some

miRNAs, such as miRNA-138, that, while having a ubiquitous expression of the pre-miRNA, are characterized by selective maturation occurring in certain cell types or at a particular developmental stage.[12] Evidence points to the presence of an inhibitor that binds to the pre-miRNA and that prevents pre-miRNA processing by Dicer. Accumulation of the pre-miRNA within the cytoplasm might not only represent an additional level of control of miRNA expression (thought to occur primarily at the transcriptional level) but also suggests the presence of a novel function for miRNA precursors (FIG. 1, 5. Accumulation of Unprocessed Precursor).

MicroRNA and Cardiovascular Biology

Muscle-specific MicroRNAs

MicroRNAs are estimated to comprise at least 1% of animal genes[36] and regulate 30% of the human genome,[37] making them one of the most abundant classes of regulators,[31] with a pattern of expression that is often perturbed in disease states.[38–40] A large array of miRNAs can be found within tissues of an organism, and at least one miRNA is specifically expressed per tissue.[41] In muscle, for example, miRNA-133 has been found to be preferentially expressed. Other miRNAs, such as miRNA-1, have also been found to be muscle specific.[42,43] To date, only miRNA-208 has been found to be purely cardiac specific.

The miRNA-1 family is one of the most highly conserved and consists of miRNA-206 (which is not expressed in cardiac muscle) and two closely related transcripts, miRNA-1-1 and miRNA-1-2.[41,44,45] The miRNA-1 family is found as part of a polycistronic unit that is transcribed together with components of the miRNA-133 family, comprised of miRNA-133a-1, miRNA-133a-2, and miRNA133-b paralogs. Chromosome 2 contains the miRNA-1-1/miRNA-133a-2 intergenic bicistron, while miRNA-206/miRNA-133b (also intergenic) is found on chromosome 1. On the other hand, miRNA-1-2/miRNA-133a-1 is intronic and is located on chromosome 18; these miRNAs are found on the opposite strand of the nonmuscle-specific protein-encoding gene, Mindbomb (Mb), between exons 12 and 13, demonstrating the complex characteristics of miRNA genetics.[46]

At the regulatory level, mammalian cardiac miRNA-1 is controlled by the serum response factor (SRF), which recruits a coactivator, myocardin, to muscle-specific genes that control differentiation.[7] This is slightly different from that occurring in skeletal muscle where miRNA-1 expression requires myogenic transcription factors, such as myogenic differentiation 1 (MyoD), myocyte enhancer factor 2, and myogenin. In addition, the presence of putative transcription factor binding sites in between miRNA-1-1 and miRNA-133a-2 suggests the possibility that the individual miRNAs contained in the polycistronic unit may be independently regulated.[46] Concurrently, similar transcriptional control has been shown for skeletal muscle-specific expression of miRNA-133.[47] However, in cardiac muscles, where expression of MyoD and myogenin is not observed, modulation of miRNA-133 levels is mainly regulated by SRF, as shown for miRNA-1. In addition, miRNA-133 has been reported to repress SRF, suggesting a possible regulatory loop.[47]

Control of Muscle Development

Skeletal and heart muscle expression of miRNA-1 and miRNA-133 has been found to be limited during early embryonic development of animals, such as zebrafish[41] and mice.[39,47] Expression then increases during the later embryonic stages and during the neonatal period and is maximal in adulthood. Analysis of mutants in *D. melanogaster* revealed that miRNA-1 is not required for specification or patterning of muscle but rather for postmitotic muscle growth of larvae.[45] Similarly, miRNA-1 spatiotemporal patterning has also been found in mice.[7] While miRNA-1-1 is found to be initially strongly expressed in the less proliferative inner curvature of the heart loop and in atria, miRNA-1-2 can be found mainly in the ventricles. The temporal pattern of expression is recapitulated in C_2C_{12} cells and human myoblasts *in vitro* in that undifferentiated myoblasts do not express these muscle-specific miRNAs, while expression increases with the differentiation into myotubes.[46,47] In skeletal muscle, overexpression experiments have demonstrated that miRNA-133 is capable of promoting proliferation by targeting SRF, while miRNA-1 promotes myogenic differentiation by targeting of HDAC4.[47] Introduction of miRNA-1 in developing *Xenopus laevis* produced an altered phenotype characterized by less cell proliferation and no development of cardiac tissue. miRNA-133 misexpression was responsible for increased proliferation and altered cardiac tissue formation with disorganized looping and chamber formation. Thus, these two opposing effects demonstrate the critical importance of the correct timing and dosing of miRNAs for heart development. The results obtained *in vitro* and in nonmammalian models have also been confirmed in mice. Overexpression of miRNA-1 in a transgenic mouse model resulted in a phenotype characterized by thin-walled ventricles from premature differentia-

tion and early withdrawal of cardiomyocytes from the cell cycle.[7] Adult miRNA-1-2 knockout mice presented thickened chamber walls from prolonged hyperplasia[48]; many of the embryos from these mice suffered from fatal septal defects. Also, Hand2, a validated target of miRNA-1 in the mammalian heart, is a transcription factor that regulates ventricular cardiomyocyte expansion by inhibiting cardiomyocyte progenitor proliferation.[48,49] In addition, HDAC4 is targeted by miRNA-1 in skeletal muscle, and this relieves inhibition of myogenesis. miRNA-133, however, is responsible for the repression of SRF in muscle cells, thereby constituting a negative regulatory loop.[47]

The question as to how the apparently opposing effects of miRNA-1 and miRNA-133 can be reconciled with the finding that this miRNA-bicistronic unit increases in expression during development highlights the need for further studies.[47]

Role in Postnatal Cardiac Pathophysiology

A recent sudden surge of reports has documented the effects of miRNAs in the postnatal heart, demonstrating the important role that miRNAs have in cardiac pathophysiology.

Hypertrophic Cardiac Growth

Several studies of skeletal muscle have shown that alterations of phenotype can occur with either the dysregulation of miRNA expression or the mutation of target mRNAs. In fact, miRNA-206 was found upregulated and miRNA-1 and miRNA-133a downregulated in mouse hindlimb muscles with functional overload-induced hypertrophy.[50] Moreover, a mutation in the 3′-UTR of the myostatin gene has been ascribed as being important for the increased muscle mass of Texel sheep in that it produces a binding site for miRNA-1/miRNA-206 that is otherwise absent.[51] The consequent gain-of-function polymorphism leads to inappropriate targeting of the myostatin gene, producing a hypertrophic phenotype similar to the myostatin-null mouse.

Dysregulation of miRNA expression during hypertrophic growth has also recently been reported to occur in the heart of mice subjected to transverse aortic constriction (TAC),[39,52–55] in transgenic calcineurin mice,[55] in mice with physiological hypertrophy,[39] and, importantly, in human heart diseases.[39,55] Arrays of miRNAs were reported to be either upregulated, downregulated, or unchanged after TAC in mice,[52–55] in cultured cardiomyocytes treated with

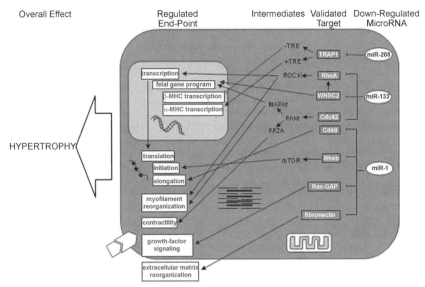

FIGURE 2. Validated miRNA targets important in hypertrophy and speculated mechanisms of action. miRNAs (ovals), targets (shaded boxes), some important end points (unshaded boxes), and intermediates (no boxes) are given.

phenylephrine (a hypertrophic agonist),[54] and in human heart failure.[55] In many of these reports, downregulated miRNAs included miRNA-1, miRNA-133, miRNA-29, miRNA-30, and miRNA-150 whereas upregulated miRNAs included miRNA-21, miRNA-23, miRNA-195, and miRNA-199. Overexpression of some of the upregulated miRNAs were individually sufficient to produce hypertrophy of cultured cardiomyocytes, whereas downregulated miRNAs could reduce cardiomyocyte size.[39,55] Hypertrophy has been reported to be induced in mice by the overexpression of transgenic miRNA-195 (an upregulated miRNA)[55] or with the inhibition of a downregulated miRNA (miRNA-133) via infusion of an miRNA-133-antisense oligonucleotide or transcoronary gene delivery of a *miRNA-133*-decoy virus.[39]

An important aspect in understanding the role of miRNAs is the correlation with specific targets, but only a few miRNAs have been ascribed validated targets (FIG. 2). Cdc42, Rho-A, and NELF-A/WHSC2, for example, have been recently confirmed as miRNA-133 targets and are characterized by increased protein levels upon establishment of cardiac hypertrophy.[39] Rho-A and Cdc42 have already been associated with cell growth, cytoskeletal and myofibrillar rearrangements, and regulation of contractility in cardiomyocytes.[56,57] Wolf-Hirschhorn Syndrome Complex 2, (WHSC2) on the other hand, has not been particularly studied, as yet, in relation to hypertrophy. A few studies have limited its identification as a repres-

sor of transcription, probably operating at the RNA elongation step.[58] In Wolf-Hirschhorn Syndrome, a congential condition characterized by mental retardation, a number of abnormalities, including cardiovascular ones, are common. Interestingly, transduction of cardiomyocytes *in vitro* and *in vivo* with an adenoviral vector containing a *Whsc2* transgene resulted in decreased protein synthesis but induced the fetal gene program, a characteristic of hypertrophic response.[39] Upregulation of Rho-A was also noted, corroborating the notion that WHSC2 could play a selective role in hypertrophy.

Four *in silico*-predicted targets of miRNA-1 have also been validated: Ras GTPase-activating protein (Ras-GAP), cyclin-dependent kinase 9 (Cdk9), Ras homolog enriched in brain (Rheb), and fibronectin.[53] In addition, miRNA-208 has been shown to control regulation of β-myosin heavy chain (MHC) in conditions of stress but not during normal development, through targeting of the thyroid hormone receptor (THR)-associated protein 1 (THRAP1), the THR coregulator.[59] The miRNA-208 sequence is encoded in intron 27 of the human and mouse α-MHC gene, and its knockout showed an apparently unremarkable phenotype. When subjected to TAC, miRNA-208 knockout mice presented, however, with blunted hypertrophic growth. Some stress markers (e.g., atrial natriuretic factor [ANF] and brain natriuretic peptide [BNP]) were increased in the hearts of these mice, as predicted, but there was no increase in β-MHC. In contrast, α-MHC

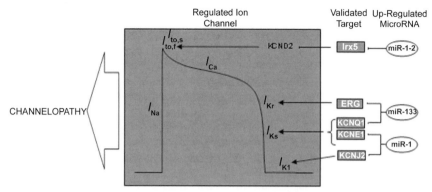

FIGURE 3. Validated miRNA targets implicated in the control of ion currents responsible for the generation of the ventricular action potential. miRNAs (ovals), targets (shaded boxes), and principal affected K^+ currents (I_{K1}, inward rectifying; $I_{Kto,f}$, transient outward fast; I_{Kr}, rapid-delayed rectifier; I_{Ks}, slow-delayed rectifier) are given.

was increased rather than decreased. A model was put forward whereby stress stimuli, usually responsible for the reduction of α-MHC transcription, consequentially reduce the level of the miRNA-208 transcript, which, in turn, relieves transcriptional repression on its target mRNA, *thrap1*. The resulting increase in THRAP1 protein influences the THR-regulated expression of α- and β-MHCs, which are inversely affected through a positive and negative thyroid hormone response element, respectively.

Conduction and Arrhythmogenesis

An important role for miRNAs has also been reported in the control of the electrical activity of the heart (Fig. 3). miRNA-1 and miRNA-133 expression was found to be spatially heterogeneous within the adult heart and with a distribution that is opposite in many aspects to that of KCNQ1 and KCNE1, two subunits that assemble in the heart to form the slow delayed rectifier K^+ current I_{Ks}.[33] Analysis of the 3′-UTRs of these mRNAs revealed putative binding sites for miRNA-1 on *KCNE1* and for miRNA-133 on *KCNQ1*. Thus, spatial differences in the expression of miRNAs can exist within an adult organ and are responsible for modulating the expression pattern of target proteins.

A relationship between altered cardiac electrical mechanisms and dysregulation of miRNAs has been shown also in disease states. miRNA-133 upregulation in hearts of a rabbit model of diabetes and in ventricular samples from diabetic patients has been directly linked to downregulation of *ERG* (ether-a-go-go-related gene), which encodes for the rapid delayed rectifier K^+ current I_{Kr}.[60] Furthermore, upregulation of miRNA-1 has been reported in individuals with coronary artery disease, the leading cause

of death in industrialized countries.[61] Many of these deaths are attributable to arrhythmias[61]; miRNA-1 ablation *in vivo* with an antisense inhibitor was sufficient to relieve arrhythmogenesis in infarcted rat hearts. Indeed, validated miRNA-1 targets were demonstrated to be *KCNJ2* (which encodes the K^+ channel subunit Kir2.1, responsible for I_{K1}) and *GJA1* (which encodes for connexin 43, involved in intercellular conductance), which are downregulated in mice after myocardial infarction and in samples from coronary artery-diseased patients. Abnormal propagation of cardiac electrical activity was a feature reported also for surviving miRNA-1-2 knockout mice where the apparently normal anatomy and function was accompanied by a slowed heart rate.[48] In that report, *Irx5* (Iroquois family of homeodomain-containing transcription factor), which regulates cardiac repolarization by repressing transcription of a key potassium channel (Kcnd2) was reported as a validated target of miRNA-1.

Angiogenesis

Vascular remodeling is a fundamental aspect of heart growth and adaptation to stress, and miRNAs have been recently documented as relevant determinants for this aspect of cardiovascular biology. Levels of important endothelial receptors, such as the vascular endothelial growth factor receptors VEGFR-2 (KDR) and VEGFR-1 (FLT); the angiopoietin receptor, Tie; and other angiogenesis and vascular remodeling genes, have been shown to be regulated by miRNAs.[62] miRNA-221 and miRNA-222, for example, indirectly regulate endothelial nitric oxide synthase (eNOS) and Tie[62] and directly control c-kit,[63] thereby contributing to angiogenesis. Interestingly, a recent finding has revealed an important role of miRNA-21 for proliferation of the intima of vessels, a common pathological lesion found in many cardiovascular diseases.[64] miRNA-133, on the other hand, was found downregulated, thus implicating a putative role of this miRNA in the vascular tissue as well. Use of a miRNA-21 antisense oligonucleotide was sufficient to inhibit neointinal formation *in vivo* and had antiproliferative and proapoptotic effects in vascular smooth muscle cells grown *in vitro*. These effects were related to a reduction of PTEN protein level, a putative miRNA-21 target that is involved in cell proliferation. In addition, as an indirect target, the apoptotic Bcl-2 protein was found upregulated upon miRNA-21 overexpression. The importance of the apoptotic gene program in the regulation of cardiac angiogenesis has been recently put forward.[65] Indeed, accumulation of the tumor suppressor gene *p53* in sustained cardiac pressure overload has been shown to suppress the vascular growth induced by hypoxia-inducible factor 1, a key regulator of angiogenic factor transcription. Intriguingly, *p53* has been demonstrated to be involved in the positive regulation of miRNAs; this *p53* activity leads to dramatic reprogramming of gene expression involved in apoptosis, cell proliferation, and angiogenesis.[66,67]

Conclusions

A complex network of mechanisms controls gene expression through gross and fine-tuning processes. Alteration of these controls is at the heart of many diseases. Because expression is also controlled at the post-transcriptional level by miRNAs, it is not surprising to find etiologies dependent on this facet of cell biology. With regard to muscle, reports have started to emerge showing how mutations that alter the susceptibility of mRNAs to miRNAs can produce abnormal phenotypes (e.g., Texel sheep).[51] Moreover, altered expression of arrays of miRNAs characterizes disease states, such as heart failure,[52–55] and may be responsible for congenital disorders, as demonstrated by the septal defects seen in knockout miRNA-1-2 mice.[48] It is possible that many of the human pathologies considered today as idiopathic will be uncovered as being miRNA based. As with any newly defined mechanism, it can be speculated that miRNAs may one day be targeted for therapeutic causes. Counteracting anomalous miRNA functioning has already been shown to reverse pathological conditions, such as hypertrophy[39] and arrhythmias,[60] in animal models of disease. However, the particular mode of action of miRNAs, including pleiotrophy (a given miRNA recognizes sites contained in multiple mRNAs), redundancy (more than one miRNA may need to target a

given mRNA to produce an effect), spatial patterning, and the fact that very small changes in their quantity may have profound effects may hamper miRNA utility. Appropriately, miRNA-mimics have been selectively designed based on the 3′-UTR of specific genes. This approach has been successful in the regulation of the expression of the cardiac pace maker genes *HCN2* and *HCN4* without disturbing other genes that would be affected with the use of an endogenous miRNA, such as miRNA-1 or miRNA-133.[68]

miRNAs are currently at the forefront of interest challenging both the scientific and clinical community. The functional role of these small molecules in the cardiovascular field certainly deserves further investigation, particularly in light of their potential use as new therapeutic agents.

Acknowledgments

G.C. is supported by the National Institute of Health (Grant RO HL078797-01A1), European Community (EU FP6 Grant LSHM-CT-2005-018833, EUGene-Heart), Italian Ministry of Health, and Italian Ministry of Research and University. D.C. is supported by a Marie Curie Outgoing International fellowship within the 6th European Framework Program.

Conflicts of Interest

The authors declare no conflicts of interest.

References

1. CHALFIE, M., H.R. HORVITZ & J.E. SULSTON. 1981. Mutations that lead to reiterations in the cell lineages of C. elegans. Cell **24:** 59–69.
2. LEE, R.C., R.L. FEINBAUM & V. AMBROS. 1993. The C. elegans heterochronic gene lin-4 encodes small RNAs with antisense complementarity to lin-14. Cell **75:** 843–854.
3. WIGHTMAN, B., I. HA & G. RUVKUN. 1993. Posttranscriptional regulation of the heterochronic gene lin-14 by lin-4 mediates temporal pattern formation in C. elegans. Cell **75:** 855–862.
4. RODRIGUEZ, A. *et al.* 2004. Identification of mammalian microRNA host genes and transcription units. Genome Res. **14:** 1902–1910.
5. BABAK, P., K.G. MAGNUSSON & S. SIGURDSSON. 2004. Dynamics of group formation in collective motion of organisms. Math. Med. Biol. **21:** 269–292.
6. BARAD, O. *et al.* 2004. MicroRNA expression detected by oligonucleotide microarrays: system establishment and expression profiling in human tissues. Genome Res. **14:** 2486–2494.

7. ZHAO, Y., E. SAMAL & D. SRIVASTAVA. 2005. Serum response factor regulates a muscle-specific microRNA that targets Hand2 during cardiogenesis. Nature **436:** 214–220.

8. O'DONNELL, K.A. *et al.* 2005. c-Myc-regulated microRNAs modulate E2F1 expression. Nature **435:** 839–843.

9. BARTEL, D.P. 2004. MicroRNAs: genomics, biogenesis, mechanism, and function. Cell **116:** 281–297.

10. LEE, Y. *et al.* 2003. The nuclear RNase III Drosha initiates microRNA processing. Nature **425:** 415–419.

11. AMBROS, V. *et al.* 2003. MicroRNAs and other tiny endogenous RNAs in C. elegans. Curr. Biol. **13:** 807–818.

12. OBERNOSTERER, G. *et al.* 2006. Post-transcriptional regulation of microRNA expression. RNA **12:** 1161–1167.

13. RUBY, J.G., C.H. JAN & D.P. BARTEL. 2007. Intronic microRNA precursors that bypass Drosha processing. Nature.

14. PILLAI, R.S. *et al.* 2005. Inhibition of translational initiation by Let-7 MicroRNA in human cells. Science **309:** 1573–1576.

15. RANA, T.M. 2007. Illuminating the silence: understanding the structure and function of small RNAs. Nat. Rev. Mol. Cell Biol. **8:** 23–36.

16. HUMPHREYS, D.T. *et al.* 2005. MicroRNAs control translation initiation by inhibiting eukaryotic initiation factor 4E/cap and poly(A) tail function. Proc. Natl. Acad. Sci. USA **102:** 16961–16966.

17. OLSEN, P.H. & V. AMBROS. 1999. The lin-4 regulatory RNA controls developmental timing in Caenorhabditis elegans by blocking LIN-14 protein synthesis after the initiation of translation. Dev. Biol. **216:** 671–680.

18. MARONEY, P.A. *et al.* 2006. Evidence that microRNAs are associated with translating messenger RNAs in human cells. Nat. Struct. Mol. Biol. **13:** 1102–1107.

19. NOTTROTT, S., M.J. SIMARD & J.D. RICHTER. 2006. Human let-7a miRNA blocks protein production on actively translating polyribosomes. Nat. Struct. Mol. Biol. **13:** 1108–1114.

20. TANG, G. 2005. siRNA and miRNA: an insight into RISCs. Trends Biochem. Sci. **30:** 106–114.

21. BEHM-ANSMANT, I. *et al.* 2006. mRNA degradation by miRNAs and GW182 requires both CCR4:NOT deadenylase and DCP1:DCP2 decapping complexes. Genes Dev. **20:** 1885–1898.

22. SCHMITTER, D. *et al.* 2006. Effects of Dicer and Argonaute down-regulation on mRNA levels in human HEK293 cells. Nucleic Acids Res. **34:** 4801–4815.

23. WU, L., J. FAN & J.G. BELASCO. 2006. MicroRNAs direct rapid deadenylation of mRNA. Proc. Natl. Acad. Sci. USA **103:** 4034–4039.

24. CHENDRIMADA, T.P. *et al.* 2007. MicroRNA silencing through RISC recruitment of eIF6. Nature **447:** 823–828.

25. GIRALDEZ, A.J. *et al.* 2006. Zebrafish MiR-430 promotes deadenylation and clearance of maternal mRNAs. Science **312:** 75–79.

26. LIU, J. *et al.* 2005. MicroRNA-dependent localization of targeted mRNAs to mammalian P-bodies. Nat. Cell Biol. **7:** 719–723.

27. HOUBAVIY, H.B., M.F. MURRAY & P.A. SHARP. 2003. Embryonic stem cell-specific MicroRNAs. Dev. Cell **5:** 351–358.

28. SUH, M.R. *et al.* 2004. Human embryonic stem cells express a unique set of microRNAs. Dev. Biol. **270:** 488–498.

29. KANELLOPOULOU, C. *et al.* 2005. Dicer-deficient mouse embryonic stem cells are defective in differentiation and centromeric silencing. Genes Dev. **19:** 489–501.

30. BERNSTEIN, E. *et al.* 2003. Dicer is essential for mouse development. Nat. Genet. **35:** 215–217.

31. STARK, A. *et al.* 2005. Animal MicroRNAs confer robustness to gene expression and have a significant impact on 3′UTR evolution. Cell **123:** 1133–1146.

32. FARH, K.K. *et al.* 2005. The widespread impact of mammalian MicroRNAs on mRNA repression and evolution. Science **310:** 1817–1821.

33. LUO, X. *et al.* 2007. Transcriptional activation by stimulating protein 1 and post-transcriptional repression by muscle-specific microRNAs of I(Ks)-encoding genes and potential implications in regional heterogeneity of their expressions. J. Cell Physiol. In press.

34. VALADI, H. *et al.* 2007. Exosome-mediated transfer of mRNAs and microRNAs is a novel mechanism of genetic exchange between cells. Nat. Cell Biol. **9**(6): 654–659.

35. HWANG, H.W., E.A. WENTZEL & J.T. MENDELL. 2007. A hexanucleotide element directs microRNA nuclear import. Science **315:** 97–100.

36. BEREZIKOV, E. *et al.* 2005. Phylogenetic shadowing and computational identification of human microRNA genes. Cell **120:** 21–24.

37. LEWIS, B.P., C.B. BURGE & D.P. BARTEL. 2005. Conserved seed pairing, often flanked by adenosines, indicates that thousands of human genes are microRNA targets. Cell **120:** 15–20.

38. LU, J. *et al.* 2005. MicroRNA expression profiles classify human cancers. Nature **435:** 834–838.

39. CARE, A. *et al.* 2007. MicroRNA-133 controls cardiac hypertrophy. Nat. Med. **13**(5): 613–618.

40. ALVAREZ-GARCIA, I. & E.A. MISKA. 2005. MicroRNA functions in animal development and human disease. Development **132:** 4653–4662.

41. WIENHOLDS, E. *et al.* 2005. MicroRNA expression in zebrafish embryonic development. Science **309:** 310–311.

42. SEMPERE, L.F. *et al.* 2004. Expression profiling of mammalian microRNAs uncovers a subset of brain-expressed microRNAs with possible roles in murine and human neuronal differentiation. Genome Biol. **5:** R13.

43. BASKERVILLE, S. & D.P. BARTEL. 2005. Microarray profiling of microRNAs reveals frequent coexpression with neighboring miRNAs and host genes. RNA **11:** 241–247.

44. BRENNECKE, J., A. STARK & S.M. COHEN. 2005. Not miRly muscular: microRNAs and muscle development. Genes Dev. **19:** 2261–2264.

45. SOKOL, N.S. & V. AMBROS. 2005. Mesodermally expressed Drosophila microRNA-1 is regulated by Twist and is required in muscles during larval growth. Genes Dev. **19:** 2343–2354.

46. RAO, P.K. *et al.* 2006. Myogenic factors that regulate expression of muscle-specific microRNAs. Proc. Natl. Acad. Sci. USA **103:** 8721–8726.

47. CHEN, J.F. *et al.* 2006. The role of microRNA-1 and microRNA-133 in skeletal muscle proliferation and differentiation. Nat. Genet. **38:** 228–233.

48. ZHAO, Y. *et al.* 2007. Dysregulation of cardiogenesis, cardiac conduction, and cell cycle in mice lacking miRNA-1-2. Cell **129**: 303–317.

49. MCFADDEN, D.G. *et al.* 2005. The Hand1 and Hand2 transcription factors regulate expansion of the embryonic cardiac ventricles in a gene dosage-dependent manner. Development **132**: 189–201.

50. MCCARTHY, J.J. & K.A. ESSER. 2007. MicroRNA-1 and microRNA-133a expression are decreased during skeletal muscle hypertrophy. J. Appl. Physiol. **102**: 306–313.

51. CLOP, A. *et al.* 2006. A mutation creating a potential illegitimate microRNA target site in the myostatin gene affects muscularity in sheep. Nat. Genet. **38**: 813–818.

52. CHENG, Y. *et al.* 2007. MicroRNAs are aberrantly expressed in hypertrophic heart. Do they play a role in cardiac hypertrophy? Am. J. Pathol. In press.

53. SAYED, D. *et al.* 2007. MicroRNAs play an essential role in the development of cardiac hypertrophy. Circ Res. **100**: 416–424.

54. TATSUGUCHI, M. *et al.* 2007. Expression of microRNAs is dynamically regulated during cardiomyocyte hypertrophy. J. Mol. Cell Cardiol.

55. VAN ROOIJ, E. *et al.* 2006. A signature pattern of stress-responsive microRNAs that can evoke cardiac hypertrophy and heart failure. Proc. Natl. Acad. Sci. USA **103**: 18255–18260.

56. BROWN, J.H., D.P. DEL RE & M.A. SUSSMAN. 2006. The Rac and Rho hall of fame: a decade of hypertrophic signaling hits. Circ. Res. **98**: 730–742.

57. KE, Y. *et al.* 2004. Intracellular localization and functional effects of P21-activated kinase-1 (Pak1) in cardiac myocytes. Circ. Res. **94**: 194–200.

58. MARIOTTI, M., M. MANGANINI & J.A. MAIER. 2000. Modulation of WHSC2 expression in human endothelial cells. FEBS Lett. **487**: 166–170.

59. VAN ROOIJ, E. *et al.* 2007. Control of stress-dependent cardiac growth and gene expression by a microRNA. Science. **316**: 575–579.

60. XIAO, J. *et al.* 2007. MicroRNA miR-133 represses HERG K+ channel expression contributing to QT prolongation in diabetic hearts. J. Biol. Chem. **282**: 12363–12367.

61. YANG, B. *et al.* 2007. The muscle-specific microRNA miR-1 regulates cardiac arrhythmogenic potential by targeting GJA1 and KCNJ2. Nat. Med. **13**: 486–491.

62. SUAREZ, Y. *et al.* 2007. Dicer dependent microRNAs regulate gene expression and functions in human endothelial cells. Circ Res. **100**: 1164–1173.

63. POLISENO, L. *et al.* 2006. MicroRNAs modulate the angiogenic properties of HUVECs. Blood **108**: 3068–3071.

64. JI, R. *et al.* 2007. MicroRNA expression signature and antisense-mediated depletion reveal an essential role of MicroRNA in vascular neointimal lesion formation. Circ. Res. **100**: 1579–1588.

65. SANO, M. *et al.* 2007. p53-induced inhibition of Hif-1 causes cardiac dysfunction during pressure overload. Nature **446**: 444–448.

66. CHANG, T.C. *et al.* 2007. Transactivation of miR-34a by p53 broadly influences gene expression and promotes apoptosis. Mol. Cell. **26**: 745–752.

67. HE, L. *et al.* 2007. A microRNA component of the p53 tumour suppressor network. Nature. In press.

68. XIAO, J. *et al.* 2007. Novel approaches for gene-specific interference via manipulating actions of microRNAs: examination on the pacemaker channel genes HCN2 and HCN4. J. Cell Physiol. **212**: 285–292.

Neonatal and Adult Cardiovascular Pathophysiological Remodeling and Repair

Developmental Role of Periostin

RUSSELL A. NORRIS,[a] THOMAS K. BORG,[b] JONATHAN T. BUTCHER,[c]
TROY A. BAUDINO,[b] INDRONEAL BANERJEE,[b] AND ROGER R. MARKWALD[a]

[a] Cardiovascular Developmental Biology Center, Medical University of South Carolina, Charleston, South Carolina, USA

[b] Department of Developmental Biology and Anatomy, University of South Carolina, Columbia, South Carolina, USA

[c] Department of Biomedical Engineering, Cornell University, Ithaca, New York, USA

The neonatal heart undergoes normal hypertrophy or compensation to complete development and adapt to increased systolic pressures. Hypertrophy and increased neonatal wall stiffness are associated with a doubling of the number of fibroblasts and *de novo* formation of collagen. Normal postnatal remodeling is completed within 3–4 weeks after birth but can be rekindled in adult life in response to environmental signals that lead to pathological hypertrophy, fibrosis, and heart failure. The signals that trigger fibroblast and collagen formation (fibrosis) as well as the origin and differentiation of the cardiac fibroblast lineage are not well understood. Using mice studies and a single-cell engraftment model, we have shown that cardiac fibroblasts are derived from two extracardiac sources: the embryonic proepicardial organ and the recruitment of circulating bone marrow cells of hematopoietic stem cell origin. Periostin, a matricellular protein, is normally expressed in differentiating fibroblasts but its expression is elevated several fold in pathological remodeling and heart failure. Our hypothesis that periostin is profibrogenic (i.e., it promotes differentiation of progenitor mesenchymal cells into fibroblasts and their secretion and compaction of collagen) was tested using isolated and cultured embryonic, neonatal, and adult wild-type and periostin-null, nonmyocyte populations. Our findings indicate that abrogation of periostin by targeted gene deletion inhibits differentiation of nonmyocyte progenitor cells or permits misdirection into a cardiomyocyte lineage. However, if cultured with periostin or forced to express periostin, they became fibroblasts. Periostin plays a significant role in promoting fibrogenesis residual stress, and tensile testings indicated that periostin played an essential regulatory role in maintaining the biomechanical properties of the adult myocardium. These findings indicate that periostin is a profibrogenic matricellular protein that promotes collagen fibrogenesis, inhibits differentiation of progenitor cells into cardiomyocytes, and is essential for maintaining the biomechanical properties of the adult myocardium.

Key words: periostin; fasciclin; myocardial remodeling; hypertrophy; fibrosis; cardiac fibroblast; myocyte; myocardial infarction; heart failure

Introduction

Heart Development Extends beyond Intrauterine Life

The neonatal heart undergoes rapid enlargement and adaptive (or physiological) "remodeling" to complete development and maturation. It is a unique transitional period that bridges the embryonic/fetal period of heart development and the fully defined adult heart. Based on pioneering studies by Borg *et al.*,[1] it is now generally recognized that the neonatal heart adapts (remodels) to sudden increased systolic pressures following birth by increasing ventricular wall thickness and stiffness (i.e., tensile strength). This is a result of a twofold increase in the number of fibroblasts and the formation, compaction, and alignment of collagen fibrils that envelop myocytes as an endomysial-like collagenous network.[2] This organization provides

Address for correspondence: Professor Roger R. Markwald, PhD, Cardiovascular Developmental Biology Center, Children's Research Institute, Medical University of South Carolina, 173 Ashley Avenue, Charleston, SC 29425. Voice: 843-792-5880; fax: 843-792-0664.

markwald@musc.edu

Ann. N.Y. Acad. Sci. 1123: 30–40 (2008). © 2008 New York Academy of Sciences.
doi: 10.1196/annals.1420.005

for potential "contact-signaling" between myocytes and fibroblasts, the fibroblasts themselves, or cell–extracellular matrix (ECM) contacts that prepare the postnatal myocardial wall to structurally adapt and respond to increases in blood pressure after birth.[3] Thus, the fibroblast is potentially a key player in neonatal development.

In the mouse, the unique, fibroblast-mediated, adaptive/remodeling period progressively diminishes after 1 week and the mature phenotype is fully established by 30 days. However, remodeling can be rekindled in adult life in response to environmental changes (e.g., pressure overload or ischemic injury) that may lead to an increase in fibroblasts and collagen content. These normal adaptive responses may eventually become pathological, resulting in fibrosis and increased wall stiffness.[3,4] Here, we address some questions that relate to the mechanisms of neonatal heart development and their possible relationship to pathological remodeling in an adult heart (e.g., if the myocardium is injured):

1. What is the origin of cardiac fibroblasts? Are they derived from a single population of progenitor cells that are carried over from intrauterine to postnatal and adult life or are new cells added or recruited into the heart?
2. Are there any signals that direct postnatal cardiac progenitor cells into a fibroblast lineage or inhibits their differentiation into other mesodermal lineages (e.g., cardiac muscle, bone, or cartilage)?
3. Are these same signals re-expressed or elevated in the adult heart following acute injuries that result in fibrosis and ventricular dysfunction?

Answers to these questions bear directly upon human heart disease, which remains one of the most prevalent and life-threatening diseases in the world and one of the fastest growing.

The Origin(s) of Cardiac Fibroblasts

Surprisingly, while there is considerable knowledge on the structure and function of neonatal cardiac myocytes, much less is known about the role(s) of fibroblasts during postnatal development or pathological remodeling. One reason is that their developmental origin is unclear. Are we born with all the fibroblasts we will ever have or are they a dynamic cell population that can be expanded or recruited from elsewhere? Information from our embryonic chick and adult mouse studies suggests there are two major sources of cardiac

fibroblasts: embryonic proepicardial organ (PEO) and circulating progenitor cells derived from adult bone marrow stem cells (as discussed and shown below and in Ref. 11).

The Proepicardial Organ

The PEO is a sac-like vascular structure that is derived from the coelomic mesothelium located at the venous inlet (sinuatrial) pole of the heart.[5] The PEO gives origin to cells that migrate as an epithelium over the surface of the entire heart to form the embryonic epicardium. The epicardium, in turn, undergoes a cell autonomous epithelial to mesenchymal transformation (EMT) to form epicardial-derived cells (EPDCs) that accumulate as mesenchyme between the epicardium and myocardium. During the late embryonic and fetal period, the EPDCs invade the atrial and ventricular walls and functionally interact with cardiomyocytes to establish a compact myocardium.[6] FIGURE 1 was prepared in collaboration with a colleague, Dr. John Burch (Fox Chase Cancer Center, Philadelphia, PA), to illustrate the invasion of EPDC into the ventricular myocardium as revealed by their expression of enhanced green fluorescent protein (EGFP) driven by the Wilms' tumor promoter, which is expressed in the epicardium and EPDCs (and not other heart cell types).[7] Failure of EPDCs to invade the myocardium results in lethality as a result of the failure to form a compact myocardium, indicating that from the beginning of cardiac embryogenesis, there is an interaction between the fibroblasts (or their progenitors) and cardiomyocytes that is necessary for myocardial growth, survival, and function.[6,8]

While the EPDCs are an immediate progenitor of cardiac fibroblasts, it is important to recognize that they are also multipotential cells. They have potential to differentiate *in vitro* into endothelial cells, smooth muscle, and cardiomyocytes.[5,6] However, *in vivo*, during normal embryonic life, most EPDC within the wall of the heart progressively differentiate into fibroblasts that oscillate or transdifferentiate between two phenotypes: one is spindle shaped; the other is more rounded and sometimes called a *myofibroblast* because it expresses α-smooth muscle actin. Over time, expression of α-smooth muscle actin is suppressed and most EPDC derivatives assume a more spindle-shaped phenotype. After birth and into adult life, the myofibroblast phenotype can be reactivated by transforming growth factor (TGF)β signaling during a pathogenic remodeling event.[9,10]

Recruited Fibrogenic Progenitor Cells

In addition to the EPDC-derived progenitor cells, we have found evidence for postnatal recruitment

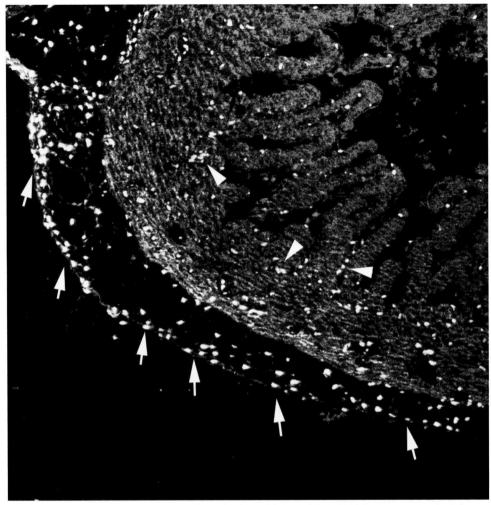

FIGURE 1. Contribution of EPDCs to the ventricular myocardium. Sagittal section of E16.5 wild-type 1 (WT1)-EGFP transgenic mouse heart showing EPDC cells invading the ventricular wall.

of circulating fibroblast progenitor cells into the ventricular myocardium.[11] This can possibly represent a continuation of a recruitment process that has begun before birth or hatching.[12] As described in Visconti *et al.*,[11,13,14] the progeny of a single (EGFP+) donor hematopoietic stem cell (HSC) injected to repopulate the bone marrow of a lethally irradiated adult mouse host will also give rise to EGFP+ cells in the adult heart. Current controversy regarding the lineage potential of any given engrafted stem cell may be because most studies evaluating stem cell potential invariably use mixed cell populations.[15,16] Thus, we believe that the definitive assignment of fate (or lineage) should be based on the analysis of a *single* donor cell's ability to achieve long-term organ engraftment and multilineage hematopoietic reconstitution of bone marrow

when injected into the tail vein of an irradiated mouse host. This permits the full potential of a single HSC to be evaluated *in vivo*. Based on our results using this method, we have concluded that circulating bone marrow HSCs engraft into the heart and give rise to fibroblasts and myofibroblasts, which, as noted above, may be the same cell type but reflect different secretory or contractile states because of different epigenetic influences.[11,17]

Whatever their origin (EPDC or HSC), it is important to recognize that undifferentiated fibroblastic progenitor cells are present in normal newborn and adult hearts. They do not express traditional fibroblast markers or only very low levels of the markers unless induced to differentiate into a mature fibroblast. Admittedly, the fibroblastic phenotype is hard

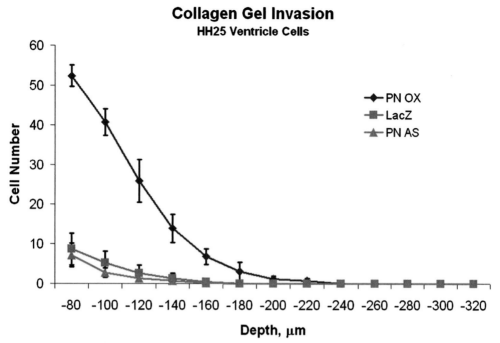

FIGURE 2. Periostin induces myocardial invasion. Ventricular apexes were isolated from HH25 chick hearts and infected with either the periostin-expressing virus (PNOX), an antisense periostin virus (PNAS), or control virus (LacZ) in hanging drop culture. The hanging drop aggregates were placed on top of type I collagen gels (1.5 mg/mL) and cultured for 3 days. The number of invasive cells, at specified depths, were counted. Forced expression of periostin in embryonic myocytes induced invasive migratory behavior not observed in empty vector controls.

to characterize satisfactorily. In part, this is because fibroblasts are heterogeneous. Arbitrarily, we assign a fibroblast phenotype to nonmyocyte cells that strongly express type I collagen (mRNA and protein), discoid domain receptor 2 (DDR2), type 3 procollagen, fibroblast surface antigen (FSA), α-smooth muscle actin, vimentin, and periostin.[4] Using established methods to isolate myocytes versus nonmyocytes,[1] we have found a small population (15%) of neonatal nonmyocytes that do not express fibroblast markers and, if plated on the surface of a collagen gel, do not exhibit invasive migratory behavior. Conversely, nonmyocyte cells that express fibroblast markers, like periostin, do invade the gel lattice. Thus, migratory potential may be a useful tool for determining differentiation status of neonatal or adult nonmyocytes and signals that might promote their expression of fibroblast markers (and not other lineage markers) and their invasive migratory activity. To test this possibility, we explanted embryonic chick ventricular myocytes, which normally do not express periostin or migrate as individual cells, onto a collagen gel lattice and assessed their potential to become invasive if infected with full-length periostin cDNA viral vectors. Forced expression of periostin in embryonic myocytes induced invasive migratory behavior that was

not observed in empty vector controls (FIG. 2). Additionally, migrating embryonic myocytes infected with periostin vectors also switched from expression of a myocyte sarcomeric marker protein to a smooth muscle cell marker protein. Collectively, these findings suggest that periostin is a candidate regulator for activating migration and/or modifying differentiation of cardiac progenitor cells.

Signal(s) that Promote Differentiation of Neonatal or Adult Cardiac Progenitor Cells into Fibroblasts

The two major adaptive changes within the left ventricular myocardium during neonatal life are an increase in fibroblast number and the progressive secretion and organization of a collagenous (endomyseal-like) network.[18] The mechanism of the increase in fibroblasts is not known but its expression profile and the phenotype of its targeted deletion[19] invoke a regulatory role for periostin. Periostin is one of several fibroblast markers that are initially expressed at low levels in the prevalvular and EPDC mesenchyme of the embryonic heart. Periostin expression peaks after

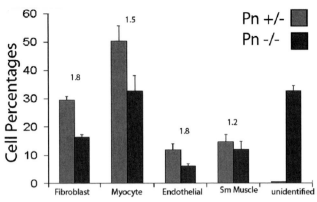

FIGURE 3. FACS analysis of wild-type and periostin-null hearts. Twelve-week-old hearts were prepared for fluorescence-activated cell sorting (FACS) using antibody markers for fibroblasts (DDR2), smooth muscle (α-SMA), myocytes (α-MHC), and endothelial cells (PECAM/CD31). In the periostin −/− mice, a large percentage of unidentified cells exist.

birth early in the neonatal period and then falls to a baseline level that is maintained throughout life; it is not expressed in neonatal or adult cardiomyocytes but only fibroblasts.[3,19,20] Thus, periostin expression correlates with increased formation of fibroblasts and collagen, presumably in response to changes in hemodynamic forces immediately after birth.

What is Periostin?

Periostin is an evolutionary-conserved ECM protein that has four domains: a signal sequence, an N-terminal domain that contains cysteine residues, a carboxyl terminus that can be alternatively spliced, and four coiled fasciclin (fas) domains that have closest homology to the *Drosophila fasciclin 1* gene.[21] In *Drosophila*, the ancestral fasciclin domain functioned as an adhesion molecule linked to axonal guidance, migration, and differentiation.[21–23] Periostin can interact *in vivo* with other ECM scaffold proteins, especially collagens.[24,25] Isoforms lacking the entire carboxyl domain inhibited cell motility and migration.[24] Known receptors for periostin are integrins (α_V/β_3 and α_V/β_5).[26,27] Periostin binding to integrins is thought to be through highly conserved H1 and H2 peptide stretches (but not RGD sequences) present within the fasciclin domain having "YH" and Asp-Ile motifs.[21,27] We have also found that periostin binding to integrins α_V/β_3 and β_1 on cardiac mesenchymal progenitor cells can initiate signaling related to migration and collagen cross-linking or compaction transduced through Rho and phosphoinositide-3 (PI3) kinases.[28] Thus, there is an integrin-based receptor mechanism for mediating the potential effects of periostin on neonatal ventricular remodeling. However, we cannot yet rule out that other receptors may exist for transducing periostin signals.

Periostin as a Matricellular Protein

Bornstein and colleagues[29,30] have proposed that there is a family of functionally related secreted proteins, called *matricellular* proteins. The latter derive their complex functions from their ability to interact with multiple cell surface receptors, especially integrins, cytokines, growth factors, and proteases; but matricellular proteins can also bind directly to structural or scaffold proteins. Examples of matricellular proteins include thrombospondin, tenascin-C, osteopontin, CCN1, and SPARC.[31] The expression of this unique family of ECM proteins is most prominent during development and growth or in response to injury.[32] Based on its known biological roles, we propose that periostin also qualifies as a matricellular protein. As such, we would expect that the contextual nature of its functions include dynamic contact-signaling between mesenchyme progenitor cells, fibroblasts and mesenchyme, or fibroblasts and cardiomyocytes.

Periostin Knockout Mice

To test whether periostin promotes differentiation of cardiac mesenchymal progenitor cells into fibroblasts *in vivo*, two periostin knockout mice were generated.[19,33] In both lines, deletion of periostin inhibited or delayed differentiation of mesenchyme into fibroblasts, which resulted postnatally in loss of fibrous tissue or abnormally differentiated cells (e.g., ectopic cardiomyocytes, bone, or cartilage). At 12 weeks, there remained a large population of undifferentiated mesenchymal-like cells (33%) in periostin-null hearts whose lineage identity could not be determined with the same marker antibodies that were sufficient to identify 100% of the cells isolated from wild-type adult hearts[34] (FIG. 3). Whether

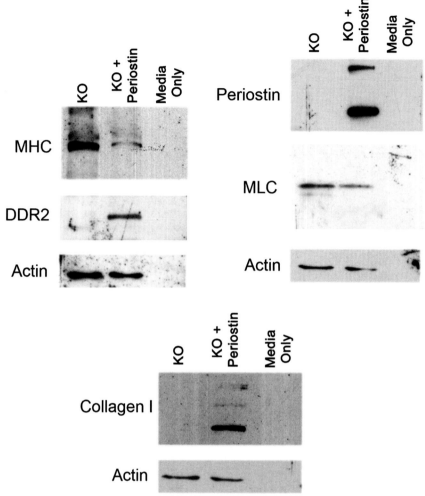

FIGURE 4. Periostin functions as a nodal differentiation switch. Western blot analysis of isolated cardiac mesenchymal cells from periostin-null mice. One group of cells was given the purified periostin protein whereas the other group was not. After 7 days in culture, protein lysates were taken and expression of fibroblasts markers (DDR2, periostin, and collagen I) or myocytes (MHC and MLC) were analyzed. Periostin expression is not only required for the expression of fibroblast markers but also for the repression of myocardial proteins.

these cells came from EPDC or HSC has not been determined, but these genetic findings and those provided by transfecting periostin genes into mesenchymal progenitor cells suggest that periostin is needed to promote full and complete differentiation of cardiac progenitor cells into postnatal fibroblasts.

Periostin as a "Nodal" Differentiation Switch

There is also evidence that periostin may act as a binary differentiation signal that determines if cardiac mesenchymal progenitor cells differentiate into fibroblastic lineages versus other mesodermal phenotypes, especially cardiac muscle. As shown in FIGURE 4, the addition of purified full-length periostin protein to cultures of periostin-null cardiac mesenchymal cells not only "rescued" collagen synthesis and expression of fibroblast markers but also suppressed the expression of myocyte marker proteins. Infarcts created in wild-type versus periostin-null hearts indicated that fibrous scar tissue is reduced in periostin nulls, replaced instead by viable cardiomyocytes.[33] Collectively, these data indicate that periostin is a signaling protein in embryonic and postnatal life that functions to promote differentiation of mesenchymal progenitor cells into fibroblasts while inhibiting their differentiation into cardiomyocytes.

Is There a Developmental Basis for Adult Cardiac Pathological Remodeling?

Remodeling can be a normal developmental process, as in neonatal life in response to environmental changes. However, if the modulating signals continue after recovery or adaptation to a changing environment, the remodeling process can become detrimental to cardiac function and ultimately lead to irreversible damage and heart failure.[35,36] As noted above, a spike in periostin expression occurs during neonatal development, which correlates with increased numbers of fibroblasts and new formation of fibrous ECM components, especially collagen.[1,2,4] While there are numerous signals that can trigger or mediate adult remodeling ranging from extracellular ligands for G protein coupled receptors to receptor tyrosine kinases,[37,38] there are also extensive data suggesting that periostin may be a key trigger. This is based on an eightfold to 40-fold increase in periostin expression when pathological remodeling is induced in adult hearts by acute or chronic injuries.[39,40] Using a TAC banding model to induce chronic pressure overload and cardiac hypertrophy, we found that periostin was upregulated over 64-fold and accompanied by extensive collagen deposition and fibrosis.[19] Katsuragi *et al.*[39,41] reported that elevated expression of periostin correlated with increased fibrosis during ventricular dilation but also noted a loss of ventricular cardiomyocytes, indicating that periostin may not only promote collagen formation and fibrosis in the adult heart but may also affect the fate of adult cardiomyocytes. Various aspects of remodeling of cardiac myocytes have been explored after infarction, hypertrophy, and congestive failure. For example, cardiomyocytes can (i) increase in size (hypertrophy),[37] (ii) undergo apoptosis,[42,43] or (iii) activate an embryonic gene program (e.g., periostin) and/or revert to a myofibroblastic phenotype.[44–46] Whether any of these changes are caused indirectly by periostin promoting formation of collagen fibrils (which, in turn, interact in some way with myocardial cell surfaces) or directly by mediating an inductive interaction between adult fibroblasts and myocytes is an important unresolved question. If embryonic development serves as a relevant precedent, then the answer would seem to be that periostin can directly signal changes in postnatal cardiac myocytes. During development, at sites where embryonic cardiac mesenchyme directly contacts myocytes (e.g., as at the atrioventricular (AV) junctional myocardium), periostin expression is elevated and the fate of the adjacent myocytes is to "disappear" or retract from the original boundary interface.[47–49] In periostin nulls or in other mouse or chick embryos where periostin expression is inhibited, myocardial cells that normally disappear persist, resulting in conduction system disturbances or the failure of AV valve leaflets to delaminate.[19,48,50] As shown in FIGURE 5, when neonatal populations of isolated cardiomyocytes are co-cultured for 48 h with periostin-positive neonatal fibroblasts, there is a progressive disruption of cardiomyocyte cell:cell associations followed by their gradual loss of myosin+ sarcomeric organization. By 72 h, periostin-positive fibroblastic cells fill spaces once occupied by an epithelial monolayer of myocytes. These findings suggest that the contact of neonatal myocytes with fibroblastic cells expressing periostin can directly modify the phenotypic stability of myocytes and possibly their survival.

Material Properties and Elevated Expression of Periostin

Because of its link to fibrosis and pathological remodeling, it is important to know if the passive biomechanical properties of the myocardium are modified as a result of enhancing or decreasing periostin. The small size of mouse hearts limits the number of *in vivo* biomechanical assays possible. Nevertheless, we have been able to perform preliminary "residual stress and tensile testing" studies. Residual stress has been previously shown to be important in the cardiovascular system for distributing loads and reducing peak myocardial wall stress. Decreased residual stress may indicate increased tissue stiffness resulting in impaired diastolic function.[51] The classical residual stress assay is the open angle measurement of the left ventricle.[52] Briefly, we performed this assay on excised wild-type versus periostin-null adult (3 months) hearts. Our results, shown in FIGURE 6, demonstrate that periostin deficiency leads to reduced residual stress. Because collagen matrix is known to have a strongly nonlinear response, we have taken these results to suggest that periostin deficiency leads to impaired collagen production and organization which, in turn, reduces wall stiffness and residual stress important for systolic and diastolic function. This conclusion is strongly supported by Oka *et al*,[33] who created a periostin overexpressor mouse and a periostin knockout mouse and subjected each to increased blood pressures or surgically created myocardial infarcts. Null animals were prone to aneurysm and rupture of the ventricular wall, whereas overexpression of periostin in the heart increased fibrosis and protected from rupture, especially after experimentally inducing a heart attack. This further indicates that a developmental protein, such as periostin, can have important regulatory effects on adaptive

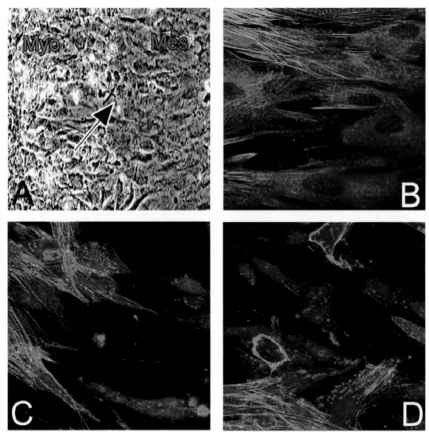

FIGURE 5. Fibroblast–myocyte co-cultures. **(A)** Bright field image of myocyte–fibroblast interaction 24 h after the spacer bar has been removed. *Arrow* points to original location of the spacer bar. **(B)** Periostin staining (red) of fibroblasts that have contacted myocytes (green) after 24 h in culture. **(C, D)** Forty-eight h in culture, myocytes in contact with fibroblasts form migratory-like cell processes coincident with reorganization of their actin cytoskeleton as seen by rings of phalloidin-positive material (green).

responses to increasing blood pressure during neonatal life and can also adjust ventricular wall stiffness in adult life in response to environmental signals.

Remodeling, Repair, and Recruitment

HSCs derived from the bone marrow circulate as cardiac progenitor cells that engraft into the heart as both valvular or ventricular interstitial cells. These engrafted cells strongly express fibroblast markers including periostin.[11,14] The question then becomes does engraftment change when the heart is injured (e.g., by a surgically created infarct or cryoinjury). In response to this question, we now know that (i) periostin is strongly upregulated in the heart following ischemic injury,[33,40] (ii) the number of fibroblasts increase at the site of the infarct,[33] and (iii) fibroblasts of the infarct intensely express periostin.[11] What is not known is (i) whether the fibroblasts of scar tissue are recruited from circulating

progenitor cells or derived from proliferation of the endogenous pool of interstitial fibroblasts and (ii) if periostin directs their differentiation into a fibroblast lineage. To answer these questions we have initiated studies using our single-cell engraftment approach in which a single wild-type EGFP+ HSC is injected into the tail vein of an irradiated wild type or host. In preliminary studies, we have found that EGFP+ cells comprise the majority of cells in the infarct that is formed in an irradiated, wild-type host. Gender mismatch controls were run and ruled out cell fusion. *This finding indicates that the fibroblasts are recruited from the HSCs of the bone marrow and thus are the most likely source of the elevated expression of periostin seen in the ischemic injury model.* Finally, when an ischemic injury is created in a periostin-null mouse, the size of the infarct is significantly reduced and many of the cells forming the infarct resemble cardiomyocytes.[33] Thus, abrogation of periostin expression by gene deletion

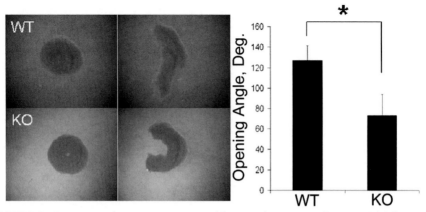

FIGURE 6. Opening angle measurements on wild-type and periostin −/− mice. Adult hearts from wild-type (WT) and periostin −/− (KO) mice were isolated and used for determining opening angle measurements. Periostin −/− mice exhibit a reduction in the opening angle measurement(*) indicating a compromise in myocardial wall stiffness.

results in reduced scar formation and inhibited and modified differentiation of progenitor cells into fibroblasts. It remains to be determined if wild-type HSCs can rescue scar formation and differentiation into fibroblasts in periostin nulls or, conversely, if a single periostin −/−, EGFP+ HSC injected into the bloodstream of a host animal, can modify scar formation and differentiation in a wild-type (irradiated) host.

Conclusions

It is clear that in either normal physiological or pathological remodeling, the usual developmentally expressed proteins, such as periostin, can be important players in the qualitative and quantitative aspects of fibrogenesis and fibrosis. Such proteins could have great potential for controlling the onset and/or progression of myocardial remodeling from normal adaptive responses to pathological changes that alter performance and possibly lead to catastrophic heart failure. Therefore, a selective spatiotemporal reduction of normal periostin expression in the infarct region following a heart attack may provide a plausible therapeutic intervention that would result in improved myocardial remodeling and overall long-term enhancement of cardiac function.

Acknowledgments

This study was partially funded by research Grants HL33756 (R.R.M) from the National Institutes of Health (NIH); COBRE P20RR016434-07 (R.R.M. and T.K.B) from NIH-National Center for Research Resources (NCRR); and FIBRE EF0526854 (R.R.M. and R.A.N.) from the National Science Foundation.

Conflicts of Interest

The authors declare no conflicts of interest.

References

1. BORG, T.K., K. RUBIN, E. LUNDGREN, *et al.* 1984. Recognition of extracellular matrix components by neonatal and adult cardiac myocytes. Dev. Biol. **104:** 86–96.
2. BORG, T.K., W.F. RANSON, F.A. MOSLEHY, *et al.* 1981. Structural basis of ventricular stiffness. Lab. Invest. **44:** 49–54.
3. BANERJEE, I., K. YEKKALA, T.K. BORG, *et al.* 2006. Dynamic interactions between myocytes, fibroblasts, and extracellular matrix. Ann. N.Y. Acad. Sci. **1080:** 76–84.
4. BAUDINO, T.A., W. CARVER, W. GILES, *et al.* 2006. Cardiac fibroblasts: friend or foe? Am. J. Physiol. Heart. Circ. Physiol. **291:** H1015–1026.
5. WESSELS, A. & J.M. PEREZ-POMARES. 2004. The epicardium and epicardially derived cells (EPDCs) as cardiac stem cells. Anat. Rec. A Discov. Mol. Cell. Evol. Biol. **276:** 43–57.
6. MANNER, J., J.M. PEREZ-POMARES, D. MACIAS, *et al.* 2001. The origin, formation and developmental significance of the epicardium: a review. Cells Tissues Organs **169:** 89–103.
7. PEREZ-POMARES, J.M., A. PHELPS, M. SEDMEROVA, *et al.* 2002. Experimental studies on the spatiotemporal expression of WT1 and RALDH2 in the embryonic avian heart: a model for the regulation of myocardial and valvuloseptal development by epicardially derived cells (EPDCs). Dev Biol. **247:** 307–326.

8. GITTENBERGER-DE GROOT, A.C., M.P. VRANCKEN PEETERS, M.M. MENTINK, *et al.* 1998. Epicardium-derived cells contribute a novel population to the myocardial wall and the atrioventricular cushions. Circ. Res. **82:** 1043–1052.

9. SHIOJIMA, I., M. AIKAWA, J. SUZUKI, *et al.* 1999. Embryonic smooth muscle myosin heavy chain SMemb is expressed in pressure-overloaded cardiac fibroblasts. Jpn. Heart J. **40:** 803–818.

10. WALKER, G.A., K.S. MASTERS, D.N. SHAH, *et al.* 2004. Valvular myofibroblast activation by transforming growth factor-beta: implications for pathological extracellular matrix remodeling in heart valve disease. Circ. Res. **95:** 253–260.

11. VISCONTI, R.P. & R.R. MARKWALD. 2006. Recruitment of new cells into the postnatal heart: potential modification of phenotype by periostin. Ann. N.Y. Acad. Sci. **1080:** 19–33.

12. EISENBERG, L.M. & R.R. MARKWALD. 2004. Cellular recruitment and the development of the myocardium. Dev. Biol. **274:** 225–232.

13. EBIHARA, Y., M. MASUYA, A.C. LARUE, *et al.* 2006. Hematopoietic origins of fibroblasts: II. In vitro studies of fibroblasts, CFU-F, and fibrocytes. Exp. Hematol. **34:** 219–229.

14. VISCONTI, R.P., Y. EBIHARA, A.C. LARUE, *et al.* 2006. An in vivo analysis of hematopoietic stem cell potential: hematopoietic origin of cardiac valve interstitial cells. Circ. Res. **98:** 690–696.

15. GOODELL, M.A., K. BROSE, G. PARADIS, *et al.* 1996. Isolation and functional properties of murine hematopoietic stem cells that are replicating in vivo. J. Exp. Med. **183:** 1797–1806.

16. ORLIC, D., J. KAJSTURA, S. CHIMENTI, *et al.* 2001. Bone marrow cells regenerate infarcted myocardium. Nature **410:** 701–705.

17. MIRONOV, V., R.P. VISCONTI & R.R. MARKWALD. 2004. What is regenerative medicine? Emergence of applied stem cell and developmental biology. Expert Opin. Biol. Ther. **4:** 773–781.

18. GOLDSMITH, E.C., A. HOFFMAN, M.O. MORALES, *et al.* 2004. Organization of fibroblasts in the heart. Dev. Dyn. **230:** 787–794.

19. SNIDER, S., J. WANG, A. LINDSLEY, *et al.* 2007. Periostin is required for maturation and ECM stabilization of the cardiac skeleton. Circulation Research (Submitted).

20. KRUZYNSKA-FREJTAG, A., M. MACHNICKI, R. ROGERS, *et al.* 2001. Periostin (an osteoblast-specific factor) is expressed within the embryonic mouse heart during valve formation. Mech. Dev. **103:** 183–188.

21. LITVIN, J., S. ZHU, R. NORRIS, *et al.* 2005. Periostin family of proteins: therapeutic targets for heart disease. Anat. Rec. A Discov. Mol. Cell. Evol. Biol. **287:** 1205–1212.

22. HORTSCH, M. & C.S. GOODMAN. 1990. Drosophila fasciclin I, a neural cell adhesion molecule, has a phosphatidylinositol lipid membrane anchor that is developmentally regulated. J. Biol. Chem. **265:** 15104–15109.

23. TAKESHITA, S., R. KIKUNO, K. TEZUKA, *et al.* 1993. Osteoblast-specific factor 2: cloning of a putative bone adhesion protein with homology with the insect protein fasciclin I. Biochem. J. **294**(Pt 1): 271–278.

24. KIM, C.J., N. YOSHIOKA, Y. TAMBE, *et al.* 2005. Periostin is down-regulated in high grade human bladder cancers and suppresses in vitro cell invasiveness and in vivo metastasis of cancer cells. Int. J. Cancer **117:** 51–58.

25. NORRIS, R.A., B. DAMON, V. MIRONOV, *et al.* 2007. Periostin regulates collagen fibrillogenesis and the biomechanical properties of connective tissues. J. Cell Biochem. **101**: 659–711.

26. BAO, S., G. OUYANG, X. BAI, *et al.* 2004. Periostin potently promotes metastatic growth of colon cancer by augmenting cell survival via the Akt/PKB pathway. Cancer Cell. **5:** 329–339.

27. GILLAN, L., D. MATEI, D.A. FISHMAN, *et al.* 2002. Periostin secreted by epithelial ovarian carcinoma is a ligand for alpha(V)beta(3) and alpha(V)beta(5) integrins and promotes cell motility. Cancer Res. **62:** 5358–5364.

28. BUTCHER, J.T., R.A. NORRIS, S. HOFFMAN, *et al.* 2007. Periostin promotes atrioventricular mesenchyme matrix invasion and remodeling mediated by integrin signaling through Rho/PI 3-kinase. Dev. Biol. **302:** 256–266.

29. BORNSTEIN, P. 2000. Matricellular proteins: an overview. Matrix Biol. **19:** 555–556.

30. YANG, Z., T.R. KYRIAKIDES & P. BORNSTEIN. 2000. Matricellular proteins as modulators of cell-matrix interactions: adhesive defect in thrombospondin 2-null fibroblasts is a consequence of increased levels of matrix metalloproteinase-2. Mol. Biol. Cell. **11:** 3353–3364.

31. MO, F.E. & L.F. LAU. 2006. The matricellular protein CCN1 is essential for cardiac development. Circ. Res. **99:** 961–969.

32. KYRIAKIDES, T.R., Y.H. ZHU, Z. YANG, *et al.* 1998. The distribution of the matricellular protein thrombospondin 2 in tissues of embryonic and adult mice. J. Histochem. Cytochem. **46:** 1007–1015.

33. OKA, T., J. XU, R.A. KAISER, *et al.* 2007. Genetic manipulation of periostin expression reveals a role in cardiac hypertrophy and ventricular remodeling. Circulation Research (Submitted).

34. BANERJEE, I., J.W. FUSELER, R.L. PRICE, *et al.* 2007. Determination of cell type and number during cardiac development in the neonatal and adult rat and mouse. Am. J. Physiol. Heart Circ. Physiol. **293:** H1883–H1891.

35. COHN, J.N. 1995. Critical review of heart failure: the role of left ventricular remodeling in the therapeutic response. Clin. Cardiol. **18:** IV4–12.

36. FEDAK, P.W., S. VERMA, R.D. WEISEL, *et al.* 2005. Cardiac remodeling and failure: from molecules to man (Part III). Cardiovasc. Pathol. **14:** 109–119.

37. HEINEKE, J. & J.D. MOLKENTIN. 2006. Regulation of cardiac hypertrophy by intracellular signalling pathways. Nat. Rev. Mol. Cell. Biol. **7:** 589–600.

38. OLSON, E.N. & M.D. SCHNEIDER. 2003. Sizing up the heart: development redux in disease. Genes Dev. **17:** 1937–1956.

39. KATSURAGI, N., R. MORISHITA, N. NAKAMURA, *et al.* 2004. Periostin as a novel factor responsible for ventricular dilation. Circulation **110:** 1806–1813.

40. STANTON, L.W., L.J. GARRARD, D. DAMM, *et al.* 2000. Altered patterns of gene expression in response to myocardial infarction. Circ. Res. **86:** 939–945.

41. IEKUSHI, K., Y. TANIYAMA, J. AZUMA, *et al.* 2007. Novel mechanisms of valsartan on the treatment of acute myocardial infarction through inhibition of the antiadhesion molecule periostin. Hypertension. **49**: 1409–1414.

42. FOO, R.S., K. MANI & R.N. KITSIS. 2005. Death begets failure in the heart. J. Clin. Invest. **115:** 565–571.

43. KITSIS, R.N. & D.L. MANN. 2005. Apoptosis and the heart: a decade of progress. J. Mol. Cell. Cardiol. **38:** 1–2.

44. DISPERSYN, G.D. & M. BORGERS. 2001. Apoptosis in the heart: about programmed cell death and survival. News Physiol. Sci. **16:** 41–47.

45. DISPERSYN, G.D., L. MESOTTEN, B. MEURIS, *et al.* 2002. Dissociation of cardiomyocyte apoptosis and dedifferentiation in infarct border zones. Eur. Heart J. **23:** 849–857.

46. LAFRAMBOISE, W.A., D. SCALISE, P. STOODLEY, *et al.* 2007. Cardiac fibroblasts influence cardiomyocyte phenotype in vitro. Am. J. Physiol. Cell. Physiol. **292**: C1799–C1808.

47. KERN, C.B., S. HOFFMAN, R. MORENO, *et al.* 2005. Immunolocalization of chick periostin protein in the developing heart. Anat. Rec. A Discov. Mol. Cell. Evol. Biol. **284:** 415–423.

48. KOLDITZ, D.P., M.C. WIJFFELS, N.A. BLOM, *et al.* 2007. Persistence of functional atrioventricular accessory pathways in postseptated embryonic avian hearts: implications for morphogenesis and functional maturation of the cardiac conduction system. Circulation **115:** 17–26.

49. NORRIS, R.A., C.B. KERN, A. WESSELS, *et al.* 2004. Identification and detection of the periostin gene in cardiac development. Anat. Rec. A Discov. Mol. Cell. Evol. Biol. **281:** 1227–1233.

50. GAUSSIN, V., G.E. MORLEY, L. COX, *et al.* 2005. Alk3/Bmpr1a receptor is required for development of the atrioventricular canal into valves and annulus fibrosus. Circ. Res. **97:** 219–226.

51. OMENS, J.H. & Y.C. FUNG. 1990. Residual strain in rat left ventricle. Circ. Res. **66:** 37–45.

52. OMENS, J.H., A.D. MCCULLOCH & J.C. CRISCIONE. 2003. Complex distributions of residual stress and strain in the mouse left ventricle: experimental and theoretical models. Biomech. Model Mechanobiol. **1:** 267–277.

The Missing Link in the Mystery of Normal Automaticity of Cardiac Pacemaker Cells

EDWARD G. LAKATTA, TATIANA M. VINOGRADOVA, AND VICTOR A. MALTSEV

National Institutes of Health, National Institute on Aging, Laboratory of Cardiovascular Science, Baltimore, Maryland 21224, USA

Earlier studies of the initiating event of normal automaticity of the heart's pacemaker cells, inspired by classical quantitative membrane theory, focused upon ion currents (I_K, I_f) that determine the maximum diastolic potential and the early phase of the spontaneous diastolic depolarization (DD). These early DD events are caused by the prior action potential (AP) and essentially reflect a membrane recovery process. Events following the recovery process that ignite APs have not been recognized and remained a mystery until recently. These critical events are linked to rhythmic intracellular signals initiated by Ca^{2+} clock (i.e., sarcoplasmic reticulum [SR] cycling Ca^{2+}). Sinoatrial cells, regardless of size, exhibit intense ryanodine receptor (RyR), Na^+/Ca^{2+} exchange (NCX)-1, and SR Ca^{2+} ATPase-2 immunolabeling and dense submembrane NCX/RyR colocalization; Ca^{2+} clocks generate spontaneous stochastic but roughly periodic local subsarcolemmal Ca^{2+} releases (LCR). LCRs generate inward currents via NCX that exponentially accelerate the late DD. The timing and amplitude of LCR/I_{NCX}-coupled events control the timing and amplitude of the nonlinear terminal DD and therefore ultimately control the chronotropic state by determining the timing of the I_{CaL} activation that initiates the next AP. LCR period is precisely controlled by the kinetics of SR Ca^{2+} cycling, which, in turn, are regulated by 1) the status of protein kinase A-dependent phosphorylation of SR Ca^{2+} cycling proteins; and 2) membrane ion channels ensuring the Ca^{2+} homeostasis and therefore the Ca^{2+} available to Ca^{2+} clock. Thus, the link between early DD and next AP, missed in earlier studies, is ensured by a precisely physiologically regulated Ca^{2+} clock within pacemaker cells that integrates multiple Ca^{2+}-dependent functions and rhythmically ignites APs during late DD via LCRs-I_{NCX} coupling.

Key words: cardiac pacemaker; calcium; ryanodine receptor; Na–Ca exchange; sarcoplasmic reticulum

Introduction

Sinoatrial nodal cell (SANC) *robustness* ("fail-safe" properties conserved during evolution of the animal kingdom) and *flexibility* (the ability to react to demands for faster or slower firing rate) require multiple interactions among intrinsic cell mechanisms. This interactive network of mechanisms intrinsic to nodal cells must also interpret and react to signals arising extrinsic to the cell (e.g., stretch or neurotransmitter or hormonal stimulation of surface membrane receptors).

During the past five decades, pacemaker theory, formulated on the basis of an ensemble of multiple voltage and time-dependent rhythms of membrane ion channels (membrane clock, FIG. 1A), has produced numerous numerical models of SANC function that are able to reproduce recurring action potentials (APs; i.e., spontaneous firing). Such membrane-delimited models, however, turned out to be fundamentally incomplete and, hence, incorrect, as they all ignored critical intracellular rhythmic signals that ensure both robustness and flexibility to SANC pacemaker function.

Recent studies in SANC suggest that the missing signal comes from a rhythmic intracellular Ca^{2+} oscillator or Ca^{2+} clock (FIG. 1B) that "crosstalks" to multiple other mechanisms that govern membrane potential and, in this way, coordinates their interaction to regulate the SANC firing rate. This Ca^{2+} clock dispenses localized, critically timed, submembrane pulses of Ca^{2+} during the SANC cycle. The timing (phase) of these rhythmic Ca^{2+} releases and their size determine their impact on the SANC firing rate and pattern (FIG. 1C). Rhythmic APs, in turn, sustain the Ca^{2+} clock by maintaining cell Ca^{2+} homeostasis via ion fluxes of membrane ion channels. Novel numerical models demonstrate that mutual entrainment of membrane and Ca^{2+} clocks confers both stability and

Address for correspondence: Victor A. Maltsev, PhD, National Institute on Aging, Gerontology Research Center, Intramural Research Program, 5600 Nathan Shock Drive, Baltimore, MD 21224. Fax: 410-558-8150.

MaltsevVi@grc.nia.nih.gov

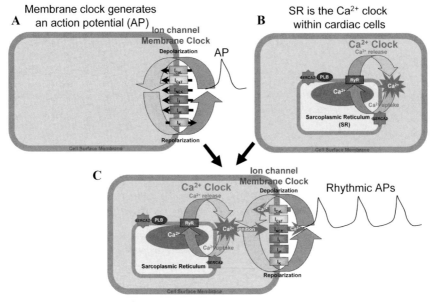

FIGURE 1. Schematic representation of an ion channel membrane clock (**A**) and Ca^{2+} clock (**B**) in cardiac pacemaker cells. Panel **C** illustrates the idea that the Ca^{2+} clock ignites the membrane clock to entrain normal automaticity in cardiac pacemaker cells (see text for details). (In color in *Annals* online.)

flexibility to the SANC pacemaker function. This review chronicles the evolution of thought that has led to this concept.

The Membrane-delimited Ion Channel Clock Concept

Historical Perspectives: Development of Theory and Mechanistic Studies of Cell Membrane Excitation

A major function of the heart's pacemaker cells is to generate spontaneous APs. In 1868, Bernstein plotted a galvanometer response to nerve excitation that was, in fact, the first record of an "action potential."[1] By 1912, based on the work of Ostwald and Nernst, Bernstein developed and experimentally proved the membrane theory of bioelectric potentials,[2] which linked the bioelectric potential (V_m) of nerve and muscle to ion gradients across their cell membrane. At the same time, he suggested that a mechanism of membrane excitability involved a sudden reduction of the K$^+$ permeability of the membrane, which reduced V_m to zero.

In 1949, testing the effect of Na$^+$ removal on V_m in large diameter squid neurons, Hodgkin and Katz[3] concluded (contrary to the original Bernstein K$^+$-driven excitability theory) that Na$^+$ was responsible for the AP upstroke and positive membrane polarization. In the same year, Ling and Gerard invented the glass microelectrode that allowed recording activity from individual cells, and Kenneth Cole suggested placing a second electrode inside the cell in order to inject the current and to "voltage clamp" the interior of the cell. In 1952, Hodgkin and Huxley applied the voltage clamp approach to nerve cells of a giant squid and established current–voltage relationships for Na$^+$ and K$^+$ currents.[4] Based on these data, they formulated a new quantitative theory of membrane excitation (i.e., of the AP generation). According to their theory, the movement of selective ions through the excitable membrane is determined by separate voltage-dependent "gates" with a simple kinetics described by differential equations.

Using the new microelectrode technique, Silvio Weidmann[5] confirmed the application of ionic theory to the cardiac muscle and explained the differences between cardiac APs and those of nerve and skeletal muscle. His findings include the genesis of the resting membrane potential, the increased permeability to Na$^+$ during membrane depolarization, high membrane resistance during the AP plateau, and the definition of threshold potential.

The Slow Diastolic Depolarization Concept

The unique feature of the APs of pacemaker cells is that they arise spontaneously because of events that occur prior to the rapid AP upstroke. A slow potential change preceding the discharge of cardiac impulses was first observed by Arvanitaki, who was working on the snail heart in 1937.[6] Later, in 1943, Bozler used monophasic potential recording and found a similar phenomenon in sinus venosus of turtle.[7] He called the slow depolarizing potential change a "prepotential" and suggested that it was "the basis underlying automaticity" including "normal rhythmicity." Bozler worked with rhythmically active strips of cardiac muscle (turtles, rabbits, cats, and dogs) and sinus venosus (turtle, high K^+)[7] and noted oscillatory monophasic subthreshold potentials of relatively (compared to nerve) low frequency of approximately 1 Hz, which were enhanced by increasing bathing $[Ca^{2+}]$ (Ca_o) or by adding adrenaline. Discharging impulses on the top of oscillatory potentials were accompanied by synchronized variations in isometric force (tonus).[8] Further, an important idea postulated at that time was that the observed oscillatory phenomena might be somehow linked to cardiac automaticity.[8] Specifically it was postulated that:

> the tonus changes and the local potentials are probably manifestations of a more fundamental process, a fluctuation in resting metabolism. . . . Their chief interest lies in their relation to the automaticity and rhythmicity of the muscle. It may be assumed that an increase in metabolism causes a rise in tonus and a decreased surface polarization. The decrease in polarization in turn may be considered as the last link in the chain of processes leading to the discharge of an impulse.

The prepotential (i.e., diastolic depolarization [DD]) was later observed with microelectrode techniques by Brady and Hecht[9] in turtle hearts and by Coraboeuf and Weidmann[10] in spontaneously active Purkinje tissue. Spontaneous APs from the sinoatrial node were first recorded using microelectrodes by West in 1955.[11] Microelectrode studies of sinoatrial node identified three important factors of pacemaker potential: the DD slope, the maximum diastolic potential, and the threshold for AP upstroke.

First Models of the Membrane-delimited Pacemaker Clock

Noble modified the Hodgkin–Huxley model to simulate both muscle (triggered) and pacemaker (spontaneous) cardiac APs.[12,13] AP shapes, known from microelectrode studies, were reproduced in the model by an interplay of hypothesized voltage-gated and time-dependent Na^+ and K^+ currents. Although specific

cardiac ion currents had not yet been measured at that time, there were some clues from earlier microelectrode findings of a membrane conductance decrease during DD in pacing Purkinje fibers (first reported by Weidman[14] in 1951) and then in beating sinus preparation of the rabbit's heart (by Dudel and Trautwein[15] in 1958). The membrane response with various bathing $[K^+]$ and $[Na^+]$ (K_o and Na_o) in the latter study suggested that the spontaneous depolarization was a result of the decay of a specific K^+ conductance (g_K). The first Purkinje fiber models[12,13] described the AP as a rapid membrane depolarization produced by a voltage-gated Na^+ current (I_{Na}) followed by a repolarization from a delayed increase in g_K. Generation of spontaneous APs in these models was thus ascribed to a slow decay of the g_K (termed g_K *decay mechanism*) so allowing any background permeability (e.g., to Na^+ ions) to drive V_m away from the K^+ equilibrium potential. The 1962 cardiac modification of the Hodgkin–Huxley model represented a mechanistic model based upon solid principles of thermodynamics of Bernstein's membrane theory and the Nernst equation. The formulation of the 1962 model was a case of "theory leading experiment."[16] The membrane-delimited pacemaker mechanism, predicted by the theory, became the subject of an extensive experimental search driven by the successful application of the voltage clamp technique to various cardiac preparations.[17–20] Many ion currents were discovered in cardiac pacemaker cells, and a variety of pacemaker mechanisms have been further suggested based on the characteristics of these currents.[16,21,22]

The Pacemaker Current

The Quest to Discover the Most Important Current that Governs Pacemaker Function Purkinje Cells

The search for the surface membrane cardiac pacemaker mechanism was initially and mainly studied experimentally and theoretically (from about 1950 to about 1980) in Purkinje fibers and not in true pacemaker cells (i.e., SANC). A slow inward current carried by Ca^{2+} (I_s, I_{si}, or I_{Ca}), activated by membrane depolarization, and increased by adrenaline was discovered in Purkinje fibers by Reuter.[23] A "pure K^+ selective" I_{K2} current (later redefined as I_f), a "funny," nonselective, monovalent cation current activated by membrane hyperpolarization,[24] was and continues to be acclaimed by some as the true "pacemaker current." But the fundamental pacemaker concept formulated in the 1962 model[13] and then revised in the

McAllister–Noble–Tsien (MNT) model of 1975,[25] based upon Purkinje fiber voltage clamp studies, did not work for all pacemakers (see review[26]).

Sinoatrial Nodal Cells

Following the invention of the patch clamp method in 1976 by Neher and Sakmann[27] and successful isolation of single SANC by Taniguchi *et al.* in 1981,[20] further extensive voltage clamp studies have identified abundant ion current components in SANC, and new components are still being discovered. Many previously identified currents are split into subcomponents; others are species dependent. For example, studies of single cardiac cells showed that the "slow" inward current I_{si} is approximately 10 times faster than previously reported in multicellular preparations (2–5 ms versus 20–100 ms) and consists of three components: T- and L-type Ca^{2+} currents (I_{CaT}, I_{CaL}) and Na^+/Ca^{2+} exchange (NCX) current (I_{NCX}) (all three in current terminology). I_{CaL} is responsible for AP upstroke in SANC. In most species, SANC had neither a fast Na^+ current nor an inward rectifying K^+ current (I_{K1}) (see review[28] for species dependency). The latter keeps the resting potential low in Purkinje fibers but not in SANC. While SANC of many species did exhibit the funny current, it plays only a minor role in pacemaker mechanism of SANC as its activation voltage is relatively low compared to maximum diastolic potential in the cells.[16,29] Furthermore, I_f has relatively slow (compared to DD duration in SANC) activation kinetics. It is not surprising that the block of I_f by Cs^+ had only a minor effect on the pacemaker rhythm.[30] It was also found that bullfrog sinus venosus cells lack I_f.[31] The delayed rectifier K^+ current (I_K) is split into slow and rapid components (I_{Ks} and I_{Kr}), with I_{Ks} being present in porcine SANC, but I_{Kr} in rat and rabbit.[28] I_{CaT} was absent in porcine SANC.[32] Other currents were observed only in subpopulations of SANC (I_{Na} and Cl^- current, I_{Cl})[33] or had no obvious molecular identity (the sustained current, I_{st}).[34] As a result, net membrane current is currently viewed as being spread among numerous components and subcomponents,[22] and the fundamental (and thus rather simple) pacemaker mechanism claimed in early theoretical models has been lost. Several early[35] and later studies,[36] which specifically compared the contribution of different ion currents to the DD, concluded that there is no single pacemaker current in the sinoatrial node.

Currently there is no consensus as to which of the SANC currents makes the major contribution to pacemaker activity. Various viewpoints on the membrane clock (FIG. 1A) include essentially combinations of previously suggested DD mechanisms: 1) original g_K decay mechanism of the 1962 model (I_{Kr} in rabbit[37]); 2) I_s (I_{CaL}) activation[38]; 3) I_f activation[39,40]; 4) I_{CaT} activation[41]; 5) I_{NCX} activation[42–44]; and 6) I_{st} activation.[34] Numerous numerical models, based upon gating schema of ensembles of SANC membrane ion currents, have been devised, and all these models featuring different contributions of the ion currents into DD can indeed generate spontaneous APs (see review[22]).

However, since the idea of a unique pacemaker current was not confirmed in the extensive experimental studies across numerous species, there was a need to formulate another fundamental principle of cardiac pacemaker function. Could it be that regardless of combination or contributions of the different ion channel types, not all of the crucial mechanisms that are implicated in spontaneous excitation of pacemaker cells are embodied in the ensemble of cell surface ion channels? Moreover, and very importantly, could it be that the formal cause of spontaneous rhythmic APs in SANC (i.e., the initiating step of their normal automaticity) is an intracellular process (e.g., similar to the one that Bozler had suggested earlier in 1943[7])?

The Existence and the Identity of Intracellular Ca^{2+} Clock

Studies in Contractile Cardiac Cells Lead the Way

Heart research had been tuned in to the crucial importance of extracellular Ca^{2+} in the cardiac muscle duty cycle since 1883 when Sidney Ringer's London tap water mistake illuminated the field.[45] Subsequent studies identified the sarcoplasmic reticulum (SR) and its Ca^{2+} pump, and the idea that cyclic Ca^{2+} release from SR is followed by pumping of Ca^{2+} back into the SR was born.

Ca^{2+} pumping into the SR is accomplished mainly by the SR Ca^{2+} ATPase (SERCA). The Ca^{2+} pump isoform in the heart, SERCA2A, is the dominant mechanism to remove Ca^{2+} from the cytosol following excitation, but its relative contribution varies among species. In most mammalian ventricular myocytes (including rabbit, guinea-pig, ferret, dog, cat, and human), SERCA function accounts for about 60–75% of Ca^{2+} extrusion, with 25–40% attributable to the sarcolemmal NCX.[46]

It was shown that ryanodine, an alkaloid that makes the SR Ca^{2+} release channel leaky and inefficient, greatly altered the contractile state of mammalian ventricular myocardium.[47,48] This extremely important pharmacological tool was used in numerous later studies to isolate, purify, and clone ryanodine receptor (RyR) permitting the assessment of SR Ca^{2+} release in the function of different cardiac cell types. Other studies[49] demonstrated a "memory" within heart muscle that produces contractions of differing amplitudes in response to changes in the rate or pattern of stimulation. It was later inferred that staircases in contraction amplitudes prior to achieving a new steady state involved changes that occur over several heart beats in Ca^{2+} loading and release from the SR[50] from beat-dependent net changes in cell Ca^{2+}. Although not articulated as such at that time, such Ca^{2+} oscillations generated by SR Ca^{2+} pumping and Ca^{2+} release are a manifestation of an intracellular Ca^{2+} clock (FIG. 1B).

In mechanically skinned cardiac cell fragments, application of graded $[Ca^{2+}]$ induced graded Ca^{2+} releases from the SR to generate graded transient increases in cytosolic Ca^{2+} that initiate graded contractions,[51,52] a process referred to as Ca^{2+}-induced Ca^{2+}-release.

The advent of intracellular Ca^{2+} indicators permitted direct demonstrations that the AP does indeed result in a release of Ca^{2+} from the SR, resulting in a transient increase in cytosolic Ca^{2+} that activates the myofilaments to drive contraction. Cytosolic Ca^{2+} transients evoked by APs were demonstrated in 1978 by Allen and Blinks[53] in atrial muscle and in 1980 by Wier[54] in Purkinje fibers by using aequorin, a Ca^{2+}-binding photoprotein. Thus, the earlier inference that Ca^{2+} releases, triggered by an AP, couples membrane excitation to contraction was correct, and the concept of "excitation-induced-Ca^{2+}-release" evolved.

It was also discovered that increases in Ca^{2+} beneath the cell surface membrane and the mechanical strain during contraction produced feedback on ion channels and transporters that underlie the AP.[55] One crucial aspect of such tuning of excitation by Ca^{2+} release in ventricular myocytes occurs during the AP and involves Ca^{2+} removal from the cell to balance the Ca^{2+} influx via L-type Ca^{2+} channels. This is achieved, in part, by the cell surface membrane NCX, which extrudes one Ca^{2+} for three Na^+. This process generates an inward current (I_{NCX}) that affects the AP repolarization phase shape[56] as well as the diastolic membrane potential.

The Potential Importance of Intracellular Ca²⁺ in Generating Spontaneous APs

Evidence that an oscillatory intracellular process can initiate both normal and abnormal cardiac rhythms stems from the turn of the last century.[57,58] The initial perspectives of the origin of cardiac automaticity focused not on membrane-delimited initiation of

the heart beat but on myogenic mechanisms. In the early 1880s, Walter Gaskell, who conducted experiments applying heat to the tortoise heart, demonstrated that heart automaticity was myogenic in origin (rather than neuorogenic). He noted that, importantly, the automaticity could occur in all the different regions of the heart including the ventricle,[59] indicating that, in contrast to skeletal muscle (which strictly followed the nerve-mediated excitation → contraction paradigm), cardiac tissues possess their own internal excitatory oscillator. The area of the turtle heart with the greatest automaticity was the sinus venosus, which emanated impulses that entrained the entire heart. Early pharmacological experimental approaches[7,8,60] towards understanding normal automaticity produced "brute force" Ca^{2+} overload. Thus, abnormal rather than normal automaticity was being explored in early studies. Nonetheless, these studies set the stage for the later rejuvenation of the hypothesis that intracellular Ca^{2+} is a partner with surface membrane channels in the process of normal automaticity.

In the late 1970s to early 1980s, the idea that surface membrane-delimited ion channel rhythms were sufficient to confer automaticity to pacemaker cells dominated the research field of cardiac automaticity. During this time, however, the idea that an intracellular oscillator drives membrane excitations was temporarily revived as the importance of intracellular Ca^{2+} transients in cardiac cells came into focus. Unfortunately, these studies of a role for Ca^{2+} in pacemaker cell automaticity continued to employ brute force Ca^{2+} overload in their approach to the problem. Pharmacological microelectrode studies of APs in the early 1970s[57] rediscovered the triggered activity described approximately 30 years earlier in the monophasic recordings.[7,8,60] It was shown that the APs of Purkinje fibers exposed to ouabain or to acetylstrophanthidin develop "delayed afterdepolarizations" and aftercontractions; the amplitude of these was enhanced by driving the fiber more rapidly,[61-64] by increasing Ca_o or by decreasing K_o, and was depressed by Mn^{2+}.[64]

Based upon experimental observations that cardiac glycosides, a reduction in Na_o, or an elevation in Ca_o cause spontaneous contractions, it was suggested that an elevation of intracellular Ca^{2+} drives these spontaneous contractions.[57] But in contrast to the idea of a spontaneous intracellular oscillator, initial interpretations from microelectrode studies suggested that spontaneous depolarization (i.e., the membrane voltage change), *per se*, initiates internal Ca^{2+} release which drives spontaneous cardiac contraction.[63-66] However, microelectrode recordings of the membrane potential do not allow assessment of the underlying ion currents, and this interpretation was disproven by voltage clamp studies that demonstrated a subcellular origin of the oscillations. It thus became apparent that intracellular Ca^{2+} could initiate triggered activity or abnormal automaticity. The concept was further supported by the knowledge of oscillatory Ca^{2+} activity in a wide variety of cells.[67] Data from studies of cardiac cells, including Ca^{2+}-dependent tension oscillations in skinned fibers (Fabiato and Fabiato[68]) and voltage noise analysis (De-Felice and DeHaan[69]) in chick heart cell aggregates led to the hypothesis that it is Ca^{2+} that provides the oscillatory substrate in the intracellular "metabolic" oscillator suggested earlier.[8]

Oscillatory Ca-dependent SR-driven Ion Current under Voltage Clamp in Cardiac Cells

Oscillatory current fluctuations under voltage clamp were observed in Purkinje fibers,[70–74] single Purkinje cells,[75] and sinoatrial node.[76] Both current and mechanical fluctuations are enhanced when intracellular free Ca^{2+} is elevated by bathing solutions containing high Ca^{2+}, low Na^+, low K^+, or toxic concentrations of digitalis. The power spectra of the currents showed peaks at oscillatory frequencies near 1 Hz at room temperature,[77] and the current oscillations were correlated with spontaneous contractile fluctuations. These oscillatory currents in Purkinje fibers disappeared after removing extracellular Ca^{2+} or chelating intracellular Ca^{2+} with injected ethylene glycol tetraacetic acid (EGTA).[74] Based on voltage clamp data, Kass and Tsien, in 1982,[74] proposed that the intracellular oscillatory mechanism involves cycles of Ca^{2+} movement between SR and myoplasm, as previously suggested for skinned cardiac preparations by Fabiato and Fabiato.[51] Since the oscillatory currents were inhibited by D600, a Ca^{2+} current blocker, a new and important idea was that Ca^{2+} oscillations represent the result of "a two-way interaction between surface membrane potential and intracellular Ca^{2+} stores".[74] The Ca^{2+} signals, including Ca^{2+} oscillations, were then indeed measured directly by aequorin in Purkinje fibers.[78,79] An important issue that needed to be addressed was how Ca^{2+} oscillatory signals are transformed into respective electrical signals of cell membrane. It turned out that major ion (K^+ and Ca^{2+}) currents in cardiac cells, including SANC, were modulated by Ca^{2+} and cyclic nucleotides. Based on this knowledge, it was speculated that intracellular Ca^{2+} oscillations could be linked to membrane potential via 1) a Ca^{2+}-dependent K^+ conductance[76]; 2) a nonselective (Na^+ and K^+) cation channel[74]; or 3) I_{NCX}.[16] Further studies of automaticity in atrial latent pacemaker cells indicated that

Ca^{2+} release interacts with cell membrane by stimulating I_{NCX}.[80]

The "Abnormal" Flavor of the Initially Discovered Internal Ca²⁺ Oscillator

As noted, in order to demonstrate Ca^{2+}-driven oscillatory currents by techniques available during the late 1970s, Ca^{2+} overload was produced in most experiments. Thus, the prevailing assumption at the time was that the oscillating current was a characteristic of abnormal function, similar to that described by Lewis, Rothberger, Scherf, Segers, and Bozler (review Ref. 57). In these earlier studies Ca^{2+} overload (not realized at that time by these investigators) was employed to disengage the intracellular Ca^{2+} clock from its entrainment by externally driven APs. Despite the fact that it had already been demonstrated that similar fluctuations can also appear quite prominently in some heart cells in the absence of overt intervention, current fluctuations were primarily interpreted as a disturbance of normal cardiac function.[74] Upon testing an integration of a simple Ca^{2+} oscillator scheme into a pacemaker model, Noble, in 1984,[16] concluded that clarification of these abnormal rhythm mechanisms required additional experiments. Other pacemaker *cognoscenti* of the day, although basically in agreement with this view, were somewhat more liberal in their vision. Cranefield[57] reasoned that the true automaticity of the sinoatrial node or of Purkinje fibers might have a close connection with triggered activity. Irisawa speculated that subthreshold oscillations contribute to a "safety margin to prevent the sinoatrial node cell from becoming quiescent."[76] However, since the spontaneous, localized, oscillatory Ca^{2+} signals in SANC could not be measured at the time, the concept that an internal Ca^{2+} oscillator provides the initiating signals for normal automaticity was abandoned again.

The Physiologic Intracellular Ca²⁺ Clock within Pacemaker Cells Is Rejuvenated

In the late 1980s, experimental investigation of the role of Ca^{2+} in pacemaker function began to shift from an experimental Ca^{2+}-overload paradigm to more physiologic conditions. Afterdepolarizations and contractions were recorded in Purkinje fibers under normal Ca^{2+} loading conditions,[81] and these fibers demonstrated localized spontaneous myofilament motion caused by spontaneous local Ca^{2+} oscillations.[82] It was discovered that ryanodine (depleting SR content) has a profound negative chronotropic effect on automaticity of subsidiary atrial pacemakers.[83] Analyses of the ryanodine effect on AP shape led to the suggestion that Ca^{2+} released from the SR contributes to DD, primarily during its latter half, and plays a prominent role in bringing the late pacemaker potential to the AP activation threshold. The importance of RyR-mediated Ca^{2+} release and I_{NCX} for normal spontaneous beating was also demonstrated in toad pacemaker cells.[84]

Additional studies provided compelling evidence that rhythmic Ca^{2+} cycling within SANC is crucial for their normal automaticity. Ca^{2+} cycling proteins, SERCA, RyR, and NCX previously identified in atrial or ventricular cells, were also identified in SANC.[85,85a] A possible importance of SR Ca^{2+} release and related I_{NCX} for SANC pacemaker function was demonstrated by the ryanodine effect in these cells.[42–44] When it became possible to routinely measure intracellular Ca^{2+} in SANC, it was observed that a cytosolic Ca^{2+} transient is evoked by the AP, and that Ca^{2+} influx via L-type Ca^{2+} channels affects Ca^{2+} loading of the SR; chelation of intracellular Ca^{2+} in rabbit SANC by bis-(o-aminophenoxy)-N,N,N',N'-tetraacetic acid (BAPTA)[43,86] (but not EGTA![86,87]), applied intracellularly, markedly slowed or abolished spontaneous beating of rabbit SANC. A strong, negative, chronotropic effect of EGTA, however, had been reported in SANC[88] and atrioventricular node cells[89] of guinea pig. Changes in the cytosolic Ca^{2+} transient in response to β-adrenergic stimulation were correlated with changes in the beating rate.[84,85] Thus, it was becoming recognized that intracellular Ca^{2+} cycling within pacemaker cells is somehow involved in normal SANC pacemaker function.

A New Pacemaker Theory Based on Integration of the Intracellular Ca²⁺ Clock and Membrane Ion Channel Clock

Ca²⁺ Clocks in SANC Ignite Rhythmic APs via Activation of NCX Current

Confocal Ca^{2+} imaging permitted detection of a more subtle form of spontaneous Ca^{2+} release than those caused by experimental Ca^{2+} overload. Recent studies in rabbit SANC (FIG. 1C) employing confocal imaging showed that, in contrast to ventricular myocytes, the SR Ca^{2+} clocks are, in part, entrained by an AP but later in the cycle become "free running."

The rhythmic local Ca^{2+} releases (LCRs) via RyRs occur spontaneously beneath the cell membrane in SANC during the later part of the spontaneous DD (FIG. 2A). Dense immunolabeling and clear colocalization of RyR and NCX beneath the surface membrane of SANC have been recently demonstrated (FIG. 2C–

E).[90] LCRs activate NCX inward currents producing miniature voltage fluctuations. The ensemble of the inward NCX currents causes the later part of the DD to increase exponentially (FIG. 2B, "Nonlinear DD"), leading to the generation of spontaneous APs.[91] LCRs are a manifestation of spontaneous SR Ca^{2+} release that does not require a trigger mechanism, such as surface membrane depolarization or inward currents; rhythmic spontaneous LCRs occur in skinned SANC and under voltage clamp.[92,93]

An argument that is often raised in an attempt to exclude submembrane pacemaker mechanisms is that in voltage-clamped SANC the membrane current does not show cyclic changes.[21,94] However, this is only true with reference to a comparison with oscillatory current scaling of more than 100 pA, as observed previously in digitalis-treated ventricular myocytes, and now interpreted as the gold standard of the oscillatory current (FIG. 1 in Ref. 94). Further, accurate perforated patch clamp studies, which preserve crucial intracellular interactions within rabbit SANC, clearly demonstrate an oscillatory nature of the net membrane current (FIG. 3).[92,93] Since the scale of the oscillatory current in SANC is only 10 pA, it likely was distorted or missed in early patch clamp whole-cell experiments.[21,94] However, it is the 10 pA current that has crucial importance during DD because membrane electrical resistance during this phase of cycle is very high; a tiny current change of approximately 3 pA is enough to drive the critical DD change in rabbit SANC.[21,40] The rhythmic oscillatory currents in SANC under voltage clamp are produced by rhythmic LCRs, since both LCRs and the currents exhibit fluctuations of the same frequency,[92,93] and both are abolished by ryanodine[95] (FIG. 4).

The Spontaneous Cycle Length is Tightly Linked to LCR Period of SANC

The Ca^{2+} clock rate in cardiac cells is reflected in the LCR period observed as a delay from the AP-triggered global Ca^{2+} transient and the subsequent LCR occurrence (FIG. 5A). The likelihood for spontaneous SR Ca^{2+} release to occur increases as a function of the SR Ca^{2+} load determined in a physiologic content by the Ca^{2+} available for pumping and the phosphorylation status of the SR Ca^{2+} cycling proteins. Thus, the convergence of multiple factors governs the LCR period. These include the rate at which Ca^{2+} is pumped into the SR, the threshold SR Ca^{2+} load required to initiate spontaneous Ca^{2+} release, and the availability of SR Ca^{2+} release channels (i.e., RyRs).

Each spontaneous cycle of the SANC LCR clock can be envisioned to initiate the occurrence of an AP via I_{NCX} activation. Ca^{2+} influx via I_{CaL} during

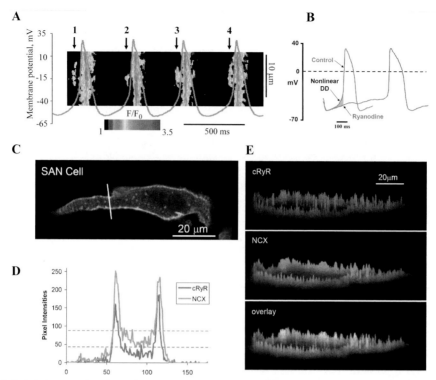

FIGURE 2. Diastolic submembrane LCRs via RyRs ignite rhythmic APs via activation of NCX imparting nonlinear (exponential) late DD in rabbit SANC. **(A)** Linescan Ca^{2+} image with superimposed spontaneous APs. *Arrows* mark LCRs. Modified from Ref. 91. **(B)** Ryanodine abolishes LCRs and thus abolishes the LCR-mediated AP ignition early in the cycle by inhibition of the nonlinear DD (gray area shown by *arrow*). Modified from Ref. 95. **(C)** Confocal whole cell image of cells doubly labeled for NCX and RyR. **(D)** Graphed pixel-by-pixel fluorescence intensities of labeling along an arbitrary line, positioned as indicated by a *thick white line* in C. The *horizontal dashed lines* report the average pixel intensity. **(E)** The topographical profiles of the pixel intensity levels of each antibody labeling and overlay of the small SANC. The maximum height represents the brightest possible pixel in the source image (using an 8-bit image intensity scale). Less bright pixels are accordingly scaled to a smaller height. C-E from Ref. 90. (In color in *Annals* online.)

AP triggers Ca^{2+}-induced Ca^{2+}-release. The resulting global SR Ca^{2+} depletion synchronizes SR throughout the cell in a Ca^{2+}-depleted state. Refilling of the SR ensures that the threshold of Ca^{2+} load required for spontaneous release is achieved at about the time when RyR inactivation following prior activation is removed; then the spontaneous LCR occurs, activating I_{NCX}, igniting next AP, and so on.

Thus, Ca^{2+} refilling of the SR to the spontaneous release threshold yields the delay determining LCR period so that the LCR period is not fixed but can vary within the kinetics of SR Ca^{2+} reloading (FIG. 5A). Variations in the LCR period are, in turn, linked to variations in the spontaneous beating rate. Thus, an extremely tight link between the LCR period and the spontaneous cycle length has been experimentally proven (TABLE 1). Note that the relationship of LCR pe-

riods to the spontaneous cycle length of SANC (slope coefficient in TABLE 1) is nearly identical and is near unity over a wide range of conditions, and the intercept indicates that, regardless of the conditions and absolute cycle length, LCRs shortly precede (within approximately 100 ms) and ignite APs.

The Crucial Role of Protein Kinase A-dependent Protein Phosphorylation in the Tightly Controlled Variability of Ca^{2+} Clock Rate and SANC Chronotropy

Protein kinase A (PKA)-dependent phosphorylation is obligatory for basal LCR occurrence; gradations in basal PKA-dependent phosphorylation result in gradations in the LCR period[93] (FIG. 5B,C). Stimulation of β-adrenergic receptors extends the range of PKA dependence of the LCR period. The PKA dependence

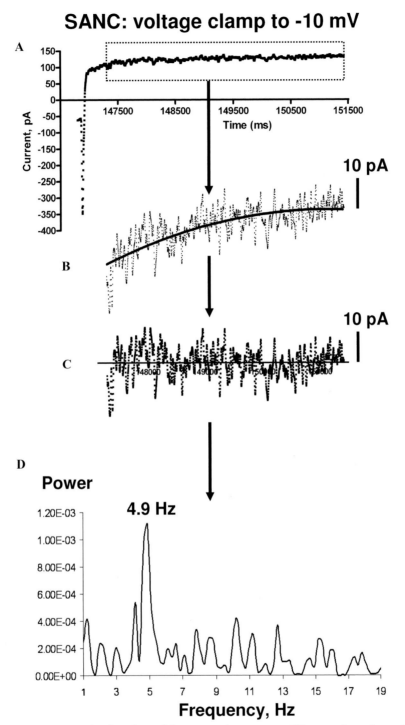

FIGURE 3. An example of analysis of the net membrane current (**A**) in a voltage-clamped rabbit SANC to detect cyclic current fluctuations. *Arrows* show the analysis sequence. The net current is first zoomed to a 10 pA scale and its slow component is fitted by a cubic polynomial function (*solid line* in (**B**). The slow component (likely from K⁺ current activation) is subtracted (**C**). The current fluctuations show a dominant oscillation frequency of 4.9 Hz in its power spectrum (**D**).

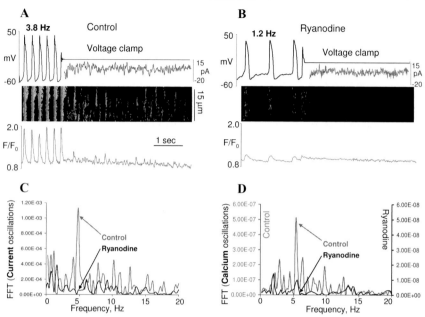

FIGURE 4. **(A,B)** Simultaneous recordings of membrane potential or current (*top*), confocal linescan image (*middle*), and normalized fluo-3 fluorescent (*bottom*) averaged over the linescan image in a representative spontaneously beating rabbit SANC with intact sarcolemma prior to and during voltage clamp to −10 mV in control (**A**) and following inhibition of RyR with 3 μmol/L ryanodine (**B**). The traces in panels A and B under voltage clamp depict membrane current fluctuations generated by LCR occurrence. **(C,D)** Fast Fourier Transform (FFT) of Ca^{2+} (**D**) and membrane current (**C**) fluctuations during voltage clamp in control and after ryanodine. Note that the rate of these current fluctuations is the same as that of LCRs and is similarly suppressed by ryanodine. From Ref. 95. (In color in *Annals* online.)

of the LCR period is attributable to at least three (and probably several more) potential mechanisms: augmentation of Ca^{2+} influx via phosphorylation of L-type Ca^{2+} channels; augmentation of Ca^{2+} pumping into the SR via phospholamban (PLB) phosphorylation; and increased availability of RyRs via their phosphorylation.

In spontaneously beating SANC, inhibition or activation of PKA results in gradations in PLB phosphorylation, paralleled by gradations in LCR period[93]; but graded phosphorylation of L-type channels and RyRs also likely occurs in such experiments. The dependence of LCR period on PKA inhibition or the addition of cyclic adenosine 3′,5′-phsophate (cAMP) to skinned SANC or during voltage clamp[93] demonstrates the role of PKA-dependent mechanisms distinct from those of L-type Ca^{2+} channels (FIG. 6A). However, there have been no experiments to date that separate the roles of Ca^{2+} pumping from RyR availability in the generation of spontaneous LCRs. As to I_{CaL}, its roles in SANC spontaneous firing rate regulation are 1) to provide AP upstroke; 2) to synchronize SR in a Ca^{2+}-depleted state; and 3) to support the increased

Ca^{2+} supply–demand balance at higher rates as ignition from lower voltages requires a larger NCX current and is thus associated with higher Ca^{2+} efflux by the NCX.

Novel Numerical Pacemaker Models Test Ca^{2+}-driven Pacemaker Mechanism

Initial tests of the ability of LCRs during the DD to command the AP firing rate used a primary rabbit SANC model developed by Zhang *et al.*[96] When experimentally measured waveforms of submembrane, LCRs were introduced into the model, spontaneous AP firing rate was easily entrained at a rate determined by the period of these LCRs.[97] Subsequently, Kurata *et al.* developed a model[98] that described Ca^{2+} concentration in submembrane space and predicted a strong negative chronotropic effect of Ca^{2+} chelation in this critical location. A novel pacemaker model[99] included LCR in the submembrane space and reproduced, for the first time, the negative chronotropic effect of ryanodine and predicted a new powerful mechanism of rate regulation by varying Ca^{2+} release rate and phase; the larger release resulted in the larger I_{NCX} that, in turn,

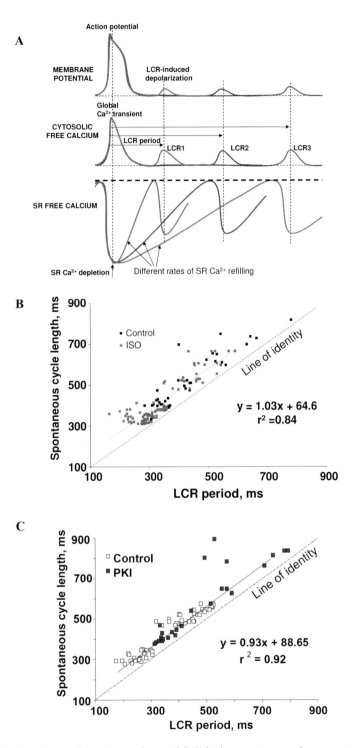

FIGURE 5. Variations in the LCR period are tightly linked to variations in the spontaneous beating. **A** schematically illustrates the idea that the rate of SR Ca²⁺ refilling controls the LCR period and the timing of the LCR-induced depolarization. **(B)** Relationship between LCR period and spontaneous cycle length is shifted to shorter periods by β-adrenergic stimulation with 0.1 μmol/L isoproterenol (ISO). **(C)** The PKI effect to increase the cycle length is linked to its effect on the LCR period. B and C are modified from Ref. 93. (In color in *Annals* online.)

TABLE 1. Variations in spontaneous cycle length are strongly correlated with variations in LCR period

Experimental paradigm	Cycle length = a·LCR Period + b
Different SANC modified from Ref. 92	CL = 0.86 · LCR period + 23 ms, $r^2 = 0.85$
SR Ca^{2+} repletion following SR Ca^{2+} depletion modified from Ref. 92	CL = 0.90 · LCR period + 85 ms, $r^2 = 0.86$
Ryanodine (Vinogradova and Lakatta)	CL = 1.00 · LCR period + 110 ms, $r^2 = 0.90$
Suppression of SR Ca pump (Vinogradova and Lakatta)	CL = 0.97 · LCR period + 98 ms, $r^2 = 0.85$
PKA inhibition (PKI)[93]	CL = 0.93 · LCR period + 89 ms, $r^2 = 0.92$
PKA inhibition (H89)[93]	CL = 0.97 · LCR period + 86 ms, $r^2 = 0.85$
Basal PDE inhibition induced reduction in cycle length (Vinogradova and Lakatta)	CL = 0.93 · LCR period + 69 ms, $r^2 = 0.89$
β-AR stimulation induced reduction in cycle length (modified from Ref. 93)	CL = 1.03 · LCR period + 64.6 ms, $r^2 = 0.84$

An extremely tight link between the LCR period and the spontaneous cycle length (CL) of SANC was identified experimentally (slope a, intercept b, and correlation r^2) over a wide range of conditions (experimental paradigm).

allowed wider rate regulation range by the phase of the release covering basically the entire physiologic range. In terms of the fine DD structure, the model showed that the LCR-activated NCX current imparts an exponentially rising part to the late DD that culminates in an AP upstroke.[99]

A subsequent upgrade of the model reproduced individual stochastic LCRs (a multicompartment SR model).[91,93] Varying frequency, size, and phase of the LCRs, according to experimental data, this model predicted the wide range of chronotropic effects that were experimentally observed with graded PKA activity; that is, basal and reserve rate regulation via a stochastic ensemble of local NCX currents induced by the LCRs.[93] The model also provided further details of fine DD structure, including its fluctuations and exponentially rising late DD component observed experimentally shortly before the excitation as well as modest beat-to-beat cycle variations from the stochastic occurrence of the LCRs.[91] This model identified substantial benefits of interactions of the rhythmic LCRs and the membrane ion channel clock in terms of overall performance of the new pacemaker mechanism. Under conditions when ion channels operating alone in numerical models fail to generate rhythmic APs, stable and rhythmic AP firing resumes when the timely and powerful prompt to the membrane sent by SR via LCRs and NCX is introduced into the model.[100]

The numerical model approaches described above illustrated that the concept of Ca^{2+} release–ignited excitation is indeed operational in primary SANC, based on the existing knowledge of LCRs and SANC-specific ion channels. However, the Ca^{2+} release function in these model simulations was either substituted by the experimental waveform or approximated by a sinusoidal function. The most recent approaches to quantitatively describe Ca^{2+} integrated SANC function has included spontaneous Ca^{2+} release mechanisms[101] and local Ca^{2+} dynamics.[102] More specifically, this approach has developed a new model[101] of primary rabbit SANC in which an integrated Ca^{2+} release mechanism is controlled by the SR Ca^{2+} load (modified from Shannon *et al.* 2004[103]). This more mechanistic model now predicts SR Ca^{2+} depletion–refilling dynamics measured by Fluo-5N[101,104] in spontaneously beating SANC and exhibits spontaneous Ca^{2+} oscillations under voltage clamp. Varying the Ca^{2+} clock model parameters results in a graded control of the timing of spontaneous Ca^{2+} release and the pacemaker rate. This result was corroborated by the experimental finding that cyclopiazonic acid, an inhibitor of SR Ca^{2+} ATPase, effects about a 50% reduction of the spontaneous beating rate of rabbit SANC.[104] These recent numerical model studies thus show that 1) periodic SR Ca^{2+} refilling and release indeed define (quantitatively rather than speculatively or phenomenologically) the Ca^{2+} clock within SANC and 2) beat-to-beat interaction of this Ca^{2+} clock with the membrane clock results in a robust pacemaker function with an effectively controlled beating rate.

The great acceleration of computational power of modern computers permits development of an extension of the presently available models to simulate local Ca^{2+} dynamics in the submembrane space of SANC, based on approximation of Ca^{2+} diffusion and stochastic activation of RyR clusters (Ca^{2+} release units).[102] These simulations predict the occurrence of local stochastic Ca^{2+} wavelets with spatiotemporal properties similar to those found for LCRs in SANC. In perspective, when the above two modeling approaches converge, a rather sophisticated powerful SANC model integrating function of membrane channels, SR Ca^{2+} release mechanism, and local Ca^{2+} dynamics will likely emerge.

Mutual Entrainment of Ca^{2+} and Membrane Clocks

The true cardiac pacemaker mechanism is complex and includes a balanced integration of both intracellular and membrane-delimited processes

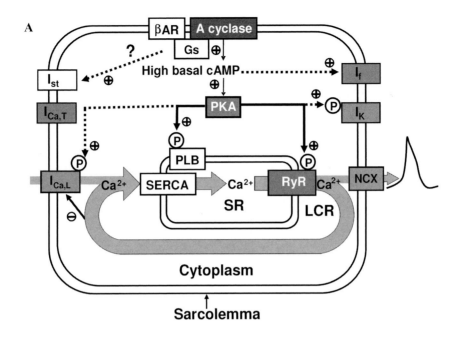

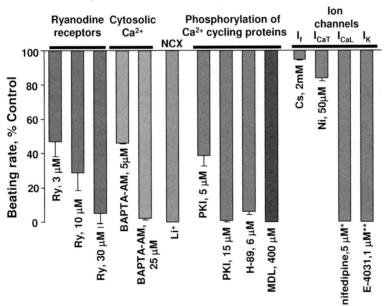

FIGURE 6. (A) Schematic illustration of functional integration and regulation of membrane and submembrane Ca^{2+} cycling to control pacemaker function via NCX-mediated ignition of rhythmic APs. The *thick line* indicates spontaneous SR Ca^{2+} cycling. Modified from Ref. 93. **(B)** Spontaneous beating of rabbit SANC critically depends upon Ca^{2+}-related mechanisms and protein phosphorylation. Bars show a decrease in the beating rate (% control) induced by different drugs that affect Ca^{2+} cycling (ryanodine receptors, cytosolic Ca^{2+}), NCX (Li^+ substitution for Na^+), protein phosphorylation (PKI, H-89, MDL), or ion channels: I_f (Cs^+), I_{CaT} (Ni^{2+}), I_{CaL} (nifedipine [105]), I_K (E-4031 [106]). PKI and H-89 are PKA inhibitors; MDL is an adenylyl cyclase inhibitor. Modified from Ref. 95. (In color in *Annals* online.)

(FIG. 6A). Interfering with NCX function or SR internal Ca^{2+} cycling directly (e.g., ryanodine) or indirectly (by inhibiting PKA-dependent phosphorylation) precludes normal pacemaker functions, just as does interfering with membrane ion channels (I_{CaL} and I_K) that generate the AP upstroke and repolarization (FIG. 6B). Note that blockade of I_{CaT} or I_f has relatively minor effects on the SANC pacemaker firing rate.

While the SANC ion channel membrane clocks are necessary to effect an AP, they are not sufficient, in our opinion, to ignite rhythmic APs at variable rates. Rather, both experimental data and numerical simulations (see above) demonstrate that the intracellular Ca^{2+} clock and its rhythmic LCRs do not merely fine tune the normal automaticity of SANC but, by igniting excitations from within the cell, represent the formal cause of APs (FIGS. 1 and 6A). The true initiation of cardiac pacemaker automaticity begins during the late part of the DD when the Ca^{2+} clock ignites the surface membrane excitation. Earlier events, such as the maximum diastolic potential and early linear DD, ensure recovery of the membrane clocks that are activated during the prior AP and need to be reset before the next AP. The spontaneous, but precisely controlled, rhythmic SR Ca^{2+} clock of SANC ensures the stability of basal rhythm and sets the basal firing rate by integrating multiple Ca^{2+}-dependent functions and rhythmically interacting with one, igniting the ensemble of surface membrane clocks to effect APs. Variable degrees of intrinsic PKA-dependent protein phosphorylation, and its attendant variations in intracellular Ca^{2+} (FIG. 6A), regulate the time that the SR Ca^{2+} clock keeps. This regulates the LCR period, which modulates the voltage- and time-dependent kinetics of the membrane clocks, thus ensuring that they produce APs with characteristics commensurate with given rhythmic ignition rates commanded by LCRs.

The rate at which the Ca^{2+} clock ticks controls rhythmic spontaneous AP firing across the broad range of physiologic firing rates. In short, variations of internal Ca^{2+} cycling within SANC result in corresponding variations in LCRs that are tightly linked to gradations in their automaticity. In addition to AP generation, surface membrane ion channels ensure the robust function of the Ca^{2+} clock by keeping intracellular ion balance, especially for Ca^{2+}. Ca^{2+} influx via I_{CaL} during AP "rewinds" the Ca^{2+} clock by balancing Ca^{2+} efflux via NCX each cycle. Maintaining this balance is indeed critical for the Ca^{2+} clock's long-term operation since LCRs become damped and cease within a few seconds following cessation of APs during voltage clamp near the maximum diastolic potential.[92] K^+ currents, in turn, regulate the AP duration commensurate with the ignition rate commanded by the Ca^{2+} clock.

Summary

The mutual entrainment of the Ca^{2+} clocks and surface membrane ion channel clocks regulate the kinetics of the SANC duty cycle. We believe that this new concept of mutually interacting clocks, governing normal automaticity of pacemaker cells, marks a new beginning of cardiac pacemaker research.

Further evidence for the crucial role of Ca^{2+} and rhythmic spontaneous Ca^{2+} releases in the initiation and regulation of normal cardiac automaticity provides the key that reunites the fields of pacemaker automaticity and cardiac muscle function.[26] The separate achievements of reductionist approaches within each of these fields of research to elucidate the strength of contraction of ventricular myocytes and the spontaneous beating rate of pacemaker cells converge and a general theory of cardiac chronotropic and inotropic mechanisms emerges: SR Ca^{2+} cycling both initiates the heartbeat in pacemaker cells and executes the heartbeat in ventricular cells.

Acknowledgment

This research was supported by the Intramural Research Program of the National Institutes of Health, National Institute on Aging.

Conflicts of Interest

The authors declare no conflicts of interest.

References

1. BERNSTEIN, J. 1868. Ueber den zeitlichen Verlauf der negativen Schwankung des Nervenstroms. Pflugers Arch. **1:** 173–207 (in German).
2. BERNSTEIN, J. 1912. Elektrobiologie. Die Lehre von den elektrischen Vorgangen im Organismus auf moderner Grundlage dargestellt: 138–141 (in German). Vieweg & Sohn. Braunschweig.
3. HODGKIN, A.L. & B. KATZ. 1949. The effect of sodium ions on the electrical activity of the giant axon of the squid. J. Physiol. **108:** 37–77.
4. HODGKIN, A.L. & A.F. HUXLEY. 1952. A quantitative description of membrane current and its application to conduction and excitation in nerve. J. Physiol. **117:** 500–544.
5. KLEBER, A.G. *et al.* 2006. The early years of cellular cardiac electrophysiology and Silvio Weidmann (1921–2005). Heart Rhythm. **3:** 353–359.

6. ARVANITAKI, A. *et al.* 1937. Reactions electroniques du myocarde en function de sur tonus initial. Compt. Rend. Sot. Biol. **124:** 165–167.

7. BOZLER, E. 1943. The initiation of impulses in cardiac muscle. Am. J. Physiol. **138:** 273–282.

8. BOZLER, E. 1943. Tonus changes in cardiac muscle and their significance for the initiation of impulses. Am. J. Physiol. **139:** 477–480.

9. BRADY, A.J. & H.H. HECHT. 1954. On the origin of the heart beat. Am. J. Med. **17:** 110.

10. CORABOEUF, E. & S. WEIDMANN. 1954. Temperature effects on the electrical activity of Purkinje fibres. Helv. Physiol. Pharmacol. Acta **12:** 32–41.

11. WEST, T.C. 1955. Ultramicroelectrode recording from the cardiac pacemakers. J. Pharmacol. Exp. Ther. **115:** 283–290.

12. NOBLE, D. 1960. Cardiac action and pacemaker potentials based on the Hodgkin-Huxley equations. Nature **188:** 495–497.

13. NOBLE, D. 1962. A modification of the Hodgkin-Huxley equations applicable to Purkinje fibre action and pacemaker potentials. J. Physiol. **160:** 317–352.

14. WEIDMANN, S. 1951. Effect of current flow on the membrane potential of cardiac muscle. J. Physiol. **115:** 227–236.

15. DUDEL, J. & W. TRAUTWEIN. 1958. The mechanism of formation of automatic rhythmical impulses in heart muscle. Pflugers Arch. **267:** 553–565.

16. NOBLE, D. 1984. The surprising heart: a review of recent progress in cardiac electrophysiology. J. Physiol. **353:** 1–50.

17. DECK, K.A., R. KERN & W. TRAUTWEIN. 1964. Voltage clamp technique in mammalian cardiac fibres. Pflugers Arch. Gesamte Physiol. Menschen Tiere **280:** 50–62.

18. POWELL, T., D.A. TERRAR & V.W. TWIST. 1980. Electrical properties of individual cells isolated from adult rat ventricular myocardium. J. Physiol **302:** 131–153.

19. IRISAWA, H. 1972. Electrical activity of rabbit sino-atrial node as studies by a double sucrose gap method. *In* Proceedings of the Satellite Symposium of the XXVth Interneational Congress of Physiological Science. The Electrical Field of the Heart. P. Rijlant, Ed.: 242–248. Press Academiques Europeenes. Bruxelles.

20. TANIGUCHI, J. *et al.* 1981. Spontaneously active cells isolated from the sino-atrial and atrio-ventricular nodes of the rabbit heart. Jpn. J. Physiol. **31:** 547–558.

21. IRISAWA, H., H.F. BROWN & W. GILES. 1993. Cardiac pacemaking in the sinoatrial node. Physiol. Rev. **73:** 197–227.

22. WILDERS, R. 2007. Computer modelling of the sinoatrial node. Med. Biol. Eng. Comput. **45:** 189–207.

23. REUTER, H. 1966. [Current-tension relations of Purkinje fibers in different extracellular concentrations of calcium and under the influence of adrenaline]. Pflugers Arch. Gesamte Physiol. Menschen Tiere **287:** 357–367.

24. DIFRANCESCO, D. & C. OJEDA. 1980. Properties of the current if in the sino-atrial node of the rabbit compared with those of the current iK, in Purkinje fibres. J. Physiol. **308:** 353–367.

25. MCALLISTER, R.E., D. NOBLE & R.W. TSIEN. 1975. Reconstruction of the electrical activity of cardiac Purkinje fibres. J. Physiol. **251:** 1–59.

26. MALTSEV, V.A., T.M. VINOGRADOVA & E.G. LAKATTA. 2006. The emergence of a general theory of the initiation and strength of the heartbeat. J. Pharmacol. Sci. **100:** 338–369.

27. NEHER, E. & B. SAKMANN. 1976. Single-channel currents recorded from membrane of denervated frog muscle fibres. Nature **260:** 799–802.

28. SATOH, H. 2003. Sino-atrial nodal cells of mammalian hearts: ionic currents and gene expression of pacemaker ionic channels. J. Smooth Muscle Res. **39:** 175–193.

29. VASSALLE, M. 1995. The pacemaker current (If) does not play an important role in regulating SA node pacemaker activity. Cardiovasc. Res. **30:** 309–310.

30. BROWN, H., J. KIMURA & S. NOBLE. 1982. The relative contributions of various time-dependent membrane currents to pacemaker activity in the sino atrial node. *In* Cardiac Rate and Rhythm. L.N. Bouman & H.J. Jongsma, Eds.: 53–68. Martinus Nijhoff. The Hague.

31. SHIBATA, E.F. & W.R. GILES. 1985. Ionic currents that generate the spontaneous diastolic depolarization in individual cardiac pacemaker cells. Proc. Natl. Acad. Sci. USA **82:** 7796–7800.

32. ONO, K. & T. IIJIMA. 2005. Pathophysiological significance of T-type Ca^{2+} channels: properties and functional roles of T-type Ca^{2+} channels in cardiac pacemaking. J. Pharmacol. Sci. **99:** 197–204.

33. VERKERK, A.O. *et al.* 2002. Ca^{2+}-activated Cl^- current in rabbit sinoatrial node cells. J. Physiol. **540:** 105–17.

34. GUO, J., K. ONO & A. NOMA. 1995. A sustained inward current activated at the diastolic potential range in rabbit sino-atrial node cells. J. Physiol. **483**(Pt 1): 1–13.

35. BROWN, H.F. *et al.* 1984. The ionic currents underlying pacemaker activity in rabbit sino-atrial node: experimental results and computer simulations. Proc. R. Soc. Lond. B Biol. Sci. **222:** 329–347.

36. ZAZA, A. *et al.* 1997. Ionic currents during sustained pacemaker activity in rabbit sino-atrial myocytes. J. Physiol. **505**(Pt 3): 677–688.

37. ONO, K. & H. ITO. 1995. Role of rapidly activating delayed rectifier K+ current in sinoatrial node pacemaker activity. Am. J. Physiol. **269:** H453–462.

38. YANAGIHARA, K., A. NOMA & H. IRISAWA. 1980. Reconstruction of sino-atrial node pacemaker potential based on the voltage clamp experiments. Jpn. J. Physiol. **30:** 841–857.

39. MAYLIE, J., M. MORAD & J. WEISS. 1981. A study of pacemaker potential in rabbit sino-atrial node: measurement of potassium activity under voltage-clamp conditions. J. Physiol. **311:** 161–178.

40. DIFRANCESCO, D. 1991. The contribution of the 'pacemaker' current (if) to generation of spontaneous activity in rabbit sino-atrial node myocytes. J. Physiol. **434:** 23–40.

41. NILIUS, B. 1986. Possible functional significance of a novel type of cardiac Ca channel. Biomed Biochim Acta **45:** K37–45.

42. RIGG, L. & D.A. TERRAR. 1996. Possible role of calcium release from the sarcoplasmic reticulum in pacemaking in guinea-pig sino-atrial node. Exp. Physiol. **81:** 877–880.

43. LI, J., J. QU & R.D. NATHAN. 1997. Ionic basis of ryanodine's negative chronotropic effect on pacemaker cells isolated from the sinoatrial node. Am. J. Physiol. **273:** H2481–2489.

44. SATOH, H. 1997. Electrophysiological actions of ryanodine on single rabbit sinoatrial nodal cells. Gen. Pharmacol. **28:** 31–38.

45. RINGER, S. 1883. A further contribution regarding the influence of the different constituents of the blood on the contraction of the heart. J. Physiol. **4:** 29–42.

46. BERS, D.M. 2001. Excitation-Contraction Coupling and Cardiac Contractile Force, 2nd ed. Kluwer Academic Publishers. Norwell, Mass.

47. SUTKO, J.L. *et al.* 1979. Ryanodine: its alterations of cat papillary muscle contractile state and responsiveness to inotropic interventions and a suggested mechanism of action. J. Pharmacol. Exp. Ther. **209:** 37–47.

48. SUTKO, J.L. & J.T. WILLERSON. 1980. Ryanodine alteration of the contractile state of rat ventricular myocardium. Comparison with dog, cat, and rabbit ventricular tissues. Circ. Res. **46:** 332–343.

49. SCHAEFER, J. *et al.* 1992. Historical note on the translation of H.P. Bowditch's paper 'Über die Eigenthümlichkeiten der Reizbarkeit, welche die Muskelfasern des Herzens zeigen' (On the peculiarities of excitability which the fibres of cardiac muscle show). *In* The Interval-Force Relationship of the Heart: Bowditch revisited. M. Noble & W.A. Seed, Eds. Cambridge University Press. Cambridge.

50. WOOD, E.H., R.L. HEPPNER & S. WEIDMANN. 1969. Inotropic effects of electric currents. I. Positive and negative effects of constant electric currents or current pulses applied during cardiac action potentials. II. Hypotheses: calcium movements, excitation-contraction coupling and inotropic effects. Circ. Res. **24:** 409–445.

51. FABIATO, A. & F. FABIATO. 1978. Calcium-induced release of calcium from the sarcoplasmic reticulum of skinned cells from adult human, dog, cat, rabbit, rat, and frog hearts and from fetal and new-born rat ventricles. Ann. N.Y. Acad. Sci. **307:** 491–522.

52. FABIATO, A. & F. FABIATO. 1972. Excitation-contraction coupling of isolated cardiac fibers with disrupted or closed sarcolemmas. Calcium-dependent cyclic and tonic contractions. Circ. Res. **31:** 293–307.

53. ALLEN, D.G. & J.R. BLINKS. 1978. Calcium transients in aequorin-injected frog cardiac muscle. Nature **273:** 509–513.

54. WIER, W.G. 1980. Calcium transients during excitation-contraction coupling in mammalian heart: aequorin signals of canine Purkinje fibers. Science **207:** 1085–1087.

55. LAB, M.J., D.G. ALLEN & C.H. ORCHARD. 1984. The effects of shortening on myoplasmic calcium concentration and on the action potential in mammalian ventricular muscle. Circ. Res. **55:** 825–829.

56. DUBELL, W.H. *et al.* 1991. The cytosolic calcium transient modulates the action potential of rat ventricular myocytes. J. Physiol. **436:** 347–369.

57. CRANEFIELD, P.F. 1977. Action potentials, afterpotentials, and arrhythmias. Circ. Res. **41:** 415–423.

58. TSIEN, R.W., R.S. KASS & R. WEINGART. 1979. Cellular and subcellular mechanisms of cardiac pacemaker oscillations. J. Exp. Biol. **81:** 205–215.

59. GASKELL, W.H. 1881. On the innervation of the heart, with special reference to the heart of the tortoise. J. Physiol. **4:** 43–127.

60. SEGERS, M. 1941. Le rôle des potentiels tardifs du coeur. Mem. Acad. R. Med. **1:** 1–30 (in French).

61. DAVIS, L.D. 1973. Effect of changes in cycle length on diastolic depolarization produced by ouabain in canine Purkinje fibers. Circ. Res. **32:** 206–214.

62. ROSEN, M.R., H. GELBAND & B.F. HOFFMAN. 1973. Correlation between effects of ouabain on the canine electrocardiogram and transmembrane potentials of isolated Purkinje fibers. Circulation **47:** 65–72.

63. FERRIER, G.R., J.H. SAUNDERS & C. MENDEZ. 1973. A cellular mechanism for the generation of ventricular arrhythmias by acetylstrophanthidin. Circ. Res. **32:** 600–609.

64. FERRIER, G.R. & G.K. MOE. 1973. Effect of calcium on acetylstrophanthidin-induced transient depolarizations in canine Purkinje tissue. Circ. Res. **33:** 508–515.

65. KAUFMANN, R., A. FLECKENSTEIN & H. ANTONI. 1963. [Causes and Conditions of Release of Myocardial Contractions without a Regular Action Potential]. Pflugers Arch. Gesamte Physiol. Menschen Tiere **278:** 435–446 (in German).

66. AKSELROD, S. *et al.* 1977. Electro-mechanical noise in atrial muscle fibres of the carp. Experientia **33:** 1058–1060.

67. RAPP, P.E. & M.J. BERRIDGE. 1977. Oscillations in calcium-cyclic AMP control loops form the basis of pacemaker activity and other high frequency biological rhythms. J. Theor. Biol. **66:** 497–525.

68. FABIATO, A. & F. FABIATO. 1975. Contractions induced by a calcium-triggered release of calcium from the sarcoplasmic reticulum of single skinned cardiac cells. J. Physiol. **249:** 469–495.

69. DEFELICE, L.J. & R.L. DEHAAN. 1975. Voltage noise and impedance from heart cell aggregates. Biophys. J. **15:** 130a (Abstract).

70. LEDERER, W.J. & R.W. TSIEN. 1976. Transient inward current underlying arrhythmogenic effects of cardiotonic steroids in Purkinje fibres. J. Physiol. **263:** 73–100.

71. ARONSON, R.S. & J.M. GELLES. 1977. The effect of ouabain, dinitrophenol, and lithium on the pacemaker current in sheep cardiac Purkinje fibers. Circ. Res. **40:** 517–524.

72. KASS, R.S. *et al.* 1978. Role of calcium ions in transient inward currents and aftercontractions induced by strophanthidin in cardiac Purkinje fibres. J. Physiol. **281:** 187–208.

73. EISNER, D.A. & W.J. LEDERER. 1979. Inotropic and arrhythmogenic effects of potassium-depleted solutions on mammalian cardiac muscle. J. Physiol. **294:** 255–277.

74. KASS, R.S. & R.W. TSIEN. 1982. Fluctuations in membrane current driven by intracellular calcium in cardiac Purkinje fibers. Biophys. J. **38:** 259–269.

75. MEHDI, T. & F. SACHS. 1978. Voltage clamp of isolated cardiac Purkinje cells. Biophys. J. **21:** 165a (Abstract).

76. IRISAWA, H. 1978. Comparative physiology of the cardiac pacemaker mechanism. Physiol. Rev. **58:** 461–498.

77. KASS, R.S., W.J. LEDERER & R.W. TSIEN. 1976. Current fluctuations in strophanthidin-treated cardiac Purkinje fibers. Biophys. J. **16**: 25a (Abstract).

78. WIER, W.G. *et al.* 1983. Cellular calcium fluctuations in mammalian heart: direct evidence from noise analysis of aequorin signals in Purkinje fibers. Proc. Natl. Acad. Sci. USA **80:** 7367–7371.

79. KORT, A.A. *et al.* 1985. Fluctuations in intracellular calcium concentration and their effect on tonic tension in canine cardiac Purkinje fibres. J. Physiol. **367:** 291–308.

80. ZHOU, Z. & S.L. LIPSIUS. 1993. Na⁺-Ca²⁺ exchange current in latent pacemaker cells isolated from cat right atrium. J. Physiol. **466:** 263–285.

81. LIPSIUS, S.L. & W.R. GIBBONS. 1982. Membrane currents, contractions, and aftercontractions in cardiac Purkinje fibers. Am. J. Physiol. **243:** H77–86.

82. KORT, A.A. & E.G. LAKATTA. 1984. Calcium-dependent mechanical oscillations occur spontaneously in unstimulated mammalian cardiac tissues. Circ. Res. **54:** 396–404.

83. RUBENSTEIN, D.S. & S.L. LIPSIUS. 1989. Mechanisms of automaticity in subsidiary pacemakers from cat right atrium. Circ. Res. **64:** 648–657.

84. JU, Y.K. & D.G. ALLEN. 1999. How does beta-adrenergic stimulation increase the heart rate? The role of intracellular Ca²⁺ release in amphibian pacemaker cells. J. Physiol. **516**(Pt 3): 793–804.

85. RIGG, L. *et al.* 2000. Localisation and functional significance of ryanodine receptors during beta-adrenoceptor stimulation in the guinea-pig sino-atrial node. Cardiovasc. Res. **48:** 254–264.

85a. MUSA, H. *et al.* 2002. Heterogenous expression of Ca²⁺ handling proteins in rabbit sinoatrial node. J. Histochem. Cytochem. **50:** 311–324.

86. VINOGRADOVA, T.M. *et al.* 2000. Sinoatrial node pacemaker activity requires Ca²⁺/calmodulin-dependent protein kinase II activation. Circ. Res. **87:** 760–767.

87. CHO, H.S., M. TAKANO & A. NOMA. 2003. The electrophysiological properties of spontaneously beating pacemaker cells isolated from mouse sinoatrial node. J. Physiol. **550:** 169–180.

88. SANDERS, L. *et al.* 2006. Fundamental importance of Na⁺-Ca²⁺ exchange for the pacemaking mechanism in guinea-pig sino-atrial node. J. Physiol. **571:** 639–649.

89. HANCOX, J.C., A.J. LEVI & P. BROOKSBY. 1994. Intracellular calcium transients recorded with Fura-2 in spontaneously active myocytes isolated from the atrioventricular node of the rabbit heart. Proc. Biol. Sci. **255:** 99–105.

90. LYASHKOV, A.E. *et al.* 2007. Calcium cycling protein density and functional importance to automaticity of isolated sinoatrial nodal cells are independent of cell size. Circ. Res. **100:** 1723–1731

91. BOGDANOV, K.Y. *et al.* 2006. Membrane potential fluctuations resulting from submembrane Ca²⁺ releases in rabbit sinoatrial nodal cells impart an exponential phase to the late diastolic depolarization that controls their chronotropic state. Circ. Res. **99:** 979–987.

92. VINOGRADOVA, T.M. *et al.* 2004. Rhythmic ryanodine receptor Ca²⁺ releases during diastolic depolarization of sinoatrial pacemaker cells do not require membrane depolarization. Circ. Res. **94:** 802–809.

93. VINOGRADOVA, T.M. *et al.* 2006. High basal protein kinase A-dependent phosphorylation drives rhythmic internal Ca²⁺ store oscillations and spontaneous beating of cardiac pacemaker cells. Circ. Res. **98:** 505–514.

94. NOMA, A. 1996. Ionic mechanisms of the cardiac pacemaker potential. Jpn. Heart J. **37:** 673–682.

95. LAKATTA, E.G. *et al.* 2006. The integration of spontaneous intracellular Ca²⁺ cycling and surface membrane ion channel activation entrains normal automaticity in cells of the heart's pacemaker. Ann. N.Y. Acad. Sci. **1080:** 178–206.

96. ZHANG, H. *et al.* 2000. Mathematical models of action potentials in the periphery and center of the rabbit sinoatrial node. Am. J. Physiol. **279:** H397–421.

97. LAKATTA, E.G. *et al.* 2003. Cyclic variation of intracellular calcium: a critical factor for cardiac pacemaker cell dominance. Circ. Res. **92:** e45–50.

98. KURATA, Y. *et al.* 2002. Dynamical description of sinotrial node pacemaking: improved mathematical model for primary pacemaker cell. Am. J. Physiol. **283:** H2074–101.

99. MALTSEV, V.A. *et al.* 2004. Diastolic calcium release controls the beating rate of rabbit sinoatrial node cells: numerical modeling of the coupling process. Biophys. J. **86:** 2596–2605.

100. MALTSEV, V.A. *et al.* 2005. Local subsarcolemmal Ca²⁺ releases within rabbit sinoatrial nodal cells not only regulate beating rate but also ensure normal rhythm during protein kinase A inhibition. Biophys. J. **88:** 89a–90a (Abstract).

101. MALTSEV, V.A. *et al.* 2007. A numerical model of a Ca²⁺ clock within sinoatrial node cells: Interactive membrane and submembrane Ca²⁺ cycling provides a novel mechanism of normal cardiac pacemaker function. Biophys. J. Supplement: 77a (Abstract).

102. MALTSEV, A.V. *et al.* 2007. A simple stochastic mechanism of a roughly periodic Ca²⁺ clock within cardiac cells. Biophys. J. Supplement: 344a (Abstract).

103. SHANNON, T.R. *et al.* 2004. A mathematical treatment of integrated Ca dynamics within the ventricular myocyte. Biophys. J. **87:** 3351–3371.

104. VINOGRADOVA, T.M. *et al.* 2007. Sarcoplasmic reticulum (SR) Ca²⁺ refilling kinetics controls the period of local subsarcolemmal Ca²⁺ releases (LCR) and the spontaneous beating rate of sinoatrial node cells (SANC). Biophys. J. Supplement: 31a (Abstract).

105. VERHEIJCK, E.E. *et al.* 1999. Contribution of L-type Ca²⁺ current to electrical activity in sinoatrial nodal myocytes of rabbits. Am. J. Physiol. **276:** H1064–1077.

106. VERHEIJCK, E.E. *et al.* 1995. Effects of delayed rectifier current blockade by E-4031 on impulse generation in single sinoatrial nodal myocytes of the rabbit. Circ. Res. **76:** 607–615.

Shear Fluid-induced Ca²⁺ Release and the Role of Mitochondria in Rat Cardiac Myocytes

STEVE BELMONTE AND MARTIN MORAD

Department of Pharmacology, Georgetown University, Washington, DC 20007, USA

Cardiac myocyte contraction occurs when Ca²⁺ influx through voltage-gated L-type Ca²⁺ channels causes Ca²⁺ release from ryanodine receptors of the sarcoplasmic reticulum (SR). Although mitochondria occupy about 35% of the cell volume in rat cardiac myocytes, and are thought to be located <300 nm from the junctional SR, their role in the beat-to-beat regulation of cardiac Ca²⁺ signaling remains unclear. We have recently shown that rapid (∼20 ms) application of shear fluid forces (∼25 dynes/cm²) to rat cardiac myocytes triggers slowly (∼300 ms) developing Ca$_i$ transients that were independent of activation of all transmembrane Ca²⁺ transporting pathways, but were suppressed by FCCP, CCCP, and Ru360, all of which are known to disrupt mitochondrial function. We have here used rapid 2-D confocal microscopy to monitor fluctuations in mitochondrial Ca²⁺ levels ([Ca²⁺]$_m$) and mitochondrial membrane potential ($\Delta\Psi_m$) in rat cardiac myocytes loaded either with rhod-2 AM or tetramethylrhodamine methyl ester (TMRM), respectively. Freshly isolated intact rat cardiac myocytes were plated on glass coverslips and incubated in 5 mM Ca²⁺ containing Tyrode's solution and 40 mM 2,3-butanedione monoxime (BDM) to inhibit cell contraction. Alternatively, myocytes were permeabilized with 10 μM digitonin and perfused with an "intracellular" solution containing 10 μM free [Ca²⁺], 5 mM EGTA, and 15 mM BDM. Direct [Ca²⁺]$_m$ measurements showed transient mitochondrial Ca²⁺ accumulation after exposure to 10 mM caffeine, as revealed by a 66% increase in the rhod-2 fluorescence intensity. Shear fluid forces, however, produced a 12% decrease in signal, suggesting that application of a mechanical force releases Ca²⁺ from the mitochondria. In addition, caffeine and CCCP or FCCP strongly reduced $\Delta\Psi_m$, while application of a pressurized solution produced a transient $\Delta\Psi_m$ hyperpolarization in intact ventricular myocytes loaded with TMRM. The close proximity of mitochondria to ryanodine receptors and large [Ca²⁺] that develop in microdomains following calcium release are likely to play a critical role in regulating cytosolic Ca²⁺ signaling. We suggest that mitochondria may accumulate and release Ca²⁺ in response to mechanical forces generated by blood flow, independent of surface membrane–regulated CICR. The extent to which such a signaling mechanism contributes to stretch-induced increase in myocardial force and pathogenesis of arrhythmias remains to be assessed.

Key words: mitochondria; Ca²⁺ signaling; shear/pressure Ca²⁺ release; TMRM; cardiac myocytes

Introduction

Mitochondria are fundamental organelles, intimately involved in many aspects of cellular physiology. Although their primary role is to provide energy (ATP) for a variety of cellular processes, mitochondria also contribute to steroid biosynthesis, free radical production, apoptosis, and regulation of cytosolic Ca²⁺ signaling pathways.[1,2] Cardiac myocytes contain an abundance of mitochondria (∼1/3 of the total cell volume), many of which are in close apposition (37–270 nm) to the SR Ca²⁺ release sites (dyadic or triadic junctions).[3–5] Such ultrastructural architecture might suggest a particularly critical role for mitochondria in cardiac Ca²⁺ signaling and contractility. Mitochondrial Ca²⁺ uptake with each heartbeat results in activation of matrix dehydrogenases that initiate ATP production, regulate Ca$_i$ transients, and suppress Ca²⁺-dependent inactivation of adjacent sarcolemmal L-type Ca²⁺ channels.[3,6,7] Here, we explore the role of mitochondria in regulation and/or generation of Ca$_i$ transients triggered by shear fluid forces in freshly isolated rat cardiac myocytes. We have previously reported that 2-s-long "pressurized puffs" (PPs) of bathing solutions activate slowly developing Ca$_i$ transients, which may reflect the activation of longitudinally propagating Ca²⁺ waves that were strongly and reversibly suppressed by short exposures to mitochondrial uncoupler, FCCP.[8,9] In order to clarify

Address for correspondence: Professor Martin Morad, Ph.D., Department of Pharmacology, Georgetown University, 3900 Reservoir Road, Washington, DC 20007. Voice: 202-687-8453; fax: 202-687-8458.
moradm@georgetown.edu

Ann. N.Y. Acad. Sci. 1123: 58–63 (2008). © 2008 New York Academy of Sciences.
doi: 10.1196/annals.1420.007

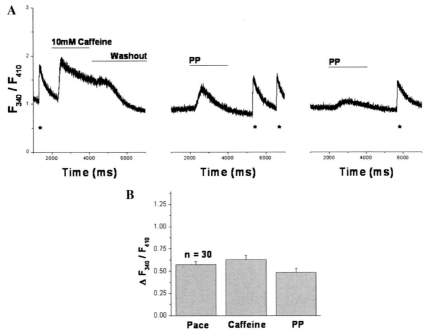

FIGURE 1. Experimental protocol for activating pressure puff–triggered Ca_i transients. (**A**) Fura-2 AM loaded ventricular myocytes were electrically stimulated (∗) and subjected to 10 mM caffeine treatment. After caffeine washout, "pressurized puff" (PP, ∼25 dynes/cm²) of Tyrode's bathing solution activated a slowly developing (∼300 ms) Ca_i transient. Application of a second PP produced a much smaller or no Ca_i transient. (**B**) Summary of data expressed as mean ± SEM.

the details of mitochondrial Ca^{2+} signaling in generation of shear flow response, we used 2-D confocal microscopy to monitor fluctuations in mitochondrial Ca^{2+} concentration ($[Ca^{2+}]_m$) and membrane potential, $\Delta\Psi_m$, in both intact and permeabilized rat ventricular myocytes. We provide evidence that mitochondria can transiently accumulate Ca^{2+} released from the SR by 10 mM caffeine. This mitochondrial Ca^{2+} pool can, in turn, be triggered by pressure/shear forces to release the accumulated Ca^{2+}, causing the slow and transient rise of cytosolic Ca^{2+}. We also show that although electrical pacing or application of caffeine depolarizes the mitochondria, pressure/shear forces hyperpolarize $\Delta\Psi_m$. Although our data strongly implicate mitochondria as the primary source for the activation of pressure/shear–induced Ca^{2+} release, the mechanism by which shear force is detected and transmitted to the mitochondria remains to be fully understood.

Methods and Results

FIGURE 1 presents the most reliable paradigm employed to trigger the pressure/shear flow–induced Ca_i transients. Freshly isolated rat ventricular my-

ocytes were incubated with 1 μM fura-2 AM for 30 min at room temperature and then perfused with a Tyrode's solution containing 5 mM Ca^{2+}. Panel A shows traces from a cell that was electrically paced at 1 Hz and exposed to a 2-s-long puff of pressurized (∼25 dynes/cm²) solution containing 10 mM caffeine. Ten seconds after washout of caffeine, the bathing solution was puffed onto the cell, producing a Ca_i transient that rises in 300 ms, and lasts ∼1.5 s. The magnitude of such Ca_i transients is comparable to that of electrically paced Ca_i transients (shown in panel B), but a second puff of solution generally failed to induce a subsequent PP-triggered Ca_i transient, despite fully loaded SR Ca^{2+} stores. The data suggest that shear/fluid forces activate Ca^{2+} release from a small Ca^{2+} store that can support only one such transient.

FIGURE 2 displays representative data from a permeabilized ventricular myocyte loaded with 15 μM of rhod-2 AM dye and imaged confocally at 7.5 frames/s. As the mitochondrial membrane potential is ∼180 mV negative to the cytosol,[10,11] cationic rhod-2 is preferentially taken up into the mitochondria. To further enhance the specificity of the mitochondrial Ca^{2+} signal, and inhibit cell motion and rise of cytosolic Ca^{2+}, myocytes were permeabilized with 10 μM digitonin

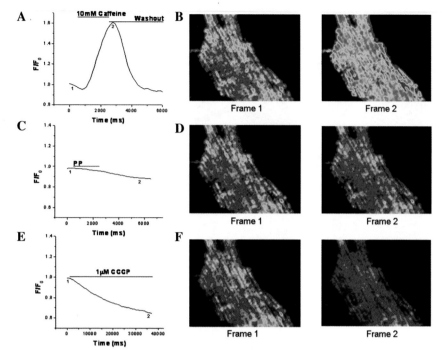

FIGURE 2. Confocal imaging of $[Ca^{2+}]_m$ in permeabilized ventricular myocytes. (**A, C,** and **E**) Ventricular myocytes were incubated with 15 μM rhod-2 AM and permeabilized with 10 μM digitonin. Traces are shown of normalized rhod-2 fluorescence intensity in response to caffeine, PP, and CCCP, respectively. (**B, D,** and **F**) Images correspond to the indicated points on the trace for caffeine, PP, and CCCP, respectively.

and perfused with an "intracellular" solution containing 5 mM EGTA and 15 mM BDM.[12] Panels A, C, and E of FIGURE 2 depict the normalized rhod-2 fluorescence signal versus time during the puffing of caffeine, bathing solution, and CCCP application, respectively. In panels B, D, and F, frames 1 and 2 are confocal images of the corresponding points for the caffeine response, PP intervention, and CCCP application, respectively. The trace in panel A shows that 10 mM caffeine caused a transient rise in normalized rhod-2 fluorescence signal (mean of 66%, $n = 6$), indicative of mitochondrial uptake and subsequent extrusion of Ca^{2+}. This is illustrated in panel B, where the fluorescence intensity increases from frame 1 to 2 at distinct points and along longitudinally arranged lines within the myocyte. Subjecting the cell to shear/fluid force of 25 dynes/cm^2 (PP) triggered a loss of mitochondrial Ca^{2+} that is not immediately replenished (mean signal decrease of 12%, $n = 6$; panels C and D). The mitochondrial uncoupler, CCCP, was used here to control the mitochondrial dye loading. CCCP specifically depolarizes $\Delta\Psi_m$ and therefore significantly diminishes $[Ca^{2+}]_m$ over 35 s of treatment (mean signal reduced by 40%, $n = 6$; panels E and F). These data confirm

the ability of mitochondria to sequester Ca^{2+} released from the SR, and suggest a critical role for the mitochondrial Ca^{2+} handling in activating PP-triggered Ca$_i$-transients.

FIGURE 3 compares the time course of global cytosolic Ca^{2+} signals in intact ventricular myocytes (fura-2 AM signal) with those of mitochondrial Ca^{2+} changes (rhod-2 AM signal) in permeabilized ventricular myocytes. The traces in panel A were obtained from an intact fura-2 AM–loaded cell, whereas panel B shows traces from a myocyte incubated with rhod-2 AM and subsequently permeabilized. The traces shown in panels A and B have identical x-axis scales, allowing easier visual comparison of the time course of events. The release of SR Ca^{2+} with 10 mM caffeine produced a large and rapid (78 ms) increase in the cytosolic Ca^{2+} of intact cells (panels A and C, $n = 30$), whereas mitochondrial Ca^{2+} rose much more slowly (1225 ms, panels B and C, $n = 9$). Intriguingly, PPs of solutions caused a decrease in $[Ca^{2+}]_m$ that began almost immediately after onset of PP force and lasted an average of 1050 ms (panels B and C). This signal has a different onset time and kinetics than the fura-2 signal induced by PPs of solutions, which increased over 138 ms and

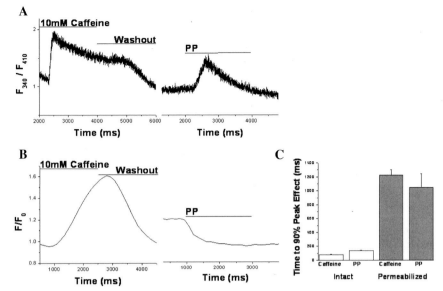

FIGURE 3. Comparison of fura-2 AM and rhod-2 AM fluorescent signals in ventricular myocytes. Global Ca^{2+} signals from an intact fura-2 AM–loaded myocyte (**A**) and $[Ca^{2+}]_m$ signals from a permeabilized myocyte incubated with rhod-2 AM (**B**). Traces in **A** and **B** are aligned vertically and the time scales are identical. (**C**) Summary of data expressed as mean ± SEM.

had a delayed onset time of ~250 ms (panels B and C). Thus, rhod-2 signals were significantly slower than fura-2 signal with respect to rise time, consistent with exclusive rhod-2 compartmentalization into mitochondria. Even though it is not certain whether the kinetics of Ca^{2+} flux in rhod-2-loaded myocytes are slowed by permeabilization and high Ca^{2+} buffering conditions, the almost immediate loss of $[Ca^{2+}]_m$ on application of shear/fluid force suggests that mitochondrial Ca^{2+} release may be the trigger for the delayed but much faster global cytosolic Ca_i-transient in intact cells.

It was of considerable interest to test whether the PP-induced changes in mitochondrial $[Ca^{2+}]_m$ were accompanied by changes in mitochondrial voltage. In order to measure $\Delta\Psi_m$, we used the lipophilic cationic dye, tetramethylrhodamine methyl ester (TMRM). Low concentrations of TMRM (50 nM) are reported to produce fluorescent signals that vary linearly with the concentration of dye, which accumulates across the mitochondrial membrane according to the electrical potential.[13,14] Confocal imaging of a ventricular myocyte incubated in 50 nM TMRM shows the mitochondria as bright intracellular structures, arranged in a regular sarcomeric pattern along longitudinal lines (FIG. 4A). Quantification of dynamic changes in $\Delta\Psi_m$ are complicated by the fact that depolarization of mitochondria results in redistribution of the dye from the mitochondria to the cytosol, often with minimal changes in the overall fluorescence intensity. The stan-

dard deviation (SD) of the signal, however, is thought to be a more sensitive measure of dye redistribution. Note that the bright staining of mitochondria is contrasted with much darker regions of the cell, representing the cytosol and nucleus, causing a high SD of signal (FIG. 4A). Depolarization will greatly reduce the SD, even if the mean fluorescence does not change much, so we attempted to quantify the $\Delta\Psi_m$ fluctuations as SD/mean.[13,15]

FIGURE 4B shows confocal data recorded from an intact ventricular myocyte perfused with Tyrode's solution and 50 nM TMRM in the bath. Caffeine caused transient depolarization of the mitochondria (12% mean decrease in SD/mean), presumably due to the large cationic influx of Ca^{2+} (FIGS. 4B and C). PP, however, produced a 4.7% mean increase in SD/mean, indicative of hyperpolarization of $\Delta\Psi_m$ (FIGS. 4B and C). FCCP or CCCP dissipated $\Delta\Psi_m$, serving as a control for mitochondrial dye loading and a reference for complete mitochondrial depolarization (FIGS. 4B and C). These data suggest that shear fluid force induces a change in $\Delta\Psi_m$ that is inconsistent with mitochondrial Ca^{2+} uptake. It remains to be worked out how PP-induced mitochondrial Ca^{2+} release hyperpolarizes $\Delta\Psi_m$.

Ultrastructural studies show that the mitochondria are highly abundant in cardiac myocytes (~35% of cellular volume) and come in close apposition with the dyadic components of SR, suggesting that

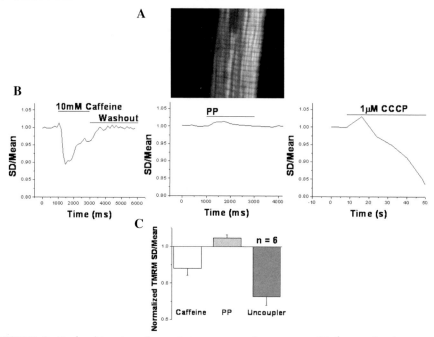

FIGURE 4. Confocal imaging of $\Delta\Psi_m$ in intact ventricular myocytes. (**A**) Raw confocal image of a ventricular myocyte incubated with 50 nM TMRM for 30 min. (**B**) Traces show changes in $\Delta\Psi_m$ (expressed as SD/mean) in response to consecutive applications of caffeine, PP, and CCCP. (**C**) Summary of data expressed as mean ± SEM.

mitochondrial outer membrane and the ryanodine receptors of SR may share the same microdomains. This close apposition may expose the mitochondria to high concentrations of Ca^{2+} during the periods of SR Ca^{2+} release and implicate mitochondria in cardiac Ca^{2+} signaling.

We have previously shown that brief pulses of shear/fluid force, termed "pressurized puffs," in cardiac myocytes activate an intracellular Ca^{2+} release that is independent of CICR and the presence of extracellular Ca^{2+}.[9] These slowly developing Ca_i transients were insensitive to stretch-activated channel blockers, Ca^{2+} channel blockers, and blockers of NO or IP_3 signaling pathways, and were inhibited by mitochondrial uncoupler, FCCP, and mitochondrial Ca^{2+} uniporter blocker, Ru360. In this communication we have assessed the role of mitochondria in activation of PP-induced Ca_i transients more directly by imaging confocally $[Ca^{2+}]_m$ and $\Delta\Psi_m$. We found that PPs of solution released mitochondrial Ca^{2+} and hyperpolarized $\Delta\Psi_m$. Since many mitochondria are positioned directly beneath the cell membrane, it is likely that some type of protein–protein interaction transduces the external mechanical force to the mitochondria. The steps through which this novel Ca^{2+} signaling pathway is propagated remain unclear. Interestingly, blocking the

TABLE 1. Summary of effects of different mitochondrial inhibitors

Pharmacologic agent	Paced Ca_i transients	PP-triggered Ca_i transients
FCCP (0.1–1 μM)	↔,↓	↓
CCCP (0.1–1 μM)	↔,↓	↓
Cyclosporine A (500 nM)	↔	↔
Ru360 (5 μM)	↔	↓
Rotenone (10 μM)	↔	↔
CGP-37157 (1 μM)	↔	↓

mitochondrial permeability transition pore (MPTP) with cyclosporine A had no effect on PP-triggered Ca^{2+} release, but selective inhibitor of the mitochondrial Na^+/Ca^{2+} exchanger (CGP-37157) did block PP-triggered Ca_i-transients (TABLE 1), suggesting that Ca^{2+} is released through the mitochondrial Na^+/Ca^{2+} exchanger.

The implications for changes in $\Delta\Psi_m$ are difficult to assess. Others have also found that mitochondrial Ca^{2+} uptake causes depolarization of $\Delta\Psi_m$, in agreement with our own findings.[16] It is less clear, though, how mitochondrial Ca^{2+} release may affect the mitochondrial membrane potential. Cellular respiration maintains the $\Delta\Psi_m$ as protons are pumped out of the matrix by the electron transport chain.[2] The dynamic nature of a number of electrogenic proteins in mitochondria (ion channels, adenine nucleotide translocase, and Na^+/Ca^{2+} exchanger), however, complicate evaluation of $\Delta\Psi_m$ in the context of PP-triggered Ca_i transients. Nevertheless, we have demonstrated the existence of a mechano-sensitive mitochondrial Ca^{2+} store that may have significant implications for the development of arrhythmias, particularly under pressure overload conditions of heart failure.

Acknowledgments

We gratefully acknowledge the detailed imaging analysis of many of our confocal images by our valued colleague, Lars Cleemann. This study was supported by NIH Grant HL-16152 (to M.M.).

Conflict of Interest

The authors declare no conflicts of interest.

References

1. DUCHEN, M.R. 2004. Roles of mitochondria in health and disease. Diabetes **53**(Suppl. 1): S96–102.
2. DUCHEN, M.R. 2004. Mitochondria in health and disease: perspectives on a new mitochondrial biology. Mol. Aspects Med. **25**: 365–451.
3. SANCHEZ, J.A., M.C. GARCIA, *et al.* 2001. Mitochondria regulate inactivation of L-type Ca^{2+} channels in rat heart. J. Physiol. **536**: 387–396.
4. MAACK, C., S. CORTASSA, *et al.* 2006. Elevated cytosolic Na^+ decreases mitochondrial Ca^{2+} uptake during excitation-contraction coupling and impairs energetic adaptation in cardiac myocytes. Circ. Res. **99**: (172–182).
5. RAMESH, V., V.K. SHARMA, *et al.* 1998. Structural proximity of mitochondria to calcium release units in rat ventricular myocardium may suggest a role in Ca^{2+} sequestration. Ann. N.Y. Acad. Sci. **853**: 341–344.
6. JO, H., A. NOMA, *et al.* 2006. Calcium-mediated coupling between mitochondrial substrate dehydrogenation and cardiac workload in single guinea-pig ventricular myocytes. J. Mol. Cell. Cardiol. **40**: 394–404.
7. SEGUCHI, H., M. RITTER, *et al.* 2005. Propagation of Ca^{2+} release in cardiac myocytes: role of mitochondria. Cell Calcium **38**: 1–9.
8. WOO, S.H., T. RISIUS, *et al.* 2007. Modulation of local Ca(2+) release sites by rapid fluid puffing in rat atrial myocytes. Cell Calcium **41**: 397–403.
9. MORAD, M., A. JAVAHERI, *et al.* 2005. Multimodality of Ca2+ signaling in rat atrial myocytes. Ann. N.Y. Acad. Sci. **1047**: 112–121.
10. BALABAN, R.S. 2002. Cardiac energy metabolism homeostasis: role of cytosolic calcium. J. Mol. Cell. Cardiol. **34**: 1259–1271.
11. DUCHEN, M.R. 2000. Mitochondria and calcium: from cell signalling to cell death. J. Physiol. **529**(Pt 1): 57–68.
12. SEDOVA, M., E.N. DEDKOVA, *et al.* 2006. Integration of rapid cytosolic Ca2+ signals by mitochondria in cat ventricular myocytes. Am. J. Physiol. Cell Physiol. **291**: C840–850.
13. DUCHEN, M.R., A. SURIN, *et al.* 2003. Imaging mitochondrial function in intact cells. Methods Enzymol. **361**: 353–389.
14. SCADUTO, R.C., JR. & L.W. GROTYOHANN. 1999. Measurement of mitochondrial membrane potential using fluorescent rhodamine derivatives. Biophys. J. **76**: 469–477.
15. BRENNAN, J.P., R.G. BERRY, *et al.* 2006. FCCP is cardioprotective at concentrations that cause mitochondrial oxidation without detectable depolarisation. Cardiovasc. Res. **72**: 322–330.
16. O'REILLY, C.M., K.E. FOGARTY, *et al.* 2004. Spontaneous mitochondrial depolarizations are independent of SR Ca^{2+} release. Am. J. Physiol. Cell Physiol. **286**: C1139–1151.

Cardiac Cell Hypertrophy In Vitro

Role of Calcineurin/NFAT as Ca^{2+} Signal Integrators

MATILDE COLELLA[a] AND TULLIO POZZAN[b,c]

[a]Department of General and Environmental Physiology, University of Bari, Bari, Italy

[b]Department of Biomedical Sciences, CNR Institute of Neurosciences, University of Padua, Padua, Italy

[c]Venetian Institute of Molecular Medicine, Padua, Italy

Various conditions were used to investigate the importance of Ca^{2+} signaling in triggering hypertrophy in neonatal rat cardiomyocytes *in vitro*. An increase in cell size and sarcomere reorganization were induced not only by treatment with receptor agonists, such as angiotensin II, aldosterone, and norepinephrine, but also by a small depolarization caused by an increase in the KCl concentration of the medium. This latter treatment has no direct effects on receptor signaling. All these hypertrophic treatments caused a long-lasting increase in the frequency of spontaneous $[Ca^{2+}]$ oscillations, causing nuclear translocation of transfected NFAT (GFP). Cyclosporine A inhibited hypertrophy and NFAT translocation, but not the increased oscillation frequency. We propose here that calcineurin–NFAT can act as integrators of the Ca^{2+} signal and can decode alterations in the frequency even of very rapid Ca^{2+} oscillations.

Key words: Ca^{2+} signaling; triggering hypertrophy; angiotensin II; aldosterone; norepinephrine; KCl; Ca^{2+} oscillations; NFAT; calcineurin

Introduction

Cardiac hypertrophy is a very common pathophysiological state that is a consequence of an increased physiological workload, as in athletes, and/or accompanies heart disease, such as that caused by genetic or congenital defects, hormonal alterations, ischemia, and hypertension. The molecular mechanisms that transform physiological into pathologic cardiac hypertrophy are still unclear and are under intense investigation, in part because of the major socioeconomic impact of such diseases.[1] At the molecular level different signaling pathways have been shown[2] to be activated during hypertrophy (e.g., mitogen-activated protein (MAP) kinases, Gp130/Stat3, and the calcineurin-regulated pathway).

One of the characteristic features of cardiac hypertrophy at the cellular level is that it somehow resembles a process of de-differentiation, whereby different hypertrophic stimuli act through the activation of fetal cardiac gene transcription, enhanced protein synthesis, and cytoskeleton reorganization.[3,4] Many of these hypertrophic stimuli are known to be coupled to alterations in Ca^{2+} homeostasis, but the key, and still unanswered, question is how the hypertrophic Ca^{2+} signal can be discriminated from the cyclic Ca^{2+} rises underlying contractions. This problem is not unique to the process of cardiac hypertrophy, but is related to a much more general question that has been addressed both at the experimental and theoretical level in many systems: Berridge and co-workers[5–7] pointed out that the decoding mechanisms of Ca^{2+} signaling are essentially of two different types: (1) brief Ca^{2+} transients that are decoded primarily through "digital tracking," whereby downstream responses closely track each Ca^{2+} transient (typical examples are contractile cells and nerve terminals); and (2) more prolonged Ca^{2+} changes, including Ca^{2+} oscillations (see below), that are primarily decoded by "integrative tracking," whereby the Ca^{2+} signal has a small effect on some dynamic process that can adopt different equilibrium positions. Small changes in individual rapid transients, or prolonged small elevations in Ca^{2+} levels, are then integrated over time to provide a change in the cellular process (in particular expression of new genes). An example of such integrative tracking is the nuclear factor of activated T cells (NFAT) shuttle. NFAT is a transcription factor that is imported into the nucleus

Address for correspondence: Prof. Tullio Pozzan, Ph.D., Venetian Institute of Molecular Medicine, Department of Biomedical Sciences, via G. Colombo 3, 35129 Padua, Italy. Voice: +39 049 7923-231; fax: +39 049 7923-260.

tullio.pozzan@unipd.it

Ann. N.Y. Acad. Sci. 1123: 64–68 (2008). © 2008 New York Academy of Sciences.
doi: 10.1196/annals.1420.008

in the dephosphorylated form that requires the activation of the Ca^{2+}-dependent phosphatase, calcineurin. The distribution of NFAT between the cytoplasm and the nucleus was found to vary, being dependent on the steady-state level of Ca^{2+} in the cytosol or on frequency of Ca^{2+} transients [8–10]

In this study we have used an *in vitro* model of cardiac hypertrophy—neonatal rat cardiomyocytes subjected to a variety of stimuli—focusing on the possibility that a modification of Ca^{2+} homeostasis and calcineurin activation could represent a common signaling pathway for *in vitro* hypertrophy. The study demonstrates that the simple modification of the frequency of spontaneous Ca^{2+} oscillations suffices to increase the cell size through the NFAT–calcineurin pathway.

Materials and Methods

Cardiomyocyte Culture and Transfection

Cultures of cardiomyocytes were prepared from ventricles of neonatal Wistar rats (0–2 days after birth) as previously described.[11] The NFATc4-GFP plasmid, a gift from Chi-Wing Chow (Department of Molecular Pharmacology, Albert Einstein College of Medicine, Bronx, NY, USA), was transfected on the second day of culture using the FuGENE 6 transfection reagent (Roche, Milan, Italy) as described previously.[11]

Ca²⁺ Measurements

Loading with the fluorescent Ca^{2+} indicators was carried out in serum-free DMEM supplemented with fura-2 AM or fluo-3 AM and 0.04% Pluronic at RT for about 15 min. To prevent probe leakage and sequestration, $250\,\mu M$ sulfinpyrazone was present throughout the loading procedure and $[Ca^{2+}]_i$ measurements. Fura-2 AM and fluo-3 AM concentrations, the exposure time and the interval between successive images, were varied in the different experiments to reduce photobleaching and phototoxicity. All the experiments were performed at room temperature.

Morphologic Analysis

For the measurement of the cell area, at least 20 randomly chosen fields were analyzed in each coverslip, measuring the area of individual cardiomyocytes using the standard algorithm provided by the manufacturer of the Till Vision program. Visualization was performed as previously described.[11]

Statistical Analysis

Data are reported as mean ± SEM of independent experiments or in cell area determination of different cells. Statistical differences were evaluated by one-tailed Student's *t*-test, a P value of <0.05 being considered statistically significant.

Materials

All chemicals were of analytical or highest available grade and, unless otherwise stated, were obtained from Sigma (Sigma-Aldrich, Milan, Italy). ZD7155 was from Tocris (Avonmouth, UK); and fura-2 AM, fluo-3 AM were obtained from Molecular Probes (Eugene, OR). When dimethyl sulfoxide or ethanol were used as solvents, the final solvent concentration never exceeded 0.01% or 0.1%, respectively.

Results and Discussion

A large body of data indicates that changes in Ca^{2+} homeostasis are pivotal in the hypertrophy of cardiac cells both *in vitro* and *in vivo* (see, for example Frey et al.[12] and Vega *et al.*[13]). In unstimulated cardiomyocytes, two types of spontaneous Ca^{2+} oscillation patterns were observed: A regular one in terms of frequency and one characterized by bursts of activity (FIG. 1). On the average, the oscillation frequency of control cells was ~0.3 Hz. Addition of angiotensin II (Ang II) induced a small Ca^{2+} concentration ($[Ca^{2+}]$) increase in ~50% of the cells that superimposed on the Ca^{2+} oscillations. These effects of Ang II on $[Ca^{2+}]$ were prevented by a classical Ang II receptor 1 (ATR1) antagonist, ZD7155 (not shown). In addition Ang II caused a prolonged increase in the frequency of Ca^{2+} spikes that lasted at least 48 h after the addition of the hormone.[14] We then tested two other hypertrophic agents, aldosterone and norepinephrine (NE). Aldosterone had no significant acute effect either on diastolic Ca^{2+} levels or on oscillation frequency, while, as expected, NE caused a rapid increase in both the amplitude and frequency of the Ca^{2+} spikes (not shown). However, prolonged (>24 h) treatment with either drug resulted in a clear increase in oscillation frequency (on average from 0.3 to 0.6 Hz).[13]

Hypertrophic agents, such as Ang II, aldosterone, endothelin, and NE, control complex signal transduction pathways. One can argue that downstream effects, other than changes in Ca^{2+} oscillation frequency, may be common among these stimuli and thus responsible for inducing cell hypertrophy. Consequently, we have incubated the cells in media where the KCl concentration was doubled to 10 mM. This increase in KCl concentration is sufficient to cause a small depolarization of membrane potential, but does not affect the membrane or intracellular receptors. The incubation of cardiomyocytes in medium containing 10 mM

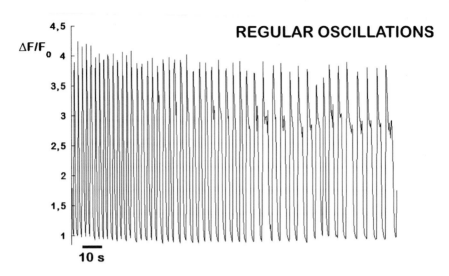

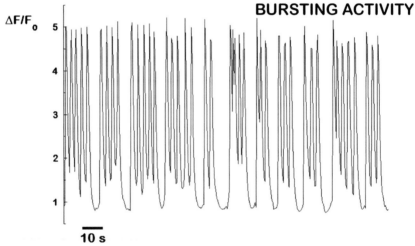

FIGURE 1. Patterns of spontaneous Ca^{2+} oscillations in neonatal cardiac cells. Ca^{2+} oscillations were monitored in neonatal cardiomyocytes loaded with the fluorescent Ca^{2+} indicator fluo-3 (see **Materials and Methods** for details). The fluorescence changes are plotted as $\Delta F/F_0$, where F_0 is the value of fluorescence at time 0 and ΔF is the change in fluorescence intensity at any given time.

KCl caused a rapid increase in the Ca^{2+} oscillation frequency by more than twofold, and this increase persisted during the whole incubation time (>48 h). The cell areas increased by ~25% in myocytes treated for 12 h with 10 mM KCl and their myofilaments presented a more regular organization (FIG. 2). A similar increase in cell size and cytoskeletal reorganization was observed in cells treated with Ang II or NE (FIG. 2). The hypertrophic effect of KCl was not modified by a cocktail of inhibitors, such as the AT1R inhibitor ZD7155 and α- and β-adrenergic antagonists (not shown).

To determine whether calcineurin is involved in the hypertrophic effect of the different treatments described above, cells stimulated with the various agents

were treated with 1 μM cyclosporine A (CsA), the classical inhibitor of calcineurin. The increase in cell area caused by the various stimuli, alone or in combination, was completely blocked, whereas the frequency of Ca^{2+} oscillations remained unaffected (not shown). In the next series of experiment the cells were transfected with NFAT-GFP, and its subcellular localization was compared in control and treated cells. All the hypertrophic agents used (Ang II, aldosterone, NE and KCl) caused an increase in NFAT-GFP translocation to the nucleus, from ~4% in control cells to ~50% in cells treated with the hypertrophic stimuli. Examples of such a translocation are presented in FIGURE 3. Last, but not least, a net translocation of NFAT-GFP was

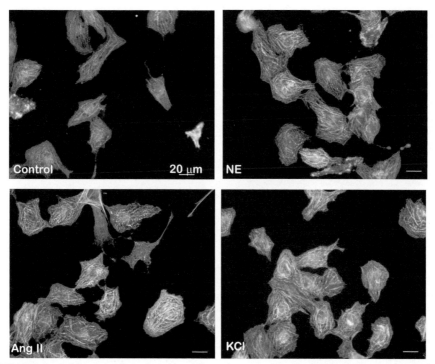

FIGURE 2. Cell hypertrophy induced by different hypertrophic stimuli. Morphology of neonatal cardiac cells incubated in the absence of serum for 48 h (Control) and incubated in the presence of nore-pinephrine 1 μM (NE), angiotensin 1 μM (Ang II), and KCl 10 mM for the last 12 h (KCl). Fixed cells were permeabilized and stained with TRITC phalloidin and examined by fluorescence microscopy (40× objective). Bar: 20 μm.

observed in cells paced with an extracellular electrode at a frequency of 3.5 Hz. Translocation of NFAT-GFP with all the stimuli was inhibited by pre-treatment with 1 μM CsA (not shown).

The question thus arises as to the molecular mechanism of such NFAT activation, in view of the very brief duration of each single Ca^{2+} spike in the cardiac cells. We propose that the "integrative tracking" model proposed by Berridge and co-workers[5-7] can also be applied to the hypertrophic signaling of cardiomyocytes, at least *in vitro*. In particular, here we propose that calcineurin/NFAT act as integrators of an oscillatory Ca^{2+} signal, independently of the duration of such oscillations. The fraction of NFAT that is dephosphorylated during a single Ca^{2+} oscillation in a cardiomyocyte is minute. If there is sufficient time between two Ca^{2+} spikes, NFAT is rephosphorylated during diastole and no net transfer of NFAT into the nucleus occurs. However, a reduction in the diastolic period reduces the time allowed for dephosphorylation, and dephosphorylated NFAT tends to slowly accumulate. If such even marginal imbalance is repeated tens of times every minute, then eventually, after hours, a significant fraction of NFAT ends up in the nucleus. According

to the model, it can be speculated that NFAT should translocate to the nucleus not only in response to an increase in Ca^{2+} oscillation frequency, but also in response to stimuli that affect the amplitude of the Ca^{2+} spikes.

Summary

As shown here, a variety of positive chronotropic agents substantially activate the calcineurin/NFAT pathway in cultured cardiac myocytes and cause an increase in the cell area. Indeed, all the hypertrophic effects of aldosterone, Ang II, NE, and KCl were inhibited by CsA, but, more importantly, all these stimuli resulted in a substantial, CsA-sensitive, translocation of NFAT into the nucleus. Most relevant, the sole increase in the frequency of Ca^{2+} oscillations evoked by an extracellular electrode or by an increase of the extracellular KCl concentration, appears sufficient to cause a substantial translocation of NFAT and an increase of the cell size. We propose that the intrinsic ability of the calcineurin/NFAT pathway to perceive and transduce into NFAT activation not only changes steady-state Ca^{2+} concentration, but also the frequency (or

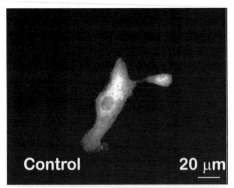

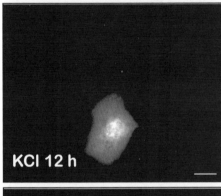

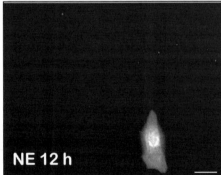

FIGURE 3. NFATc4-GFP translocation induced by hypertrophic stimuli. Representative images are shown of neonatal cardiomyocytes expressing NFATc4-GFP and incubated for 12 h with KCl (10 mM) or NE (1 μM).

amplitude) of Ca^{2+} oscillations, independently of the duration of the single episode of Ca^{2+} increase.

Acknowledgments

This work was supported by grants from Telethon–Italy (1226/02), from the Italian Association for Cancer Research–AIRC (51/2004), from the Italian Ministry of University and Scientific Research (MIUR) Cofin Project 2002 and 2003, FIRB 2001 (RBNE01ERXR), and from the Veneto Region Biotech grant (to T.P.). We are grateful to Prof. S. Schiaffino for helpful discussions and for the kind gift of the NFAT gene reporter.

Conflict of Interest

The authors declare no conflicts of interest.

References

1. AMMAR, K.A. *et al.* 2007. Prevalence and prognostic significance of heart failure stages: application of the American College of Cardiology/American Heart Association heart failure staging criteria in the community. Circulation **115:** 1563–1570.
2. FREY, N. *et al.* 2004. Hypertrophy of the heart: a new therapeutic target? Circulation **109:** 1580–1589.
3. FREY, N. & E.N. OLSON. 2003. Cardiac hypertrophy: the good, the bad, and the ugly. Annu. Rev. Physiol. **65:** 45–79.
4. MOLKENTIN, J.D. 2000. Calcineurin and beyond: cardiac hypertrophic signaling. Circ. Res. **87:** 731–738.
5. BERRIDGE, M.J. 2006. Remodelling Ca^{2+} signalling systems and cardiac hypertrophy. Biochem. Soc. Trans. **34:** 228–231.
6. BERRIDGE, M.J. 2006. Calcium microdomains: organization and function. Cell Calcium **40:** 405–412.
7. BERRIDGE, M.J., M.D. BOOTMAN & H.L. RODERICK. 2003. Calcium signalling: dynamics, homeostasis and remodelling. Nat. Rev. Mol. Cell Biol. **4:** 517–529.
8. TIMMERMAN, L.A. *et al.* 1996. Rapid shuttling of NF-AT in discrimination of Ca2+ signals and immunosuppression. Nature **383:** 837–840.
9. DOLMETSCH, R.E., K. XU & R.S. LEWIS. 1998. Calcium oscillations increase the efficiency and specificity of gene expression. Nature **392:** 933–936.
10. LI, W. *et al.* 1998. Cell-permeant caged InsP3 ester shows that Ca^{2+} spike frequency can optimize gene expression. Nature **392:** 936–941.
11. ROBERT, V. *et al.* 2001. Beat-to-beat oscillations of mitochondrial [Ca2+] in cardiac cells. EMBO J. **20:** 4998–5007.
12. FREY, N., T.A. MCKINSEY & E.N. OLSON. 2000. Decoding calcium signals involved in cardiac growth and function. Nat. Med. **6:** 1221–1227.
13. VEGA, R.B., R. BASSEL-DUBY & E.N. OLSON. 2003. Control of cardiac growth and function by calcineurin signaling. J. Biol. Chem. **278:** 36981–36984.
14. COLELLA, M. *et al.* 2007. Calcineurin and NFAT as Ca^{2+} signal integrators in cardiomyocyte hypertrophy. In preparation.

The Role of Ca^{2+} in Coupling Cardiac Metabolism with Regulation of Contraction

In Silico *Modeling*

Yael Yaniv,[a] William C. Stanley,[b] Gerald M. Saidel,[c] Marco E. Cabrera,[c] and Amir Landesberg[a]

[a]*Faculty of Biomedical Engineering, Technion – Israel Institute of Technology, Haifa, Israel*

[b]*Division of Cardiology, School of Medicine, University of Maryland, Baltimore, Maryland, USA*

[c]*Faculty of Biomedical Engineering and Center for Modeling Integrated Metabolic Systems, Case Western Reverse University, Cleveland, Ohio, USA*

The heart adapts the rate of mitochondrial ATP production to energy demand without noticeable changes in the concentration of ATP, ADP and Pi, even for large transitions between different workloads. We suggest that the changes in demand modulate the cytosolic Ca^{2+} concentration that changes mitochondrial Ca^{2+} to regulate ATP production. Thus, the rate of ATP production by the mitochondria is coupled to the rate of ATP consumption by the sarcomere cross-bridges (XBs). An integrated model was developed to couple cardiac metabolism and mitochondrial ATP production with the regulation of Ca^{2+} transient and ATP consumption by the sarcomere. The model includes two interrelated systems that run simultaneously utilizing two different integration steps: (1) The faster system describes the control of excitation contraction coupling with fast cytosolic Ca^{2+} transients, twitch mechanical contractions, and associated fluctuations in the mitochondrial Ca^{2+}. (2) A slower system simulates the metabolic system, which consists of three different compartments: blood, cytosol, and mitochondria. The basic elements of the model are dynamic mass balances in the different compartments. Cytosolic Ca^{2+} handling is determined by four organelles: *sarcolemmal* Ca^{2+} influx and efflux; *sarcoplasmic reticulum* (SR) Ca^{2+} release and sequestration (SR); binding and dissociation from *sarcomeric* regulatory troponin complexes; and *mitochondrial* Ca^{2+} flows. Mitochondrial Ca^{2+} flows are determined by the Ca^{2+} uniporter and the mitochondrial $Na^{+}Ca^{2+}$ exchanger. The cytosolic Ca^{2+} determines the rate of ATP consumption by the sarcomere. Ca^{2+} binding to troponin regulates the rate of XBs recruitment and force development. The mitochondrial Ca^{2+} concentration determines the pyruvate dehydrogenase activity and the rate of ATP production by the F_1-F_0 ATPase. The workload modulates the cytosolic Ca^{2+} concentration through feedback loops. The preload and afterload affect the number of strong XBs. The number of strong XBs determines the affinity of troponin for Ca^{2+}, which alters the cytosolic Ca^{2+} transient. Model simulations quantify the role of Ca^{2+} in simultaneously controlling the power of contraction and the rate of ATP production. It explains the established empirical observation that significant changes in the metabolic fluxes can occur without significant changes in the key nucleotide (ATP and ADP) concentrations. Quantitative investigations of the mechanisms underlying the cardiac control of biochemical to mechanical energy conversion may lead to novel therapeutic modalities for the ischemic and failing myocardium.

Key words: cardiac energetics; mitochondria; sarcomere; calcium; cooperativity; cross-bridge; pyruvate dehydrogenase; F_0-F_1 ATPase

Introduction

The cardiac muscle generates more power at higher heart rate or in response to activation of the sympathetic system. The increase in mechanical power originates from higher rate of ATP consumption by the cardiac cross-bridges (XBs). Despite the increase

Address for correspondence: Prof. Amir Landesberg, M.D., Ph.D., Faculty of Biomedical Engineering, Technion – IIT, Haifa 32000 Israel. Fax: 972-4-8294599.

amir@bm.technion.ac.il

Ann. N.Y. Acad. Sci. 1123: 69–78 (2008). © 2008 New York Academy of Sciences.
doi: 10.1196/annals.1420.009

in the rate of ATP consumption, the ATP and ADP concentrations[1] do not change noticeably under normal aerobic physiological conditions and normal coronary perfusion. Thus, the cardiac myocyte has a comprehensive control system that enables a change in demand to be immediately accommodated by a change in mitochondrial ATP production. The mechanisms underlying this control system are still not well elucidated.

We hypothesize that the cytosolic and mitochondrial calcium concentrations ($[Ca^{2+}]_i$, and $[Ca^{2+}]_m$) play key roles in regulating the cytosolic ATP consumption rate and the mitochondrial ATP production rate. We further suggest that the relationship between cytosolic and mitochondrial Ca^{2+} can provide a fast adaptive response of ATP production rate to quick changes in ATP consumption. Most Ca^{2+} which is released into the cytosol during the action potential, through the sarcolemma or from the SR, binds to troponin ($56 \, \mu M$).[2] Less than 1% ($0.35 \, \mu M$) enters into the mitochondria, and about 1% remains as free cytosolic Ca^{2+}. Although free cytosolic Ca^{2+} and mitochondrial Ca^{2+} are only a fraction of the total released Ca^{2+}, they can play major regulatory roles.[1,2]

The cytosolic Ca^{2+} regulates the rate of ATP consumption by the actin–myosin XBs, which accounts for most of the cytosolic ATP consumption in the cardiac muscle. The cytosolic Ca^{2+} also determines ATP consumption by SR-ATPase, the second cytosolic ATP consumer. Calcium binding to sarcomeric troponin regulates the hydrolysis of ATP by actomyosin ATPase and allows XB recruitment, that is, the turnover from the weak to the strong XB conformation. The myocytes control the rate of XB recruitment and the generated power by modulating either the free Ca^{2+} transient (amplitude and duration) or the affinity of Ca^{2+} binding to troponin.

Previous studies[3–10] have established that the affinity of Ca^{2+} binding to troponin is determined by two sarcomeric feedback control mechanisms. By these adaptive control mechanisms, myocytes can accommodate the rate of energy consumption to changes in the mechanical loading condition and demand. At higher workloads (preload or afterload), during which more Ca^{2+} is bound to troponin, myocytes consume more energy and generate more work. The sarcomeric control mechanisms involve (1) *mechanical feedback*,[3] whereby sarcomere-shortening velocity determines the rate of XB cycling and (2) *cooperativity*,[4] whereby calcium affinity depends on the number of strong XBs. An increase in the afterload decreases the sarcomere-shortening velocity.[5] Slowing

the sarcomere-shortening velocity decreases the XB weakening rate (i.e., rate of XB turnover from strong to weak conformation) through *mechanical feedback*. The ensuing increase in the number of strong XBs increases calcium affinity and the amount of Ca^{2+} bound to troponin via the *cooperativity*. Consequently, the load affects the amount of Ca^{2+} bound to troponin, the energy consumption, and the generated work.[6]

These feedback mechanisms suffice to explain the adaptive control of cardiac mechanics by the loading conditions. The *cooperativity* mechanism determines the force–length relationship of isolated cardiac fiber[4] and the related Frank–Starling law of the heart.[7] This mechanism can account for the length-dependent Ca^{2+} sensitivity[4] and the fast switching mechanism of cardiac activation, whereby small changes in the free Ca^{2+} concentration yield huge changes in the number of recruited XBs and the mechanical output.[3] The *mechanical feedback* mechanism accounts for the muscle force–velocity relationship in accordance with the well-established, but empirical, Hill equation.[3] This mechanism leads to a linear relationship between energy consumption and the generated mechanical energy.[6] Thus, *mechanical feedback* determines the generated power and cardiac efficiency.[6]

The validity of the *cooperativity* and *mechanical feedback* mechanisms was strengthened by our recent studies in isolated rat cardiac trabeculae,[8–10] which established the existence of different patterns of hysteresis in the force response to length oscillations. During each cycle at low oscillation frequency, hysteresis occurs in the counterclockwise direction and the muscle generates external work. In each force–length hysteresis loop, the force is larger during shortening than during lengthening at the same sarcomere length (SL). This observation is explained by the *cooperativity*, whereby lengthening increases the force, Ca^{2+} affinity, and amount of Ca^{2+} bound to troponin. This augments XB recruitment that generates external work. During each cycle at high oscillation frequency, hysteresis occurs in the clockwise direction and the muscle absorbs energy. In these hysteretic loops, the muscle behaves as a damper so that the force is lower during shortening than during lengthening at the same SL. This phenomenon is dominated by *mechanical feedback* since the dependence of XB weakening rate on the velocity resembles the behavior of a viscous element.

The mitochondrial ATP production is regulated by the mitochondrial calcium ($[Ca^{2+}]_m$),[1,11,12] which is determined by Ca^{2+} influx from the cytosol through the uniporter[13] and the efflux through the mitochondrial Na^+–Ca^{2+} exchanger. The Ca^{2+} influx through the uniporter, which is driven by the electrochemical gradient across the mitochondrial membrane, does not require ATP hydrolysis or ion exchange. The $[Ca^{2+}]_m$ modulates the activities of pyruvate dehydrogenase and several enzymes in the tricarboxylic acid (TCA) cycle (namely, isocitrate dehydrogenase and α-oxoglutarate dehydrogenase).[11] It was recently shown that $[Ca^{2+}]_m$ modulates the F_1F_0-ATPase activity in the cardiac mitochondria.[12] The F_1F_0-ATPase is attached to the mitochondrial inner membrane and synthesizes ATP via the potential energy of the transmembrane proton gradient.

The aim of this study was to develop a mechanistic mathematical model that integrates cardiac metabolism with the regulation of muscle contraction, allowing us to describe biochemical conversion to mechanical energy. The study focuses on intracellular control mechanisms and the role of Ca^{2+} in the regulation of cardiac energetics. The study investigates whether Ca^{2+} can simultaneously modulate ATP production and consumption without noticeable changes in the key nucleotides (ATP, ADP). The model must simulate cardiac metabolism, mitochondrial activity, and sarcomere dynamics together with the regulation of cytosolic and mitochondrial calcium handling. The model integrates sarcomere dynamics with mitochondrial activity and highlights the role of cytosolic and mitochondrial calcium in the regulation of cardiac energetics. The $[Ca^{2+}]_i$, and $[Ca^{2+}]_m$ are the mediators, or "state variables" in the control system terminology, which provide the signal transduction between the intracellular organelles (sarcomere, mitochondria, and glycolytic compartment). The model describes the functional control of three organelles: mitochondria, sarcomere, and sarcoplasmic reticulum, as well as the regulation of the metabolism in the cytosol. The model includes two interlaced systems that run simultaneously: (1) The faster control system describes the control of excitation contraction coupling with the sharp cytosolic Ca^{2+} transient, the associated fluctuations in the mitochondrial Ca^{2+}, and the twitch mechanical contraction. (2) The slower system simulates the metabolic reactions in blood, cytosol, and mitochondria. With this model, we investigate the effects of changing the extracellular Ca^{2+} concentrations. The results support the hypothesis that Ca^{2+} modulates the biochemical-to-mechanical energy conversion in the cardiac muscle in response to demand. Regulation by Ca^{2+} can explain the increase in the rate of energy consumption without noticeable concentration changes of key cellular nucleotides (ATP, ADP).

Methods

Integration of Cardiac Myocyte Mechanics and Metabolism

The model couples energy production with sarcomeric energy consumption by relating to two types of important mediators: nucleotide concentrations (ATP, ADP and Pi) and calcium in the cytosol and mitochondria as shown schematically in FIGURE 1.

Free Cytosolic Calcium Dynamics

The free cytosolic calcium ($[Ca^{2+}]_i$) dynamics are determined by Ca^{2+} fluxes from four compartments: (1) Ca^{2+} influx into (I_s) and efflux from (I_o) the sarcolemma; (2) Ca^{2+} release from (I_i) and uptake (I_u) by the sarcoplasmic-reticulum (SR); (3) Ca^{2+} flow into (I_{mi}) and out of (I_{mo}) the mitochondria; and (4) Ca^{2+} binding rate (B_t) to the sarcomeric regulatory protein complexes:

$$\frac{d[Ca^{2+}]_i}{dt} = I_s - I_o + I_i - I_u - I_{mi} + I_{mo} - B_t \quad t \geq o \quad (1)$$

The Ca^{2+} sarcolemmal influx through the L-type calcium channel and SR calcium release through the ryanodine channel are described by phenomenologic equations[14]:

$$I_s = [Ca^{2+}]_0 \cdot Q_s \cdot \left(1 - e^{\frac{-t}{\tau_{SR}}}\right) \cdot e^{\frac{-t}{\tau_{SF}}} \quad (2)$$

$$I_i = [Ca^{2+}]_{SR} \cdot Q_i \cdot \left(1 - e^{\frac{-t}{\tau_{IR}}}\right) \cdot e^{\frac{-t}{\tau_{IF}}}, \quad t \geq 0 \quad (3)$$

where $[Ca^{2+}]_0$ and $[Ca^{2+}]_{SR}$ are the extracellular and the SR concentrations; and Q_s and Q_i are maximal Ca^{2+} flow parameters.

Calcium sequestration through the sarcolemma, mainly by the Na^+–Ca^{2+} exchanges (NCX), is approximated by a first-order reaction:

$$I_0 = Q_0 \cdot [Ca^{2+}]_i \quad (4)$$

where Q_o is rate coefficient.

Calcium uptake into the SR through Ca^{2+} ATPase is determined by saturation Michaelis–Menten kinetics:

$$I_u = Q_u \cdot [Ca^{2+}]_i/(K_{mu} + [Ca^{2+}]_i) \quad (5)$$

where Q_u is the maximal rate of uptake.

Calcium enters into the mitochondria through an electrogenic uniporter that behaves like an ion channel and is thus described by a Goldman–Hodgkin–Katz equation[15]:

$$I_{mi} = Q_{mi} \frac{z\psi_m F}{RT} \cdot \frac{\alpha_m[Ca^{2+}]_m e^{\frac{-z\psi_m F}{RT}} - \alpha_i[Ca^{2+}]_i}{e^{\frac{-z\psi_m F}{RT}} - 1} \quad (6)$$

where ψ_m is the mitochondrial membrane potential; and α_m and α_i are the mitochondrial and cytosolic Ca^{2+} activity coefficients, respectively. The coefficient Q_{mi} is determined by the uniporter maximal permeability, uniporter density, and the ratio of mitochondrial volume to cellular volume.

The main Ca^{2+} outflow from the mitochondria via the Na^+–Ca^{2+} exchanger (NCX_m), on the mitochondrial membrane, is proportional to mitochondrial Ca^{2+}:

$$I_{mo} = Q_{mo}[Ca^{2+}]_m \qquad (7)$$

The coefficient Q_{mo} depends on the NCX_m maximal permeability and density, mitochondrial membrane potential, cytosolic and mitochondrial Na^+ concentrations, and the ratio of the mitochondrial volume to the cellular volume. The mitochondrial potential and the cytosolic and mitochondrial Na^+ concentrations are constant for a wide range of mitochondrial activity under normal physiological conditions[13] and therefore Q_{mo} is constant.

Calcium Binding to Troponin

The rate of Ca^{2+} binding to troponin (Tn) is related to force developed through the feedback mechanisms.[4–10] Force kinetics depend on (1) Ca^{2+}–Tn binding and dissociation and (2) XB cycling between weak and strong conformations. The Tn regulatory units are distributed between four different states: (1) **state R**, a resting state with weak XBs and no Ca^{2+}–Tn binding; (2) **state A** "activated" by Ca^{2+} binding to Tn, while the adjacent XBs are weak; (3) **state T** with strong XBs and Ca^{2+}–Tn binding; and (4) **state U** with strong XBs and no Ca^{2+}–Tn binding. The density of the Tn-regulatory units in each state (i.e., the number of regulatory units per filament unit length) denoted by R, A, T and U sums to a constant total density (TRo):

$$R + A + T + U = TRo \qquad (8)$$

The sarcomeric system dynamics can be represented as:

$$\frac{dA}{dt} = k_\ell [Ca^{2+}]_i (TRo - A - T - U) - (f + k_{-\ell})A + gT \qquad (9)$$

$$\frac{dT}{dt} = fA - (g + k_{-\ell})T + k_\ell [Ca^{2+}]_i U \qquad (10)$$

$$\frac{dU}{dt} = k_{-\ell}T - (g + k_\ell [Ca^{2+}]_i)U \qquad (11)$$

where k_ℓ and $k_{-\ell}$ represent Ca^{2+} binding and dissociation from troponin. **f** and **g** are rates of XB cycling between the weak and the strong conformations. f is the rate of ATP hydrolysis by the actomyosin ATPase.

The amount of Ca^{2+} bound to Tn is depicted by states A and T. Hence, the rate of Ca^{2+} binding to Tn (B_t) is given by:

$$B_t = \frac{d(A + T)}{dt} = k_\ell [Ca^{2+}]_i TRo$$
$$- (k_\ell [Ca^{2+}]_i + k_{-\ell})(A + T) \qquad (12)$$

where the rate coefficient k_ℓ is a constant[14,16] and diffusion-limited, whereas the dissociation rate coefficient ($k_{-\ell}$) is determined by the affinity of troponin for calcium, $k_{-\ell} = k_\ell / K_{[Ca]}$. Calcium affinity ($K_{[Ca]}$) is determined by the number of force generating XBs (N_{XB}), that is, the number of regulatory units in state T and U according to *cooperativity*. $N_{XB} = L_s(T + U)$, where L_s is the single overlap length between the actin and myosin filaments. Calcium affinity is describes here by a sigmoidal function of N_{XB}:

$$K_{[Ca]} = K_0 + K_1 \frac{N_{XB}^n}{K_{0.5}^n + N_{XB}^n} \qquad (13)$$

where K_0 is the calcium affinity at rest and K_1 is the *cooperativity* coefficient.

Free Mitochondrial and SR Calcium

The $[Ca^{2+}]_m$ is determined by the inflow (I_{mi}) through the uniporter and the efflux out (I_{mo}) through the mitochondrial Na^+–Ca^{2+} exchanger[1,12]:

$$\frac{d[Ca^{2+}]_m}{dt} = r_m(I_{uni} - I_{Na-Ca}) \qquad (14)$$

where the coefficient r_m is determined by the ratio between cytosolic and mitochondrial volumes.

The SR calcium content depends on Ca^{2+} uptake (I_u) and release (I_i) from the SR:

$$\frac{d[Ca^{2+}]_{SR}}{dt} = r_{SR}(I_u - I_i) \qquad (15)$$

where the coefficient r_{SR} reflects the ratio between the cytosolic and the SR volumes.

Calcium Regulation of ATP Production and Consumption

Calcium regulates the activity of several mitochondrial enzymes including pyruvate dehydrogenase[11] and the activity of the F_1F_0-ATPase.[20] The dependence of the F_1F_0-ATPase activity ($\mathcal{J}_p$) on mitochondrial calcium concentration has been described previously[21]:

$$\mathcal{J}_p = K_p(1 - e^{-[Ca]_m / K_{Ca,ATP}}) \qquad (16)$$

The mean ATP production by the average F_1F_0-ATPase $(\bar{\mathcal{J}}_p)$ over the twitch duration (TD) is calculated by:

$$\bar{\mathcal{J}}_p = \frac{1}{TD} \int_0^{TD} \mathcal{J}_p \, dt \qquad (17)$$

The rate of ATP consumption by the sarcomere is determined by the rate of XB turnover from the weak to the strong conformation. Each transition from the weak to the strong conformation requires one ATP. The XB can turn to strong conformation only between state **A** and **T**. Hence, the instantaneous rate of ATP consumption by the sarcomere is given by $\mathcal{J}_{XB} = f\,A$ and the mean sarcomeric ATP consumption rate $(\bar{\mathcal{J}}_{XB})$ over the TD is:

$$\bar{\mathcal{J}}_{XB} = \frac{1}{TD} \int_0^{TD} f \cdot A \, dt \qquad (18)$$

The energetic variables $\bar{\mathcal{J}}_p$ and $\bar{\mathcal{J}}_p$, together with the mean $[Ca^{2+}]_m$, link the faster system, which describes the excitation–contraction coupling, with the slower system, which simulates the metabolic dynamics.

Metabolic Dynamics of Cardiac Substrates

The model of cardiac metabolism describes the concentrations of the various metabolites, considers the different rates of the metabolic processes, and formulates the mass balances for the metabolites that pass between different compartments[17–19]

The model describes passive and carrier-mediated transports across the different compartments: blood, cytosol, and mitochondria. It also describes the comprehensive chemical reactions that are involved in energy transduction in the different cellular compartments: glycolysis in the cytosol, ATP production by the mitochondria, and ATP consumption by the sarcomeres within the cytosol.[17]

The dynamic mass balances in the capillaries describe the changes in the metabolite and oxygen concentrations in the capillaries due to transport between the blood and the cytosol $(\mathcal{J}_{b-c,j})$. The change in the capillary blood content (subscript b) of a given metabolite (C_{bj}) is determined by the amount that enters the capillary from the arterial side, the amount that leaves the capillary at the venous side, and the transport across the capillary wall:

$$V_b \frac{dC_{bj}}{dt} = Q(C_{aj} - C_{vj}) - \mathcal{J}_{b-c,j} \qquad (19)$$

where C_{aj} and C_{vj} are the arterial and venous concentrations of metabolite j, respectively. V_b is the volume

of the capillary domain, and Q is flow through the capillary blood domain.

In the cytosolic domain (subscript c), the concentration of metabolite j, C_{cj}, may change on account of metabolic reaction (R_{cj}) and mass transfer across the cellular membrane $(\mathcal{J}_{b-c,j})$ and mitochondrial membrane $(\mathcal{J}_{c-m,j})$:

$$V_c \frac{dC_{cj}}{dt} = R_{cj} + \mathcal{J}_{b-c,j} - \mathcal{J}_{c-m,j} \qquad (20)$$

where V_c is the volume of the cytosol.

The change in concentration of metabolite j, C_{mj} in the mitochondrial domain is given by the following mass balance equation:

$$V_m \frac{dC_{mj}}{dt} = R_{mj} + \mathcal{J}_{c-m,j} \qquad (21)$$

where V_{mj} is the mitochondrial volume and R_{mj} is the net reaction rate inside the mitochondria.

Transport Fluxes

There are two types of transport between the different compartments, passive transport and metabolic reaction. The passive diffusion $(\mathcal{J}^p_{x \to y,j})$ between domains $\boldsymbol{x}$ and $\boldsymbol{y}$ is given by:

$$\mathcal{J}^p_{x \to y,j} = \lambda_{x \to y,j}(C_{x,j} - \sigma_{x \to y,j} C_{y,j}) \qquad (22)$$

where $\lambda_{x \to y,j}$ is the membrane transport coefficient that incorporates membrane permeability and effective surface area and $\sigma_{x \to y,j}$ is the partition coefficient.

The rate of carrier-mediated transport is expressed by the Michaelis–Menten equation characterizing enzyme kinetics:

$$\mathcal{J}^f_{x \to y,j} = \frac{T_{x \to y,j} \, C_{x,j}}{M_{x \to y,j} + C_{x,j}} \qquad (23)$$

where $T_{x \to y,j}$ is the transport rate coefficient and $M_{x \to y,j}$ is the affinity coefficient.

Metabolic Fluxes

The net reaction rate of species j in domain $\boldsymbol{x}$ (R_{xj}) can be represented as the difference of rates of species production (P_{xj}) and utilization U_{xj}:

$$R_{xj} = P_{xj} - U_{xj} = \sum_{k=1}^{n} \beta_{k,j} \, \phi_{k,j} - \sum_{k=1}^{m} \beta_{j,k} \phi_{j,k} \qquad (24)$$

where $\beta_{k,j}$ is the corresponding stoichiometric, n is the number of reaction fluxes forming species j from species k, m is the number of reaction fluxes forming species k from species j and $\phi_{k,j}$ is the reaction flux from species k to species j. In general for a reactants X and Y with products Z and W the reaction flux is:

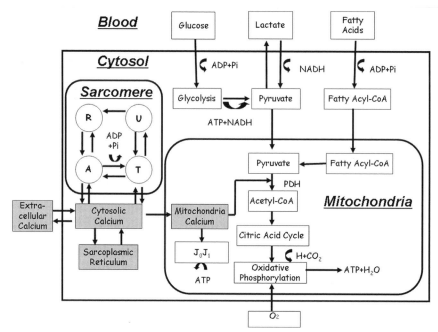

FIGURE 1. A schematic diagram of the integrated regulation of cardiac metabolism, biochemical to mechanical energy conversion, and calcium handling.

$$\phi_{X-Y,Z-W} = V_{X-Y,Z-W} \frac{C_X C_Y}{K_{X-Y,Z-W} + C_X C_Y} \quad (25)$$

where $V_{x-y,z-w}$ and $K_{x-y,z-w}$ are Michaelis–Menten coefficients specific to the reaction process. In general, the rate coefficients $V_{x-y,z-w}$ are nonlinear function of the metabolite concentration:

$$V_{x-y,z-w} = V_{MAX,\,x-y,z-w} \left(\frac{PS^{\pm}}{\mu^{\pm} + PS^{\pm}} \right) \left(\frac{RS^{\pm}}{v^{\pm} + RS^{\pm}} \right) \quad (26)$$

where $V_{MAX,\,x-y,z-w}$ is the maximum reaction rate and PS and RS are metabolite concentration ratios that relate to phosphorylation states ($PS^{+} = C_{ATP}/C_{ADP}$, $PS^{-} = C_{ADP}/C_{ATP}$) and the redox state ($RS^{+} = C_{NADH}/C_{NAD}, RS^{-} = C_{NAD}/C_{NADH}$). Changes in these metabolites accelerate or inhibit the biochemical reaction, thereby providing feedback mechanisms to any reaction in which the metabolites participate.

Model Simulations

The simulations include two interlaced systems that run simultaneously, using two different integration steps: The faster control of the excitation contraction coupling and the slower control of cardiac metabolism. The faster system simulates the transient $[Ca^{2+}]_i$, the transient $[Ca^{2+}]_m$, and the twitch force. This fast system determines the mean rate of ATP consumed by the XB (Eq. 17) and the mean rate of ATP produc-

tion by F_1F_0-ATPase (Eq. 18). These values are used by the slower metabolic system (Eqs. 19–25) that describes cardiac metabolism.

The model was used to investigate the changes in cardiac nucleotide concentrations in response to variations in the extracellular calcium concentration. These simulations relate to the effects of changes in the demand or in the humoral stress. The model utilizes the MATLAB Simulink toolbox.

Results and Discussion

Insignificant Changes in Nucleotide Concentrations with Different Metabolic Rates

The mathematical model allows us to simulate calcium activity, force development, and the regulation of cardiac metabolism at any mechanical loading conditions, metabolite concentrations, and metabolic demand. It provides the transient responses to changes in the mechanical loading conditions or metabolite concentrations, and yields the new steady state in response to any change. To validate the utility of the model we tested the effects of increasing the ATP consumption and production rates on the concentration of key nucleotides (ATP, ADP). An increase in the extracellular calcium ($[Ca^{2+}]_0$) increases cytosolic Ca^{2+} and,

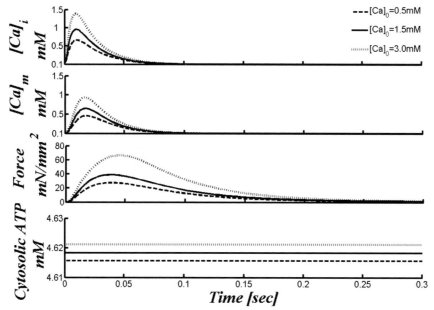

FIGURE 2. The effect of changes in the extracellular Ca²⁺ ([Ca²⁺]₀) on the cytosolic ([Ca²⁺]ᵢ) and mitochondrial ([Ca²⁺]ₘ) Ca²⁺ transients, twitch force, and cytosolic ATP concentration.

thereby, force generation and ATP consumption. An increase in the transient cytosolic calcium is associated with a parallel increase in the mitochondrial calcium and in the F_1-F_0 ATPase and pyruvate dehydrogenase (PDH) activities. The parallel increase in the ATP consumption and production yields only an insignificant change in the ATP concentration, as shown in FIGURE 2. A sixfold increase in $[Ca^{2+}]_0$, from 0.5 to 3 mM, increases the peak $[Ca^{2+}]_i$ by 100%, from 0.7 to 1.4 μM, and the peak force by 260%, from 29.36 to 76 mN/mm². However, the changes in the ATP concentration were less than 0.1%. The results are consistent with well-established observations.[1]

Cortassa *et al.*[22] simulated the effects of increasing the stimulation rate on cardiac metabolism. Their simulation suggests that an increase in the heart rate from 0.5 Hz to 2 Hz increases the $[Ca^{2+}]_m$ but also has a significant effect on the ADP concentration. In their model and in our model, calcium regulates ATP consumption; however, our model predicts only minor changes in the nucleotide (ATP, ADP) concentration in accord with extensive experimental evidence.[1,12]

The results thus support the hypothesis that calcium modulates the biochemical to mechanical energy conversion in the cardiac muscle in response to demand. Regulation by Ca²⁺ can explain the increase in the rate of energy consumption without noticeable concentration changes in the cellular key nucleotides (ATP, ADP).

Relationship of Cytosolic and Mitochondrial Calcium

To investigate how calcium provides the observed tight correlation between the changes in the ATP consumption and ATP production, without noticeable changes in the nucleotide concentrations (ATP, ADP), we have evaluated the relationship between $[Ca^{2+}]_i$, and $[Ca^{2+}]_m$. The relationship between $[Ca^{2+}]_i$, and $[Ca^{2+}]_m$ was assessed at different extracellular calcium concentrations.

An increase in the extracellular Ca²⁺ increases the cytosolic Ca²⁺ and is associated with a parallel increase in mitochondrial Ca²⁺ (FIG. 3). A linear relationship is obtained between the peak $[Ca^{2+}]_i$, and the peak $[Ca^{2+}]_m$, with similar relative changes in the Ca²⁺ concentrations in the two compartments. An increase in the extracellular Ca²⁺ from 0.5 to 3 mM doubles the peak $[Ca^{2+}]_i$ from 0.69 to 1.4 μM and is associated with doubling the peak $[Ca^{2+}]_m$ from 0.47 to 0.96 μM.

This model prediction is of great importance. It indicates that changes in the ATP consumption that are regulated by the cytosolic Ca²⁺ are accommodated by similar changes in the mitochondrial ATP production, which is regulated by the mitochondrial Ca²⁺. Since identical changes occur in the production and consumption of ATP, no changes occur in the metabolite concentrations. Although it is well established that no significant changes occur in the nucleotide

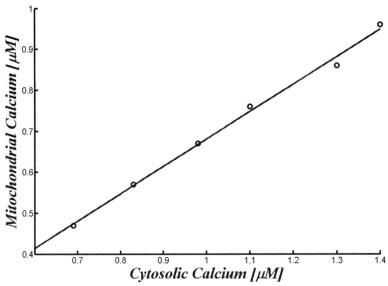

FIGURE 3. A tight linear relationship is obtained between the peak mitochondrial ($[Ca^{2+}]_m$) and peak cytosolic ($[Ca^{2+}]_i$) calcium transients, with various extracellular Ca^{2+} concentrations.

concentrations,[1] our model predicts a tight relationship between $[Ca^{2+}]_i$ and $[Ca^{2+}]_m$ with various extracellular Ca^{2+} concentrations. This prediction was not investigated before, and new experimental studies are needed to validate this prediction.

Calcium may enter the mitochondria via several mechanisms. The present model assumes that the dominant mechanism is inflow through the uniporter (Eq. 6). Ruthenium red, which inhibits the uniporter, decreases the mitochondrial Ca^{2+} and diminishes the mitochondrial Ca^{2+} transient.[23] This observation supports our assumption that Ca^{2+} enters the mitochondria mainly through the uniporter under normal conditions. Although Ca^{2+} may also enter into the mitochondria through the rapid Ca^{2+} transport mechanism, denoted as the Ram channels,[24] this mechanism is effective only at very low Ca^{2+} levels. For Ca^{2+} concentrations above $200\,nM$, that is, above the normal diastolic Ca^{2+} concentration in the cytosol, these Ram channels do not contribute significantly to the Ca^{2+} inflow.[24] An increase in the cytosolic Ca^{2+} above $200\,nM$ increases the mitochondrial Ca^{2+} mainly through the uniporter.[24] Under normal physiological conditions, Ram channels do not play an important role in the regulation of mitochondrial Ca^{2+}.

Clinical Merits

The study suggests that calcium may provide the fast response mechanism whereby ATP production can ac-

commodate the changes in demands. Therefore, any pathology in the regulation of calcium handling may impair the regulation of cardiac biochemical to mechanical energy conversion. This likely occurs in postischemic reperfusion, heart failure, and genetic diseases that alter calcium handling (cardiomyopathies).

The main calcium storage in the cell is in the sarcoplasmic reticulum. The second main calcium binding site is the sarcomeric troponin regulatory complex. Most of the calcium that is released during the action potential binds to troponin.[2] Thus, mutations and abnormalities in the SR or the sarcomere function can have significant effects on calcium handling and the regulation of cardiac energetics. Heart failure impairs the SR function (e.g., leaky ryanodine channel, decrease in the SR Ca^{2+} pump activity) and is associated with an increase in the diastolic Ca^{2+}, a decrease in the peak Ca^{2+}, and prolongation of the Ca^{2+} transient. Such changes in calcium dynamics can impair the ability of failing heart to accommodate the changes in demands since calcium plays a key role in this control mechanism. Mutations in the troponin regulatory complex or in other functionally associated sarcomeric proteins (e.g., myosin light chains) also can affect the intracellular Ca^{2+} though the *cooperativity* and *mechanical feedback*. Javadpour *et al.*[25] have shown that mutation R92Q in the troponin-T, which causes severe hypertrophic cardiomyopathy in patients, affects cardiac mechanics and energetics. Troponin-T does not directly interact with the ATP hydrolysis site on the myosin head, but is essential for coupling calcium binding to troponin with the regulation of XB cycling. It is also essential for the cooperative activation of the contractile protein. Interestingly, this mutation alters the ability of the muscle to accommodate the increase in the extracellular calcium,[24] and an increase in the extracellular calcium is associated with a decrease in the ATP concentration and in the free energy available from ATP hydrolysis.

Summary

This study strongly supports the hypothesis that a tight interaction exists between the cytosolic and mitochondrial calcium concentrations. This interaction plays an important role in the regulation of cardiac energetics and the adaptive control of ATP production to changes in the ATP consumption rate. Changes in cytosolic calcium that modulate ATP consumption also affect the mitochondrial calcium. The ensuing changes in the mitochondrial calcium regulate the pyruvate dehydrogenase and other Krebs cycle enzymes that modulate the F_1-F_0 ATPase activity. Consequently, parallel changes in the ATP production and consumption rates can exist without significant changes in key cardiac nucleotides (ATP, ADP). Understanding the intracellular regulation of cardiac mechanics and metabolism under normal conditions will allow further investigation of various cardiac disorders and the development of novel therapeutic strategies.

Acknowledgments

This study was supported by the Fund for the Promotion of Research at the Technion–ITT (A.L.) and a grant from the United States–Israel Binational Science Foundation (BSF Research Project No. 2003399).

Conflict of Interest

The authors declare no conflicts of interest.

References

1. BALABAN, R.S. 2002. Cardiac energy metabolism homeostasis: role of cytosolic calcium. J. Mol. Cell. Cardiol. **34:** 1259–1271.
2. BERS, D.M. 2001. Cardiac Excitation–Contraction Coupling and Contractile Force. Kluwer. Dordrecht, the Netherlands.
3. LANDESBERG, A. & S. SIDEMAN. 1994. Mechanical regulation in the cardiac muscle by coupling calcium binding to troponin-C and crossbridge cycling: a dynamic model. Am. J. Physiol. **267:** H779–H795.
4. LANDESBERG, A. & S. SIDEMAN. 1994. Coupling calcium binding to troponin-C and crossbridge cycling kinetics in skinned cardiac cells. Am. J. Physiol. **266:** H1261–H1271.
5. DANIELS, M., M.I.M. NOBLE, H.E.D.J. TER KEURS & B. WOHLEART. 1984. Velocity of sarcomere shortening in rat cardiac muscle: relationship to force, sarcomere length, calcium and time. J. Physiol. **355:** 367–381.
6. LANDESBERG, A. & S. SIDEMAN. 2000. Force-velocity relationship and biochemical-to-mechanical energy conversion by the sarcomere. Am. J. Physiol. **278:** H1274–H1284.
7. LANDESBERG, A. 1996. End–systolic pressure-volume relationship and intracellular control of contraction. Am. J. Physiol. **270:** H338–H349.
8. LEVY, C. & A. LANDESBERG. 2004. Hysteresis in the force-length relation and the regulation of cross-bridge recruitment in the tetanized rat trabeculae. Am. J. Physiol. **286:** H434–H441.
9. LEVY, C. & A. LANDESBERG. 2006. Cross-bridge dependent cooperativity determines the cardiac force-length relationship. J. Mol. Cell. Cardiol. **40:** 639–647.
10. YANIV, Y., R. SIVAN & A. LANDESBERG. 2005. Analysis of hystereses in force length and force calcium relations. Am. J. Physiol. **288:** H389–H399.
11. HARRIS, D.A & A.M. DAS. 1991. Control of mitochondrial ATP synthesis in the heart. Biochem. J. **280:** 561–573.
12. TERRITO, P.R., V.G. MOOTHA, S.A. FRENCH & R.S. BALABAN. 2000. Ca²⁺ activation of heart mitochondrial oxidative phosphorylation: role of F0/F1ATPase. Am. J. Physiol. **278:** c423–c435.
13. KIRICHOK, Y., G. KRAPIVINSKY & D.E. CLAPHAM. 2004. The mitochondrial calcium uniporter is a highly selective ion channel. Nature **427:** 360–364.
14. LEE, J.A. & D.G. ALLEN. 1991. EMD 53998 sensitizes the contractile proteins to calcium in intact ferret ventricular muscle. Circ. Res. **69:** 927–936.
15. NGUYEN, M.T. & M.S. JAFRI. 2005. Mitochondrial calcium signaling and energy metabolism. Ann. N. Y. Acad. Sci. **1047:** 127–137.
16. ROBERTSON, S.P., J.D. JOHNSON & J.D. POTTER. 1981. The time course of calcium exchange with calmodulin, troponin, parvalbumin and myosin in response to transient increases in calcium. Biophys. J. **34:** 559–569.
17. ZHOU, L., W.C. STANLEY, G.M. SAIDEL, X. YU & M. CABRERA. 2005. Regulation of lactate production at the onset of ischemia is independent of mitochondrial NADH/NAD+: insights from in silico studies. J. Physiol. **562:** 593–603.
18. ZHOU, L., J.E. SALEM, G.M. SAIDEL, W.C. STANLEY & M. CABRERA. 2005. Mechanistic model of cardiac energy metabolism predicts localization of glycolysis to cytosolic subdomain during ischemia. Am. J. Physiol. **88:** 2400–2411.
19. ZHOU, L., M. CABRERA, I.C. OKERE, N. SHARMA & W.C. STANLEY. 2006. Regulation of myocardial substrate metabolism during increased energy expenditure: insights from computational studies. Am. J. Physiol. **91:** 1036–1046.

20. DAS, A.M. & D.A. HARRIS. 1989. Reversible modulation of the mitochondrial ATP synthase with energy demand in cultured rat cardiomyocytes. FEBS Lett. **256:** 97–100.

21. BALABAN, R.S., S. BOSE, S.A. FRENCH & P.R. TERRITO. 2003. Role of calcium in metabolic signaling between cardiac sarcoplasmic reticulum. Am. J. Physiol. **278:** C285–C293.

22. CORTASSA, A. & M.A AON, *et al.* 2006. A computational model integrating electrophysiology, contraction and mitochondrial bioenergetics in the ventricular myocytes. Biophys. J. **91:** 1564–1589.

23. TROLLINGER, D.R., W.E. CASCIO & J.J. LEMASTERS. 2000. Mitochondrial calcium transients in adult rabbit cardiac myocytes: Inhibition by ruthenium red and artifacts caused by lysosomal loading of calcium indicating fluorophores. J. Biophys. **79:** 39–50.

24. BUNTINAS, L., K.K. GUNTER, G.C. SPARAGNA & T.E. GUNTER. 2001. The rapid mode of calcium uptake into heart mitochondria(RaM): comparison to RaM in liver mitochondria. Biochim. Biophys. Acta **1504:** 248–261.

25. JAVADPOUR, M.M. & J.C. TARDIFF, *et al.* 2003. Decreased energetics in murine hearts bearing the R92Q mutation in cardiac troponin T. J. Clin. Invest. **112:** 768–775.

Sarcomere Mechanics in Uniform and Nonuniform Cardiac Muscle

A Link between Pump Function and Arrhythmias

HENK E.D.J. TER KEURS,[a] TSUYOSHI SHINOZAKI,[b] YING MING ZHANG,[a] YUJI WAKAYAMA,[b] YOSHINAO SUGAI,[b] YUTAKA KAGAYA,[b] MASAHITO MIURA,[b] PENELOPE A. BOYDEN,[c] BRUNO D.M. STUYVERS,[a] AND AMIR LANDESBERG[a]

[a]Department of Physiology, School of Medicine, University of Calgary, Calgary, Alberta, Canada

[b]Tohoku University Graduate School of Medicine, Sendai, Japan

[c]Columbia University, New York, New York, USA

Starling's law and the end-systolic pressure–volume relationship (ESPVR) reflect the effect of sarcomere length (SL) on the development of stress (σ) and shortening by myocytes in the uniform ventricle. We show here that tetanic contractions of rat cardiac trabeculae exhibit a σ–SL relationship at saturating [Ca^{2+}] that depends on sarcomere geometry in a manner similar to that of skeletal sarcomeres and the existence of opposing forces in cardiac muscle shortened below slack length. The σ–SL –[Ca^{2+}]$_{free}$ relationships (σ–SL–Ca relationships) at submaximal [Ca^{2+}] in intact and skinned trabeculae were similar, although the sensitivity for Ca^{2+} of intact muscle was higher. We analyzed the mechanisms underlying the σ–SL–Ca relationship by using a kinetic model assuming that the rates of Tn-C Ca^{2+} binding and/or cross-bridge (XB) cycling are determined by either the SL, [Ca^{2+}], or σ. We analyzed the correlation between the model results and steady-state σ measurements at varied SL at [Ca^{2+}] from skinned rat cardiac trabeculae to test the hypotheses that the dominant feedback mechanism is SL-, σ-, or [Ca^{2+}]-dependent, and that the feedback mechanism regulates Tn-C Ca^{2+} affinity, XB kinetics, or the unitary XB force. The analysis strongly suggests that the feedback of the number of strong XBs to cardiac Tn-C Ca^{2+} affinity is the dominant mechanism regulating XB recruitment. Using this concept in a model of twitch-σ accurately reproduced the σ–SL–Ca relationship and the time courses of twitch σ and the intracellular [Ca^{2+}]$_i$. The foregoing concept has equally important repercussions for the nonuniformly contracting heart, in which arrhythmogenic Ca^{2+} waves arise from weakened areas in the cardiac muscle. These Ca^{2+} waves can reversibly be induced with nonuniform excitation–contraction coupling (ECC) by the cycle of stretch and release in the border zone between the damaged and intact regions. Stimulus trains induced propagating Ca^{2+} waves and reversibly induced arrhythmias. We hypothesize that rapid force loss by the sarcomeres in the border zone during relaxation causes Ca^{2+} release from Tn-C and initiates Ca^{2+} waves propagated by the sarcoplasmic reticulum (SR). Modeling of the response of the cardiac twitch to rapid force changes using the feedback concept uniquely predicts the occurrence of [Ca^{2+}]$_i$ transients as a result of accelerated Ca^{2+} dissociation from Tn-C. These results are consistent with the hypothesis that a force feedback to Ca^{2+} binding by Tn-C is responsible for Starling's law and the ESPVR in the uniform myocardium and leads to a surge of Ca^{2+} released by the myofilaments during relaxation in the nonuniform myocardium, which initiates arrhythmogenic propagating Ca^{2+} release by the SR.

Key words: Starling's law; arrhythmias; Ca^{2+} waves; nonuniform mechanics

Address for correspondence: Prof. Henk E.D.J. ter Keurs, M.D., Ph.D., School of Medicine, Department of Physiology, University of Calgary, 3330 Hospital Dr., N.W., Calgary, Alberta T2N 4N1, Canada. Voice: 403-220-4521; fax: 403-210-9739.

terkeurs@ucalgary.ca

Introduction

It is well known that during systole the ventricle can be characterized by a unique end-systolic pressure–volume relationship (ESPVR), which depends little on the volume change during ejection.[1–7] Hence, the

Ann. N.Y. Acad. Sci. 1123: 79–95 (2008). © 2008 New York Academy of Sciences.
doi: 10.1196/annals.1420.010

ESPVR dictates the interrelationship between stroke volume and end-diastolic volume, which has given rise to Starling's law of the heart.[7] Since the ESPVR is a geometric transform of the force–length relationship of the cardiac sarcomere, it follows that Starling's law is based on the properties of the myofilaments. Sarcomere length (SL) dependence on force output during steady activation at controlled levels of free calcium $[Ca^{2+}]$[8] stems from the length-dependent sensitivity of the contractile system to Ca^{2+}[5,9–11] and has initiated the search for the mechanism for this length dependence of Ca^{2+} sensitivity by showing that the force–SL relationships in intact and skinned cardiac muscle were similar. The latter appeared to be to due the effect of stretch in the skinned fiber by the increase in the maximal force as well as the shift of the sigmoidal force– $[Ca^{2+}]$ relation leftward, suggesting that stretch increases Ca^{2+} binding to troponin-C (Tn-C). The observation that a rapid ($<10\,ms$) stretch of skinned fibers causes both a slow increase in force and a slow decrease of $[Ca^{2+}]_i$ surrounding the myofilaments suggested that the force development rather than the SL by itself, determines Ca^{2+} binding.[12] Hence, SL and/or force have to feed back onto the properties of cardiac Tn-C (cTn-C). Mammalian cardiac muscle operates only on the ascending limb of the force–SL relationship at lengths (*in vivo* usually 1.7–2.15 μm[13]; *in vitro* maximally 1.5–2.3 μm[2,3,14,15]) at which the actin filaments (1.125 μm long) are always in double overlap. Still, the relationship between force and SL at a saturating $[Ca^{2+}]$ in intact fibers is unknown. This knowledge is required if one is to distinguish whether the length-dependent sensitivity of the contractile system for Ca^{2+} may be due to SL-dependent Ca^{2+} affinity of Tn-C with Tn-C as the length sensor[16] or the variation of Ca^{2+} affinity of Tn-C as a result of a feedback of force generated by the cross-bridge (XB). The observation that, after the expression of skeletal muscle Tn-C (sync) in mouse cardiac muscle, SL still regulates the Ca^{2+} sensitivity of force development[17] makes it unlikely that cTn-C is the length sensor for Ca^{2+} sensitivity. Similarly, the fact that cTn-C present in soleus muscle does not confer length-dependent Ca^{2+} sensitivity to the soleus muscle suggests that other properties of the contractile filaments may be involved.[18,19] Fuchs[20] and collaborators have shown that the amount of Ca^{2+} bound to the thin filament is sensitive to lattice spacing of the myofilaments[20] and consequently XB formation and force development.[21] These studies created a foundation for models which suggest a role for force development by the XBs in length dependence of Ca^{2+} activation of the mammalian cardiac sarcomere, and which faithfully predict mechanical properties of

the cardiac ventricle.[22]

The current study presents data on the effect of filament geometry on maximal force development and on the effect of stretch of uniform intact and skinned muscle on the sensitivity of the contractile system for Ca^{2+} ions. We have used a model wherein the kinetic rates of Ca^{2+} binding to Tn-C and XB cycling were assumed to be determined by either SL, $[Ca^{2+}]$, or stress. The tested hypotheses consider whether the dominant mechanism is SL (interfilament spacing)-, force-, or $[Ca^{2+}]$-dependent, and whether the dominant mechanism regulates Ca^{2+} affinity, XB kinetics, or the unitary force per XB. The different predictions were tested by use of the skinned muscle data. SL- and Ca^{2+}-dependent mechanisms were unable to explain the data. However, for all the trabeculae, an identical relationship between Ca^{2+} affinity and stress demonstrated that an increase in the stress is associated with a proportional increase in apparent Ca^{2+} affinity. The analysis strongly supports the hypothesis that the dependence of Ca^{2+} affinity on the number of strong XBs is the dominant mechanism that regulates XB recruitment in cardiac muscle. This mechanism regulates cardiac mechanics and energetics.

The analysis clearly favours a model where the XB force exerts a feedback on binding of Ca^{2+} to Tn-C. Application of this model to contraction of uniform and non-uniform muscle suggests that this feedback mechanism plays an important role in the genesis of arrhythmias in non-uniform muscle.

Methods

Experiments

All experiments were performed at 26°C. The methods for isolating the trabeculae, skinning and measurement of SL, and stress have been published before.[23] Both the dissecting and perfusion solutions, containing (in mM) 137.2 NaCl, 5.0 KCl, $CaCl_2$ as specified, 1.2 $MgCl_2$, 2.0 sodium acetate, 10 taurine, 10 glucose, 10 N-2-hydroxyethylpiperazine-N'-2-ethanesulfonic acid (HEPES), were equilibrated with 100% O_2 and adjusted to a pH of 7.35 by 1 M NaOH. The dissection solution contained 0.2 mM $CaCl_2$ and additional 15 mM KCl. The trabeculae were mounted between force transducers using a silicon strain gauge (AE801, AME; Sensonor, Horten, Norway) and a pin attached to a servomotor controller (300S; Cambridge Technology, Cambridge, MA) in a bath on the stage of an inverted microscope (TMD; Nikon, Tokyo, Japan). Three interrelated studies are reported in this overview: the σ–SL–Ca relationship during steady

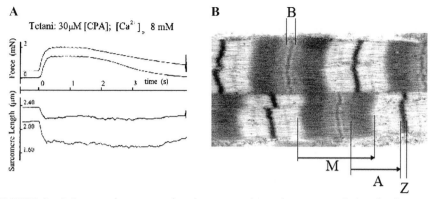

FIGURE 1. A Force and sarcomere length tracings obtained at two muscle lengths during 1.5 s lasting tetani. The muscle was stimulated with 8 Hz pulse of 40 ms duration in the presence of 30 μM CPA. $[Ca^{2+}]_o$ was 8 mM. Tetani were repeated at varied initial muscle length so that SL during the tetani ranged from 1.6 to 2.2 μm. Panel **B** shows and electron micrograph of a trypsin treated trabecula which had been fixed at a SL of 2.86 μm.

state in uniform intact and skinned fibers and studies of nonuniformly contracting trabeculae. The force was measured by using the silicon strain gauge.

Tetanic Contractions and Skinned Fibers

Tetanic contractions, as shown in Figure 1, were reproducibly elicited by trains of 40-ms pulses at 4–6 Hz for 2 s with perfusion solution containing 30 μM/L CPA (Sigma Chemical, St. Louis, MO)[24] and varied $[Ca^{2+}]_o$ (0.3–6 mM). Final concentrations of DMSO in perfusion solution were 0.05%. Data were analyzed after 2 s when a stable steady state of force, $[Ca^{2+}]_i$ (corrected for changes in autofluorescence) and SL (<2.10 μm) had been attained.

A "jump" bath was used for rapid sequential change between skinned fiber solutions[25] with seven $[Ca^{2+}]$s: 1.04, 1.48, 1.82, 2.33, 4.8, 9.04, and 27.67 μM. SL was continuously measured by laser diffraction technique.[26] SL was manipulated by changing muscle length using a micromanipulator in each activation solution. The SLs used at the various $[Ca^{2+}]$ levels were not identical, since the SL was determined by the micromanipulator and there was shortening of the center of the fiber at the expense of the tricuspid valve and the right-ventricle remnant, before the steady-state stress was reached. A set of five skinned trabeculae was treated with trypsin (0.15 mg/mL for 5 min), fixed in the bath, and processed for electron microscopy.

Nonuniform Trabeculae

To produce the nonuniform excitation–contraction coupling (ECC), a restricted region was exposed to a small "jet" of solution (≈0.06 mL/min) that had been directed perpendicularly to a small muscle segment (300 μm) using a syringe pump connected to a glass pipette.[27,28] The jet solution contained standard HEPES solution as well as either (1) caffeine to deplete the sarcoplasmic reticulum (SR) of Ca^{2+} through opening of the SR Ca^{2+} release channels[3,29]; (2) 2,3-butanedione monoxime (BDM; 20 mmol/L) to suppress myosin ATPase and, thus, activation of XBs[29] while reducing the SR Ca^{2+} content modestly; or (3) low $[Ca^{2+}]$ to reduce Ca^{2+} availability for ECC.[3] The Ca^{2+} concentration in the jet ($[Ca^{2+}]_{jet}$) was usually identical to the bath solution ($[Ca^{2+}]_o$), except for low-$[Ca^{2+}]_{jet}$ solution or unless mentioned otherwise. During exposure to the jet solutions, Ca^{2+} waves underlying triggered propagated contractions (TPCs) were induced by stimulation of the muscle at 2.5 Hz for 7.5 s, repeated every 15 s at $[Ca^{2+}]_o$ of 2–3 mmol/L at 24°C.[30] Measurement of $[Ca^{2+}]_i$ commenced within 10 min, as soon as the amplitude of stimulated twitches, TPCs, and underlying Ca^{2+} waves were constant.

Analysis of the Force–[Ca²⁺]–SL Relationships

The model used here couples the kinetics of Ca^{2+} binding to a single binding site on uniformly distributed Tn-C on actin filaments with the regulation of asynchronous XB cycling via the troponin–tropomyosin complex.[22] The model does not assume any dependence of the rate kinetics on the input variables (Ca^{2+} or SL) or on the stress (state variables T and U below). The XBs shuttle between a "weak" non-force-generating conformation and a "strong" force-generating conformation due to binding and hydrolysis of ATP. The rate of XB return from the strong to a weak conformation (XB weakening) is independent of the presence of bound Ca^{2+}.[22] The cardiac sarcomere may exhibit three forms of overlap between thin and thick filaments: a *non-overlap* region, a *single-overlap* region, and a *double-overlap* region.

The regulatory units are distributed between four "states." Regulated actin is turned "off" only at rest (*state R*) when XBs are "weak" and Ca^{2+} is not bound to Tn-C. Ca^{2+} binding to Tn-C turns actin "on" and leads to mechanical activation (*state A*), while XBs are still weak.[22] ATP hydrolysis and phosphate release allow XB turnover to the "strong" conformation, leading to tension development (*state T*). Ca^{2+} dissociation from Tn-C in state T leads to *state U (unbound Ca^{2+}*, in which the XBs are still strong. Kinetics between states are bi-directional except that transition from the state R to $\underline{U}$ is negligible[31] and formation of force-generating XBs is allowed only through state A. R, A, T, and U reflect the density of the corresponding regulatory unit states along the thin filament. The coefficients k_l and k_{-l} represent the rate constants of Ca^{2+} binding to, and dissociation from, the Tn-C regulatory site; f and g_0 represent the rate constants of XB turnover during isometric contraction. These rate coefficients may be a function of stress (state variables T and U), or of input variables such as the SL or $[Ca^{2+}]$. It follows that stress is given by:

$$\sigma = TR_0 \cdot \bar{F}_{XB} \cdot \frac{1}{2} \cdot (SL - L_0) \cdot P_{XB} \cdot$$
$$\frac{K \cdot [Ca]}{(1 - P_{XB}) + K \cdot [Ca]} \quad (1)$$

where TR_0 is the density of the regulatory units; L_0 determines the apparent single overlap length; P_{XB} is the probability of strong XB at full Ca^{2+} activation, $P_{XB} = f / (f + g_0)$; $\bar{F}_{XB}$ is the XB unitary force, and K is the affinity of Ca^{2+} binding to Tn-C, $K = k_\ell / k_{-\ell}$.

The Input Limb of the Cooperativity Mechanism

We have studied how L_0, P_{XB}, K, M, and $\bar{F}_{XB}$ depend on SL, $[Ca^{2+}]$, and the number of regulatory units associated with force development (A.J.P.; submitted for publication). The approach is summarized here. First, we have defined the input limb of the cooperativity mechanism by determining whether any of the parameters depend on the variables SL, $[Ca^{2+}]$, and stress. In order to do so, we test each possible variable separately, analyzing data with the same value of the tested variable. By doing this we have to estimate only a set of constant parameters, since the independent variable is set to be constant. This approach yielded a series of functions denoted as $\Psi_{variable}$, where the sub-

script "variable" refers to the tested variable. Using, as an example, the analysis of the data at constant SL, Ψ_{SL} yields:

$$\Psi_{SL} = \frac{[Ca]_2 \cdot \sigma_1 - [Ca]_1 \cdot \sigma_2}{[Ca]_1 \cdot [Ca]_2 \cdot (\sigma_2 - \sigma_1)}$$
$$= \frac{K(SL)}{1 - P_{XB}(SL)} = const \quad (2)$$

The left side of the equation includes only measured data, whereas the right side reflects Ca^{2+} affinity and the probability of XB attachment in the model. Equation 2 predicts that Ψ_{SL} should be positive $[P_{XB}(SL) < 1]$ and a monotonically increasing function of SL.[32] Using the hypotheses that Ca^{2+} affinity (K),[32] XB dynamics (P_{XB}),[21] or the apparent SL (L_0)[21] are SL-dependent led to the same predicted dependence of Ψ_{SL} on SL (FIG. 2). All these hypotheses were tested by calculating the predicted relationship (Ψ_{SL}) between the stresses and Ca^{2+} levels at the same SL.

Constant $[Ca^{2+}]$

We derived the stress–SL relationship at constant $[Ca^{2+}]$ by means of the same approach. It yielded linear stress–SL relationships with increasing slope for increasing $[Ca^{2+}]$.

Constant Stress

Using this approach, again we derive the relationship between SL–$[Ca^{2+}]$ combinations at the same stress level, yielding:

$$\Psi_\alpha = L_0(\sigma) \cdot \Psi_\beta + \frac{K(\sigma)}{1 - P_{XB}(\sigma)} \quad (3)$$

Equation 3 predicts that the relationship between Ψ_α and Ψ_β calculated from the experimental data should be linear for any σ. Both the slope $[L_0(\sigma)]$ and the intercept $[K(\sigma)/(1 - P(\sigma)]$ of this relationship depend on the stress. Different slopes will appear only if L_0 depends on σ. All hypotheses that suggest dependence of Ca^{2+} affinity[22] or XB cycling[33] on the number of strong XBs yield Equation 3 and were tested against the $\Psi_\alpha - \Psi_\beta$ relationship.

The Output Limb of the Cooperativity Mechanism

The output limb of the cooperativity mechanism was defined subsequently, that is, which parameters (Ca^{2+} affinity (K) and/or stress (σ)) are affected by the input variable (Ca^{2+} affinity, XB dynamics, unitary force, or apparent SL). It can be shown that if K depends on stress, Equation 1 yields:

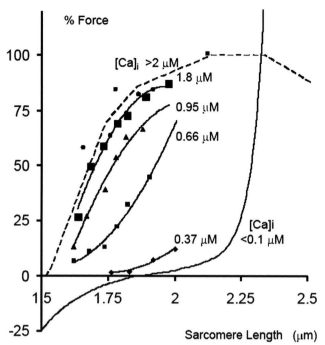

FIGURE 2. The force sarcomere length relationships of passive, active, and maximally activated cardiac muscle. The dashed line is the force length relationship predicted by the cross-bridge theory of muscle contraction and was constructed by using the sarcomere measures from Figure 1. At sarcomere lengths below 1.86 μm forces opposing shortening below slack length[44] were subtracted from the predicted actively generated force. Note that cardiac muscle cannot reach the SL range of partial overlap of actin and myosin because of the presence of a stiff parallel elastic element. (For further explanation see text.) The data have been obtained using tetanized cardiac trabeculae at varied $[Ca^{2+}]_o$, which the $[Ca^{2+}]_i$ was measured using fura-2 (levels of $[Ca^{2+}]_i$ are indicated next to the graphs. Activation of contraction was observed at $[Ca^{2+}]_i$ exceeding 200 nM and saturated at $[Ca^{2+}]_i$ of ~2μM.

$$\Psi_K = \frac{2}{TR_0 \cdot \bar{F}_{XB} \cdot P_{XB}} \cdot \sigma + L_0 \qquad (4)$$

ψ_K, calculated from the measured SLs and $[Ca^{2+}]$s, is a linear function of σ. Moreover, a single unique dependence of ψ_K on σ is predicted for all pairs of data taken at the same stress level. L_0 represents the apparent SL and may reflect the effect of changes in sarcomere geometry including the filament spacing.[21] It can again be shown that if the apparent geometry is a function of σ, Equation 1 yields Ψ_L (not shown here), which should be a linear function of σ, passing through the origin.

XB Dynamics

Equation 1 predicts that if the output limb of the model consists of the dynamics of the XBs, the $\Psi_\alpha - \Psi_\beta$ relationship should have a constant intercept with the Ψ_α axis which equals the Ca^{2+} affinity, K.

XB Force

In this case, the slope of the $\Psi_\alpha - \Psi_\beta$-relationship should be a constant.

Parameter estimation for this analysis was performed using the Gauss–Newton method of minimizing the least-mean–square error of the fit of the raw experimental data from individual muscles and from all experimental data as a group.

Model Simulations of Ca²⁺ and Force Kinetics

We have numerically simulated the experimentally observed Ca^{2+} transients and twitch kinetics as well as the F–pCa relationship to gain insight into the parameters of Ca^{2+} release and the removal processes that dictate their time course.[6,27] The model assumes a Ca^{2+} release flux with an exponential rise and fall; ligand binding; Ca^{2+} extrusion dominated by the SR, which exhibits Hill kinetics.[34] Ca^{2+} binding to Tn-C is diffusion limited[35]; hence we chose to create a feedback of force to Ca^{2+}–Tn-C kinetics by assuming that deformation of actin is accompanied by a structural change of Tn-C which reduces the rate of dissociation of bound Ca^{2+}. The cytosolic $[Ca^{2+}]_i$ was assumed to be 70 nM[36,37] and the calculations were started with the buffers in equilibrium. For simplification we have only incorporated Ca^{2+} binding to Tn-C, although other ligands are known.[3,38] The calculations were performed with an integration interval of 1 µs. Rise times of $[Ca^{2+}]_i$ transients were simulated by fitting the rate constants of Ca^{2+} channel opening and closing, their open time (Δt), and R to the experimentally observed rising phase of the Ca^{2+} transient. The decline of a Ca^{2+} transient was simulated by fitting the SR Ca^{2+} uptake parameters to the observed decline of the transients in the absence of F.[39] For the steady-state F–pCa relationship, we assumed Hill behavior, while the interaction between F and the K_D of Ca·Tn-C was assumed to be strain dependent. F_{max} at saturating $[Ca^{2+}]_i$ is a function of acto-myosin overlap, which depends on the geometry of the sarcomere.[24] The feedback of F to the dissociation rate of Ca^{2+} from Tn-C was identical for simulations of each of the relationships.

Results

Tetanic Force, [Ca²⁺], and SL in Uniform Cardiac Muscle

It is well known that the twitch σ–SL–Ca relationship of cardiac muscle is curvilinear with convexity toward the ordinate at high $[Ca^{2+}]_o$ and convexity toward the abscissa at low $[Ca^{2+}]_o$.[25] Less known is the nature of the relationship during steady-state activation. We used here two approaches to activate the cardiac myofilaments in a steady manner: tetanic stimulation after paralysis of the SR using the drug cyclopiazonic acid (CPA), which blocks the SERCA2 Ca^{2+} pump of the SR,[3] and direct activation of the myofilaments in skinned fibers.[40]

The effect of 30 µM [CPA] on Ca^{2+} sensitivity of the contractile filaments is negligible; hence, we used this concentration of CPA to induce tetani in intact muscle (FIG. 1A). Typically, the sarcomeres shortened initially; then SL remained constant for the duration of the plateau (FIG. 1A). The initial shortening of the sarcomeres is known to take place at the expense of the series elastic element in the muscle (usually the tricuspid valve attachment in these trabeculae). FIGURE 3A shows the increase of F_{max} with increasing SL, during tetany in [CPA] = 30 µM and at $[Ca^{2+}]_o = 8$ mM. This increase in F_{max} with SL paralleled the force increment that is predicted by the assumption that F_{max} generation depends on the mechanical constraints on the interaction between the XBs and the actin fila-

ments. Measurement of the mechanical constraints was derived from electron micrographs of trypsin-treated trabeculae.[41] We have employed trypsin treatment (see the Methods section) in order to break the titin filaments and allow stretch of the muscle to a SL where the edge of the thin filament array is visible on the electron micrographs. When this approach was employed the edge of the thin filaments always paralleled the contour of the Z line (FIG. 1B). The average myosin length, actin length, bare zone, and Z-band width were: 1.65 µm, 1.125 µm, 0.18 µm, and 0.14 µm, respectively.[42] These dimensions predict a force plateau from 2.25 to 2.45 µm and an ascending limb of the F–SL relationship between 1.5 and 2.26 µm (FIG. 2, dashed line). Increasing the double-overlap would proportionally reduce force at shorter lengths so that SL = 1.65 µm (~70% of the maximal force) is generated. Myosin distortion would cause a further and steeper decline with an intercept at the SL axis near 1.25 µm. The same assumptions have generated the ascending limb of the classical force–length relationship for the skeletal muscle. In addition, it has been shown that cardiac muscle generates a force that opposes shortening below slack length (SL = 1.86 µm[43]). The opposing force has been shown to increase the cardiac trabeculae in proportion to shortening; at SL = 1.5 µm the opposing force was 25% of the maximal active force.[44] The opposing force is generated in part by the titin filament in the sarcomere itself[45] and by the collagen meshwork around the myocytes.[46] These opposing forces (drawn line in FIG. 2 marked $[Ca^{2+}]_i < 0.1$ µM) are probably substantially larger than the same in skeletal muscle fibers, which are the source of the textbook force–length relationship, because titin in skeletal muscle fibers is more compliant and single skeletal muscle fibers do not contain a collagen meshwork like myocardium. FIGURE 2 shows that the actually measured maximal force during the tetanus at saturating $[Ca^{2+}]$ coincided exactly with the force predicted on the basis of these mechanical considerations. Hence, we conclude that there is no significant difference between the factors that dictate maximal force development in skeletal and in cardiac muscle. The figure also shows that the passive forces prevent the SL of the cardiac muscle from ever being stretched beyond the plateau on account of the presence of collagen strands in the fascicles of myocardium.[47]

Effect of [Ca²⁺]ᵢ on the Steady State σ-SL-Ca Relationship in Intact Cardiac Muscle

FIGURE 2 shows the typical shape of the σ–SL–Ca relationships at both maximal and submaximal

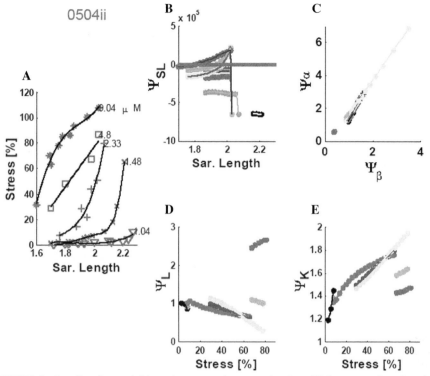

FIGURE 3. Feedback model hypotheses tested against the data (**A**) from a single trabecula. **B** and **C** suggest that the afferent limb of the cooperativity is not length (**B**) but stress (**C**) dependent. **C, D,** and **E** suggest that the efferent limb does not affect cross-bridge kinetics (**C**) apparent sarcomere length (**D**) or the unitary force per cross-bridge (**E**) but determines Ca^{2+} affinity (**E**) (for further explanation see text).

$[Ca^{2+}]_i$ measured using fura-2. The σ at any SL increases with $[Ca^{2+}]_i$ yielding a series of σ–SL–Ca relationships, which are convex to the abscissa at low $[Ca^{2+}]_i$ and convex to the ordinate at high $[Ca^{2+}]_i$. The figure also shows that the force saturated at free $[Ca^{2+}]_i$ above 2 μM. The relationship between σ and $[Ca^{2+}]_i$ fitted Hill equations[40] in which the Hill coefficient increased with SL (3–5.5) and EC_{50} decreased with SL (1 to 0.5 μM at SL 1.65 and 1.9 μm, respectively) (FIG. 1B). The Ca^{2+} sensitivity in the intact trabeculae was ~4.5 times greater than in skinned cardiac muscle.[24] The σ–SL relationship in skinned muscles at high free $[Ca^{2+}]$ was again convex to the ordinate. whereas at lower $[Ca^{2+}]$ it was concave. The results shown for a typical skinned trabecula in FIGURE 3 resemble the data of Kentish *et al.*[40] We used data from nine muscles studied in this manner to identify the nature of the cooperativity during Ca^{2+}-activated force development in the cardiac muscle.

Analysis of the Steady-State σ–SL–Ca Relationship in Skinned Fibers

Cooperativity between Stress Development and Tn-C Ca^{2+} Affinity

FIGURE 3 summarizes the analysis of the stress–SL relationship (FIG. 3A) obtained from a typical trabecula. The linear dependence of Ψ_α on Ψ_β (FIG. 1C) is in accordance with the hypothesis that the stress is the afferent limb of the cooperativity mechanism. The calculated Ψ_{SL} (Eq. 3) has no defined dependence on SL and has unrealistic negative values (FIG. 3B), which voids the hypothesis that the SL is the input variable of the cooperativity mechanism. As for the efferent limb, the negative intercepts of extension of the lines in FIGURE 3C with the Ψ_α axis and the identical linear dependence of Ψ_α on Ψ_β, with the same slope for the various stress levels rule out the hypothesis that stress affects the rates of XB cycling (P_{XB}). The decrease in Ψ_L with the increased stress (FIG. 3D) conflicted with

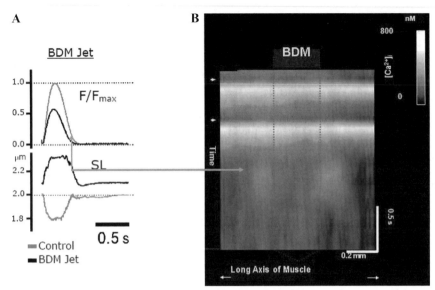

FIGURE 4. Nonuniform cardiac muscle and [Ca^{2+}]$_i$Panel **A** shows SL changes in a muscle during the twitch during exposure to a jet containing BDM (20 mmol/L). Normal contraction in the normal segments is accompanied by stretch in the jet exposed region. Panel **B** shows the [Ca^{2+}]$_i$ (in gray scale; see calibration bar) along the muscle (ordinate) as a function of time (abscissa) during and after two electrically driven contractions. The region subjected to the jet is delineated by the dashed lines. The panel shows the initiating events of Ca^{2+} waves induced by local BDM (20 mmol/L) exposure at normal [Ca^{2+}] in the bath (1 mmol/L). Only a local Ca^{2+} surge (starting 360 ms) after stimulation is observed.

the predicted behavior based on the assumption that the apparent SL is stress dependent. The monotonic increase of Ψ_K with increasing stress (FIG. 3E) rules out the hypothesis that the unitary force per XB is a function of the number of strong XBs and is consistent with the hypothesis that Ca^{2+} affinity is stress dependent.

Contraction of a Nonuniform Muscle and Arrhythmogenesis

The cooperativity mechanism identified above may be of importance as a potential source of arrhythmias in nonuniformly contracting cardiac muscle. Nonuniformity is the hallmark of diseased myocardium. While it is generally accepted that electrical nonuniformity of diseased myocardium is arrhythmogenic, nonuniformity of contraction has received far less attention. Hence, we set out to investigate the effects of nonuniformity by rendering a small segment of trabeculae weak by localized superfusion of a segment of the muscle with a jet of solution from a micropipette which contained a modified HEPES solution in order to reduce force development in the ~300-μm-long segment near the pipette.[27] The fluid flow from the pipette using a solution with the same HEPES solution as in the bath had

no effect on σ or SL by itself. We have studied the effects of both BDM (FIGS. 4 and 5) and caffeine as well as low and high [Ca^{2+}] in the jet. Here we will focus of the effects of BDM. When a jet containing BDM (FIG. 4) was applied to the stimulated trabeculae, sarcomere stretch replaced rapidly the normal active shortening of the sarcomeres in the exposed segment, while peak force (F/F$_{max}$) succinctly decreased. All effects were rapidly reversible. Sarcomere dynamics along muscles exposed to a jet revealed three distinct regions during the stimulated twitch[27,48]: (1) a segment exposed to the jet where sarcomeres were stretched; (2) a region located >200 μm from the jet where sarcomeres exhibited typical shortening; and (3) a "border zone" (BZ) between these two segments, where sarcomeres shortened early during the twitch and then were stretched, although less than in the jet-exposed segment. The BZ extended 1–2 cell lengths beyond the jet-exposed region. Similar changes in regional sarcomere dynamics were observed with caffeine and low as well as high [Ca^{2+}]$_{jet}$ experiments.

Initiation of Ca^{2+} Waves

FIGURES 4 and 5 show initiation of Ca^{2+} waves in the BZ of a BDM-exposed trabeculae. All muscles responded similarly to the three interventions: a localized transient in [Ca^{2+}]$_i$ (≈300 nmol/L)—denoted as

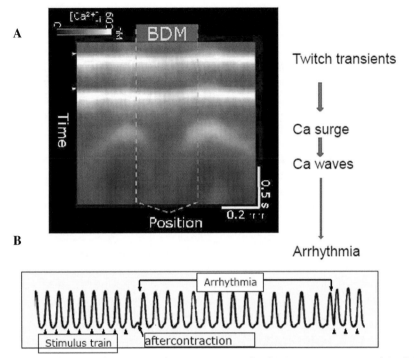

FIGURE 5. A Propagated Ca²⁺ waves cause arrhythmias (A) At increased $[Ca^{2+}]_o$ (2 mmol/L) bi-directional Ca^{2+} waves are initiated at the site of the Ca^{2+} surge and propagate into the segment inside the jet and into the normal muscle. Both amplitude of the initial and propagating transient as well as propagation velocity increased with increase of $[Ca^{2+}]_o$, while the latency of onset of the Ca^{2+} transient decreased (300 ms). Arrows indicate initiation sites of propagating waves. **(B)** Nonuniform ECC caused by the jet containing BDM (20 mmol/L) is arrhythmogenic. Recording of Force showing that stimulus trains during local exposure to BDM consistently induced arrhythmias. The force tracing showing that spontaneous contractions were both preceded and followed by after-contractions induced by the stimulus train. When the jet was turned off, the contractile nonuniformity and its arrhythmogenic effects disappeared rapidly.

Ca²⁺ surge occurred along ~100–150 μm of the BZs without apparent propagation of Ca^{2+} waves. The Ca^{2+} surge took place ~325 ms after the stimulus, during the relaxation phase of the twitch, when force had declined by 70 to 80% (FIG. 4). Increasing $[Ca^{2+}]_o$ led to a decrease of the latency and an increase of the Ca^{2+} surge (FIG. 4) and led to development of bi-directional propagating Ca^{2+} waves. The earliest Ca^{2+} surge with a caffeine jet was also observed in the BZ. The Ca^{2+} surge in the high $[Ca^{2+}]_{jet}$ occurred in the jet-exposed region itself. These observations suggest strongly that the Ca^{2+} surge was the initiating event of Ca^{2+} waves. Ca^{2+} waves started systematically after the decline of the last stimulated Ca^{2+} transient (FIG. 5). These waves propagated into the regions outside and—in the cases of BDM and low $[Ca^{2+}]_{jet}$—inside the jet-exposed region. FIGURE 5 (BDM jet) clearly shows two initiation sites of four Ca^{2+} waves in the BZ and symmetric propagation into re-

gions outside and inside the jet. In caffeine-exposed muscles, the Ca^{2+} waves did not propagate into the jet region. Propagation velocity of the Ca^{2+} waves as in FIGURE 5 outside and inside the jet region, ranged from 0.2 to 2.8 mm/s, that is, comparable to Ca^{2+} waves observed in studies of damaged muscle.[49,50] Jets of caffeine, BDM, or modified $[Ca^{2+}]_o$ solution had distinct effects on $[Ca^{2+}]_i$. Robust, electrically driven $[Ca^{2+}]_I$ transients occurred in the regions outside the jet independent of the composition of the jet solution (FIG. 4). Both the caffeine-jet and low-$[Ca^{2+}]_{jet}$ decreased the peak of the stimulated $[Ca^{2+}]_i$-transient in the jet region. BDM decreased the $[Ca^{2+}]_I$ transient only slightly (FIGS. 4 and 5). Caffeine increased diastolic $[Ca^{2+}]_i$; low $[Ca^{2+}]_{jet}$ and BDM both decreased diastolic $[Ca^{2+}]_i$ (FIG. 5). High $[Ca^{2+}]_{jet}$ increased diastolic $[Ca^{2+}]$ in the jet, but reduced the $[Ca^{2+}]$ and force development in the jet region. The $[Ca^{2+}]_I$ changes were smaller in the BZ, consistent with a gradient between the regions on account of the mixing of the contents of the jet with the main solution in the bath.

Nonuniformity and Arrhythmias

Our experiments with muscles rendered nonuniform by a jet of solution containing an inhibitor of ECC prove that mechanical nonuniformity is a sufficient requirement for arrhythmogenicity. FIGURE 5B shows that nonuniformity of ECC created by the jet induced rapidly reversible non-driven rhythmic activity. The arrhythmia consisted of spontaneous twitches at regular intervals starting after an after-contraction that followed the last stimulated contraction. The arrhythmias could persist for several minutes in a nonuniform muscle with intervals between non-driven contractions that were usually slightly longer than those of the preceding stimulus train. These arrhythmias were no longer inducible shortly after the jet was turned off and the uniformity of ECC restored.

Simulation of ECC with Feedback between Force and Ca^{2+} Dissociation from Tn-C

We have tested whether a feedback of the force to Tn-C Ca^{2+} kinetics can reproduce the Ca^{2+} surge that appears to trigger the Ca^{2+} waves (see **Methods**). A full report is given elsewhere.[45] In summary, it is well known that an action potential in cardiac muscle leads to an inward current of Ca^{2+} ions which triggers release of Ca^{2+} ions by Ca^{2+}-induced Ca^{2+} release (CICR) from the SR. The released Ca^{2+} ions bind to Tn-C in a manner that appears to depend on the resultant force, leading to a brisk response of force to activation. The normal cardiac contraction is then terminated by extrusion of Ca^{2+} ions from the cytosol, predominantly by the SR, although the rate of extrusion is influenced by the feedback mechanism. We generated a simplified model of the twitch by assuming a single release pulse and we simplified the relaxation phase by assuming a single Ca^{2+} removal process. These assumption generated a time course of $[Ca^{2+}]_i$ similar to that of unloaded cardiac myocytes.[35] Intracellular Ca^{2+} was assumed to bind to intracellular ligands, including Tn-C, and ATP. Ca^{2+} was assumed to bind to a single low-affinity site on Tn-C with a fixed on-rate constant, but to dissociate with an off-rate that was inversely proportional to the $Ca^{2+} \bullet$Tn-C deformation that is caused by XB force that strains the actin filament (κ) (FIG. 6). The rate of formation (f) of force developing XBs was fit to the

rate of force development of trabeculae after a quick release, but the ratio of f and g was taken from the literature.[51,52]

Simulation of the feedback of force to Ca^{2+} dissociation from $Ca^{2+} \bullet$Tn-C through the deformation of the actin filament and Tn-C on the filament by a single feedback of actin strain κ predicts fundamental properties known for the cardiac muscle, a realistic reproduction of σ–pCa–SL relationship.[40] An increase of SL reproduced an apparent shift of the F–pCa relationship to lower pCa as a result of an increase of the apparent Hill coefficient, although Tn-C still is assumed to bind only one Ca^{2+} ion. The feedback predicted realistic F–SL relationships at varied activation levels for $[Ca^{2+}]_i$. FIGURE 6B shows that introduction of a tight relationship between the actin strain κ and a decrease of the dissociation rate of Ca from $Ca^{2+} \bullet$Tn-C reproduces the experimental finding that increased force development progressively increases the duration of the twitch.[43,53] This observation holds both in the model and the experiments, irrespective of whether the increased force is caused by varied SL, $[Ca^{2+}]_o$, or by other interventions.[54] FIGURE 6 shows that twitch prolongation is accompanied by a proportional prolongation of the time during which Tn-C is occupied by Ca^{2+} ions (FIG. 6B, middle pannel). FIGURE 6 shows the implication of this feedback to the kinetics of the cytosolic $[Ca^{2+}]$ transient and replicates our observation,[42] and the recent observation by Julian's group, that the $[Ca^{2+}]_i$ transient during the twitch exhibits a prominent plateau at higher force levels.[55] The plateau is completely lost in the absence of a feedback of force to Ca^{2+} binding to Tn-C (FIG. 6).

Discussion

A Mechanism that Links Contraction in the Uniform Heart with Arrhythmias in Nonuniform Myocardium

This study describes the stress–SL interrelationship of a maximally Ca^{2+} activated intact cardiac muscle. This relationship is almost identical to the ascending limb of the force–length relationship based on the XB theory of muscle contraction for the skeletal muscle, provided that the dimensions of the filaments for the rat are used and provided that forces opposing shortening below slack length are incorporated in the relationship. The correspondence between the observed data and the theoretical relationship suggests that there is no difference in the mechanisms which dictate force at saturating $[Ca^{2+}]$ between cardiac and

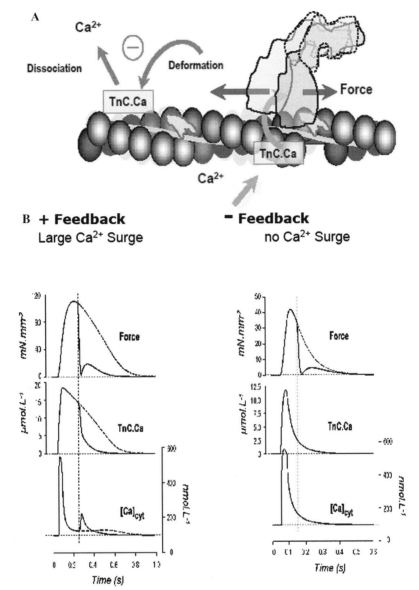

FIGURE 6. Cross-bridge feedback on Ca²⁺ affinity of Tn-C may explain arrhythmo-genic Ca²⁺ waves Panel **A** shows a cartoon of the cross-bridge-actin-troponin-C (Tn-C) interaction. We assume in the model that that force development by XBs feeds back on dissociation of Ca²⁺ ions from Ca²⁺•Tn-C by inducing a deformation of Ca²⁺•Tn-C complex, which reduces the off-rate of Ca²⁺ from Ca²⁺•Tn-C. Panel B (*left side*) shows the simulated time course of F, free [Ca²⁺]ᵢ and [Ca²⁺•Tn-C] during a twitch *in the presence* of feedback of F to the off-rate of Ca²⁺ from Ca²⁺•Tn-C. The time course of F is identical to the time course of F during the published sarcomere isometric contraction.[26] The time course of [Ca²⁺]ᵢ is identical to the published time course of [Ca²⁺]ᵢ[55] including the plateau of the [Ca²⁺]ᵢ transient. The effect of feedback of F on dissociation of Ca²⁺ ions from Ca²⁺•Tn-C leads to prolonged binding of Ca²⁺ to Tn-C, which exceeds the time course of the [Ca²⁺]ᵢ transient. The effect of a quick release of the muscle 200 ms after onset of contraction (indicated by the vertical dashed line through the tracings) causes a rapid drop of F and a rapid and large decrease of the amount of Ca²⁺ bound to Ca²⁺•Tn-C, which causes a substantial increase in free [Ca²⁺]ᵢ , denoted here as the Ca²⁺ surge. Panel **B** (*right side*) shows the time course of F, free [Ca²⁺]ᵢ and [Ca²⁺Tn-C] during the twitch *in the absence* of feedback of F to the off-rate of Ca²⁺ from Ca²⁺•Tn-C. Peak F is smaller and the [Ca²⁺•Tn-C] now follows the [Ca²⁺]ᵢ transient. A quick release (at the dashed line) still causes a rapid drop of force but fails to elicit a [Ca²⁺]ᵢ transient during relaxation.

skeletal muscle. Furthermore, this study in the intact muscle confirms that stretch of cardiac sarcomeres over the range of lengths, where only double-overlap of the actin filaments plays a role, leads to increase in the apparent sensitivity of the contractile system to Ca^{2+} ions. Comparison of the data in this study with the data from skinned fiber experiments shows that the mechanisms underlying the length dependence of Ca^{2+} sensitivity are the same, despite the loss of soluble cytosolic proteins from the skinned fiber. These data and the analysis of skinned fiber data are consistent with the hypothesis that the degree of stretch of the sarcomeres determines the force development by the XB on account of physical intra-sarcomeric factors. In turn, developed force affects the sensitivity of the contractile system to Ca^{2+} ions, observed at submaximal $[Ca^{2+}]$. Taken together, the data provide a rational framework of mechanisms in intact cardiac muscle underlying Frank–Starling's law of the heart. In the following, we will show that the same mechanism can explain the initiation of arrhythmias in the nonuniform myocardium.

Experimental Non-uniform ECC

By creating nonuniformity in ECC by exposing a small segment of the muscle to caffeine, BDM, or low $[Ca^{2+}]_o$, we have expected that (1) low $[Ca^{2+}]_{jet}$ would reduce Ca^{2+} currents and thereby SR–Ca^{2+} content[3]; (2) caffeine would open SR–Ca^{2+} release channels and thereby deplete the SR[1,3,56,57]; and (3) BDM would modestly affect Ca^{2+} transport[2,58] and inhibit XB cycling.[29] Consistent with these expectations, the amplitude of stimulated Ca^{2+} transients, which reflects the SR–Ca^{2+} load,[3] decreased dramatically in regions exposed to caffeine and low $[Ca^{2+}]_{jet}$, but only slightly to BDM (FIG. 4). The effects of BDM are emphasized here because the drug has apparently only reduced force development—at this concentration—in the jet-exposed region significantly and left the Ca^{2+} transients relatively unaffected.

Each of these perturbations reduced muscle force due to creation of a muscle segment which developed less twitch force than the normal cells remote from the jet, as is witnessed by the stretch of the weakened sarcomeres in the jet by the fully activated sarcomeres outside the exposed region. These regions were connected mechanically by a BZ of one to two cells, where the sarcomeres first contracted, and then were stretched. The diffraction pattern of sarcomeres in the BZ showed a clear single peak during both shortening and lengthening, strongly suggesting that sarcomere contraction in the BZ was also partially suppressed, probably because of diffusion of the contents of each jet solution.

Nonuniform ECC and the Ca^{2+} Surge

The common effect of the three protocols was to suppress contraction and reduce sarcomere force. The observation that the effect of force was common to all three interventions, while the effect on $[Ca^{2+}]_i$ and on the Ca^{2+}-transient was dramatically different between the interventions, makes it reasonable to conclude that Ca^{2+} waves are initiated as a result of nonuniformity of the sarcomere force generation and the resultant sequence of stretch and quick release of sarcomeres in the BZ by contraction of normal cells in the region remote from the jet. A Ca^{2+} surge initiating Ca^{2+} waves took place late during the relaxation phase of the twitch when both force and free Ca^{2+} in the cytosol had decayed by 70 to 80% (FIG. 4D). By this time the SR–Ca^{2+} channels have partially recovered[59] and were able to support CICR and Ca^{2+} wave generation.[3] However, the delay between the start of the stimulus and the Ca^{2+} surge makes it highly unlikely that Ca^{2+} entry via L-type Ca^{2+} channels causes CICR from the SR to be involved in the initial Ca^{2+} surge. Furthermore, Ca^{2+} waves never started in jet-exposed regions, where sarcomeres were maximally stretched even if the amplitude of the stimulated Ca^{2+} transients witnessed a robust SR Ca^{2+} load.[3] Ca^{2+} waves never started simultaneously with the peak of the stretch (FIG. 4), making it unlikely that a stretch-related mechanism such as activation of Gd^{3+}-sensitive stretch-activated channels[6,42,43,60,61] is involved in the initial Ca^{2+} surge. Finally, Ca^{2+} waves started from the BZ with reduced SR Ca^{2+} release—(CF and low $[Ca^{2+}]_{jet}$ and/or reduced $[Ca^{2+}]_i$ low $[Ca^{2+}]_{jet}$, and BDM)—suggests that another mechanism of wave initiation, different from "spontaneous SR Ca^{2+}release," was involved.

Initiation and Propagation of Ca^{2+} Waves

We propose an explanation on the basis of these experiments and the results of model studies. It is likely that quick-release-induced Ca^{2+}dissociation from the myofilaments, demonstrated in uniform cardiac muscle, is applicable to the chain of cells in the nonuniform muscle exposed to the jet. Rapid sarcomere shortening during the force decline occurred both in the jet region and in the BZ, but led only to a Ca^{2+} surge and Ca^{2+} wave initiation in the BZ (FIG. 1C), making it probable that quick-release-induced Ca^{2+} dissociation from Tn-C caused by the decline of force in the shortening BZ sarcomeres led to the local Ca^{2+}surge.[32] The region inside the jet, where ECC was all but abolished, probably contained either little Tn-C–Ca^{2+} (caffeine or low $[Ca^{2+}]_{jet}$) or only few Ca^{2+} activated force-generating XBs (BDM), which would render a quick release of this region unable to generate a Ca^{2+} surge and Ca^{2+} wave . The BZ, on the other hand, could generate a Ca^{2+} surge[14,32,42,45,56,62] that is large enough to induce local CICR and thus a Ca^{2+} wave even if only a fraction of Tn-C[38,63] were occupied with Ca^{2+}.[37,64] The region inside the jet of high $[Ca^{2+}]$ is still able to contract actively, but is clearly weakened by spontaneous chaotic Ca^{2+} release, witnessed by spontaneous sarcomere activity,[65] but has a robust Ca^{2+} content of the SR and elevated diastolic $[Ca^{2+}]$. Consistent with the above criteria for generation of the Ca^{2+} surge, we observed that Ca^{2+} waves started inside this region.

The Role of Feedback between XB Force and Ca^{2+} Dissociation from Tn-C

FIGURE 2 shows the fundamental property of XB–Tn-C interaction that underlies the model: feedback of F to Ca^{2+} dissociation from $Ca^{2+} \cdot$Tn-C through the strain κ of the actin filament and subsequent deformation of Tn-C on the filament predicts fundamental properties known for cardiac muscle. In view of the realistic reproduction of the steady state F–pCa relationship,[40] is it not surprising that the feedback predicts realistic σ–SL relationships at varied activation levels for $[Ca^{2+}]_i$. The prediction by the model that the σ–SL exhibits little sensitivity to shortening, as has been observed,[43] needs to be tested in order to ascertain its merit in explaining both the ESPVR described by Suga and Sagawa and Starling's law of the heart.

Twitches generated by the model are similar to twitches generated by rat cardiac trabeculae at 25°C. The increased time of Ca^{2+} binding to Tn-C as a result of the feedback of force development explains that progressive prolongation of the twitch has indeed been shown for cardiac muscle twitches at varied length, while SL was held constant during the twitch.[43] This property was reproduced accurately by introducing a close relationship between the actin strain κ and a decrease of the dissociation rate of Ca from $Ca^{2+} \cdot$Tn-C. This observation holds both in model and experiments irrespective of whether the increased force is caused by varied SL, or $[Ca^{2+}]_o$, or varied pH.[54] Conversely, any force decrease is expected to cause accelerated dissociation of Ca^{2+} from $Ca^{2+} \cdot$Tn-C and to a decrease to subsequent force. This phenomenon is the well known as the deactivation response to shortening and is realistically replicated by the model (FIG. 6B). The degree of force reduction during shortening deactivation depends on the redistribution of Ca^{2+} over ligands and Ca^{2+} extrusion.

The simulations also reproduced our observation,[42] and the recent observation by Julian's group, that the $[Ca^{2+}]_i$ transient during the twitch exhibits a prominent plateau at higher force levels.[55] These investigators explained their observation on the basis of cooperativity between force development and affinity of Tn-C for Ca^{2+} ions, but we are not aware of previous quantitative modeling explaining this observation. The plateau is obvious in the $[Ca^{2+}]_i$ simulation of FIGURE 7 (left part of panel) as well, and is completely lost in the absence of the feedback (FIG. 7). The observation was made that the simulated $[Ca^{2+}]_i$ transient does not last longer than the fluorescence transient in the two aforementioned studies.[42,55] The predicted fluorescent transients in our simulations (data not shown) were similar to those measured experimentally; the difference in time course was explained by the limited dissociation rate of Ca^{2+} from fura-2 (80–130 $\mu M^{-1}\,s^{-1}$).[42,55,66] The combination of these predictions is no proof for the postulated σ-$Ca^{2+} \cdot$Tn-C feedback, but makes it a useful working model for the cardiac muscle and should stimulate further studies of the dynamics of the interaction between Tn-C and Ca^{2+}.

Reverse Excitation–Contraction Coupling (RECC)

FIGURE 6 shows the possible importance of this mechanism of the response of F and $[Ca^{2+}]_i$ to a quick reduction of force in that the model predicts large $[Ca^{2+}]_i$ transients upon a quick release on account of Ca^{2+} dissociation from the $Ca^{2+} \cdot$Tn-C. This phenomenon is well known as the deactivation response upon a quick release of cardiac muscle and is realistically replicated by the model (FIG. 6). The resultant cytosolic $[Ca^{2+}]$ transient is well above the $[Ca^{2+}]_i$ level that is required to induce propagation of CICR.[37] The effect of a quick release in nonuniform muscle is illustrated in FIGURE 7, where strong segments have stretched a weak segment. This process reduces the force generated by the strong segments, but both enhances and prolongs force development by the weak segment and permits a balance of forces. Early relaxation by the shortened strong segments will now impose a quick release of force on the weak segments, which are still activated as a result of their enhanced force development during the tug of war earlier during the twitch. We postulate that it is this quick release that induces a $[Ca^{2+}]_i$ surge in the nonuniform muscle and causes RECC, which is responsible for Ca^{2+} waves, TPCs, delayed afterdepolarizations (DADs), and arrhythmias as illustrated in FIGURE 7.

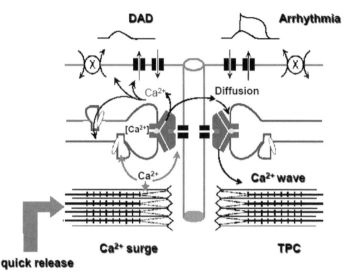

FIGURE 7. Reverse excitation–contraction coupling The figure shows the postulated events in non-uniform muscle during triggering of the TPC. Nonuniform muscle contains weak segments (symbolized by sarcomere on the left) which are stretched by strong segments (sarcomere on the right) during the twitch. (*left side*) Quick release of the weak sarcomeres leads to dissociation of Ca^{2+} from the contractile filaments during the relaxation phase, which induces SR-Ca^{2+} release. The SR is enough recovered to respond to the increase in $[Ca^{2+}]_i$ by Ca^{2+} induced Ca^{2+} release. The resultant elevation of $[Ca^{2+}]_i$ causes diffusion of Ca^{2+} to adjacent sarcomeres. The *right side* shows that arrival of diffusing Ca^{2+} after SR-Ca^{2+} release in the weak region leads to Ca^{2+} release by the SR in the adjacent sarcomeres. Ca^{2+} diffuses again the next sarcomere causing a propagating contraction as well as an arrhythmogenic delayed afterdepolarization (DAD) due to electrogenic Na^+/Ca^{2+} exchange and activation of Ca^{2+} sensitive non selective channels in the sarcolemma. Diffusion of Ca^{2+} along its gradient maintains the propagation of the Ca^{2+} wave.

Implications: Nonuniform ECC, Ca^{2+} Waves, and Arrhythmias

Our simulations, which assume that XB force retards Ca^{2+} dissociation from the $Ca^{2+} \bullet Tn$-C complex on actin, predict a large $[Ca^{2+}]_i$ transient upon a quick release on account of Ca^{2+} dissociation from the $Ca^{2+} \bullet Tn$-C. The Ca^{2+} transient is well above the $[Ca^{2+}]_i$ level that is required to induce propagating CICR.[37] It should be emphasized here that no other feedback mechanism is able to explain the Ca^{2+} surge observed in these studies. We postulate that these $[Ca^{2+}]_i$ transients occur in nonuniform muscle and cause RECC (FIG. 7), which is responsible for Ca^{2+} waves. Such Ca^{2+} waves are known to cause TPCs and DADs[6,30,67–69] and cause arrhythmias[70] (FIG. 7). Whether this mechanism contributes to arrhythmias in the diseased heart, where nonuniform segmental wall motion[71,72] may result from ischemia, nonuniform electrical activation, or nonuniform adrenergic activation[71,73] remains to be proven, although the arrangement of the cardiac wall in muscle fascicles, which transmit force longitudinally and therefore are subject to comparable constraints as the trabeculae in this study, makes this possibility highly likely.

It has been shown that whenever Ca^{2+} waves occur in muscles rendered nonuniform by damage, DADs accompany them with a duration that correlates exactly with the time during which the TPCs travel through the trabeculae. The amplitude of the DADs also correlates exactly with the amplitude of the TPCs.[30] This tight correspondence between TPCs and DADs suggests that the DAD is elicited by a Ca^{2+}-dependent current, as proposed by Kass and Tsien.[74] Hence, if the $[Ca^{2+}]$ transient is large enough, it is expected to lead to action potential generation. Clearly, nonuniformity due to damage initiates a chain of subcellular events leading to arrhythmogenic oscillations of $[Ca^{2+}]_i$ (FIG. 5B).

Importantly, our model of RECC also predicts that the initial Ca^{2+} surge from Tn-C will lead more readily to arrhythmias in hearts with abnormal SR Ca^{2+} storage[75–77] and/or with increased open probability of the SR Ca^{2+}-release channels[78] because of either mutation of the channel gene[69,71,75] or post-translational changes of the channel in heart failure.[76,79]

Acknowledgments

These studies were supported by Grants HL-58860 and HL-66140 from the National Heart, Lung and Blood Institute of the NIH, by the Canadian Institutes for Health Research, and by a NATO Collaborative Research Grant. Professor Henk ter Keurs is a Medical Scientist of the Alberta Heritage Foundation for Medical Research. We thank Drs. B.U. Suzuki and M. Obayashi for their discussions and for comments on the manuscript.

Conflict of Interest

The authors declare no conflicts of interest.

References

1. KONISHI, M., S. KURIHARA & T. SAKAI. 1984. The effects of caffeine on tension development and intracellular calcium transients in rat ventricular muscle. J. Physiol. **355:** 605–618.

2. HERMANN, C., M. HOUADJETO, F. TRAVERS, *et al.* 1992. Early step of the Mg-ATPase of relax myofibrils: a comparison with Ca-activated myofibril and myosin subfragment 1. Biochemistry **31:** 8036–8042.

3. BERS, D.M. 2001. Excitation-Contraction Coupling and Cardiac Contractile Force, 2nd ed. Kluwer Academic Publishers. Dordrecht, the Netherlands.

4. WANNENBURG, T., S.P. SCHULMAN & D. BURKHOFF. 1992. End systolic pressure-volume and MVO2-pressure-volume area relations of isolated rat hearts. Am. J. Physiol. **262:** H1287–H1293.

5. SELLIN, L.C. & J.J. MCARDLE. 1994. Multiple effects of 2,3-butanedione monoxime. Pharmacol. Toxicol. **74:** 305–313.

6. WAKAYAMA, Y., Y. SUGAI, Y. KAGAYA, *et al.* 2001. Stretch and quick release of cardiac trabeculae accelerates Ca2+ waves and triggered propagated contractions. Am. J. Physiol. Circ. Physiol. **281:** H2133–2142.

7. SAGAWA, K., L. MAUGHAN, H. SUGA, *et al.* 1988. Cardiac Contraction and the Pressure-Volume Relationship. Oxford University Press. New York; Oxford.

8. JEWELL, B.R. 1977. A reexamination of the influence of muscle length on myocardial performance. Circ. Res. **40:** 221–230.

9. CHILIAN, W.M. & G.J. GROSS. 1999. Prologue: ischemic preconditioning in cardiac vascular muscle. Am. J. Physiol. **277:** H2416–2417.

10. HIBBERD, M.G. & B.R. JEWELL. 1982. Calcium- and length-dependent force production in rat ventricular muscle. J. Physiol. **329:** 527–540.

11. KENTISH, J.C., H.E.D.J. TER KEURS, L. RICCIARDI, *et al.* 1986. Comparison between the sarcomere length-force relations of intact and skinned trabeculae from rat right ventricle. Circ. Res. **58:** 755–768.

12. ALLEN, D.G. & S. KURIHARA. 1982. The effects of muscle length on intracellular calcium transients in mammalian cardiac muscle. J. Physiol. **327:** 79–94.

13. RODRIGUEZ, E.K., W.C. HUNTER, M.J. ROYCE, *et al.* 1992. A method to reconstruct myocardial sarcomere lengths and orientations at transmural sites in beating canine hearts. Am. J. Physiol. Heart Circ. Physiol. **263:** H293–H306.

14. ALLEN, D.G. & S. KURIHARA. 1982. The effects of muscle length on intracellular calcium transients in mammalian cardiac muscle. J. Physiol. **327:** 79–94.

15. TER KEURS, H.E.D.J., J.J.J. BUCX, P.P. DE TOMBE, *et al.* 1988. The effects of sarcomere length and Ca++ on force and velocity of shortening in cardiac muscle. *In* Molecular Mechanism of Muscle Contraction. H. Sugi & G.H. Pollack, Eds.: 581–593. Plenum Press Publishing. New York.

16. BABU, A., E.H. SONNENBLICK & J. GULATI. 1988. Molecular basis for the influence of muscle length on myocardial performance. Science **240:** 74–76.

17. MCDONALD, K.S., L.J. FIELDS, M.S. PARMACEK, *et al.* 1995. Length dependence of Ca^{2+} sensitivity of tension in mouse cardiac myocytes expressing skeletal troponin C. J. Physiol. **483:** 131–139.

18. WANG, Y.-P. & F. FUCHS. 1994. Length, force, and Ca^{2+}-troponin C affinity in cardiac and slow skeletal muscle. Am. J. Physiol. **266:** C1077–C1082.

19. BRANDT, P.W., M.S. DIAMOND & J.S. RUTCHIK. 1987. Co-operative interactions between troponin-tropomyosin units extend the length of the thin filament in skeletal muscle. J. Mol. Biol. **195:** 885–896.

20. WANG, Y.-P. & F. FUCHS. 1995. Osmotic compression of skinned cardiac and skeletal muscle bundles: effects of force generation, Ca^{2+} sensitivity and Ca^{2+} binding. J. Mol. Cell Cardiol. **27:** 1235–1244.

21. FUCHS, F. & Y.-P. WANG. 1996. Sarcomere length versus interfilament spacing as determinants of cardiac myofilament Ca^{2+} sensitivity and Ca^{2+} binding. J. Mol. Cell. Cardiol. **28:** 1375–1383.

22. LANDESBERG, A. & S. SIDEMAN. 1999. Regulation of energy consumption in cardiac muscle: analysis of isometric contractions. Am. J. Physiol. **276:** H998–H1011.

23. KONHILAS, J.P., T.C. IRVING & P.P. DE TOMBE. 2002. Myofilament calcium sensitivity in skinned rat cardiac trabeculae: role of interfilament spacing. Circ. Res. **90:** 59–65.

24. TER KEURS, H.E.D.J., E.H. HOLLANDER & M.H.C. TER KEURS. 2000. The effects of sarcomere length on the force-cytosolic Ca^{2+} relationship in intact rat trabeculae. *In* Skeletal Muscle Mechanics. W. Herzog, Ed.: 53–70. John Wiley and Sons. Toronto.

25. KENTISH, J.C., H.E.D.J. TER KEURS & D.G. ALLEN. 1988. The contribution of myofibrillar properties to the sarcomere length-force relationship of cardiac muscle. *In* Starling's Law of the Heart Revisited. H.E.D.J. ter Keurs & M.I.M. Noble, Eds.: 5–18. Kluwar Academic Publishers. New York.

26. VAN HEUNINGEN, R., W.H. RIJNSBURGER & H.E.D.J. TER KEURS. 1982. Sarcomere length control in striated muscle. Am. J. Physiol. **242:** H411–H420.

27. WAKAYAMA, Y., M. MIURA, B.D. STUYVERS, *et al.* 2005. Spatial nonuniformity of excitation-contraction coupling

causes arrhythmogenic Ca^{2+} waves in rat cardiac muscle. Circ. Res. **96**: 1266–1273.

28. TER KEURS, H.E., Y. WAKAYAMA, M. MIURA, *et al.* 2006. Arrhythmogenic Ca(2+) release from cardiac myofilaments. Prog. Biophys. Mol. Biol. **90**: 151–171.

29. BACKX, P.H., W.D. GAO, M.D. AZAN-BACKX, *et al.* 1994. Mechanism of force inhibition by 2,3-butanedione monoxime in rat cardiac muscle: roles of [Ca2+]i and cross-bridge kinetics. J. Physiol. **476**: 487–500.

30. DANIELS, M.C.G., D. FEDIDA, C. LAMONT, *et al.* 1991. Role of the sarcolemma in triggered propagated contractions in rat cardiac trabeculae. Circ. Res. **68**: 1408–1421.

31. CHALOVICH, J.M. & E. EISENBERG. 1982. Inhibition of actomyosin ATPase activity by troponin-tropomyosin without blocking the binding of myosin to actin. J. Biol. Chem. **257**: 2431–2437.

32. ALLEN, D.G. & J.C. KENTISH. 1988. Calcium concentration in the myoplasm of skinned ferret ventricular muscle following changes in muscle length. J. Physiol. **407**: 489–503.

33. GRABAREK, Z., R.Y. TAN, J. WANG, *et al.* 1990. Inhibition of mutant troponin C activity by an intra-domain disulphide bond. Nature **345**: 132–135.

34. DAVIDOFF, A.W., P.A. BOYDEN, K. SCHWARTZ, *et al.* 2004. Congestive heart failure after myocardial infarction in the rat: cardiac force and spontaneous sarcomere activity. Ann. N.Y. Acad. Sci. **1015**: 84–95.

35. MICHAILOVA, A., F. DELPRINCIPE, M. EGGER, *et al.* 2002. Spatiotemporal features of Ca^{2+} buffering and diffusion in atrial cardiac myocytes with inhibited sarcoplasmic reticulum. Biophys. J. **83**: 3134–3151.

36. STUYVERS, B.D., M. MIURA, & H.E. TER KEURS. 1997. Dynamics of viscoelastic properties of rat cardiac sarcomeres during the diastolic interval: involvement of Ca^{2+}. J. Physiol. **502**: 661–677.

37. STUYVERS, B.D., W. DUN, S. MATKOVICH, *et al.* 2005. Ca^{2+} sparks and waves in canine purkinje cells: a triple layered system of Ca^{2+} activation. Circ. Res. **97**: 35–43.

38. FABIATO, A. 1983. Calcium-induced release of calcium from the cardiac sarcoplasmic reticulum. Am. J. Physiol. **245**: C1–C14.

39. ISHIDE, N., M. MIURA, M. SAKURAI, *et al.* 1992. Initiation and development of calcium waves in rat myocytes. Am. J. Physiol. **263**: H327–H332.

40. KENTISH, J.C., H.E.D.J. TER KEURS, L. RICCIARDI, *et al.* 1986. Comparison between the sarcomere length-force relations of intact and skinned trabeculae from rat right ventricle. Circ. Res. **58**: 755–768.

41. GRANZIER, H.L. & T.C. IRVING. 1995. Passive tension in cardiac muscle: contribution of collagen, titin, microtubules, and intermediate filaments. Biophys. J. **68**: 1027–1044.

42. BACKX, P.H. & H.E.D.J. TER KEURS. 1993. Fluorescent properties of rat cardiac trabeculae microinjected with fura-2 salt. Am. J. Physiol. **264**: H1098–H1110.

43. TER KEURS, H.E.D.J., W.H. RIJNSBURGER, R. VAN HEUNINGEN, *et al.* 1980. Tension development and sarcomere length in rat cardiac trabeculae: evidence of length-dependent activation. Circ. Res. **46**: 703–714.

44. BACKX, P.H.M. 1989. Force Sarcomere Relation in Cardiac Myocardium. The University of Calgary: Ph.D. Thesis, 1989.

45. HOUSMANS, P.R., N.K.M. LEE & J.R. BLINKS. 1983. Active shortening retards the decline of the intracellular calcium transient in mammalian heart muscle. Science **221**: 159–161.

46. ROBINSON, T.F. & S. WINEGRAD. 1981. A variety of intercellular connections in heart muscle. J. Mol. Cell. Cardiol. **13**: 185–195.

47. HANLEY, P.J., A.A. YOUNG, I.J. LEGRICE, *et al.* 1999. 3-Dimensional configuration of perimysial collagen fibres in rat cardiac muscle at resting and extended sarcomere lengths. J. Physiol. **517**: 831–837.

48. TER KEURS, H.E., Y. WAKAYAMA, M. MIURA, *et al.* 2006. Arrhythmogenic Ca(2+) release from cardiac myofilaments. Prog. Biophys. Mol. Biol. **90**: 151–171.

49. MIURA, M., P.A. BOYDEN & H.E.D.J. TER KEURS. 1998. Ca^{2+} waves during triggered propagated contractions in intact trabeculae. Am. J. Physiol. **274**: H266–H276.

50. MIURA, M., P.A. BOYDEN & H.E. TER KEURS. 1999. Ca^{2+} waves during triggered propagated contractions in intact trabeculae: determinants of the velocity of propagation. Circ. Res. **84**: 1459–1468.

51. WOLEDGE, R.C., N.A. CURTIN & E. HOMSHER. 1985. Basic Facts and Ideas. Energetic Aspects of Muscle Contraction. pp. 1–26. London: Academic Press Inc.

52. WOLEDGE, R.C., N.A. CURTIN & E. HOMSHER. 1985. Energetics of Aspects of Muscle Contraction, pp. 85–117. London: Academic Press Inc.

53. LANDESBERG, A. & S. SIDEMAN. 1994. Coupling calcium binding to troponin-C and cross-bridge cycling in skinned cardiac cells. Am. J. Physiol. **266**: H1260–H1271.

54. BUCX, J.J.J. 1995. Ischemia of the heart: a study of sarcomere dynamics and cellular metabolism. Ph.D. thesis. Eramus University Rotterdam, the Netherlands.

55. JIANG, Y., M.F. PATTERSON, D.L. MORGAN, *et al.* 1998. Basis for late rise in fura 2 R signal reporting $[Ca^{2+}]_i$ during relaxation in intact rat ventricular trabeculae. Am. J. Physiol. **274**: C1273–C1282.

56. KURIHARA, S. & K. KOMUKAI. 1995. Tension-dependent changes of the intracellular Ca^{2+} transients in ferret ventricular muscles. J. Physiol. **489**: 617–625.

57. SITSAPESAN, R., R.A.P. MONTGOMERY, K.T. MACLEOD, *et al.* 1990. Increased open probability of sheep cardiac sarcoplasmic reticulum calcium-release channels induced by low temperatures. J. Physiol. **426**: 21 P.

58. BACKX, P.H.M., W.D. GAO, M.D. AZAN-BACKX, *et al.* 1994. Mechanism of force inhibition by 2,3-butanedione monoxime in rat cardiac muscle: roles of $[Ca^{++}]_i$ and cross-bridge kinetics. J. Physiol. **476**: 487–500.

59. BANIJAMALI, H., W.D. GAO, B. MACINTOSH, *et al.* 1998. Force-interval relations of twitches and cold contractures in rat cardiac trabeculae: effect of ryanodine. Circ. Res. **69**: 937–948.

60. ZHANG, Y., M. MIURA & H.E.D.J. TER KEURS. 1996. Triggered propagated contractions in rat cardiac trabeculae; inhibition by octanol and heptanol. Circ. Res. **79**: 1077–1085.

61. ZHANG, Y. & H.E.D.J. TER KEURS. 1996. Effects of gadolinium on twitch force and triggered propagated contractions in rat cardiac trabeculae. Cardiovasc. Res. **32**: 180–188.

62. LAB, M.J., D.G. ALLEN & C.H. ORCHARD. 1984. The effects of shortening on myoplasmic calcium concentration and on the action potential in mammalian ventricular muscle. Circ. Res. **55:** 825–829.

63. BACKX, P.H., P.P. DE TOMBE, J.H. VAN DEEN, *et al.* 1989. A model of propagating calcium-induced calcium release mediated by calcium diffusion. J. Gen. Physiol. **93:** 963–977.

64. TER KEURS, H.E.D.J., Y.M. ZHANG & M. MIURA. 1998. Damage induced arrhythmias: reversal of excitation-contraction coupling. Cardiovasc. Res. **40:** 444–455.

65. LAKATTA, E.G. & D.L. LAPPE. 1981. Diastolic scattered light fluctuation, resting force and twitch force in mammalian cardiac muscle. J. Physiol. **315:** 369–394.

66. KAO, J.P. & R.Y. TSIEN. 1988. Ca^{2+} binding kinetics of fura-2 and azo-1 from temperature-jump relaxation measurements. Biophys. J. **53:** 635–639.

67. LAKATTA, E.G. 1992. Functional implications of spontaneous sarcoplasmic reticulum Ca^{2+} release in the heart. Cardiovasc. Res. **26:** 193–214.

68. MIURA, M., N. ISHIDE, H. ODA, *et al.* 1993. Spatial features of calcium transients during early and delayed afterdepolarizations. Am J Physiol. **265:** H439–H444.

69. SCHLOTTHAUER, K. & D.M. BERS. 2000. Sarcoplasmic reticulum $Ca(2+)$ release causes myocyte depolarization. Underlying mechanism and threshold for triggered action potentials. Circ. Res. **87:** 774–780.

70. DANIELS, M.C., D. FEDIDA, C. LAMONT, *et al.* 1991. Role of the sarcolemma in triggered propagated contractions in rat cardiac trabeculae. Circ. Res. **68:** 1408–1421.

71. SIOGAS, K., S. PAPPAS, G. GRAEKAS, *et al.* 1998. Segmental wall motion abnormalities alter vulnerability to ventricular ectopic beats associated with acute increases in aortic pressure in patients with underlying coronary artery disease. Heart **79:** 268–273.

72. YOUNG, A.A., S. DOKOS, K.A. POWELL, *et al.* 2001. Regional heterogeneity of function in nonischemic dilated cardiomyopathy. Cardiovasc. Res. **49:** 308–318.

73. JIANG, Y. & F.J. JULIAN. 1997. Pacing rate, halothane, and BDM affect fura 2 reporting of $[Ca^{2+}]_i$ in intact rat trabeculae. Am. J. Physiol. **273:** C2046–C2056.

74. KASS, R.S., W.J. LEDERER, R.W. TSIEN, *et al.* 1978. Role of calcium ions in transient inward currents and aftercontractions induced by strophanthidin in cardiac purkinje fibres. J. Physiol. **281:** 187–208.

75. LAITINEN, P.J., K.M. BROWN, K. PIIPPO, *et al.* 2001. Mutations of the cardiac ryanodine receptor (RyR2) gene in familial polymorphic ventricular tachycardia. Circulation **103:** 485–490.

76. MARKS, A.R. 2001. Ryanodine receptors/calcium release channels in heart failure and sudden cardiac death. J. Mol. Cell Cardiol. **33:** 615–624.

77. POSTMA, A.V., I. DENJOY, T.M. JOORNTJE, *et al.* 2002. Absence of calsequesterin 2 causes severe forms of catecholaminergic polymorphic ventricular tachycardia. Circ. Res. **91:** e21–e26.

78. BOYDEN, P.A. & H.E.D.J. TER KEURS. 2001. Reverse excitation contract coupling: Ca^{2+} ions as initiators of arrhythmias. J. Cardiovasc. Electrophysiol. **12:** 382–385.

79. TER KEURS, H.E.D.J. & P.A. BOYDEN. 2007. Calcium and arrhythmogenesis. Physiol. Rev. **87:** 457–506.

Modulation of Cardiac Performance by Motor Protein Gene Transfer

Todd J. Herron,[a] Eric J. Devaney,[b] and Joseph M. Metzger[a]

[a] Department of Molecular and Integrative Physiology,
and
[b] Department of Surgery, Division of Pediatric Cardiovascular Surgery,
University of Michigan Medical School, Ann Arbor, Michigan 48109, USA

Cardiac muscle performance can be determined by factors intrinsic to each cardiac muscle cell, such as protein isoform expression. One protein whose expression plays a major role in determining cardiac performance is myosin. Myosin is the heart's molecular motor which transduces the chemical energy from ATP hydrolysis into the mechanical energy of each heartbeat. Alterations of myosin isoform expression are routinely associated with acquired and inherited cases of cardiomyopathy. For example, human heart failure is consistently associated with increased expression of a slow myosin motor isoform and a concomitant decreased expression of the heart's fast myosin motor isoform. Further, mutations of the cardiac myosin gene are the most common cause of inherited hypertrophic cardiomyopathy. Transgenic animal studies have provided insight into cardiac functional effects caused by myosin isoform gene switching (fast-to-slow myosin or slow-to-fast myosin) or by expression of a disease-related mutant motor. More direct structure–function analysis using acute gene transfer of myosin motors provides evidence that the inotropic state of cardiac muscle can be affected by motor protein isoform shifting independent of intracellular calcium handling. Because most therapies for the diseased heart target intracellular calcium handling, acute gene transfer of cardiac molecular motors to modulate heart performance offers a novel therapeutic strategy for the compromised heart. Although the development of safe vectors for therapeutic myosin gene delivery are in their infancy, studies focused on acute genetic engineering of the heart's molecular motor will provide a foundation for therapeutic vector development and insight into mechanisms that contribute to cardiomyopathy.

Key words: myosin; molecular motor; acute genetic engineering; cardiac performance

Introduction

Oxygen and nutrient-rich blood is pumped to the body's tissues with each heartbeat. The force and pressure development of each heartbeat are due to the contraction and shortening of cardiac muscle. At the molecular level cardiac muscle contraction involves complex dynamic interactions between calcium and the myofilament proteins. Cardiac muscle force generation and shortening are initiated by calcium released from intracellular stores of the sarcoplasmic reticulum (SR) and executed by the activity of molecular motors within the myofilaments. A popular perspective is that the level of force (or pressure) generation of cardiac muscle is determined primarily by the amount of calcium released from the SR and delivered to the myofilaments. However, mounting evidence suggests that cardiac muscle performance can also be altered independent of the calcium signal by the activity of the heart's molecular motor. In fact, a recent review article[1] has proposed a dominant role for molecular motors in the intrinsic regulation of cardiac performance.

Heart failure is a major health problem, affecting over 5 million people in the United States, and the numbers continue to grow. In cases of heart failure, cardiac performance is diminished and the pumping action of the heart is poor. As a result, the heart is unable to supply a sufficient amount of oxygenated and nutrient-rich blood to the tissues and organs that demand it. Common symptoms of heart failure include dyspnea, exercise intolerance, and edema. Current therapies manage the symptoms of heart failure

Address for correspondence: Prof. Joseph M. Metzger, Ph.D., Department of Molecular and Integrative Physiology, University of Michigan, Ann Arbor, MI 48109. Voice: +734 647-6460; fax: +734 647-6461.

metzgerj@umich.edu

and surgical intervention or cardiac transplantation is ultimately necessary. The development of novel therapies for heart failure is clearly necessary, and some investigators have predicted a shift in emphasis toward the development of gene-based therapy.[2] One closely related pair of genes consistently associated with heart failure is the one that encodes the heart's molecular motors.[3–5]

The importance of the heart's molecular motor, myosin, in the regulation of cardiac performance has been underscored in the past two decades. Since the first discovery that a myosin gene missense mutation can cause inherited hypertrophic cardiomyopathy (HCM)[6] nearly 200 other disease-associated myosin mutations have been identified. Further, myosin missense mutations have also been implicated in other inherited cardiovascular diseases, including dilated cardiomyopathy (DCM),[7,8] and in skeletal muscle myopathies.[9] Since the cardiac phenotype of inherited cardiomyopathies can be similar to the phenotype of acquired heart failure, it is likely that alterations of myosin function contribute to the pathogenesis of more common forms of heart failure. The clinical importance of cardiac myosin makes it a good target for gene transfer experiments using viral vectors to study the molecular mechanisms that contribute to heart failure and for the development of novel gene-based therapies.

Molecular Motors: Cardiac Myosin Isoforms

Cardiac myosin belongs to the myosin II family of conventional myosins and is the molecular motor that drives myocardial contraction. Cardiac myosin is the most abundant protein within cardiac muscle and is the primary consumer of cellular energy, in the form of ATP. Muscular contraction is due primarily to cyclical interactions between interdigitating myosin and actin filaments. Myosin is an asymmetric molecule made up of two myosin heavy chains (MyHC) and two pairs of myosin light chains. The crystal structure of myosin has been solved[10] and at its N terminus MyHC has a globular head domain which contains both an actin-binding site and an ATP binding and hydrolysis site (FIG. 1A). At its C terminus MyHC contains a rod domain made up of an α-helical tail that interacts with the α-helical tail of neighboring MyHC molecules to form the bipolar thick-filament backbone. Thus, in addition to hydrolyzing ATP as the heart's molecular motor, the MyHC also contributes to the structure of cardiac muscle. Two structurally and functionally distinct iso-

forms of myosin are differentially expressed in cardiac muscle, namely, α- and β-MyHC isoforms (FIG. 1A). These isoforms can form hetero- and homodimers and these dimers have been separated electrophoretically and are referred to as myosin V_1 (α-α homodimer), V_2 (α-β heterodimer), and V_3 (β-β homodimer).[11,12] The α- and β-myosin genes, *MYH6* and *MYH7*, respectively, extend over 51 kb on chromosome 14 in humans[13] and chromosome 11 in mouse separated by 4 kb of sequence between them.[13] Transcription of each gene is independently controlled, but coordinately regulated,[14] and changes throughout mammalian development. The V_3 (β-β homodimer) myosin isoform is predominantly expressed during early development in the ventricles of all mammals and remains the predominant isoform throughout the lifetime of humans and larger mammals. The V_1 (α-α homodimer) isoform, on the other hand, predominates soon after birth in most small mammals and is the predominant isoform throughout the lifetime of mice and rats.[15]

Despite sharing >90% amino acid homology, each molecular motor isoform functions in a distinct manner. For example, biochemical experiments have demonstrated that β-MyHC hydrolyzes ATP ~3–7 times slower than α-MyHC,[16,17] the "fast" motor in the heart. Using the *in vitro* motility assay, where actin propulsion by single myosin motors is measured, it has been demonstrated that α-MyHC propels actin filaments 2–3 times faster than β-MyHC.[17,18] These biochemical functional differences between the isoforms do not manifest as differences in force-generating capacity between the two isoforms. Using the laser trap assay, where the amount of isometric force produced by a single myosin molecule is measured, it has been demonstrated that α- and β-MyHC generate identical amounts of force, displace actin by the same extent, and only differ in their kinetics of ATP hydrolysis product release.[19] Although differences in peak twitch tension are not apparent, the energetic economy and tension-time integral for preparations containing predominantly V1 myosin are less than that of preparations with predominantly V3 myosin.[20,21] These kinetic differences in the myosin molecular motor isoforms manifest functionally as differences in the kinetics of force and power development in the context of single cardiac myocytes,[22] cardiac muscle strips, and whole heart function.[23] FIGURE 1B, for example, demonstrates that the power output of single cardiac myocytes[24] is much lower in cardiac myocytes that exclusively express the β-MyHC isoform compared to cardiac myocytes that only express the faster α-MyHC molecular motor isoform. Because maximum force production is not different between the

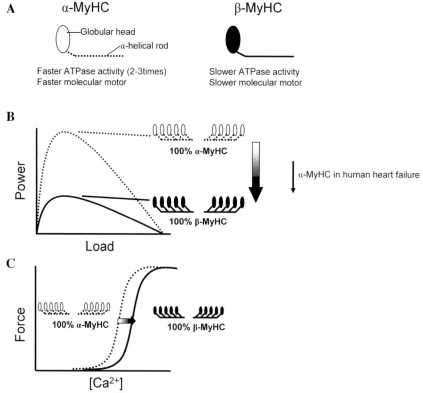

FIGURE 1. (A) Schematic representation of the cardiac MyHC molecule. Cardiac myosin is made up of a globular head domain and an α-helical rod domain. Two distinct myosin isoforms are expressed in the heart: α- and β-MyHC. **(B)** Cardiac myosin isoforms determine the power output of single cardiac muscle cells (reproduced from Herron *et al.* [24] with permission). **(C)** Myofilament calcium sensitivity is reduced in cardiac myocytes that express predominantly β-MyHC.

isoforms, loaded shortening velocities (myocyte shortening against a clamped load as the heart does with each stroke) are faster in cardiac myocytes that express the faster α-MyHC molecular motor. It has also been reported that cardiac myofilament responsiveness to calcium is depressed and the force–calcium relationship is shifted to the right in cardiac myocytes expressing the slower β-MyHC isoform[25] (FIG. 1C). These effects of myosin isoform switching are also apparent at the level of the whole working heart[23,26] (FIG. 2). Collectively, the literature supports a role for myosin isoforms as an important determinant of cardiac performance.

Myosin Isoform Expression and Cardiovascular Disease

Expression of the slow β-MyHC motor increases relative to α-MyHC in rodent models of cardiovascular disease, including diabetes,[27] hypothyroidism,[22,25]

cardiac hypertrophy,[28] and in aging.[29] In these instances the total amount of myosin remains unchanged, but the relative expression of each isoform changes. In rat models of cardiovascular disease, for example, the rat cardiac ventricle transforms and expression of the α-MyHC isoform is replaced by expression of the slower β-MyHC motor.[29,30] Since this switch of myosin isoform expression occurs in a plethora of models of heart disease, increased expression of the β-MyHC motor expression is generally accepted as a molecular biomarker of cardiac disease in rodents.

Alterations of cardiac MyHC isoform expression may play a role in the pathogenesis of human heart failure as well. The "normal" human cardiac ventricles express predominantly β-MyHC (85–90% of total myosin) and a small amount of the faster α-MyHC (10–15% of total myosin) motor.[4] The failing human heart is consistently associated with a loss of α-MyHC and exclusive expression of the slow β-MyHC motor. Patients undergoing successful treatment of heart failure with pharmacologic or surgical interventions have

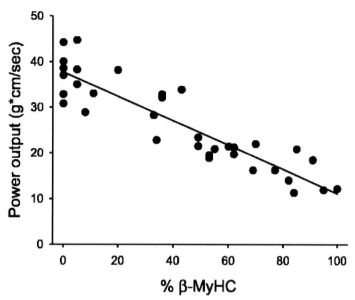

FIGURE 2. Power output of the whole working heart is reduced as the relative expression of β-MyHC increases. (From Korte *et al.*[26] Reproduced by permission).

been shown to develop upregulation of levels of α-MyHC.[31–33] These data suggest that levels of α-MyHC serve at the least as a surrogate marker for cardiac function in heart failure, but this hypothesis has been difficult to test directly. In animal models, even a small increase in expression of α-MyHC results in an increase in power output by the heart.[34] In a rabbit model of induced cardiomyopathy, overexpression of an α-MyHC transgene was found to be protective.[35] Also, in another transgenic model, expression of β-MyHC, with concomitant downregulation of α-MyHC, was found to have a detrimental effect on mice that were placed under cardiovascular stress.[36] Taken together, the evidence suggests that acute upregulation of α-MyHC expression may lead to improved cardiac function, while increases in β-MyHC expression may be detrimental to cardiac performance. Study of the direct effects of increased fast α-MyHC motor expression will require the use of acute gene transfer of MYH6 (the gene that encodes α-MyHC) to failing cardiac myocytes *in vitro* and to failing hearts *in vivo*.

Strategies that have been employed to investigate the functional significance of myosin isoform switching have inherent limitations. The most commonly used model has been the rodent hypothyroid animal model. Thyroid hormone is a potent inducer of α-MyHC expression and in the absence of thyroid hormone, by thyroidectomy or propyl-thio-uracil (PTU) treatment, there is a well documented reduction of α-MyHC expression accompanied by an increase of

β-MyHC expression.[14,37] Studies using hypothyroid rodents have provided evidence that increased β-MyHC expression depresses contractile function of single cardiac myocytes[24,25,34,38] and in the whole heart.[23,39] However, a key limitation of the hypothyroid model is that in concert with alterations in the MyHC isoform profile, the expression of key calcium handling proteins, including the SR Ca^{2+}-ATPase (SERCA2a), are also affected.[40,41] Thus, interpretation of functional data is confounded because SERCA2a activity itself is well known to markedly affect cardiac performance.[42–44] To address this confounding issue, which is attributed to altered calcium handling, the permeabilized (skinned) myocyte preparation, a preparation that eliminates any contribution of calcium-handling proteins (like SERCA2a) to myocyte functional studies, has been enlisted.[24,25] These studies have demonstrated that mechanical properties of single myocytes (e.g., power output and unloaded shortening velocity) are attenuated by increased relative expression of β-MyHC. While these studies provided important information about the contribution of myosin isoforms to myofilament function, they do not address the role of myosin isoform switching in the more physiologically relevant setting of the electrically stimulated intact cardiac muscle cell where excitation–contraction coupling mechanisms are intact and fully functional.

Chronic transgenic animal models have also been used to manipulate MyHC isoform expression in the mammalian heart *in vivo*.[35,36,45] Under baseline conditions, the near full replacement of α-MyHC with β-MyHC in the hearts of transgenic mice had no detected effects on echocardiography-derived shortening fraction, an *in vivo* measure of systolic function.[36] Isolated permeabilized muscle preparations from these mice, however, did show adverse effects on contractile performance at baseline. It is possible that chronic forced genetic transition from α-MyHC to β-MyHC throughout the development of these animals may have caused other adaptations in the mouse heart to compensate for effects of β-MyHC expression *in vivo*. This is conceivable, considering that these transgenic mice would have sustained nearly 150 million contractile cycles *in vivo* prior to cardiac performance assessment in adult mice. In transgenic animal models, therefore, it may be difficult to distinguish primary effects caused by forced β-MyHC expression from secondary compensatory changes that may occur throughout the development and lifetime of a transgenic animal.[46] The utilization and development of acute gene transfer of myosin molecular motors circumvents many of the complication(s) posed by secondary compensatory alterations of protein expression that may have altered or masked a direct effect of MyHC isoform switching on cardiac performance.

Myosin Isoform Switching: Acute Gene Transfer

We recently used acute genetic engineering to study the direct effect of increased relative β-MyHC expression on single membrane intact cardiac myocyte function.[47] Experimentally we designed and generated a recombinant adenovirus, AdMYH7, to express the full-length human MYH7 (full-length β-MyHC molecule) gene in α-MyHC-dominant rat myocytes *in vitro*. The efficiency of gene transfer was ∼100% and stoichiometric replacement of the endogenous α-MyHC motor with the full length human β-MyHC motor was shown by Western blot analysis. Proper sarcomeric incorporation of virally directed β-MyHC expression was detected in the A-band of the sarcomere by using indirect immunofluorescence and high-resolution confocal imaging. Functional effects of β-MyHC gene transfer were determined in electrically paced myocytes by measuring changes in sarcomere spacing and intracellular calcium transients using fura-2. β-MyHC gene transfer attenuated myocyte contractility just 1 day after gene transfer (FIG. 3), when β-MyHC accounted for just 18% of the total myosin (FIG. 3). This functional effect was transgene dose-dependent as the effect on contraction amplitude was greater on day 2, when β-MyHC accounted for ∼40% of the total myosin. The amplitude of the electrically stimulated intracellular calcium transient, however, was not affected by β-MyHC expression. This study demonstrated for the first time that increased expression of the heart's slow β-MyHC molecular motor can have calcium-independent negative inotropic effects on single cardiac myocyte contractility. Thus, in addition to being a common molecular marker for heart failure, increased β-MyHC expression can directly contribute to cardiac dysfunction. Acute gene transfer of molecular motors in cardiac muscle offers a novel means of calcium independent inotropy.

Cardiac Myosin and Inherited Cardiomyopathies

Inherited cardiomyopathy can be caused by mutations of sarcomere proteins. The one most commonly affected sarcomeric protein in familial cardiomyopathy is β-MyHC. Mutations of the MYH7 gene, the gene that encodes β-MyHC, have been associated with both HCM and DCM. Although the human heart expresses predominantly β-MyHC, mutations of MYH6, the gene that encodes α-MyHC, have recently been reported to also cause either HCM or DCM.[8] The clinical importance of MYH7 mutations is underscored by the finding that MYH7 mutations have also been linked to skeletal muscle myopathies, including Laing distal myopathy,[9] myosin storage myopathy,[48] and hyaline body myopathy.[49] In fact, a report from the American Heart Association has suggested that for cardiovascular genetic counseling, genetic screening should start with examination of the MYH7 gene.[50]

The first HCM-causing myosin mutation was discovered in the motor domain of the β-MyHC molecule.[6] This missense mutation results in an arginine-to-glutamine substitution at position 403 (R403Q), which is located in the actin-binding interface of the N-terminal globular head.[51] This mutation results in the loss of a positively charged amino acid residue in the actin-binding site. All affected members of this large kindred were heterozygous for the R403Q mutation, and unaffected family members did not carry this mutation. The R403Q mutation is associated with early onset of symptoms and a high mortality rate, with an average life span of 30–35 years.[52,53] Since the discovery of this myosin mutation, much effort has been devoted to determining how it affects

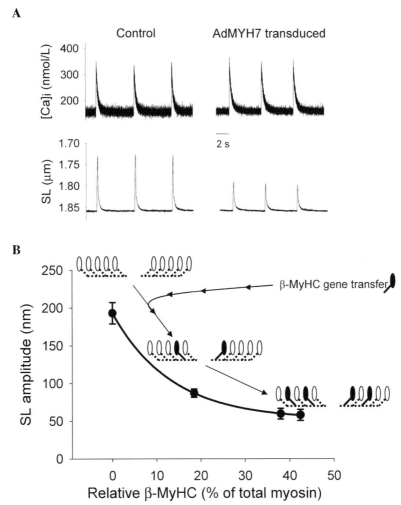

FIGURE 3. (A) Gene transfer of the slow β-MyHC motor directly attenuates cardiac myocyte contractility independent of the intracellular calcium transient.[47] **(B)** Increasing relative β-MyHC expression by acute gene transfer attenuates contractility in a dose-dependent manner. SL = sarcomere length.

myosin motor function and triggers cardiomyopathy, but the issue remains unresolved as some studies have indicated that motor function is impaired,[54–57] while others suggest a gain of function for the myosin motor.[58–61] The links between mutated myosin (R403Q) function and cardiac hypertrophy are unclear, though some studies have suggested a mechanism involving altered calcium homeostasis.[62–64] The role of calcium in the pathogenesis initiated by R403Q myosin is unclear as it has been reported that transgenic animals expressing R403Q have elevated calcium levels,[63] while other reports suggest depressed calcium levels in heart cells from these animals.[62,64] One limitation of these elegant studies is that the R403Q mutation was studied in the context of the mouse α-MyHC molecule

(MYH6 gene), while the disease-causing mutation in humans occurs in the context of the MYH7 gene (β-MyHC molecule). As outlined above, α- and β-MyHC are structurally and functionally distinct, thus making it conceivable that a point mutation in one molecule may not affect the other in the same way. A transgenic rabbit, whose myosin isoform expression profile is similar to that of humans, has been developed which expresses the R403Q mutant myosin in the context of the human MYH7 gene.[65] This model recapitulates much of the phenotype observed in humans (e.g., fibrosis, cardiac hypertrophy, whole-organ dysfunction), but the acute effect of this mutant myosin on cardiac performance (single-cell contractility and calcium homeostasis) has not been studied in this model.

Acute Gene Transfer of HCM Mutant Myosin

Two studies from different laboratories have used acute gene transfer technology to study the effect of the HCM disease-causing R403Q mutant myosin on cardiac tissue. In 1995, Marian *et al.*[66] used a replication-deficient recombinant adenoviral construct to deliver wild-type and mutated (R403Q) human β-MyHC to feline cardiac myocyte *in vitro*. The advantage of this approach is that the feline species is known to develop HCM with a phenotype identical to that in humans and that the investigators generated this mutant myosin in the context of the *bona fide* disease-causing gene (MYH7). The efficiency of *in vitro* gene transfer was reportedly high (>95%), and the mutant myosin caused sarcomere disruption 120 hours after gene transfer. While this study indicated that disruption of sarcomere assembly and myofibrillar organization due to mutant β-MyHC is a primary defect in HCM, it did not assess the acute functional effect of this mutant myosin on cardiac performance (i.e., single-cell contractility or intracellular calcium homeostasis.).

In 2003, Wang *et al.*[67] investigated the functional impact of mutant myosin on cardiac myofibril organization and myosin motor activity. Using embryonic chicken myosin fused to a GFP reporter, Wang *et al.* found that expression of the R403Q mutant myosin accelerated actin filament sliding, but decreased myosin's affinity for actin in an *in vitro* motility assay. This is consistent with previous reports that suggest augmentation of myosin function due to the R403Q mutation.[58–61] Insight from this study, however, is limited since the myosin used was tagged with GFP, a fluorescent molecule that can have direct effects on actomyosin interactions.[68] Further, this study performed acute gene transfer of mutant myosin in chicken embryonic cardiac myocytes and in a mouse myogenic cell line (C2C12) and not in adult mammalian cardiac myocytes. The direct effects of mutant myosin on adult cardiac myocyte contractility and calcium homeostasis using acute gene transfer technology are largely unexplored. Determination of the acute effects of mutant myosins on adult myocyte function will provide important insight into molecular triggers that initiate and contribute to cardiac disease.

Summary

Cardiac molecular motors are important regulators of cardiac performance. Cardiac muscle contraction is initiated by calcium binding to and activating the thin filament, thus promoting molecular motor (myosin) activity and force generation. While the amount of force generation can be graded with calcium, myocyte contractility can also be altered by the motor protein isoform expression profile. Experimentally, acute gene transfer of the slow β-MyHC molecular motor provides a means of calcium-independent negative inotropy in single cardiac myocytes. Acute gene transfer of the fast α-MyHC molecular motor may provide a means of calcium-independent positive inotropy, though this remains to be tested in the laboratory. While acute gene transfer of molecular motors is in its infancy, future research aimed at using myosin gene transfer as a therapeutic agent for the failing heart will require development of safe and effective vectors for delivery of the myosin gene (~6 kb) to the heart *in vivo*. The study of acute functional effects of disease-causing myosin mutations that trigger cardiomyopathy will provide mechanistic insight into the molecular mechanisms that initiate heart disease.

Conflict of Interest

The authors declare no conflicts of interest.

References

1. Hinken, A.C. & R.J. Solaro. 2007. A dominant role of cardiac molecular motors in the intrinsic regulation of ventricular ejection and relaxation. Physiology **22:** 73–80.

2. Mitka, M. 2006. Do lackluster trial findings mean new avenues are needed for heart research? JAMA **295:** 611–612.

3. Lowes, B.D., W. Minobe, W.T. Abraham, *et al.* 1997. Changes in gene expression in the intact human heart: downregulation of alpha-myosin heavy chain in hypertrophied, failing ventricular myocardium. J. Clin. Invest. **100:** 2315–2324.

4. Miyata, S., W. Minobe, M.R. Bristow & L.A. Leinwand. 2000. Myosin heavy chain isoform expression in the failing and nonfailing human heart. Circ. Res. **86:** 386–390.

5. Nakao, K., W. Minobe, R. Roden, *et al.* 1997. Myosin heavy chain gene expression in human heart failure. J. Clin. Invest. **100:** 2362–2370.

6. Geisterfer-Lowrance, A.A., S. Kass, G. Tanigawa, *et al.* 1990. A molecular basis for familial hypertrophic cardiomyopathy: a beta cardiac myosin heavy chain gene missense mutation. Cell **62:** 999–1006.

7. Fatkin, D. & R.M. Graham. 2002. Molecular mechanisms of inherited cardiomyopathies. Physiol. Rev. **82:** 945–980.

8. Carniel, E., M.R.G. Taylor, G. Sinagra, *et al.* 2005. α-Myosin heavy chain: a sarcomeric gene associated with dilated and hypertrophic phenotypes of cardiomyopathy. Circulation **112:** 54–59.

9. Meredith, C., R. Herrmann, C. Parry, *et al.* 2004. Mutations in the slow skeletal muscle fiber myosin heavy chain

gene (MYH7) cause Laing early-onset distal myopathy (MPD1). Am. J. Hum. Genet. **75:** 703–708.

10. RAYMENT, I., W.R. RYPNIEWSKI, K. SCHMIDT-BASE, *et al.* 1993. Three-dimensional structure of myosin subfragment-1: a molecular motor. Science **261:** 50–58.

11. HOH, J.F.Y., G.P.S. YEOH, M.A.W. THOMAS, *et al.* 1979. Structural differences in the heavy chains of rat ventricular myosin isoenzymes. FEBS Lett. **97:** 330–334.

12. POPE, B., J.F.Y. HOH & A. WEEDS. 1980. The ATPase activities of rat cardiac myosin isoenzymes. FEBS Lett. **118:** 205–208.

13. SAEZ, L.J., K.M. GIANOLA, E.M. MCNALLY, *et al.* 1987. Human cardiac myosin heavy chain genes and their linkage in the genome. Nucleic Acids Res. **15:** 5443–5459.

14. HADDAD, F., P.W. BODELL, A.X. QIN, *et al.* 2003. Role of antisense RNA in coordinating cardiac myosin heavy chain gene switching. J. Biol. Chem. **278:** 37132–37138.

15. LOMPRE, A.M., B. NADAL-GINARD & V. MAHDAVI. 1984. Expression of cardiac alpha and beta-myosin heavy chain genes is developmentally and hormonally regulated. J. Biol. Chem. **259:** 6437–6446.

16. HARRIS, D.E., S.S. WORK, R.K. WRIGHT, *et al.* 1994. Smooth, cardiac, and skeletal muscle myosin force and motion generation assessed by cross-bridge mechanical interactions in vitro. J. Muscle Res. Cell Motil. **15:** 11–19.

17. VAN BUREN, P., D.E. HARRIS, N.R. ALPERT. 1995. Cardiac V1 and V3 myosins differ in their hydrolytic and mechanical activities *in vitro.* Circ. Res. **77:** 439–444.

18. MALMQVIST, U.P., A. ARONSHTAM & S. LOWEY. 2004. Cardiac myosin isoforms from different species have unique enzymatic and mechanical properties. Biochemistry **43:** 15058–15065.

19. PALMITER, K.A., M.J. TYSKA, D.E. DUPUIS, *et al.* 1999. Kinetic differences at the single molecule level account for the functional diversity of rabbit cardiac myosin isoforms. J. Physiol. (Lond.) **519:** 669–678.

20. HASENFUSS, G., L.A. MULIERI, E.M. BLANCHARD, *et al.* 1991. Energetics of isometric force development in control and volume- overload human myocardium: comparison with animal species. Circ. Res. **68:** 836–846.

21. HOLUBARSCH, C., R.P. GOULETTE, R.Z. LITTEN, *et al.* 1985. The economy of isometric force development, myosin isoenzyme pattern and myofibrillar ATPase activity in normal and hypothyroid rat myocardium. Circ. Res. **56:** 78–86.

22. RUNDELL, V.L.M., V. MANAVES, A.F. MARTIN, *et al.* 2005. Impact of β-myosin heavy chain isoform expression on cross-bridge cycling kinetics. Am. J. Physiol. Heart Circ. Physiol. **288:** H896–H903.

23. TANG, Y.D., J.A. KUZMAN, S. SAID, et al. 2005. Low thyroid function leads to cardiac atrophy with chamber dilatation, impaired myocardial blood flow, loss of arterioles, and severe systolic dysfunction. Circulation **112:** 3122–3130.

24. HERRON, T.J., F.S. KORTE & K.S. MCDONALD. 2001. Loaded shortening and power output in cardiac myocytes are dependent on myosin heavy chain isoform expression. Am. J. Physiol. Heart Circ. Physiol. **281:** H1217–H1222.

25. METZGER, J.M., P.A. WAHR, D.E. MICHELE, *et al.* 1999. Effects of myosin heavy chain isoform switching on Ca^{2+}-activated tension development in single adult cardiac myocytes. Circ. Res. **84:** 1310–1317.

26. KORTE, F.S., T.J. HERRON, M.J. ROVETTO, *et al.* 2005. Power output is linearly related to MyHC content in rat skinned myocytes and isolated working hearts. Am. J. Physiol. Heart. Circ. Physiol. **289:** H801–H812.

27. RUNDELL, V.L.M., D.L. GEENEN, P.M. BUTTRICK, *et al.* 2004. Depressed cardiac tension cost in experimental diabetes is due to altered myosin heavy chain isoform expression. Am. J. Physiol. Heart Circ. Physiol. **287:** H408–H413.

28. MERCADIER, J.J., A.M. LOMPRE, C. WISNEWSKY, *et al.* 1981. Myosin isoenzymic changes in several models of rat cardiac hypertrophy. Circ. Res. **49:** 525–532.

29. FITZSIMONS, D.P., J.R. PATEL & R.L. MOSS. 1999. Aging-dependent depression in the kinetics of force development in rat skinned myocardium. Am. J. Physiol. Heart Circ. Physiol. **276:** H1511–H1519.

30. MASAKI, H., T. IMAIZUMI, S. ANDO, *et al.* 1993. Production of chronic congestive heart failure by rapid ventricular pacing in the rabbit. Cardiovasc. Res. **27:** 828–831.

31. WANG, J., X. GUO & N.S. DHALLA. 2004. Modification of myosin protein and gene expression in failing hearts due to myocardial infarction by enalapril or losartan. Biochim. Biophys. Acta **1690:** 177–184.

32. LOWES, B.D., E.M. GILBERT, W.T. ABRAHAM, *et al.* 2002. Myocardial gene expression in dilated cardiomyopathy treated with beta-blocking agents. N. Engl. J. Med. **346:** 1357–1365.

33. RASTOGI, S., S. MISHRA, R. GUPTA, *et al.* 2005. Reversal of maladaptive gene program in left ventricular myocardium of dogs with heart failure following long-term therapy with the Acorn cardiac support device. Heart Failure Rev. **10:** 157–163.

34. HERRON, T.J. & K.S. MCDONALD. 2002. Small amounts of α-myosin heavy chain isoform expression significantly increase power output of rat cardiac myocyte fragments. Circ. Res. **90:** 1150–1152.

35. JAMES, J., L. MARTIN, M. KRENZ, *et al.* 2005. Forced expression of α-myosin heavy chain in the rabbit ventricle results in cardioprotection under cardiomyopathic conditions. Circulation **111:** 2339–2346.

36. KRENZ, M. & J. ROBBINS. 2004. Impact of beta-myosin heavy chain expression on cardiac function during stress. J. Am. Coll. Cardiol. **44:** 2390–2397.

37. IZUMO, S., B. NADAL-GINARD & V. MAHDAVI. 1986. All members of the MHC multigene family respond to thyroid hormone in a highly tissue-specific manner. Science **231:** 597–600.

38. FITZSIMONS, D.P., J.R. PATEL & R.L. MOSS. 1998. Role of myosin heavy chain composition in kinetics of force development and relaxation in rat myocardium. J. Physiol. (Lond.) **513:** 171–183.

39. KORTE, F.S., T.J. HERRON, M.J. ROVETTO, *et al.* 2005. Power output is linearly related to myosin heavy chain content in rat skinned myocytes and isolated working hearts. Am. J. Physiol. Heart Circ. Physiol. **289:** H801–H812.

40. SAYEN, M.R., D.K. ROHRER & W.H. DILLMANN. 1992. Thyroid hormone response of slow and fast sarcoplasmic reticulum Ca^{2+} ATPase mRNA in striated muscle. Mol. Cell. Endocrinol. **87:** 87–93.

41. CARR, A.N, & E.G. KRANIAS. 2002. Thyroid hormone regulation of calcium cycling proteins. Thyroid **12**: 453–457.

42. ARAI, M., N.R. ALPERT, D.H. MacLENNAN, *et al.* 1993. Alterations in sarcoplasmic reticulum gene expression in human heart failure: a possible mechanism for alterations in systolic and diastolic properties of the failing myocardium. Circ. Res. **72**: 463–469.

43. HASENFUSS, G., H. REINECKE, R. STUDER, *et al.* 1994. Relation between myocardial function and expression of sarcoplasmic reticulum Ca(2+)-ATPase in failing and nonfailing human myocardium. Circ. Res. **75**: 434–442.

44. TEUCHER, N., J. PRESTLE, T. SEIDLER, *et al.* 2004. Excessive sarcoplasmic/endoplasmic reticulum Ca^{2+}-ATPase expression causes increased sarcoplasmic reticulum Ca^{2+} uptake but decreases myocyte shortening. Circulation **110**: 3553–3559.

45. TARDIFF, J.C., T.E. HEWETT, S.M. FACTOR, *et al.* 2000. Expression of the beta (slow)-isoform of MHC in the adult mouse heart causes dominant-negative functional effects. Am. J. Physiol. Heart Circ. Physiol. **278**: H412–H419.

46. MICHELE, D.E. & J.M. METZGER, 2000. Contractile dysfunction in hypertrophic cardiomyopathy: elucidating primary defects of mutant contractile proteins by gene transfer. Trends Cardiovasc. Med. **10**: 177–182.

47. HERRON, T.J., R. VAN DEN BOOM, E. FOMICHEVA, *et al.* 2007. Calcium-independent negative inotropy by β-myosin heavy chain gene transfer in cardiac myocytes. Circ. Res. **100**: 1182–1190.

48. TAJSHARGHI, H., L.E. THORNELL, C. LINDBERG, *et al.* 2004. Myosin storage myopathy associated with a heterozygous missense mutation in MYH7. Ann. Neurol. **54**: 494–500.

49. BOHLEGA, S., S.N. BU-AMERO, S.M. WAKIL, *et al.* 2004. Mutation of the slow myosin heavy chain rod domain underlies hyaline body myopathy. Neurology **62**: 1518–1521.

50. RICHARD, P., P. CHARRON, L. CARRIER, *et al.* 2003. Hypertrophic cardiomyopathy: distribution of disease genes, spectrum of mutations, and implications for a molecular diagnosis strategy. Circulation **107**: 2227–2232.

51. RAYMENT, I., H.M. HOLDEN, J.R. SELLERS, *et al.* 1995. Structural interpretation of the mutations in the β-cardiac myosin that have been implicated in familial hypertrophic cardiomyopathy. Proc. Natl. Acad. Sci. USA **92**: 3864–3868.

52. FANANAPAZIR, L., M.C. DALAKAS, F. CYRAN, *et al.* 1993. Missense mutations in the β-myosin heavy-chain gene cause central core disease in hypertrophic cardiomyopathy. Proc. Natl. Acad. Sci. USA **90**: 3993–3997.

53. WATKINS, H., A. ROSENZWEIG, D.S. HWANNG, *et al.* 1992. Characteristics and prognostic implications of myosin missense mutations in familial hypertrophic cardiomyopathy. N. Engl. J. Med. **326**: 1108–1114.

54. BLANCHARD, E., C. SEIDMAN, J.G. SEIDMAN, *et al.* 1999. Altered crossbridge kinetics in the α MHC403/+ mouse model of familial hypertrophic cardiomyopathy. Circ. Res. **84**: 475–483.

55. CUDA, G., L. FANANAPAZIR, W.S. ZHU, *et al.* 1993. Skeletal muscle expression and abnormal of beta-myosin in hypertrophic cardiomyopathy. J. Clin. Invest. **91**: 2861–2865.

56. CUDA, G., L. FANANAPAZIR, N.D. EPSTEIN, *et al.* 1997. The *in vitro* motility activity of beta-cardiac myosin depends on the nature of the beta myosin heavy chain gene mutation in hypertrophic cardiomyopathy. J. Muscle Res. Cell Motil. **18**: 275–283.

57. SWEENEY, H.L., A.J. STRACESKI, L.A. LEINWAND, *et al.* 1994. Heterologous expression of a cardiomyopathic myosin that is defective in its actin interaction. J. Biol. Chem. **269**: 1603–1605.

58. PALMER, B.M., D.E. FISHBAUGHER, J.P. SCHMITT, *et al.* 2004. Differential cross-bridge kinetics of FHC myosin mutations R403Q and R453C in heterozygous mouse myocardium. Am. J. Physiol. Heart Circ. Physiol. **287**: H91–H99.

59. TYSKA, M.J., E. HAYES, M. GIEWAT, *et al.* 2000. Single-molecule mechanics of R403Q cardiac myosin isolated from the mouse model of familial hypertrophic cardiomyopathy. Circ. Res. **86**: 737–744.

60. YAMASHITA, H., M.J. TYSKA, D.M. WARSHAW, *et al.* 2000. Functional consequences of mutations in the smooth muscle myosin heavy chain at sites implicated in familial hypertrophic cardiomyopathy. J. Biol. Chem. **275**: 28045–28052.

61. DEBOLD, E.P., J.P. SCHMITT, J.R. MOORE, *et al.* 2007. Hypertrophic and dilated cardiomyopathy mutations differentially affect the molecular force generation of mouse α-cardiac myosin in the laser trap assay. Am. J. Physiol. Heart Circ. Physiol. 00128.

62. FATKIN, D., B.K. McCONNELL, J.O. MUDD, *et al.* 2000. An abnormal Ca^{2+} response in mutant sarcomere protein-mediated familial hypertrophic cardiomyopathy. J. Clin. Invest. **106**: 1351–1359.

63. GAO, W.D., N.G. PEREZ, C.E. SEIDMAN, *et al.* 1999. Altered cardiac excitation–contraction coupling in mutant mice with familial hypertrophic cardiomyopathy. J. Clin. Invest. **103**: 661–666.

64. SEMSARIAN, C., I. AHMAD, M. GIEWAT, *et al.* 2002. The L-type calcium channel inhibitor diltiazem prevents cardiomyopathy in a mouse model. J. Clin. Invest. **109**: 1013–1020.

65. MARIAN, A.J., Y. WU, D.S. LIM, *et al.* 1999. A transgenic rabbit model for human hypertrophic cardiomyopathy. J. Clin. Invest. **104**: 1683–1692.

66. MARIAN, A.J., Q.T. YU, D.L. MANN, *et al.* 1995. Expression of a mutation causing hypertrophic cardiomyopathy disrupts sarcomere assembly in adult feline cardiac myocytes. Circ. Res. **77**: 98–106.

67. WANG, Q., C.L. MONCMAN & D.A. WINKELMANN. 2003. Mutations in the motor domain modulate myosin activity and myofibril organization. J. Cell Sci. **116**: 4227–4238.

68. AGBULUT, O., A. HUET, N. NIEDERLANDER, *et al.* 2007. Green fluorescent protein impairs actin-myosin interactions by binding to the actin-binding site of myosin. J. Biol. Chem. **282**: 10465–10471.

Atrial-Selective Sodium Channel Block as a Strategy for Suppression of Atrial Fibrillation

ALEXANDER BURASHNIKOV,[a] JOSÉ M. DI DIEGO,[a] ANDREW C. ZYGMUNT,[a] LUIZ BELARDINELLI,[b] AND Charles ANTZELEVITCH[a]

[a]Masonic Medical Research Laboratory, Utica, New York, USA

[b]CV Therapeutics Inc., Palo Alto, California, USA

Antiarrhythmic drug therapy remains the principal approach for suppression of atrial fibrillation (AF) and flutter (AFl) and prevention of their recurrence. Among the current strategies for suppression of AF/AFl is the development of antiarrhythmic agents that preferentially affect atrial, rather than ventricular electrical parameters. Inhibition of the ultrarapid delayed rectifier potassium current (I_{Kur}), present in the atria, but not in the ventricles, is an example of an atrial-selective approach. Our recent study examined the hypothesis that sodium channel characteristics differ between atrial and ventricular cells and that atrial-selective sodium channel block is another effective strategy for the management of AF. We have demonstrated very significant differences in the inactivation characteristics of atrial versus ventricular sodium channels and a striking atrial selectivity for the action of ranolazine, an inactivated-state sodium channel blocker, to produce use-dependent block of the sodium channels, leading to depression of excitability, development of post-repolarization refractoriness (PRR), and suppression of AF. Lidocaine and chronic amiodarone, both predominantly inactivated-state sodium channel blockers, also produced a preferential depression of sodium channel–dependent parameters (V_{Max} conduction velocity, diastolic threshold of excitation, and PRR) in the atria. Propafenone, a predominantly open-state sodium channel blocker, produced similar changes of electrophysiological parameters, which were was not atrial-selective. The ability of ranolazine, chronic amiodarone, and propafenone to prolong the atrial action potential potentiated their ability to suppress AF in coronary-perfused canine atrial preparations. In conclusion: Our data demonstrate important differences in the inactivation characteristics of atrial versus ventricular sodium channels and a striking atrial selectivity for the action of agents like ranolazine to produce use-dependent block of sodium channels leading to suppression of AF. Our findings suggest that atrial-selective sodium channel block may be a valuable strategy to combat AF.

Key words: electrophysiology; cardiac arrhythmias; lidocaine; amiodarone; ranolazine; propafenone

Introduction

Antiarrhythmic drug therapy remains the principal approach for suppression of atrial fibrillation (AF) and flutter (AFl) and their recurrences. Among the current strategies for suppression of AF/AFl is the development of antiarrhythmic agents that preferentially affect atrial, rather than ventricular electrical parameters. Inhibition of the ultrarapid delayed rectifier potassium current (I_{Kur}), present in atria but not in ventricles, is an example of an atrial-selective approach.[1] We recently examined the hypothesis that sodium channel characteristics differ between atrial and ventricular cells and that atrial-selective sodium channel block is another effective strategy for the management of AF.[2,3] Biophysical characteristics of sodium channels were measured in single myocytes isolated from canine atria and ventricles. Four agents capable of blocking cardiac sodium channels (ranolazine, lidocaine, propafenone, and chronically administered amiodarone) were compared with regard to their ability to alter the electrophysiology of canine coronary artery–perfused atrial and ventricular preparations as well as their ability to suppress AF. This review contrasts the effects of these open- and closed-state channel blockers.

Address for correspondence: Charles Antzelevitch, Ph.D., Masonic Medical Research Laboratory, 2150 Bleecker Street, Utica, NY 13501. Fax: 315-735-5648.

ca@mmrl.edu

Ann. N.Y. Acad. Sci. 1123: 105–112 (2008). © 2008 New York Academy of Sciences.
doi: 10.1196/annals.1420.012

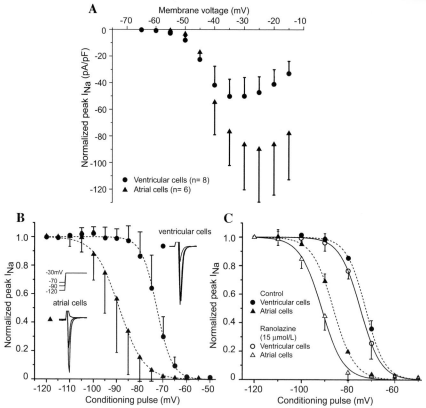

FIGURE 1. Activation and steady-state inactivation in canine atrial versus ventricular myocytes. (**A**) Current–voltage relation in ventricular and atrial myocytes. Voltage of peak I_{Na} is more positive and current density is larger in atrial versus ventricular myocytes. (**B**) Summarized steady-state inactivation curves. The half-inactivation voltage ($V_{0.5}$) is -88.80 ± 0.19 mV in atrial cells ($n = 9$) and -72.64 ± 0.14 mV in ventricular cells ($P < 0.001$, n = 7). *Insets* show representative atrial and ventricular traces after 1-s conditioning pulses to the indicated potentials. (**C**) Steady-state inactivation curves before and after addition of 15 μM ranolazine. Ranolazine shifts $V_{0.5}$ from -72.53 ± 0.16 mV to -74.81 ± 0.14 mV ($P < 0.01$) in ventricular myocytes ($n = 4$) and from -86.35 ± 0.19 to -91.38 ± 0.35 mV ($P < 0.001$) in atrial myocytes ($n = 5$). (From Burashnikov *et al.*[2] Reproduced with permission.)

Methods and Results

Sodium Channel Inactivation Characteristics in Isolated Atrial versus Ventricular Myocytes

Whole-cell peak sodium currents were recorded at 37°C in low-sodium external solution from myocytes isolated from the right atrium and left ventricle (LV) of adult mongrel dogs. The half inactivation voltage ($V_{0.5}$) in atrial myocytes was about 15 mV more negative than that recorded in ventricular myocytes, and the differences were increased after exposure to ranolazine (FIG. 1). These data indicate that a greater percentage of atrial versus ventricular sodium channels would be inactivated at a given resting or take-off potential and that inactivated-state sodium channel blockers may, therefore, be more effective in blocking

sodium channels in atria than in ventricles. An intrinsically more positive resting membrane potential (RMP) in atria (-83 mV) versus ventricles (-87 mV) would further reduce the availability of sodium channels in atria and accentuate the atrial selectivity of sodium channel blockers.

Because ranolazine has recently been identified as an inactivated sodium channel state blocker[4] with little effect on peak I_{Na} or I_{Na}-mediated parameters in ventricular myocardium at therapeutic concentrations,[5,6] we hypothesized that this agent may act as an atrial-selective sodium channel blocker. We contrasted the effects of ranolazine with those of other inactivated-state sodium channel blockers, such as lidocaine and amiodarone, as well as an open-state sodium channel blocker, propafenone, in atria and ventricles.

TABLE 1. The effect of ranolazine, lidocaine, propafenone, and chronic amiodarone on sodium channel–dependent parameters in canine isolated coronary artery–perfused atrial and ventricular preparations

		Ranolazine (10 μM) 0.5/0.3 s	Lidocaine (21 μM) 0.5/03 s	Propafenone (1.5 μM) 0.5/0.3 s	Chronic Amiodarone 0.5/0.3 s
V_{max} (% Δ)	Atria	−26/43	−31/40	−46/78	−42/67
	Ventricles	−9/15	−16/23	−40/51	−9/16
DTE (% Δ)	Atria	+18/139	+30/105	+112/172	+109/148
	Ventricles	+3/8	+8/40	+84/125	NA
CV (% Δ)	Atria	−14/46	−29/57	−55/97	+25/56
	Ventricles	−5/11	−12/36	−44/71	+6/21
PRR (ms)	Atria	51/79	71/84	68/94	48/107
	Ventricles	3/7	47/69	52/83	31/36

NOTE: Data recorded at pacing cycle lengths of 0.5 and 0.3 s. V_{max} = maximum rate of rise of the action potential upstroke; DTE = diastolic threshold of excitation; CV = conduction velocity; PRR = post-repolarization refractoriness. PRR was determined as the difference between ERP and APD_{75} in atria and APD_{90} in ventricles. (ERP is coincident with APD_{75} in atria and APD_{90} in ventricles.) CV was approximated from the duration of the P wave complex in atria and QRS complex in ventricles on the pseudo-ECG recordings. $n = 3–18$. (Modified from Burashnikov *et al.*[2,3])

Sodium Channel–Dependent Parameters in Multicellular Atrial and Ventricular Preparations

Experiments were performed using isolated arterially perfused canine right atrial preparations and left ventricular arterially perfused wedge preparations.[5,7,8] Therapeutic plasma concentrations of ranolazine (1–10 μM), lidocaine (2.1–21 μM), and propafenone (0.3–3.0 μM) were examined. Amiodarone was chronically administered at a dose of 40 mg/kg/day for 6 weeks.

Sodium channel–mediated parameters, such as the maximum rate of rise of the AP upstroke (V_{max}), conduction velocity (CV), diastolic threshold of excitation (DTE), and post-repolarization refractorines (PRR) were evaluated. PRR was defined as the difference between action potential duration (APD) and atrial effective refractory period (ERP). ERP normally coincides with $APD_{70–90}$, but may extend well beyond $APD_{70–90}$ or even APD_{100} (causing the appearance of PRR) under conditions associated with a reduction of excitability (ischemia, sodium channel block, etc.[9]).

Ranolazine and propafenone prolong APD_{90} selectively in atria (by 11% and 13%, respectively), with little change of APD_{90} in the ventricles (+ 2% and +3%, respectively; CL = 500 ms). Chronic amiodarone produced a greater prolongation of APD_{90} in atria than in ventricles (22 versus 12%, respectively; CL = 500 ms). In contrast, lidocaine abbreviates APD_{90} in both the atria and ventricles (6% and 9%, respectively; CL = 500 ms). Ranolazine, lidocaine, and chronic amiodarone lengthened ERP selectively (ranolazine) or predominantly (amiodarone and lido-

caine) in atria in a rate-dependent manner, leading to the development of greater PRR in atria versus ventricles In contrast, propafenone induced prominent PRR in both the atria and ventricles (TABLE 1).

Ranolazine and chronic amiodarone caused a much greater rate-dependent reduction in V_{max}, increase in DTE, and slowing of CV in atrial than ventricular preparations (FIG. 2; TABLE 1). Lidocaine also preferentially suppressed these parameters in atria, although to a lesser extent (TABLE 1). Propafenone depressed sodium channel–mediated parameters more potently than ranolazine, lidocaine, or chronic amiodarone, but without a sizable chamber selectivity at normal pacing rates (CL = 500 ms; TABLE 1). At a pacing CL of 300 ms, propafenone produced a potent depression of I_{Na}-mediated parameters in both atria and ventricles, but the effect in atria was more pronounced (TABLE 1). This atrial selectivity of propafenone at rapid activation rates was associated with atrial-selective prolongation of APD_{90}, leading to elimination of diastolic intervals in atria but not in ventricles. Atrial selectivity of these agents to depress I_{Na}-dependent parameters derives in part from the agents' ability to prolong atrial APD and thus leads to more positive take-off potential and elimination of the diastolic interval at rapid rates of activation (FIG. 3), potentiating the actions of the drug to depress I_{Na}.

Antiarrhythmic Effects of Ranolazine, Lidocaine, Propafenone, and Chronic Amiodarone in a Model of AF

Persistent AF is induced in 100% of canine coronary arterially perfused atrial preparations in the presence of acetylcholine (0.5 μM).[8,10] Ranolazine was found to

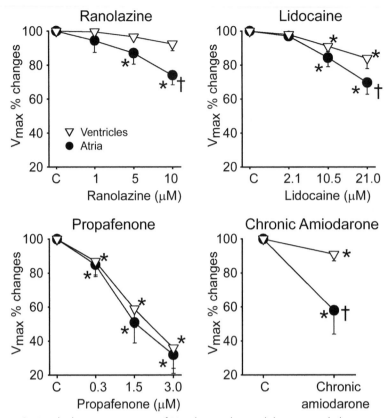

FIGURE 2. Atrial-selective suppression of V_{max} by ranolazine, lidocaine, and chronic amiodarone, but not propafenone in canine coronary artery–perfused atrial and ventricular preparations. $*P < 0.05$ versus respective control (C); $†P < 0.05$ versus respective ventricular values. $n = 8$–15. CL = 500 ms.

be more effective than lidocaine, but less effective than propafenone, in terminating acetylcholine-mediated persistent AF in coronary-perfused atria as well as in preventing the initiation of AF (TABLE 2). Persistent acetylcholine-mediated AF could be induced in only 1 of 6 atria isolated from dogs chronically treated with amiodarone (versus 10 of 10 untreated atria). Anti-AF actions of ranolazine, lidocaine, propafenone, and amiodarone were associated with the development of significant rate-dependent PRR.

Ranolazine (5–10 µM) also prevented the induction of AF in 4 of 5 atria in which self-terminating AF was induced by exposure to ischemia and β-adrenergic agonists.[2,11] Ischemia/reperfusion coupled with iso-proterenol mimics the conditions that prevail during acute myocardial infarction or the substrate encountered postsurgically.

Discussion

Our recent studies demonstrate very significant differences in the inactivation characteristics of atrial versus ventricular sodium channels and a striking atrial selectivity for the action of ranolazine, an inactivated-state sodium channel blocker, to produce use-dependent block of the sodium channels, leading to depression of excitability, development of PRR, and suppression of AF.[2,3] Lidocaine and amiodarone, two other predominantly inactivated-state sodium channel blockers,[12] depressed sodium channel–dependent parameters (V_{max}, CV, DTE, and PRR) predominantly in atria, providing support for the hypothesis that inactivated-state sodium channel blockers are likely to be atrial-selective. It is noteworthy that lidocaine was much less atrial-selective than ranolazine or chronic amiodarone (TABLE 1). Further evidence in support of the hypothesis derives from the demonstration that propafenone, a predominantly open-state sodium channel blocker,[12] is not atrial-selective. Of interest, lidocaine causes tonic block in atrial, but not ventricular preparations, while quinidine and prajmaline (predominantly open-state sodium channel blockers) produce tonic block in both atrial and ventricular preparations.[13] The atrioventricular

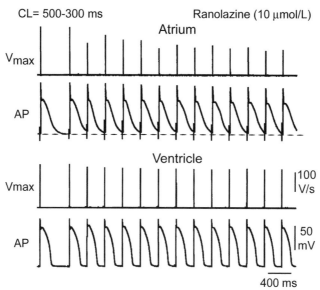

FIGURE 3. Mechanisms contributing to atrial selectivity of ranolazine in depressing V_{max} at fast pacing rates: the role of atrial-selective APD prolongations. Shown are action potential tracings and corresponding V_{max} values recorded during acceleration of pacing rate from a CL of 500 to 300 ms in atrial and ventricular preparations in the presence of ranolazine. Ranolazine prolongs late repolarization in atria, but not in ventricles. Acceleration of rate leads to elimination of the diastolic interval, resulting in a more positive take-off potential in atrium. The diastolic interval remains relatively long in ventricles. (Modified from Burashnikov et al.[2])

TABLE 2. Effectiveness of ranolazine, lidocaine, propafenone, and chronic amiodarone in terminating and preventing induction of acetylcholine-mediated AF in coronary artery–perfused right atrial preparations

	Ranolazine (10.0 μM)	Lidocaine (21.0 μM)	Propafenone (1.5–3.0 μM)	Chronic Amiodarone
Termination of AF	66% (4/6)	33% (2/6)	100% (7/7)	NA
Prevention of induction of AF	80% (8/10)	57% (4/7)	100% (6/6)	83% (5/6)

NOTE: Persistent AF was inducible in 100% of atria under baseline conditions. (Modified from Burashnikov et al.[2,3])

differences in RMP are likely to contribute to the atrial selectivity of inactivated-state sodium channel blockers. When RMP is depolarized by high potassium in ventricular tissues (increasing the fraction of inactivated sodium channels), inactivated-state sodium channel blockers produce a greater reduction of V_{max} than do open-state sodium channel blockers.[14]

In canine ventricular myocytes, ranolazine has been shown to inhibit late I_{Na} with an IC_{50} of 6 μM,[5] but to inhibit peak I_{Na} with an IC_{50} of 294 μM.[4] Consistent with the latter, ranolazine has been reported to suppresses V_{max} with an IC_{50} of >100 μM in ventricular Purkinje fibers and M cell preparations paced at a CL of 500 ms.[5,6] In sharp contrast, ranolazine causes a prominent use-dependent reduction of I_{Na} (estimated based on changes in V_{max}) in atrial preparations at con-

centrations within the therapeutic range of ranolazine (2–10 μM).[2]

Sodium channel blockers generally bind more effectively to open and/or inactivated sodium channels (i.e., during the action potential) than to resting sodium channels (i.e., during the diastolic interval). Unblocking occurs largely during the resting state.[12] Rapid activation rates contribute to the development of sodium channel block by increasing the proportion of time that the sodium channels are in the open/inactivated state and reducing the time that the channels are in the resting state. Agents that prolong APD selectively in atria but not ventricles are expected to display atrial-selective I_{Na} block, particularly at rapid activation rates on account of their ability to reduce or eliminate the diastolic interval and depolarize take-off potential in an atrial-selective manner (FIG. 3). The more depolarized RMP in atria potentiates the effects of I_{Na} blockers by increasing the fraction of channels in the inactivated state, which reduces the availability of sodium channels and prolongs the time needed for the sodium channels to recover from inactivation.

Ranolazine was more atrial-selective than was lidocaine and more effective than lidocaine in terminating and preventing recurrence of AF. This may be due to the fact that ranolazine prolongs only atrial APD because of its ability to also block the rapidly activating delayed rectifier potassium current (I_{Kr}, $IC_{50} = 12\,\mu M$),[5] whereas lidocaine, a more selective I_{Na} blocker, abbreviates both atrial and ventricular APD. It is noteworthy that I_{Kr} blockers preferentially prolong atrial versus ventricular APD (see below). The selective prolongation of APD in atria by ranolazine leads to elimination of diastolic intervals and more depolarized take-off potentials at rapid rates in atria but not ventricles (FIG. 3). The more negative h-curve in atria and acceleration-induced depolarization of take-off potential act in concert to increase the fraction of channels in the inactivated state, making sodium channels less available and more sensitive to block by ranolazine. The result is a greater atrial versus ventricular suppression of I_{Na}-dependent parameters such as V_{max}, DTE, and CV, and the development of use-dependent PRR. The effect of ranolazine to prolong atrial repolarization potentiates but does not appear to be a determining factor in ranolazine's atrial specificity and in antiarrhythmic efficacy. Propafenone (I_{Na} and I_{Kr} blocker), like ranolazine, selectively prolongs atrial APD_{90} but suppresses I_{Na}-dependent parameters in both the atrial and the ventricular preparations to a similar extent at a CL 500 ms,[2] as does GE 68, a propafenone analogue.[15] At faster pacing rates, propafenone more effectively depresses V_{max} and CV in atria on account of atrial-selective APD_{90} prolongation (leading to elimination of diastolic interval in atria). Lidocaine abbreviates both atrial and ventricular APD_{90}, but shows atrial specificity in depression of I_{Na}-dependent parameters. Chronic amiodarone produces depression of I_{Na}-dependent parameters predominantly in atria via a similar mechanism, which includes preferential prolongation of atrial APD.

These results suggest that the I_{Kr} blocking effect of ranolazine, chronic amiodarone, and propafenone potentiates sodium channel inhibitory effect of these drugs in atria at fast pacing rates. Interestingly, I_{Kr} blockers generally produce a much greater APD prolongation in atria than in ventricles. Selective inhibition of I_{Kr} prolongs atrial ERP more than ventricular ERP at normal or moderately rapid activation rates,[16,17] but not at slow rates. At relatively slow activation rates or following long pauses, I_{Kr} block preferentially prolongs ventricular versus atrial APD, leading to development of early afterdepolarization (EAD) and torsade de pointes arrhythmias in the ventricles, but not in atria.[18–20]

A number of antiarrhythmic agents have been shown to be effective in terminating and/or preventing clinical AF/AFl. Most of these agents have as a primary action the ability to reduce I_{Na} (e.g., propafenone or flecainide) and I_{Kr} (e.g., dofetilide) or to inhibit multiple ion channels, as in the case of amiodarone. An important limitation of these antiarrhythmic agents is their potential ventricular proarrhythmic actions and/or organ toxicity at therapeutically effective doses.[6,18,21] This has prompted to the development of atrial-selective antiarrhythmic agents, such as those that block I_{Kur}.[1,27,28] However, block of I_{Kur} alone may not be sufficient for the suppression of AF.[22] In remodeled atria, I_{Kur} block selectively prolongs atrial APD_{90} (but only slightly) and, when combined with I_{to} (perhaps with I_{K-ACh}) and/or I_{Na} inhibition, can suppress AF/AFl.[22–24] In nonremodeled healthy atria, I_{Kur} inhibition abbreviates APD_{90},[7,22,25] and can promote AF.[22] These data are consistent with the results of a recent study showing an association of loss-of-function mutations in KCNA5, which encodes the α-subunit of I_{Kur} channel, with familial AF.[26] Our results suggest that atrial-selective sodium channel block may be another effective approach for the management of AF.

Summary

Recent studies point to major distinctions in the inactivation characteristics of atrial and ventricular cardiac sodium channels. Sodium channel blockers such as ranolazine and chronic amiodarone take advantage of these distinctions and are capable of inhibiting sodium channels more effectively in the atria of the heart. Our findings suggest that inactivated-state sodium channel blockers have a proclivity towards producing atrial-selective sodium channel inhibition. The results point to atrial-selective sodium channel block as a novel strategy for the management of AF and suggest that the additional presence of I_{Kr} block and APD prolongation can potentiate the atrial selectivity of I_{Na} blockers and thus enhance their effectiveness in suppressing and preventing the development of AF.

Acknowledgments

This study was supported by grants from CV Therapeutics (to C.A.), the NIH (HL-47687; to C.A.), and the New York State and Florida Grand Lodges of the Free and Accepted Masons.

Conflict of Interest

The authors declare no conflicts of interest.

References

1. NATTELS, S., C. MATTHEWS, E. DE BLASIO, *et al.* 2000. Dose-dependence of 4-aminopyridine plasma concentrations and electrophysiological effects in dogs: potential relevance to ionic mechanisms *in vivo*. Circulation **101**: 1179–1184.

2. BURASHNIKOV, A., J.M. DI DIEGO, A.C. ZYGMUNT, *et al.* 2007. Atrial-selective sodium channel block as a strategy for suppression of atrial fibrillation. Differences in sodium channel inactivation between atria and ventricles and the role of ranolazine. Circulation **116**: 1449–1457.

3. BURASHNIKOV, A., L. BELARDINELLI & C. ANTZELEVITCH. 2007. Ranolazine and propafenone both suppress atrial fibrillation but ranolazine unlike propafenone does it without prominent effects on ventricular myocardium [abstract]. Heart Rhythm **4**: S163.

4. UNDROVINAS, A.I., L. BELARDINELLI, N.A. UNDROVINAS & H.N. SABBAH. 2006. Ranolazine improves abnormal repolarization and contraction in left ventricular myocytes of dogs with heart failure by inhibiting late sodium current. J. Cardiovasc. Electrophysiol. **17**: S161–S177.

5. ANTZELEVITCH, C., L. BELARDINELLI, *et al.* 2004. Electrophysiologic effects of ranolazine: a novel anti-anginal agent with antiarrhythmic properties. Circulation **110**: 904–910.

6. ANTZELEVITCH, C., L. BELARDINELLI, L. WU, *et al.* 2004. Electrophysiologic properties and antiarrhythmic actions of a novel anti-anginal agent. J. Cardiovasc. Pharmacol. Therapeut. **9**(Suppl 1): S65–S83.

7. BURASHNIKOV, A., S. MANNAVA & C. ANTZELEVITCH. 2004. Transmembrane action potential heterogeneity in the canine isolated arterially-perfused atrium: effect of IKr and Ito/IKur block. Am. J. Physiol. **286**: H2393–H2400.

8. BURASHNIKOV, A. & C. ANTZELEVITCH. 2003. Reinduction of atrial fibrillation immediately after termination of the arrhythmia is mediated by late phase 3 early afterdepolarization-induced triggered activity. Circulation **107**: 2355–2360.

9. DAVIDENKO, J.M. & C. ANTZELEVITCH. 1986. Electrophysiological mechanisms underlying rate-dependent changes of refractoriness in normal and segmentally depressed canine Purkinje fibers: the characteristics of post-repolarization refractoriness. Circ. Res. **58**: 257–268.

10. BURASHNIKOV, A. & C. ANTZELEVITCH. 2005. Role of repolarization restitution in the development of coarse and fine atrial fibrillation in the isolated canine right atria. J. Cardiovasc. Electrophysiol. **16**: 639–645.

11. BURASHNIKOV, A. & C. ANTZELEVITCH. 2005. Beta-adrenergic stimulation is highly arrhythmogenic following ischemia/reperfusion injury in the isolated canine right atrium [abstract]. Heart Rhythm **2**: S179.

12. WHALLEY, D.W., D.J. WENDT & A.O. GRANT. 1995. Basic concepts in cellular cardiac electrophysiology. Part II: Block of ion channels by antiarrhythmic drugs. Pacing Clin. Elecrophysiol. **18**: 1686–1704.

13. LANGENFELD, H., J. WEIRICH, C. KOHLER & K. KOCHSIEK. 1990. Comparative analysis of the action of class I antiarrhythmic drugs (lidocaine, quinidine, and prajmaline) in rabbit atrial and ventricular myocardium. J. Cardiovasc. Pharmacol. **15**: 338-345.

14. CAMPBELL, T.J., K.R. WYSE & P.D. HEMSWORTH. 1991. Effects of hyperkalemia, acidosis, and hypoxia on the depression of maximum rate of depolarization by class-I antiarrhythmic drugs in guinea pig myocardium: differential actions of class-Ib and class-Ic agents. J. Cardiovasc. Pharmacol. **18**: 51–59.

15. LEMMENS-GRUBER, R., H. MAREI & P. HEISTRACHER. 1997. Electrophysiological properties of the propafenone-analogue GE 68 (1-[3-(phenylethyl)-2-benzofuryl]-2-(propylamino)-ethanol) in isolated preparations and ventricular myocytes of guinea-pig hearts. Naunyn Schmiedebergs Arch. Pharmacol. **355**: 230–238.

16. SPINELLI, W., R.W. PARSONS & T.J. COLATSKY. 1992. Effects of WAY-123,398, a new Class-III antiarrhythmic agent, on cardiac refractoriness and ventricular fibrillation threshold in anesthetized dogs: a comparison with UK-68798, e-4031, and DL- sotalol. J. Cardiovasc. Pharmacol. **20**: 913–922.

17. WIESFELD, A.C., C.D. DE LANGEN, *et al.* 1996. Rate-dependent effects of the class III antiarrhythmic drug almokalant on refractoriness in the pig. J. Cardiovasc. Pharmacol. **27**: 594–600.

18. ANTZELEVITCH, C., W. SHIMIZU, G.X. YAN, *et al.* 1999. The M cell: its contribution to the ECG and to normal and abnormal electrical function of the heart. J. Cardiovasc Electrophysiol. **10**: 1124–1152.

19. BURASHNIKOV, A. & C. ANTZELEVITCH. 2006. Late-phase 3 EAD: a unique mechanism contributing to initiation of atrial fibrillation. Pacing Clin. Electrophysiol. **29**: 290–295.

20. VINCENT, G.M. 2003. Atrial arrhythmias in the inherited lLong QT syndrome. J. Cardiovasc. Electrophysiol. **14**: 1034–1035.

21. Cardiac Arrhythmia Suppression Trial (CAST) Investigators. 1989. Preliminary report: effect of encainide and flecainide on mortality in a randomized trial of arrhythmia suppression after myocardial infarction. N. Engl. J. Med. **321**: 406–412.

22. BURASHNIKOV, A. & C. ANTZELEVITCH. 2007. IKUR block promotes atrial fibrillation in healthy canine atria [abstract]. Heart Rhythm **4**: S112.

23. BLAAUW, Y., H. GOGELEIN, R.G. TIELEMAN, *et al.* 2004. "Early" class III drugs for the treatment of atrial fibrillation: efficacy and atrial selectivity of AVE0118 in remodeled atria of the goat. Circulation **110**: 1717–1724.

24. GOLDSTEIN, R.N., C. KHRESTIAN, L. CARLSSON & A.L. WALDO. 2004. Azd7009: a new antiarrhythmic drug with predominant effects on the atria effectively terminates and prevents reinduction of atrial fibrillation and flutter in the sterile pericarditis model. J. Cardiovasc. Electrophysiol. **15**: 1444–1450.

25. WETTWER, E., O. HALA, T. CHRIST, *et al.* 2004. Role of IKur in controlling action potential shape and contractility in the human atrium: influence of chronic atrial fibrillation. Circulation **110:** 2299–2306.

26. OLSON, T.M., A.E. ALEKSEEV, X.K. LIU, *et al.* 2006. Kv1.5 channelopathy due to KCNA5 loss-of-function mutation causes human atrial fibrillation. Hum. Mol. Genet. **15:** 2185–2191.

27. WANG, Z.G., B. FERMINI & S. NATTEL. 1993. Sustained depolarization-induced outward current in human atrial myocytes: evidence for a novel delayed rectifier K+ current similar to Kv1.5 cloned channel currents. Circ. Res. **73:** 1061–1076.

28. AMOS, G.J., E. WETTWER, F. METZGER, *et al.* 1996. Differences between outward currents of human atrial and subepicardial ventricular myocytes. J. Physiol **491**(Pt 1): 31–50.

Molecular Basis of Cardiac Action Potential Repolarization

YORAM RUDY

Cardiac Bioelectricity and Arrhythmia Center, Washington University in St. Louis, St. Louis, Missouri, USA

The action potential (AP) is generated by transport of ions through transmembrane ion channels. Rate dependence of AP repolarization is a fundamental property of cardiac cells, and its modification by disease or drugs can lead to fatal arrhythmias. Using a computational biology approach, we investigated the gating kinetics of the rapid (I_{Kr}) and slow (I_{Ks}) K^+ currents during the AP in order to provide insight into the molecular basis of their role in AP repolarization. Results show that I_{Kr} intensifies during the late AP plateau by progressively recovering from inactivation and generating a pronounced late peak of open-state occupancy. The delayed peak makes I_{Kr} an effective determinant of AP repolarization. I_{Ks} builds an available reserve of channels in closed states near the open state that can open rapidly to generate current during the AP repolarization phase. By doing so, I_{Ks} can provide repolarizing current when other currents (e.g., I_{Kr}) are compromised by disease or drugs, thus preventing excessive AP prolongation and arrhythmic activity.

Key words: cardiac action potential; cardiac repolarization; ion channels; cardiac arrhythmias

Introduction

The cardiac action potential (AP) is generated by complex nonlinear interactions between membrane ion channels, the membrane voltage, and the dynamic ionic environment of the cell. These processes determine the AP rate-dependent properties and are subject to various regulatory mechanisms.

The process of normal AP generation is dominated by a single type of ion channel, the fast sodium channel, which generates a large and fast inward current, I_{Na}, to depolarize the membrane during the fast AP upstroke, the rising phase. In contrast, the AP repolarization phase is a much slower process that is determined by a delicate balance between inward depolarizing and outward repolarizing membrane currents carried by various types of ion channels. These different mechanisms are consistent with the cell's electrophysiologic function. AP depolarization is a robust "all-or-none" process that is executed with a large current reserve (i.e., high safety factor), while repolarization is a precisely controlled process that determines with great

exactness the rate-dependent AP morphology and duration (APD). APD shortening with increasing heart rate, a process termed *APD adaptation*, is a property of cardiac AP that is essential for normal heart function. The multi-current mechanism of repolarization provides many "control points" for this process and is therefore highly suitable for precise control of APD and its rate dependence. Unfortunately, the dependence on a delicate balance between various currents that provides high level of control also exposes AP repolarization to easy perturbation by disease or drugs, which can affect any of the ion channels involved. Examples of such perturbations that lead to abnormal AP repolarization and cardiac arrhythmias include hereditary channelopathies (e.g., the long QT, Brugada, and Timothy syndromes)[1–3] and acquired syndromes caused by drug modification of ion-channel function.[4]

The balance of currents that controls the repolarization process is determined by the gating kinetic properties of the ion channels that participate in this process. It is important, therefore, to understand the channels' gating processes that underlie the ionic currents during the AP. In this article we use a computational biology "mathematical modeling" approach to study ion-channel gating during the AP. We use single-channel-based Markov models to compute kinetic state occupancy and transitions between these states in a detailed model of the cardiac ventricular cell that is paced at different (slow and fast) rates. The focus is on the

Address for correspondence: Prof. Yoram Rudy, Ph.D., Cardiac Bioelectricity Center, Washington University in St. Louis, 290 Whitaker Hall, Campus Box 1097, One Brookings Drive, St. Louis, MO 63130. Fax: 314-935-8168.

Rudy@wustl.edu

Ann. N.Y. Acad. Sci. 1123: 113–118 (2008). © 2008 New York Academy of Sciences.
doi: 10.1196/annals.1420.013

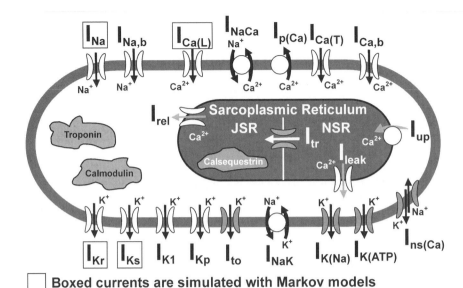

☐ **Boxed currents are simulated with Markov models**

FIGURE 1. A schematic of the Luo–Rudy dynamic (LRd) model of a cardiac ventricular cell. The model is based mostly on guinea-pig data. Details of the model can be found in Refs. 10 and 12 and at <http://rudylab.wustl.edu>, where the model code is also available. Boxed currents are simulated with Markov models.

two major repolarizing currents—I_{Kr} (rapid delayed rectifier) and I_{Ks} (slow delayed rectifier)—carried by K^+ ions. The results presented here summarize our earlier reports.[5–7] Kinetic transitions during the AP of the depolarizing currents I_{Na} and $I_{Ca(L)}$ (L-type calcium current) can be found in our recent publications.[7,8]

Methodology

We have developed an approach for incorporating Markov models of cardiac ion channels into the Luo–Rudy dynamic (LRd) model of a cardiac ventricular myocyte.[9] With this approach, single-channel kinetic properties can be studied during the AP. The LRd model (FIG. 1) includes all the major ion channels, pumps, and exchangers that generate the AP and determine its rate-dependent properties. The cell model accounts for dynamic concentration changes of Na^+, Ca^{2+}, and K^+ during the AP over long periods of pacing. It includes a two-compartment model of the sarcoplasmic reticulum (SR) and simulates Ca^{2+} cycling by the cell and the resulting Ca^{2+} transient during the AP. Details of the LRd model and its formulation can be found in previous publications[10–12] and in the research section of <http://rudylab.wustl.edu>, where the source program code (in C^{++}) is also available. Prior to stimulation at various pacing rates, cells were kept quiescent for 10 min to achieve steady-state resting conditions before all pacing protocols. Results

shown are during steady state at the indicated pacing rate.

Results

Role of I_{Kr}

I_{Kr} is a tetrameric channel formed by four identical α subunits that are encoded by the HERG gene. Each subunit contains six transmembrane-spanning segments (S1–S6). Similar to other ion channels, the S5–S6 linker ("P-loop") forms the channel pore and confers ion-selectivity to the channel. S4, a positively charged segment, serves as the voltage sensor; it moves in response to changes in the membrane potential (V_m), thereby changing the channel conformation. Upon V_m depolarization, the conformation change results in channel opening. The Markov model of I_{Kr} (FIG. 2A) consists of three closed states (C3, C2, C1), an open state (O), and an inactivated state (I).

FIGURES 2B and C show the AP (V_m) and I_{Kr} (top), and I_{Kr} kinetic-states occupancies (bottom) during the AP for slow (cycle length CL = 1000 ms; panel B) and fast (CL = 300 ms; panel C) pacing. Guinea-pig I_{Kr} activation and inactivation are rapid. During the AP upstroke, channels move rapidly from the deepest closed state, C3, through C2 to C1, from which they can open (O) or inactivate directly (I). Channels that open inactivate very rapidly (O to I) and the balance between activation and inactivation favors inactivation during most of the AP (FIGS. 2B and C). As the AP plateau

A I_{Kr} Model

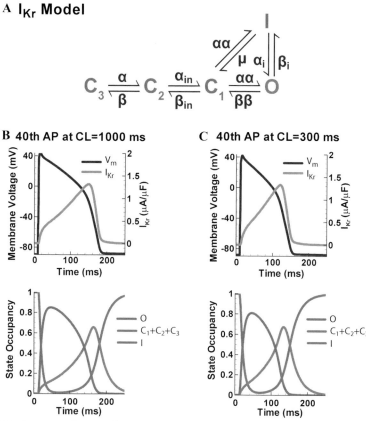

FIGURE 2. Transitions between kinetic states of I_{Kr} during the AP at slow and fast rates. **(A)** Markov model of I_{Kr}; closed states are in blue, inactivated in purple, and open in red. **(B)** V_m and I_{Kr} (*top*), and channel-state occupancies (*bottom*) during the AP at the slow rate (cycle length CL = 1000 ms). Traces are color-coded to the state types in panel **A**. **(C)** as in panel B, except for fast rate, CL = 300 ms. (From Rudy and Silva.[7] Reproduced by permission.) (In color in *Annals* online.)

repolarizes, channels recover gradually from inactivation (transition from I to O). The balance between inactivation and recovery progressively favors recovery as the AP plateau repolarizes, generating a pronounced peak of open-state occupancy (O) late during the AP plateau. This dynamic interplay ensures that I_{Kr} intensifies late during the AP plateau, when it is most effective in influencing AP repolarization and APD. At this phase, the AP is determined by a very delicate balance between depolarizing and repolarizing currents and is therefore easily modulated by the increasing I_{Kr}. Following the open-state peak, channels slowly deactivate (transition from O to C1). Note that peak I_{Kr} is similar at slow and fast rates (panels B and C, respectively). This behavior reflects the much stronger dependence of recovery from inactivation on voltage than on time during the AP. Consequently, in the guinea-pig, I_{Kr} plays a secondary role in the rate dependence of APD.

Role of I_{Ks}

Like I_{Kr}, I_{Ks} has a tetrameric structure of α-subunits (KCNQ1). A β-subunit (KCNE1) is also incorporated in the channel assembly. During membrane depolarization, all four voltage sensors of the channel (one in each α-subunit) move to cause channel opening. It has been recently shown that during activation each voltage sensor undergoes at least two transitions before channel opening. Thus, each sensor starts from a resting position (R1) and moves to an intermediate position (R2), from which it moves to the activated position (A). Considering all combinations of possible voltage sensor positions during this gating process, one obtains 15 closed states (C1–C15) prior to channel opening. For example, C1 represents all sensors in their R1 position; in C3 two are in R1 and two are in R2; in C12 two are in R2 and two in A, etc. The 15 closed states are shown in FIGURE 3 (the transitions are defined in the inset). A left–right transition represents movement of

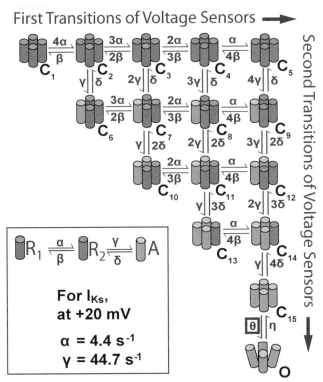

FIGURE 3. Possible combinations of voltage-sensor positions during I_{Ks} activation. Blue, red, and green indicate a voltage sensor in position R1, R2, or A, respectively (see text). All combinations are represented by 15 closed states (C1–C15). Once all four voltage sensors are in the activated position (C15, all green) the channel can open (state O). Note that the first transitions (horizontal in the diagram) are an order of magnitude slower than the second (vertical) transitions (transition rates are given below the diagram). (From Rudy and Silva.[7] Reproduced by permission.) (In color in *Annals* online.)

a voltage sensor from R1 to R2. A top-down transition represents movement from R2 to A. The first (R1 to R2) transitions are relatively slow (transition rate $\alpha = 4.4\,\mathrm{s}^{-1}$), while the second (R2 to A) transitions are an order of magnitude faster ($\gamma = 44.7\,\mathrm{s}^{-1}$). When all voltage sensors are in the activated position (C15), a cooperative voltage-independent transition occurs to the open state, O. Note that channels in closed states C5, C9, C12, C14 (right column in FIG. 3) have to only undergo the second fast transition from R2 to A before channel opening. Thus, channels in these closed states are available to open rapidly, while channels in more remote ("deeper") closed states must go through a slow transition first, which makes them much less available.

The diagram in FIGURE 3 provides the basis for the I_{Ks} Markov model in FIGURE 4A, which contains 15 closed states (C1 to C15) and two open states (O1 and O2). We classify the closed states into two zones: zone 2 (green) contains channels with at least one voltage sensor at its R1 position, requiring a first transition to R2 before channel opening; and zone 1 (blue) contains

channels for which all four voltage sensors have already completed the first transition to R1 and require only the second transition to A. Zone 1 includes the closed states in the right column of the diagrams in FIGURES 3 and 4A and constitutes the zone of readily available channels. As demonstrated below, this division into two zones with different kinetic properties makes I_{Ks} an effective repolarizing current and determinant of APD and its rate dependence.

FIGURE 4B shows the AP and I_{Ks} (top) and channel-state occupancies (bottom) during the AP at slow pacing (CL = 1000 ms). FIGURE 4C provides similar data for fast pacing (CL = 300 ms). During slow pacing, 60% of channels reside in zone 2 before the AP depolarization (panel B, bottom) and must undergo a slow transition to zone 1 before channel opening; only 40% of channels are available for fast opening in zone 1 at this pacing rate. In contrast (panel C, bottom), 75% of channels at the fast rate accumulate in zone 1 before the AP upstroke, as they do not have sufficient time between beats for transition to deeper closed states of

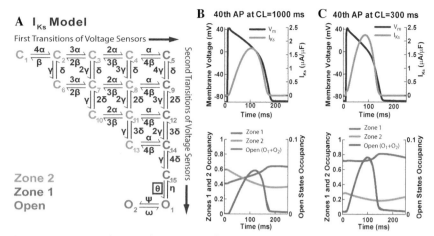

FIGURE 4. Transitions between kinetic states of I_{Ks} during the AP at slow and fast rate. (**A**) Markov model of I_{Ks}, based on the diagram of FIGURE 3. Closed states are grouped into two zones: zone 2 (green), for which not all voltage sensors completed the slow first transition, and zone 1 (blue), for which all first transitions have been completed and only fast second transitions remain before channel opening (open states O, red). (**B**) V_m and I_{Ks} (*top*), and channel-state occupancies (*bottom*) during the AP at slow rate CL = 1000 ms. (**C**) Vm and I_{Ks} (*top*), and channel-state occupancies (*bottom*) during the AP at fast rate CL = 300 ms. Note accumulation in zone 1 at AP onset during fast pacing. This creates an "available reserve" (AR) of channels that can open rapidly to shorten the AP (see text). (From Rudy and Silva.[7] Reproduced by permission.) (In color in *Annals* online.)

zone 2 at this fast rate. The accumulation in zone 1 facilitates fast channel openings and rapid rise of I_{Ks} to a larger peak than at slow rate during the AP repolarization phase (compare panels B and C). Thus, at fast rate channels accumulate in zone 1 between beats, creating an "available reserve" (AR) of channels that can open rapidly "on demand" to generate a large I_{Ks} late during the AP plateau, when the current can effectively shorten the APD.

Summary

The cardiac AP is generated by complex, nonlinear interactions between ion channels, the membrane voltage, the dynamic ionic environment of the cell, and a variety of regulatory molecules. Computational biology is a powerful approach that can provide insight into the mechanistic link between ion-channel kinetics and the AP. The simulations presented here explore the mechanisms by which two major repolarizing currents, I_{Kr} and I_{Ks}, exert their effects on AP repolarization. The precise timecourse of the AP is determined by a delicate balance between several voltage- and time-dependent inward depolarizing and outward repolarizing ionic currents. The dominant depolarizing current during the AP plateau is $I_{Ca(L)}$, which progressively decreases during the plateau be-

cause of voltage- and calcium-dependent inactivation processes. To gradually shift the balance of currents in the outward repolarizing direction, I_{Kr} and I_{Ks} increase during the plateau, reaching a peak at the late plateau phase. The two currents accomplish this via very different mechanisms: I_{Kr} inactivates almost instantaneously after activating rapidly during the AP early depolarization phase. It then gradually recovers from inactivation during the plateau phase to reach a late peak in open channel occupancy, which generates a large current when it most effectively affects AP repolarization. Thus, I_{Kr} relies on its recovery from inactivation kinetic properties for effective functioning during the AP. I_{Ks} also intensifies to a late peak during the AP plateau, but as a result of delayed activation due to two-stage voltage sensor movement. At fast rate, I_{Ks} channels accumulate between beats at zone 1 of closed states, from which they can open rapidly. Consequently, I_{Ks} increases faster during the plateau to reach a larger peak, causing AP shortening at fast rate ("APD adaptation"). This build-up of an AR of channels in closed states is very different from channels accumulating in the open state between beats, the traditionally accepted mechanism of APD adaptation. AP-clamp experiments[13] confirm the AR mechanism, showing a rapid increase in I_{Ks} at fast rate, but no instantaneous current at AP initiation, which would be indicative of open-state accumulation. The I_{Ks}

activation kinetic properties that generate the AR depend on the interaction between the α and β subunits of the channel. As a consequence of the AR buildup, I_{Ks} can provide "repolarization reserve" when other repolarizing currents (e.g., I_{Kr}) are compromised by mutations or drugs and prevent arrhythmogenic behavior of the cellular AP (see Figure 7 in Silva and Rudy[6]).

Acknowledgments

This work was supported by NIH-NHLBI Grant RO1 HL49054 and Merit Award R37 HL33343.

Conflict of Interest

The author declares no conflicts of interest.

References

1. KEATING, M.T. & M.C. SANGUINETTI. 1996. Molecular genetic insights into cardiovascular disease. Science **272:** 681–685.

2. ANTZELEVITCH, C. 2001. The Brugada syndrome: ionic basis and arrhythmia mechanisms. J. Cardiovasc. Electrophysiol. **12:** 268–272.

3. SPLAWSKI, I., K.W. TIMOTHY, N. DECHER, *et al.* 2005. Severe arrhythmia disorder caused by cardiac L-type calcium channel mutations. Proc. Natl. Acad. Sci. USA **102:** 8089–8096.

4. RODEN, D.M. 2004. Drug-induced prolongation of the QT interval. N. Engl. J. Med. **350:** 1013–1022.

5. CLANCY, C.E. & Y. RUDY. 2001. Cellular consequences of HERG mutations in the long QT syndrome: precursors to sudden cardiac death. Cardiovasc. Res. **50:** 301–313.

6. SILVA, J. & Y. RUDY. 2005. Subunit interaction determines IKs participation in cardiac repolarization and repolarization reserve. Circulation **112:** 1384–1391.

7. RUDY, Y. & J.R. SILVA. 2006. Computational biology in the study of cardiac ion channels and cell electrophysiology. Q. Rev. Biophys. **39:** 57–116.

8. FABER, G.M., J. SILVA, L. LIVSCHITZ & Y. RUDY. 2007. Kinetic properties of the cardiac L-type Ca2+ channel and its role in myocyte electrophysiology: a theoretical investigation. Biophysical. J. **92:** 1522–1543.

9. CLANCY, C.E. & Y. RUDY. 1999. Linking a genetic defect to its cellular phenotype in a cardiac arrhythmia. Nature **400:** 566–569.

10. LUO, C.H. & Y. RUDY. 1994. A dynamic model of the cardiac ventricular action potential. I. Simulations of ionic currents and concentration changes. Circ. Res. **74:** 1071–1096.

11. ZENG, J., K.R. LAURITA, D.S. ROSENBAUM & Y. RUDY. 1995. Two components of the delayed rectifier K+ current in ventricular myocytes of the guinea pig type: theoretical formulation and their role in repolarization. Circ. Res. **77:** 140–152.

12. FABER, G.M. & Y. RUDY. 2000. Action potential and contractility changes in [Na$^+$]$_i$ overloaded cardiac myocytes: a simulation study. Biophys. J. **78:** 2392–2404.

13. ROCCHETTI, M., A. BESANA, G.B. GURROLA, *et al.* 2001. Rate dependency of delayed rectifier currents during the guinea-pig ventricular action potential. J. Physiol. **534:** 721–732.

Intracellular Transport

From Physics to . . . Biology

AURÉLIEN ROUX,[a,b] DAMIEN CUVELIER,[b] PATRICIA BASSEREAU,[b] AND BRUNO GOUD[a]

[a]UMR 144 CNRS, Department of Cell Biology, and
[b]UMR 168 CNRS, Department of Physics, Institut Curie, Paris, France

Considerable effort over the past three decades has allowed the identification of the protein families that control the cellular machinery responsible for intracellular transport within eukaryotic cells. These proteins are estimated to represent about 10–20% of the human "proteome." The complexity of intracellular transport makes useful the development of model membranes. We describe here experimental systems based on lipid giant unilamellar vesicles (GUVs), which are attached to kinesin molecules. These systems give rise to thin membrane tubes and to complex tubular networks when incubated *in vitro* with microtubules and ATP. This type of assay, which mimics key events occurring during intracellular transport, allows physicists and biologists to understand how the unique mechanical properties of lipid membranes could be involved in the budding process, the sorting of cargo proteins and lipids, and the separation of the buds from a donor membrane.

Key words: intracellular transport; model membranes; phase separation; lipid sorting

Introduction

A fundamental feature of eukaryotic cells is the compartmentalization of their cytoplasm into distinct membrane-limited organelles that have a unique composition of proteins and lipids. This structural and functional compartmentalization has provided the basis for the development of a wide array of cellular processes and the rise of the complex multicellular organization of metazoans.

Organelles are intimately connected through specific transport mechanisms that are required not only to direct macromolecules to defined locations, but also to maintain the protein and lipid composition of a given organelle and therefore its identity. Transport is a multi-step process, involving the formation of transport carriers loaded with defined sets of cargo (budding), the movement of these carriers between compartments, and their specific fusion with target membranes. Real-time analysis of cargo movement by videomicroscopy has revealed that most transport carriers exist as pleiomorphic structures, such as vesicular tubular clusters and even membrane tubules, rather than the "classical" round transport vesicle. For instance, tubules have been shown to mediate the bulk of cargo movement from the Golgi complex to the cell surface.[1–3] From the development of methods based on the use of green fluorescent protein and its derivatives, it is also now clear that intracellular membrane compartments are not only highly dynamic, but are also composed of a collection of domains or subdomains that fulfill different functions.[4]

Considerable effort over the past three decades has allowed the identification of the protein families (e.g., coat proteins, SNAREs, molecular motors, Ras-like GTPases) that control the complex cellular machinery responsible for intracellular transport. These proteins are estimated to represent about 10–20% of the human "proteome." However, a clear picture of how really intracellular transport works is lacking. For instance, whether vesicular transport occurs between Golgi stacks or transport is associated with a maturation of the stacks is still a matter of hot debate! Therefore, the complexity of intracellular transport makes the development of model systems useful to understand the physical basis of membrane remodeling required for the formation of transport carriers. Here, we review some of the most recent *in vitro* findings that have allowed physicists and biologists to understand how the unique mechanical properties of lipid membranes could be involved in the budding process, the sorting of cargo proteins and lipids, and the separation of the buds from the donor membrane.

Address for correspondence: Prof. Bruno Goud, Ph.D., UMR 144 CNRS/Institut Curie, Institut Curie, 26 rue d'Ulm, 75248 Paris Cedex 05, France. Voice: 33 1 42 34 63 98; fax: 33 1 42 34 63 82.
bruno.goud@curie.fr

Ann. N.Y. Acad. Sci. 1123: 119–125 (2008). © 2008 New York Academy of Sciences.
doi: 10.1196/annals.1420.014

119

Physics of Membrane Tube Extraction

When a force is applied on a lipid vesicle, it responds by changing its shape. In most conditions, it is energetically more favorable to pull a membrane tube, rather than inducing an overall deformation of the membrane. Tubes can be extracted from vesicles by a shear flow[5] or by applying localized forces on the membrane. This latter approach was developed at the Institut Curie by combining an optical tweezer set-up with a micropipette technique[6,7] (FIG. 1).

The force (F) that has to be applied on the vesicle depends on membrane tension (σ) and on bending rigidity (κ): $f_0 = 2\pi\sqrt{2\kappa\sigma}$. At equilibrium, the radius of the tube is thus given by $r_0 = \sqrt{(\kappa/2\sigma)}$. The bending rigidity of a membrane directly depends on its lipid composition. For instance, a vesicle made of pure dioleyolphosphatidylcholine (DOPC) has a κ of about $10\,\kappa T$. For a typical surface tension of 5×10^{-5} N/m, one calculates that the radius of the tube is about 20 nm and the force F exerted on the tube is 12.6 pN. It should be pointed out that the initial formation of a tube requires a force slightly higher than the one required to elongate it (FIG. 1).

Membrane Tube Extraction by Molecular Motors

A few years ago, we showed that the attachment of motors of the kinesin family to the membrane of giant unilamellar vesicles (GUVs) is sufficient to promote the formation of membrane tubes.[8] In the original assay, biotinylated kinesins (consisting of the motor domain of the well-characterized *D. melanogaster* kinesin) were attached to the GUV membranes made of 95% egg phosphatidylcholine and 5% biotinylated dioleyol phosphoethanolamine (Biot-DOPE) via 100-nm polystyrene beads coated with streptavidin. GUVs and the kinesin-coated beads were injected into a small chamber coated with taxol-polymerized microtubules, and the chamber was filled with a buffer containing 1 mM ATP. Approximately 10 min after injection, membrane tubes started to form (FIG. 2). Once generated, tubes continued to grow along the microtubules and generated a tubular network until consumption of the membrane reservoir provided by the GUV.

Subsequently we and others showed that kinesins could be directly attached to the lipid bilayer without bead intermediates.[9,10] An interesting development of the assay came from the availability of a lipid with both a biotin function and a fluorophore attached to the head group (rhodamin-biotin-di-hexadecanoyl-phosphatidylethanolamine,

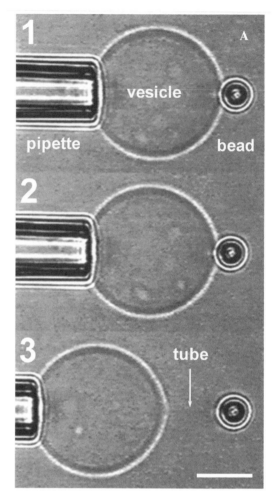

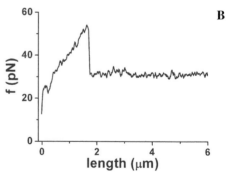

FIGURE 1. (A) Set-up to study membrane forces during tubule extraction. (**A**) Giant unilamellar vesicle (GUV) is held with a micropipette, which controls membrane tension by aspiration, and a streptavidin-coated bead held in optical tweezers is gently applied to the membrane. Then, after a lapse of time allowing the bead to stick to biotinylated lipids, the micropipette is moved away from the optical trap, pulling a tubule connecting the bead to the vesicle. (**B**) Typical force displacement diagram obtained from tracking the position of the bead. Bar, 10 microns.

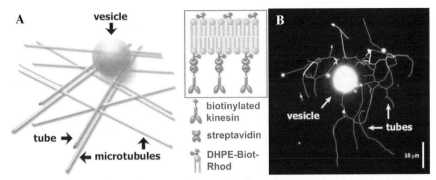

FIGURE 2. (A) *In vitro* assay allowing the growth of membrane tubules pulled by kinesins out of GUVs. A GUV coated with kinesins (via a streptavidin–biotin link described on *left*) is deposited on a network of microtubules coating a glass surface. **(B)** A typical confocal image of a membrane tubule network grown out of a GUV containing a fluorescent lipid.

DHPE-Biot-Rhod) (FIG. 2A). The number of kinesins was directly controlled by fixing the concentration of DHPE-Biot-Rhod lipid in the membrane. A protocol was set up so that one kinesin molecule binds to one biotinylated lipid. First, kinesin and streptavidin concentrations were adjusted so that at most one kinesin binds to one streptavidin on account of the large excess of streptavidin compared to kinesin. Then, by immobilizing the streptavidin–kinesin complexes on the microtubules before vesicle injection, it was possible to thoroughly rinse the chamber and get rid of free streptavidins and nonactive motors in solution. Because the total number of biotinylated lipids in the vesicle was much lower than the number of available streptavidin–kinesin complexes (by at least one order of magnitude), every binding site for motors was occupied on account of the high affinity of streptavidin for biotin, and the streptavidin–kinesin complexes not attached to DHPE-Biot-Rhod lipids remained in solution. The site saturation was achieved faster than the time required to pull the first tube (∼1 min). The number of motors attached to the membrane is therefore equal to the number of biotinylated lipids. Moreover, for concentrations above 0.01 mol% DHPE-Biot-Rhod, it was also checked that the number of streptavidin molecules per biotinylated lipids remains constant when varying the DHPE-Biot-Rhod concentration. Besides, it was also verified that the quantity of streptavidin–kinesin complexes attached to the lipids through nonspecific interactions can be neglected for concentrations of DHPE-Biot-Rhod above 0.001 mol%. The measurement of fluorescence intensity of DHPE-Biot-Rhod along the tube gives the motor distribution as the experimental protocol was adjusted to have one motor per DHPE-Biot-Rhod. With this protocol, it is then possible to simultaneously visualize the motor distribution and control the concentration of motors on the vesicle.

This assay was used for an experimental and theoretical analysis of the dynamics of motors on both vesicle and tube surfaces. In particular, it revealed the existence of an initial minimal surface density of motors on the vesicle below which no tubes can be pulled and a dynamic cluster of motors at the tips of growing tubes.[10]

Lipid Sorting

During the formation of transport intermediates (budding process), sorting of lipids occurs, some of them being incorporated into transport intermediates while others are being excluded.[11,12] *In vitro* experiments using model membranes made of dioleylphosphatidylcholine (DOPC), cholesterol (Chol), and sphingomyelin (SM) reveal, under appropriate conditions, the coexistence of two types of fluid membrane organization called liquid-ordered L_o and liquid-disordered L_d phases.[13] The composition of L_o and L_d phases is different: compared to the global average composition, L_o is enriched in SM whereas L_d is enriched in DOPC.[14] Thus, the ability of L_o versus L_d phases to bud could be a critical parameter in sorting. For instance, transport intermediates may form from a preexisting lipid domain on the donor membrane.[12] This possibility adds up to two other possible mechanisms: sorting according to molecular shape of lipids and dynamical sorting coupled to the formation of transport intermediates.[15]

To test the ability of L_o and L_d phases to form membrane tubes, we have prepared GUVs from various mixtures of BSM, Chol, and DOPC. The vesicles were fluorescently labeled by incorporation of a fluorescent lipid BODIPY$_{FL}$-C5-hexadecanoyl phosphatidylcholine (BODIPY$_{FL}$-C5-HPC) at a concentration

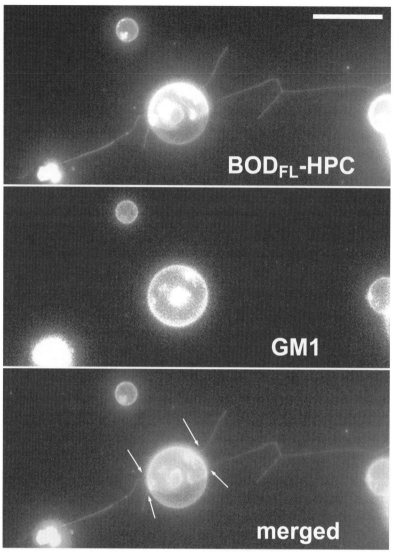

FIGURE 3. Growth of tubules out of a segregated vesicle. Tubules are mainly connected and filled up with the green phase (BOD$_{FL}$-C$_5$-HPC) that corresponds to the L$_d$ phase. *Arrows* point to connections between the tubules and the GUV. Bar, 10 microns. (In color in *Annals* online.)

of 0.5% mol/mol. A total of 11 different compositions of BSM:Chol:DOPC mixtures were tested. The vesicles displayed domains of different phases corresponding to segregation of lipids to various degrees depending on the relative ratio of BSM, Chol, and DOPC. Vesicles with a homogeneous fluorescence, reflecting an absence of lipid segregation, were observed at molar ratios of 1:1:0, 0:1:1, 1:2:1 and 1:2:3 of BSM:Chol:DOPC, respectively. These values correspond to "high" cholesterol content (over 30%). Vesicles with both highly and weakly fluorescent domains of various sizes were observed at lower Chol concen-

trations (molar ratios of 3:2:1 and 3:1:3). GUVs made of the 1:1:1 lipid mixture showed particular properties. The majority exhibited a uniform fluorescence phase, whereas the others (typically 10–30% of the population) showed a fluorescent domain covering only one hemisphere. In addition, small changes in cholesterol concentration lead to the formation of small nonfluorescent domains, suggesting that the 1:1:1 vesicles represent a frontier situation in which lipids can be segregated or not, depending on small changes in cholesterol concentration. In the vesicles of various lipid compositions, BODIPY$_{FL}$-C5-HPC segregated into the L$_d$

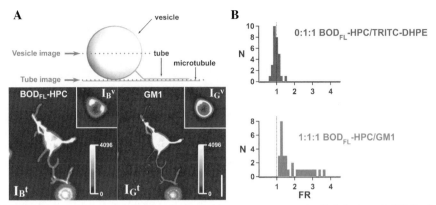

FIGURE 4. Lipid sorting in tubules pulled out of homogeneous vesicles (1:1:1 mixture). (**A**) Confocal images of tubules pulled out of membranes labeled with BODIPY$_{FL}$-C5-HPC lipids (BOD$_{FL}$-HPC) and Cy3-cholera toxin–GM1 complexes (GM1). Images were recorded at two levels: one at the vesicle equator (vesicle image) and one on the substrate (tube image). Left fluorescence image corresponds to the BOD$_{FL}$-HPC channel, whereas the right image refers to the GM1 channel. Tubule images show that the (BOD$_{FL}$-HPC) intensity is higher than that of Cy3-cholera toxin (GM1), whereas it is the opposite in the vesicle image (see *insets*). Fluorescence intensities of BOD$_{FL}$-HPC and GM1, respectively, in the tubules (IB$_t$, IG$_t$) and in the vesicle (IB$_v$, IG$_v$) were measured from tubule and vesicle images. Highly fluorescent dots on the vesicle images correspond to the connection between the tubules and the vesicle. Bar, 10 mm. (**B**) The fluorescence ratio FR = (IB$_t$/IG$_t$)/(IB$_v$/IG$_v$) was calculated for each network. Two compositions were tested: 0:1:1 and 1:1:1. For 0:1:1 (L$_d$ phase), the vesicles contained two L$_d$ phase fluorophores (BOD$_{FL}$-HPC and TRITC-DHPE) as a control experiment; the FR histogram (dark gray) calculated from 30 different networks was centered on the value 1, indicating that no relative sorting occurs under these conditions. For 1:1:1, vesicles contained 1% GM1 and 0.5% BOD$_{FL}$-HPC; FR histogram (light gray) shows that values obtained from 30 different networks were always superior to 1, reflecting a relative depletion of GM1 in tubules or equivalently a relative enrichment in BOD$_{FL}$-HPC. (In color in *Annals* online.)

phase, where the L$_o$ phase of a segregated vesicle (3:1:3) was labeled with the ganglioside GM1.

Biotinylated lipid used to anchor kinesins to membrane are equally distributed within the L$_d$ and L$_o$ phases. Therefore, motors are able to pull on both phases. However, in segregated vesicles labeled with BODIPY$_{FL}$-C5-HPC, the majority of the tubes were not only fluorescent, but also connected to the fluorescent domains (FIG. 3). In addition, GM1 was essentially excluded from tubes, suggesting that tubes were essentially composed of membranes in L$_d$ phase enriched in DOPC. Importantly, sorting between fluorescent lipids used as markers also occurred in tubes grown from nonsegregated vesicles of 1:1:1 lipid composition. Direct evidence came from the comparison of GM1 (labeled by Cy3-cholera toxin) and BODIPY$_{FL}$-C5-HPC amounts present in tubes and in vesicles (FIG. 4). Tubes pulled out of homogeneous 1:1:1 vesicles contained 1% GM1 (labeled by Cy3-cholera toxin) and 0.5% BODIPY$_{FL}$-C5-HPC. The fluorescence ratio between BODIPY$_{FL}$-C5-HPC and GM1 was increased in the tube as compared to the donor vesicle (FIG. 4).

As indicated above, the force F required to pull tubes is proportional to the square root of the bending rigidity κ and of the membrane tension σ. Using the optical tweezers set-up coupled to the micropipette system, we calculated that an L$_o$ phase is about 2.2 times more rigid than a L$_d$ phase. Molecular motors should then preferentially pull tubes out of the L$_d$ phase in a segregated vesicle since the force required is lower. This is exactly what we obtained experimentally in the experiment described in FIGURE 3. The observation that L$_d$ lipids are enriched in tubes pulled out of segregated vesicles indicates that differences in the ability of phases to form curved structures can lead to lipid sorting.

Phase Separation and Tube Fission

Interestingly, the induction of phase separation on tubes pulled out of, for instance, 1:1:1 homogeneous GUVs, provoked numerous fission events, as illustrated in FIGURE 5. At 4 s after photoactivation, a small domain of L$_d$ phase appeared at the tip of the tube. It is characterized both by an intense fluorescence and a tube diameter smaller than that of the initial tube. About 20 s later, the tube broke at the limit between the

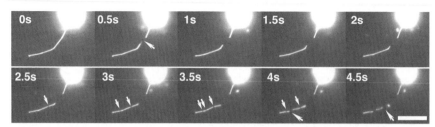

FIGURE 5. Phase separation of lipids leads to fission of tubules. Induction of phase separation by photoactivation of fluorescent lipids leads to the appearance of small domains less fluorescently labeled along the tubules (*small arrows*). Fission occurs in the vicinity of these newly formed domains (*bigger arrows*). Time is in seconds. Bar, 10 microns.

strongly and the weakly fluorescent domains. This led to the formation of an almost spherical vesicle. In some cases, the fission process led to complete fragmentation of the tubes into vesicles. The time required for tube fission after domain formation was observed to rank statistically between less than 100 ms and more than 10 s, depending on lipid composition. These observations are consistent with a theoretical analysis in which rupture originates both from line tension at the domain interfaces and Gaussian curvature discontinuity.[16]

Our results point out a (relatively simple) mechanism based on phase separation that could couple sorting and fission events during the formation of transport intermediates. Indeed, it is a characteristic of many systems in biology to work in the vicinity of a phase transition in order to increase their sensitivity.[17,18] One of the main roles of the numerous proteins that have been implicated in sorting and fission events[19] of biological membranes could then be to trigger phase separation of membrane lipids, either by clustering specific lipids or by inducing membrane tubulation. Other proteins (e.g., dynamin) might locally change the lipid composition in order to induce a phase separation that will promote the local fission of the membrane.

Future Directions

Much can still be learned by using model membranes and minimal systems to understand the physical laws that underlie the mechanisms of intracellular transport. One way to go is to progressively increase the complexity of the assays, such as the one described here. For instance, we are currently reconstituting the ARF1-dependent binding of the COPI coat (involved in transport events at the Golgi–endoplasmic reticulum interface) to GUVs. This allows us to quantitatively address the role of physical parameters such as membrane composition, membrane curvature, or membrane tension in COPI binding. Preliminary results indicate that membrane tension plays a critical role in membrane deformation induced by COPI binding.

Conflict of Interest

The authors declare no conflicts of interest.

References

1. HIRSCHBERG, K., C.M. MILLER, *et al.* 1998. Kinetic analysis of secretory protein traffic and characterization of Golgi to plasma membrane transport intermediates in living cells. J. Cell Biol. **143:** 1485–503.
2. TOOMRE, D., P. KELLER, *et al.* 1999. Dual-color visualization of trans-Golgi network to plasma membrane traffic along microtubules in living cells. J. Cell Sci. **112:** 21–33.
3. POLISHCHUK, R.S., E.V. POLISHCHUK, *et al.* 2000. Correlative light-electron microscopy reveals the tubular-saccular ultrastructure of carriers operating between Golgi apparatus and plasma membrane. J. Cell Biol. **148:** 45–58.
4. PFEFFER, S. 2003. Membrane domains in the secretory and endocytic pathways. Cell **112:** 507–517.
5. ROSSIER, O., D. CUVELIER, *et al.* 2003. Giant vesicles under flows: extrusion and retraction of tubes. Langmuir **19:** 575–584.
6. CUVELIER, D., N. CHIARUTTINI, P. BASSEREAU & P. NASSOY. 2005. Pulling long tubes from firmly adhered vesicles. Europhys. Lett. **71:** 1015–1021.
7. CUVELIER, D., I. DERENYI, P. BASSEREAU & P. NASSOY. 2005. Coalescence of membrane tethers: experiments, theory, and applications. Biophys. J. **88:** 2714–2726.
8. ROUX, A., G. CAPPELLO, J. CARTAUD, *et al.* 2002. A minimal system allowing tubulation with molecular motors pulling on giant liposomes. Proc. Natl. Acad. Sci. USA **99:** 5394–5399.
9. KOSTER, G., M. VANDUIJN, B. HOFS & M. DOGTEROM. 2003. Membrane tube formation from giant vesicles by dynamic association of motor proteins. Proc. Natl. Acad. Sci. USA **100:** 15583–15588.
10. LEDUC, C., O. CAMPAS, *et al.* 2004. Cooperative extraction of membrane nanotubes by molecular motors. Proc. Natl. Acad. Sci. USA **101:** 17096–17101.
11. BRUGGER, B., R. SANDHOFF, *et al.* 2000. Evidence for segregation of sphingomyelin and cholesterol during formation of COPI-coated vesicles. J. Cell Biol. **151:** 507–518.

12. VAN MEER, G. & Q. LISMAN. 2002. Sphingolipid transport: rafts and translocators. J. Biol. Chem. **277:** 25855–25858.

13. DIETRICH, C., L.A. BAGATOLLI, *et al.* 2001. Lipid rafts reconstituted in model membranes. Biophys. J. **80:** 1417–1428.

14. EDIDIN, M. 2003. Lipids on the frontier: a century of cell-membrane bilayers. Nat. Rev. Mol. Cell Biol. **4:** 414–418.

15. MUKHERJEE, S. & F.R. MAXFIELD. 2000. Role of membrane organization and membrane domains in endocytic lipid trafficking. Traffic **1:** 203–211.

16. ALLAIN, J.M., C. STORM, A. ROUX, *et al.* 2004. Fission of a multiphase membrane tube. Phys. Rev. Lett. **93:** 158104.

17. DUKE, T.A. & D. BRAY. 1999. Heightened sensitivity of a lattice of membrane receptors. Proc. Natl. Acad. Sci. USA **96:** 10104–10108.

18. CAMALET, S., T. DUKE, F. JULICHER & J. PROST. 2000. Auditory sensitivity provided by self-tuned critical oscillations of hair cells. Proc. Natl. Acad. Sci. USA **97:** 3183–3188.

19. SLEPNEV, V.I. & P. DE CAMILLI. 2000. Accessory factors in clathrin-dependent synaptic vesicle endocytosis. Nat. Rev. Neurosci. **1:** 161–172.

Linking Cellular Energetics to Local Flow Regulation in the Heart

JAMES B. BASSINGTHWAIGHTE

Department of Bioengineering, University of Washington, Seattle, Washington, USA

A mathematical model has been developed to explain the metabolic and energetic responses induced by abnormal routes of cardiac excitation. For example, in left bundle branch block (LBBB), both glucose uptake and flow are reduced in the septal region, similar to the situation in dogs paced at the right ventricular outflow tract. In these conditions the septum is activated early, the sarcomere lengths shorten rapidly against low left ventricular (LV) pressure, and the blood flow to the interventricular septum diminishes. In contrast, the work load and the blood flow increases in the later-activated LV free wall. To provide a logical, quantitatively appropriate representation, the model links: (1) the processes of excitation–contraction coupling; (2) regional ATP utilization for force development at the cross-bridge, for ion pumping, and for cell maintenance; (3) the regulation of demands on local fatty acid and glucose metabolism for ATP generation by glycolysis and oxidative phosphorylation; and (4) feedback regulation of blood flow to supply substrate and oxygen. The heart is considered as a cylinder composed of two parts: an early-activated region and a late-activated region in tandem, but activated separately with the time delay representing the time for excitation to spread from septum to free wall. The same model equations and parameter sets are used for the two regions. The contraction of the early-activated region stretches the other region, with the result that the early-stimulated region has diminished oxygen requirements compared to those found with simultaneous stimulation. The late-activated region has increased work and increased oxygen consumption, as seen in the intact heart. Integrating the modeling of cardiac energy metabolism with local blood flow regulation and capillary–tissue substrate exchange provides a quantitative description, an hypothesis formulated to stimulate further experimentation to test its validity. The hypothesis "explains" observations of contraction and metabolism in LBBB, but whether this concept can be extended to explain the normal flow heterogeneity in the heart remains unknown.

Key words: myocardial blood flow; heterogeneity; cardiac pacing; excitation–contraction coupling; cellular energetics; cardiac cell model; bundle branch block; shortening deactivation; ionic regulation; smooth muscle receptors; branch point competition; apparent cooperativity

Introduction

Left bundle branch block (LBBB) in humans[1] and right ventricular (RV) outflow tract pacing in dogs[2] result in changes in regional blood flows and metabolism. In the early-activated regions glucose uptake and flow were both reduced. The early activation initiates rapid local contraction, shortening regional sarcomere lengths without initially raising left ventricular (LV) pressure significantly. In LBBB septal blood flow was reduced, while in the late-activated LV free wall, flow increased. Further, in both the humans with LBBB and

the dogs chronically paced at the RV outflow, the septal wall became thinner, while the late-activated LV regions thickened by as much as 40%.[3] In the paced dogs, the LV free wall was initially passively stretched by the contraction of the early-activated regions before it was electrically excited, and it contracted against the developing LV pressure. The result was that the pre-stretch was followed by a prolonged, more forceful contraction, presumably due to the combination of the Starling effect and the increased afterload.

The present model links an integrated set of events: (1) the processes of excitation–contraction coupling; (2) the regional ATP utilization for force development at the cross-bridge, ion pumping, and cell maintenance; (3) the regulation of demands on local fatty acid and glucose metabolism for ATP generation by glycolysis and oxidative phosphorylation; and (4) feedback regulation of blood flow to supply substrate and oxygen.

Address for correspondence: Prof. James B. Bassingthwaighte, M.D., Ph.D., Department of Bioengineering, University of Washington, Seattle WA 98195-5061. Voice: 206-685-2012; fax: 206-685-3300.

jbb@bioeng.washington.edu

Ann. N.Y. Acad. Sci. 1123: 126–133 (2008). © 2008 New York Academy of Sciences.
doi: 10.1196/annals.1420.015

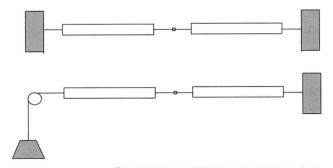

FIGURE 1. Two versions of a two-sarcomere heart. *Top:* two segments in tandem, but constrained to constant length (isometric total length). In this case, early activation of one of the two, any shortening causes lengthening of the other. *Bottom:* situation allowing the total length to shorten against load, performing external work.

This model is of a severely reduced form, focusing on the relationship between the shortening strain, strain velocity, stress, and the local blood flow in a two-element model of heart. The two elements represent the early- and late-activated portions of the LV, arranged in tandem so that each pulls upon the other. The link between the contractile events and regional flow is mediated through local metabolic demands, consumption of substrates and oxygen used to form ATP. The inference is that in regions where there is high demand there is, at least transiently, sufficient ATP breakdown in excess of the mitochondrial capacity for the rephosphorylation of ADP to provide a vasodilatory signal. A feature of the hypothesized relationship takes into account the ultrasensitivity of vascular vasodilatory responses to interstitial adenosine,[4] that is, the observation that the coronary blood flow increase versus the adenosine concentration is switch-like, having a Hill coefficient of around 7.5.

The Model

The Regional Contraction Model

The ventricular contraction model is a two-segment heart, as if the ventricle were made up of two half cylinders, one activated before the other. No parametric differences between early and late-activated regions were used to portray the steady state of asynchronous activation in this two part "ventricle." Instead, the model uses a single parameter set for the two regions linked in tandem, but activates them separately. The time delay from one to the other represents the time for excitation to spread from the LV septum to the free wall. The early activation of one half stretches the other, and the stronger contraction of the late-activated, prestretched region opposes continued contraction of the

early-activated region. The model allows exploration of these questions: (*a*) What reduction of ATP consumption occurs with shortening deactivation in the early-activated part? Is the reduction enough to explain the observed diminutions in flow and substrate uptake? (*b*) What is the magnitude of the Starling effect in increasing force, ATP consumption, and blood flow in the late-activated part? (*c*) What does it take to make a contractile two-element model stable? The incentive for this last question is that the standard Huxley contractile sarcomere is inherently unstable in the situation diagrammed in FIGURE 1, even though it is normally found that sarcomere lengths in stimulated heart or skeletal muscle shorten together. The Huxley model increases the power of its contraction with increasing overlap of thick and thin filaments, so that if one segment shortens an iota stronger or a millisecond earlier than the other, then it wins the tug of war and shortens by lengthening the delayed sarcomere in tandem with it.

Of the several models that could be chosen for the contractile element, we chose the four-state model of Landesberg and Sideman,[5] in which the rate of formation of the strong ATP-bound state decreases as a function of the shortening velocity, and where the rate of transformation from the strong to the weak state is increased with faster shortening. If shortening is very rapid, against little load (i.e., near maximum shortening velocity), then ATP hydrolysis is reduced compared to the rate of hydrolysis with a slower velocity of shortening when there is a higher load. This phenomenon is known as shortening deactivation. Shortening deactivation, which occurs in the early-activated septum in LBBB, then provides an explanation for the observed diminution in local blood flow and reduced glucose consumption in the septum. The corollary is that the more highly stressed, slower contracting LV free wall works harder, consuming more substrate and ATP, and demands more flow.

The Model for Purine Metabolism in Cardiomyocytes

The cardiac responses to asynchronous activation in the two regions of the model are different because of the differential in ATP turnover rates at the cross-bridge. Presumably, activation and ion fluxes are similar in the two regions since the whole cardiac syncytium is activated. The cross-bridge hydrolysis of ATP drives the demand for production of ATP and the utilization of substrates and oxygen to supply ATP. While Beard[6] has built the best detailed model for mitochondrial oxidative phosphorylation, we have chosen, for reasons of simplicity, to use the stoichiometric kinetic model of cellular ATP regulation of van Beek,[7] which is based on the fact that the ADP levels in the cytosol and mitochondrial intermembrane space are the primary drivers for mitochondrial ATP generation. Accepting this premise as being quantitatively descriptive, even though it lacks Beard's mechanistic biochemical detail, we use the muscle model described in FIGURE 1 to drive the rate of conversion of ATP to ADP and the resultant ADP levels to drive the demand for oxygen and substrate, and to provide a byproduct critically important to the model behavior, namely the production and release of adenosine into the interstitial space. This is then used in the model module describing the receptor activation of smooth muscle vasodilatation.

van Beek's model for the mitochondrial response to ADP in the intermembrane space accounts for ATP and ADP in cytosol and intermembrane space, for the permeation of the outer mitochondrial membrane by ATP, ADP, Pi, creatine and phosphocreatine and H^+ ion. Taking one ADP across the inner membrane produces one ATP in return. A fixed stoichiometric relationship to the number of oxygens used completes the stoichiometric relationships, but the rate of adenosine production at low oxygen tension (low PO_2) is then a function of the myokinase reaction and the activity of 5′-nucleotidase in hydrolyzing AMP to adenosine and Pi (inorganic phosphate). Since 5′-nucleotidase appears to be downregulated in ischemia[8,9] and the dependency of downregulation on the level of hypoxia and its duration is not at all well defined, this is an area of model and experimental exploration. Fortunately, there are data on the permeability–surface area products for adenosine in myocardial endothelial cells and cardiomyocytes,[10,11] both of which are needed for constraining the estimates of the intracellular concentrations.

The Model for Capillary–Tissue Exchange and Metabolism

The minimal capillary–tissue exchange unit is composed of blood, endothelial cells, interstitial space, and the parenchymal cells of the organ, in this case the cardiomyocytes. Endothelial cells lining myocardial capillaries comprise about 1% of tissue volume, yet impede transport of blood solutes to the contractile cells and they takeup and release substrates in competition with myocytes. Solutes permeating this barrier exhibit concentration gradients along the capillary. We use here a generic model, GENTEX, to characterize blood–tissue exchanges.[12] GENTEX is a whole-organ model of the vascular network providing intra-organ flow heterogeneity and accounts for substrate transmembrane transport, binding, and metabolism in erythrocytes, plasma, endothelial cells, interstitial space, and cardiomyocytes. Its primary use has been in the analysis of data from positron tomographic imaging and magnetic resonance imaging for the estimation of regional flows and metabolism. The model for purine nucleoside metabolism has been characterized via the analysis of multiple tracer indicator-dilution data from the isolated Krebs–Henseleit-perfused non-working hearts, accounting for uptake and metabolism in myocytes and endothelial cells.

The blood–tissue exchange unit is described by partial differential equations for each of the five regions, solved by special numerical techniques[13] to account for the convection along the capillary, diffusional spreading of the solutes within capillary and interstitium, transmembrane transport though specialized transporters,[10,11] and the intracellular reactions. For this "heart model," the standard GENTEX model[12] has been modified to allow for variable flow, complicating the numerical methods, but essential to defining the differential responses in the two halves of the cardiac contractile model. For the variable flow, a Mac-Cormack solver was chosen from amongst the solvers available within JSim.

Modeling the Regulation of Arteriolar Resistance

The transmembrane transport of adenosine is directly related to the regulation of the capillary blood flow during hypoxia, where it is a powerful vasodilator acting through A_2 receptors on vascular smooth muscle cells of the arterioles. Presumably this is a direct local affect since cardiomyocytes surround the terminal arterioles, and the diffusion distances to the smooth muscle cells are less than a micron.

While it is understood that vasomotor control is still a research forefront, specific metabolites (lactate,

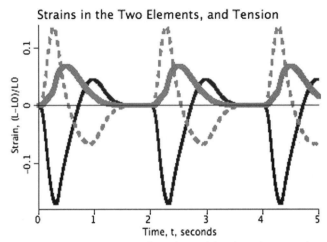

FIGURE 2. Two elements contracting in tandem: The model is driven by a stimulus releasing a wave of increased calcium concentration, thereby allowing the formation of cross-bridges between actin and myosin, the hydrolysis of ATP, and the development of contractile force and shortening. Three cycles are shown. Shortening of the first-stimulated segment (*continuous thin trace*) is slowed by stretching the series elastic element in the second (*dashed trace*) until it begins to contract at about 0.2 s after the first. The series elasticity prevents the stretching of the second segment from being so great that actin–myosin overlap goes to zero, so the second element does develop force. Parameters for the two elements are identical, the only difference being that one is stimulated 200 ms earlier than the other. Development of pressure is represented by the rise in tension (*thick curve*), lifting the weight in the diagram in bottom part of FIGURE 1, resulting in a tension waveform analogous to that of LV cavity pressure.

H^+ ion, CO_2), K^+ released from myocytes, and heightened interstitial osmolarity all play a role. Under conditions of increased cellular work and energy demand, there is also release of purine nucleosides and nucleotides that reach the adenosine receptors on smooth muscle cells to cause vasodilatation. The interesting peculiarity of the situation is that competition between the high-affinity receptors and high-capacity, lower-affinity transporters on the neighboring endothelial and myocardial cells results in the relationship between blood flow and adenosine concentration being shifted dramatically. The result is the shift to a higher apparent K_d (lower apparent affinity) and a marked steepening of the slope of the flow versus [adenosine], both of these phenomena being exhibited for a single site receptor without any cooperativity; appropriate parameters result in the Hill coefficient being about 7.5, as in the data of Stepp *et al.*[4] Thus the modeling can and does bring the varied aspects of microcirculatory regulation together with the local ATP use.

Mathematical Methods

Ordinary differential equations were used for the muscle and metabolic reactions. Partial and ordinary differential equations were used for the capillary–tissue

exchange, and the receptor activation modeling used partial and ordinary differential equations. These were written in a convenient mathematical modeling language (MML) and the model solutions were run and displayed under JSim, our simulation interface system. The models and JSim itself are available for free download <http://www.physiome.org>. Modules of the transporters, pumps, ionic exchanges, capillary–tissue exchange, and some components of the metabolic components are available at this site, and are described in detail there.

Results

Given that the likely initiating events defining the regional flow heterogeneities in the heart are at the level of the metabolic requirements to support contraction, the magnitude of the effects of asynchronicity is critical. Our model is focused on timing events, and it will be important later to take it to a second level to account for differences regionally in initial sarcomere lengths and in the degree of pre-stretch induced by contraction elsewhere.

The behavior of the two-element heart model is shown in FIGURE 2 for 200-ms asynchrony in calcium release. Each element consists of a contractile element

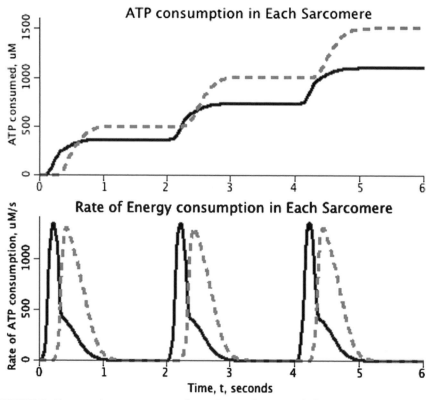

FIGURE 3. Energy utilization in a pair of asynchronously activated elements in series. (Parameters identical to those used for FIG. 2.) *Lower panel:* Rate of ATP hydrolysis by the first-stimulated (*continuous curve*) and second-stimulated (*dashed*) elements of the two-element heart. Although the contraction of the second element prolongs the ATP turnover of the earlier-contracting element, the area under the energy consumption curve is greater for the second element. *Upper panel:* The cumulative ATP use is higher in the late-activated element (*dashed curve*) compared to that of the early-activated element (*continuous curve*), even though both have exactly the same parameters.

in parallel with a viscous damping resistance and an elastic element. The elastic element is itself sufficient to effect stability if the Young's modulus is sufficiently high, although it cannot be so high that the muscle cannot contract, and it must be fairly low in order to allow the observed levels of shortening of 10% or so. The onset of contraction in the second element quickly brings the shortening of the first element to a stop, then forcefully lengthens it toward its rest length. As the second element has the same duration of calcium release, it continues to contract after the first element's contraction ceases, with the result that the first element is stretched to a length greater than its rest length, shown by the thinner black curve lying above 0.0 from about 0.7 to 1.3 s in the plot. The overall shortening of the second, pre-stretched element, is greater than that of the first element, even though this model's equations do not include any Starling effect. The maximum strain of element two is about 0.2, while that of element 1 is

only about 0.15. (Strain is defined as the fractional length change relative to the initial length.)

These changes of length, represented by the velocities of shortening and lengthening, were translated into ATP use through the equations for the reaction rates for the Landesberg–Sideman model, a four-state model for the Ca-bound and unbound forms of troponin-C and the activation (or not) of the strong versus weak form of the Ca-bound form. The assumption was that the rate of ATP use depended not only on the concentrations of the reactants, but also on the shortening velocity itself, as Landesberg and Sideman[5] also considered, the ATP hydrolysis rate decreasing with increasing shortening velocity, describing the decreased likelihood of cross-bridge attachment and ATP turnover when the myosin heads move too rapidly to bind to the attachment sites on the actin. The result is that more ATP, and therefore oxygen, is used by the late-activated element, as is shown in FIGURE 3.

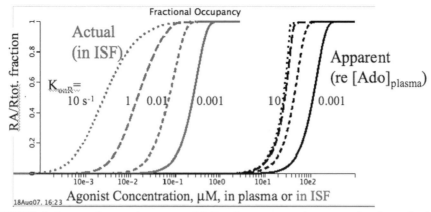

FIGURE 4. Receptor fractional occupancy, RA/Rtot, for an interstitial receptor, when there is hindered delivery of the agonist and competition for it from lower-affinity transporters and enzymes in the interstitial space (ISF). Agonist was delivered at high flow into the plasma space with a slowly rising concentration and permeated the capillary with a PS of 0.2 mL/(g·min). The actual receptor $K_D = 1 \, \mu$M and the competing enzyme dissociation constant $K_E = 10$ mM, 10,000 times lower affinity. The four curves to the *left* show RA/Rtot versus interstitial concentration [agonist]$_{ISF}$. The group of four curves to the *right* show RA/Rtot versus plasma concentration [agonist]$_{pl}$.

Intermediate in the system, lying between the contractile events and the flow responses, is the metabolic network and its regulation. This concerns the utilization of substrates, the shifts to downregulate glucose utilization in the early-activated regions and upregulate it in late-activated regions, the balance between glucose and fatty acid use, and, over the longer term, the regulation of transcription of contractile proteins, and of the growth of capillaries in the late-activated regional subject to regional hypertrophy.

Linking Local Metabolism to Local Flow

There are multitudinous factors involved in vasoregulation, and in this modeling approach we attempt only to give an example of the kind of influences at play, focusing on the observations of Stepp et al.[4] that the relationship between plasma adenosine concentration and coronary blood flow is switch-like: the Hill coefficient for the plot of flow versus plasma [adenosine] is about 7.5.

FIGURE 4 shows the fractional occupancy of a receptor with a single binding site, and no cooperativity, as a function of the concentration in the ISF and in the capillary plasma. The geometry of the model is normal: a capillary bounded by a permeable capillary wall, a surrounding interstitial fluid region (ISF), and a receptor within the ISF. The receptor has a high affinity for adenosine, and is in competition with low-affinity reaction sites consuming adenosine, for example, ISF adenosine deaminase or the adenosine transporters on the neighboring cells, which are low-affinity sites but in

high concentrations. The interesting thing is that the presence of the low-affinity competitors transforms the receptor activation from a standard Michaelis–Menten relationship with a Hill coefficient of unity into almost switch-like activation.

FIGURE 4 shows in the four curves to the left, the relationship of site occupancy to the local adenosine concentration at four different rates of association of agonist to receptor. At a fast rate of binding, $K_{onR} = 10$/sec, and the binding curve is close to the curve for equilibrium dissociation, but the apparent K_D is about $2.10-3$ or twice the actual K_D (dotted leftmost curve). At slower binding rates (with the same K_D), the binding curves are shifted strongly rightward (apparent lower affinities) and the curve become steeper (higher effective Hill coefficients).

The four curves to the right show the relationships between fractional occupancy, RA/R$_{tot}$, and the driving concentration in the plasma. What is particularly interesting is that *all* four curves have a high Hill coefficient, and that binding to the receptor is resulting in switch-like behavior from the point of view of the plasma concentrations. Since the ISF concentration cannot normally be observed, this is what counts. The results show that the data of Stepp et al.[4] showing the Hill coefficient of 7.5 can be explained in terms of a noncooperative receptor binding of a single agonist molecule. Secondly, the rate of agonist binding has relatively little influence; at the fastest association rate, 10/s, the apparent K_D is 25 mM, and with K_{onR} 10^4 slower, the apparent K_D is only 4 times higher, not 10^4.

The general result follows from this illustration of behavior in FIGURE 4: whenever a high-affinity binding site (enzyme, transporter, receptor, simple binding site) is situated behind a membrane or in a location where there is slow access to the site, the presence of a lower-affinity, high-capacity site in the neighborhood will shift the behavior of the *system* in a fashion that the binding site appears to demonstrate cooperativity (a high Hill coefficient) and to be operating with a much reduced affinity (a higher K_D). This obviously is a warning concerning the interpretation of data suggesting cooperativity.

Discussion

At this point in the development of the concept, the computer modeling shows that a rational conceptual basis for the normal heterogeneity of flow may be based on local differences in a variation in local work loads within the myocardium. There are many further steps to be taken before this can be regarded as the *accepted* working hypothesis.

The evidence from the heterogeneity of fatty acid uptake[14] is strong because those data can only be explained by specialized mechanisms for either uptake or retention of isotopically labeled fatty acid, and an alternative hypothesis of passive uptake and retention is rejected by the modeling analysis. High-flow regions required much greater transport or reaction capacity than low-flow regions, implying that specific proteins (transporters or enzymes) had much higher expression levels in the high-flow regions.

The same case might be made for the oxygen data— that is, high-flow regions required more oxygen, except in this case it is not proven that expression levels of mitochondrial proteins involved in oxidative phosphorylation are expressed to higher levels in high-flow regions. This is something that might be examined.

Another test of the hypothesis would be to obtain at higher spatial resolution the evidence relating local work load to substrate uptake. Ultrasound measures of local strain, coupled into three-dimensional model interpretation in terms of work, coupled with evidence of local oxygen consumption, would strengthen the evidence in favor of the idea.

The modeling discussed here is not definitively described, in the sense that the detailed equations, the justifications for them, the validation against experimental data on a time-point by time-point basis, have not been provided. Thus the requirements set forth for modeling, defined as a set of proposed standards for publishing models and put forward on the website <www.physiome.org/Models>, are not fulfilled. However, the individual components are available as modules on that website, and can be built upon by other investigators.

Conclusions

The anatomic–physiologic basis for the broad heterogeneity of regional blood flows observed in the normally functioning healthy heart is based on innate heterogeneity of contractile workload. More work requires more ATP formation, more substrate, and the delivery of more oxygen and substrate to the higher-flow regions. The modeling predicts that increases in ATP turnover rates result in increased local flow, which makes good sense with transitions from rest to exercise or exerting stronger contractions in any particular region. The model, lacking any description for the regulation of transcription, fails of course to predict the long-term adaptations that occur in the LV free wall subsequent to initiating LBBB, namely hypertrophy and an increased ratio in the number of capillaries per cardiocyte. Further data to test these concepts is needed.

Acknowledgments

The author's research has been supported by NIH Grants HL19139 (for the heterogeneity studies) and RR01243 and BE01973 (for the development of JSim), and by NSF Grant 0506477 and an NIH grant for the multiscale model designing.

Conflict of Interest

The author declares no conflicts of interest.

References

1. ALTEHOEFER, C. 1998. Editorial: LBBB: challenging our concept of metabolic heart imaging with fluorine-18-FDG and PET. J. Nucl. Med. **39:** 263–265.
2. PRINZEN, F.W., C.H. AUGUSTIJN, T. ARTS, *et al.* 1990. Redistribution of myocardial fiber strain and blood flow by asynchronous activation. Am. J. Physiol. (Heart Circ. Physiol. 28) **259:** H300–H308.
3. VAN OOSTERHOUT, M.F.M., T. ARTS, J.B. BASSINGTH-WAIGHTE, *et al.* 2002. Relation between local myocardial growth and blood flow during chronic ventricular pacing. Cardiovasc. Res. **53:** 831–840.
4. STEPP, D.W., R. VAN BIBBER, K. KROLL & E.O. FEIGL. 1996. Quantitative relation between interstitial adenosine concentration and coronary blood flow. Circ. Res. **79:** 601–610.
5. LANDESBERG, A. & S. SIDEMAN. 2000. Force-velocity relationship and biochemical-to-mechanical energy

conversion by the sarcomere. Am. J. Physiol. Heart Circ. Physiol. **278:** H1274–H1284.

6. BEARD, D.A. 2006. A biophysical model of the mitochondrial respiratory system and oxidative phosphorylation. PLoS Comput. Biol. **1:** e36 (0252–0264).

7. VAN BEEK, J.H.G.M. 2007. Adenine nucleotide-creatine-phosphate module in myocardial metabolic system explains fast phase of dynamic regulation of oxidative phosphorylation. Am. J. Physiol. Cell Physiol. **293:** C815–C829.

8. GUSTAFSON, L.A. & KROLL, K. 1998. Down regulation of 5′ nucleotidase in rabbit heart during coronary underperfusion. Am. J. Physiol. Heart Circ. Physiol. **274:** H529–H538.

9. GUSTAFSON, L.A., C.J. ZUURBIER, J.E. BASSETT, et al. 1999. Increased hypoxic stress decreases AMP hydrolysis in rabbit heart. Cardiovasc. Res. **44:** 333–343.

10. SCHWARTZ, L.M., T.R. BUKOWSKI, J.H. REVKIN & J.B. BASSINGTHWAIGHTE. 1999. Cardiac endothelial transport and metabolism of adenosine and inosine. Am. J. Physiol. Heart Circ. Physiol. **277:** H1241–H1251.

11. SCHWARTZ, L.M., T.R. BUKOWSKI, J.D. PLOGER & J.B. BASSINGTHWAIGHTE. 2000. Endothelial adenosine transporter characterization in perfused guinea pig hearts. Am. J. Physiol. Heart Circ. Physiol. **279:** H1502–H1511.

12. BASSINGTHWAIGHTE, J.B., G.R. RAYMOND, et al. 2006. GENTEX, a general multiscale model for *in vivo* tissue exchanges and intraorgan metabolism. Phil. Trans. Roy. Soc. A: Math. Phys. Eng. Sci. **364:** 1423–1442.

13. BASSINGTHWAIGHTE, J.B., C.Y. WANG & I.S. CHAN. 1989. Blood-tissue exchange via transport and transformation by endothelial cells. Circ. Res. **65:** 997–1020.

14. CALDWELL, J.H., G.V. MARTIN, G.M. RAYMOND & J.B. BASSINGTHWAIGHTE. 1994. Regional myocardial flow and capillary permeability-surface area products are nearly proportional. Am. J. Physiol. Heart Circ. Physiol. **267:** H654–H666.

Regulation of Endothelial Junctional Permeability

EMILY VANDENBROUCKE, DOLLY MEHTA, RICHARD MINSHALL, AND ASRAR B. MALIK

Department of Pharmacology and Center for Lung and Vascular Biology, The University of Illinois College of Medicine, Chicago, Illinois, USA

The endothelium is a semi-permeable barrier that regulates the flux of liquid and solutes, including plasma proteins, between the blood and surrounding tissue. The permeability of the vascular barrier can be modified in response to specific stimuli acting on endothelial cells. Transport across the endothelium can occur via two different pathways: through the endothelial cell (transcellular) or between adjacent cells, through interendothelial junctions (paracellular). This review focuses on the regulation of the paracellular pathway. The paracellular pathway is composed of adhesive junctions between endothelial cells, both tight junctions and adherens junctions. The actin cytoskeleton is bound to each junction and controls the integrity of each through actin remodeling. These interendothelial junctions can be disassembled or assembled to either increase or decrease paracellular permeability. Mediators, such as thrombin, TNF-α, and LPS, stimulate their respective receptor on endothelial cells to initiate signaling that increases cytosolic Ca^{2+} and activates myosin light chain kinase (MLCK), as well as monomeric GTPases RhoA, Rac1, and Cdc42. Ca^{2+} activation of MLCK and RhoA disrupts junctions, whereas Rac1 and Cdc42 promote junctional assembly. Increased endothelial permeability can be reversed with "barrier stabilizing agents," such as sphingosine-1-phosphate and cyclic adenosine monophosphate (cAMP). This review provides an overview of the mechanisms that regulate paracellular permeability.

Key words: interendothelial junctions; paracellular permeability; endothelial regulation

Introduction

The endothelium is a semi-permeable barrier that lines the vasculature and regulates fluid and solute exchange between the blood and interstitial space.[1] Endothelial cells (ECs) arise from the mesoderm and line all blood and lymphatic vessels. In pathological states, endothelial barrier dysfunction is responsible for protein-rich tissue edema and is a significant pathogenic component in multiple diseases, such as atherosclerosis[2] and diabetes-associated vascular disease.[3] There are two pathways that allow solutes to traverse the endothelium: transcellular and paracellular. The transcellular pathway, also known as transcytosis, is defined as vesicle-mediated transport of macromolecules, e.g., plasma proteins, across the endothelial barrier in a caveolae-dependent manner.[4] The paracellular pathway is formed by the minute intercellular space between contacting cells. Its primary function is to restrict free passage of macromolecules in the range of 3 nm and above through interendothelial junctions (IEJs), while allowing the convective and diffusive transport of molecules of less than 3 nm in diameter.[5] Permeability of the IEJs is determined by the adhesive properties of the proteins that comprise the tight junctions (TJs) and adherens junctions (AJs) (FIG. 1). However, the junctional barrier is a dynamic structure. It responds to permeability-increasing agonists and migrating leukocytes with disassembly of IEJs and to barrier-stabilizing mediators by increasing the surface expression and adhesiveness of junctional proteins in order to strengthen IEJs.[1,6,7]

Paracellular Permeability

The primary function of AJs and TJs is to maintain tissue fluid homeostasis. The Starling equation explains the physical forces governing the rate of fluid exchange across the endothelium (for review, see Ref. 8). The net fluid filtration rate is: $\mathcal{J}_v = L_P S[(P_c - P_i) - \sigma(\pi_c - \pi_i)]$, where L_p is hydraulic conductivity, S is vessel surface area, P is capillary hydrostatic pressure, σ is the osmotic reflection coefficient, and π is

Address for correspondence: Prof. Asrar B. Malik, PhD, Department of Pharmacology, 835 S. Wolcott (m/c 868) Chicago, IL 60612. Fax: (312) 996-1225.

abmalik@uic.edu

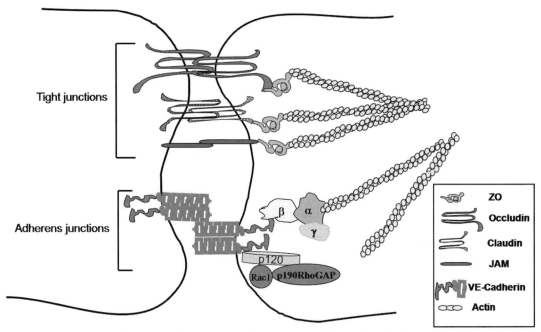

FIGURE 1. Structure of IEJs. A typical arrangement of IEJs consisting of TJs and AJs which generates endothelial cell–cell adhesion is shown in this figure. Occludins, claudins, and JAMs are the backbones of TJs, whereas VE-cadherin is required for formation of AJs. While the extracellular domains of occludins, claudins, JAMs, and VE-cadherin maintain cell–cell contact, intracellular domains provide junctional stability through their linkages with the actin cytoskeleton via α, β, and γ-catenins or zona occludin-1 protein (ZO-1). p120 can regulate endothelial permeability changes through association with p190RhoGAP and Rac1. p190RhoGAP activates Rac1 and decreases RhoA activity to combat permeability increases and facilitate re-annealing of the IEJs. (Modified from Mehta and Malik[1]; used with permission.)

oncotic pressure, with subscripts i and c for interstitial and capillary compartments, respectively. Net filtration is at equilibrium between proximal and distal capillaries for a given tissue. Tissue edema ensues when equilibrium is disrupted.[9] An increase in endothelial permeability caused by disassembly of IEJs is attenuated once P_i is equal to P_c.[8] Thus, equalizing the pressure can be regarded as a short-term protective mechanism that limits tissue edema formation.[9]

Experimental manipulation of forces in the Starling equation allows the measurement of permeability of blood vessels.[10,11] The capillary filtration coefficient ($K_{f,c}$), the value of which is equivalent to L_pS,[8,11] can be experimentally determined (for methodology, see Ref. 11). An increase in the $K_{f,c}$ signifies an increase in endothelial permeability. Determining the σ of a specific molecule, such as albumin, is a means (based on the Starling equation) of quantifying endothelial permeability.[10] When σ of albumin equals zero, the membrane is fully permeable to albumin, and if σ equals 1, the membrane is fully impermeable.[1]

Structure of the Interendothelial Junctions

The Adherens Junctions

AJs represent the majority of junctions comprising the endothelial barrier,[1] in contrast to the epithelium where TJs predominate.[12] AJs are composed of vascular endothelial (VE) cadherin consisting of 5 cadherin-like repeats that associate homotypically with VE-cadherin on the adjacent cell in a Ca^{2+}-dependent manner (FIG. 1). The juxtamembrane domain (JMD) of VE-cadherin binds p120-catenin (p120), whereas the C-terminal domain binds with β-catenins and α-catenins, which in turn link AJs to the actin cytoskeleton.[13] Although adhesion between cadherin and catenins is required for maintaining AJ integrity, p120 binding to cadherins is perhaps the most important determinant of AJ stability.[14] The scaffold function of p120 serves to regulate the interactions between cadherins, kinases, phosphatases, and RhoGTPases, which in turn control the phosphorylation state and stability of cadherin interaction with one another, as well as with catenins.[1] In addition, p120 via its interaction with the molecular motor kinesin and transcription factor kaiso may regulate surface expression of cadherins.[15,16] This conclusion is supported by the findings that deletion of p120 from ECs using small interfering RNA (siRNA) causes a marked decrease in cellular levels of VE-cadherin,[17,18] whereas overexpression of p120 increases VE-cadherin surface expression.[18] Thus, p120 is an important modulator of VE-cadherin expression and thereby can control endothelial permeability by multiple mechanisms.

Dynamic interactions between the AJ proteins and the actin cytoskeleton are crucial to understanding the regulation of junctional permeability. Whereas the cortical actin band stabilizes AJs, re-organization of actin into contractile stress fibers disrupts AJs. Actin-mediated EC contraction is the result of myosin light chain (MLC) phosphorylation,[19,20] which drives myosin-actin cross-bridge cycling.[20,21] The MLC is phosphorylated by the endothelial isoform of MLC kinase (MLCK) in a Ca^{2+}/calmodulin-dependent manner.[22,23] RhoA, through its downstream effector Rho kinase (ROCK), additionally potentiates MLC phosphorylation by inhibiting MLC phosphatase (MLCP or PP1) activity.[24,25] Actin/myosin-driven contraction generates a centripetal contractile force that pulls VE-cadherin inward, forcing it to dissociate from its adjacent partner thereby producing interendothelial gaps.[26] The disruption of VE-cadherin homotypic interaction is followed by internalization of AJ proteins and requires signaling cues to re-localize them to the lateral membrane for reformation of AJs.[1]

Another possible mechanism of AJ disassembly and interendothelial gap formation involves microtubule disassembly.[1] Permeability-increasing agonists, such as TNF-α, cause destabilization of microtubules as a means of inducing barrier dysfunction.[27] Interestingly, microtubule destabilization increases endothelial contraction through a RhoA/ROCK-dependent, but MLCK-independent, pathway.[19,27,28] LimK domain-containing kinase 1 (LIMK1), a downstream effector of ROCK, may be an important mediator of regulating the state of actin and microtubule assembly during a MLCK-independent hyperpermeability response. It has been shown to coordinate both microtubule disassembly and actin stress fiber formation in response to thrombin in ECs.[29] Thus, microtubule destabilization may serve to amplify EC contraction (beyond that induced by myosin-actin cross-bridging), thereby inducing a profound increase in EC permeability. Stabilization of microtubules inhibited EC contraction and prevented the redistribution of AJs and interendothelial gap formation in response to permeability agonists,[28] supporting the contention that microtubule

assembly is crucial in maintaining actin-induced AJ integrity.

Integrity of AJs can be regulated by changes in phosphorylation of AJ proteins.[30] Thrombin signals through protein kinase C (PKC) α-induced modification of VE-cadherin and β-catenin phosphorylation to cause disassembly of AJs.[30] Vascular endothelial growth factor (VEGF)-mediated phosphorylation of VE-cadherin at Ser665 residue induced internalization of VE-cadherin and loss of endothelial barrier function.[31] In addition, p120 function is regulated by phosphorylation on its amino terminus.[14,32] Cadherin internalization in response to an agonist (an endothelial permeability increasing mediator, such as thrombin) increased the level of cytoplasmic p120, which induced a decrease in active RhoA and an increase in levels of the barrier-stabilizing GTPases, Rac1 and Cdc42.[33] Decrease in tyrosine phosphorylation of VE-cadherin strengthens homotypic cell–cell adhesion by increasing the amount of p120 associating with the cadherin.[34] Thus, phosphorylation-induced increase in cytosolic p120 may be a crucial feedback signaling mechanism promoting an increase in cell–cell contact needed for an intact endothelial barrier.

Tight Junctions

In comparison to AJs, the regulation of TJs in the endothelial barrier is far less well understood. Although only one-fifth of the cell junctions in the endothelium are TJs,[1] they are essential in maintaining the integrity of the endothelial barrier. Endothelial TJs are formed by the homotypic adhesion of occludin or claudins (for review, see Refs. 35, 36) (Fig. 1). Occludin is expressed in ECs and epithelial cells and requires the co-expression of junctional protein zona occludin-1 (ZO-1) for cell surface expression.[36,37] Zona occludins 1–3, present in ECs, are intracellular components of TJs that associate with cortical actin.[36] Of the 24-member claudin family only claudin-5 appears to be an endothelial-specific isoform.[38] ZOs bind claudins and occludin to the actin cytoskeleton through cingulin.[39] These interactions allow actin rearrangement to regulate TJ stability through a similar mechanism as described for AJs.[36] As for specific mechanisms governing TJ regulation, the research is extremely limited. PKC activation clearly affects the stability of the junctions; however, how it affects stability can vary with different PKC activators.[36,40,41] PKC activation generally supports assembly of TJs but impairs the integrity of established TJs,[35] suggesting phosphorylation of TJ proteins has a different effect on endothelial permeability depending on whether TJs are in the assembled or disassembled state.

Junctional Adhesion Molecules

Junctional adhesion molecules (JAMs) members of the IgG superfamily of proteins, are transmembrane proteins that also form homotypic interactions with neighboring ECs,[42] further allowing them to stabilize the endothelial barrier. JAM expression is upregulated and they are redistributed away from junctions in response to inflammation and ischemia.[42] However, JAMs are the least understood member of the IEJ adhesive protein complex regulating paracellular permeability.[42] JAM-A, JAM-B, and JAM-C are expressed in ECs[42]; JAM-A colocalizes with occludin, claudins, ZOs, and cingulin at the level of TJs.[43] It is postulated that JAM-A regulates junctional assembly through recruiting and binding these proteins to its intracellular C terminus in order to colocalize junctional proteins with the nascent junctions.[36] Evidence in fact shows that JAM-A facilitates the localization of both ZO-1 and occludin at sites of developing TJs in the membrane.[43] JAM-A responds to permeability-increasing cytokines by dissociating from the actin cytoskeleton and destabilizing junctions.[35,44,45] JAM-C regulates paracellular permeability by inducing inhibition of small GTPase Rap1 to allow disassembly of AJs.[46] Inhibition of JAM-C function blocked inflammation-associated as well as angiogenesis-associated increases in endothelial permeability.[46] JAM-B is upregulated in chronic inflammation,[47] and both JAM-A and JAM-C are upregulated in ECs obtained from atherosclerotic vessels[48,49]; however, the role of JAMs in the mechanism of increased endothelial permeability associated with these diseases is not understood.

Regulating Junctional Permeability

Role of Small GTPases

Small RhoGTPases regulate IEJ stability and hence endothelial permeability through the control of actin dynamics.[50] As described above, activated RhoA disrupts IEJs by inducing MLC-dependent actin stress fibers formation and initiating cytoskeleton retraction.[19,27,51] Inhibition of RhoA blocked the permeability response to thrombin.[52] Downstream of RhoA, ROCK also disrupts junctions by phosphorylating occludin and promoting disassembly of TJs.[53] ROCK may also phosphorylate ERM (ezrin, radixin, and moesin) proteins, which in turn may induce actin stress fiber formation leading to increased EC permeability.[54–56]

While RhoA decreases barrier function, Rac1 and Cdc42 signal to repair IEJs.[50] RhoA is rapidly activated in response to permeability agonists (such as thrombin) whereas Rac1 and Cdc42 are activated only after RhoA activity subsides.[57] This sequential activation of RhoGTPases suggests that a cross-talk exists between GTPases that controls the opposing activities of these GTPases.[50] Perhaps spatial cues induced by p120 and upstream proteins regulating RhoA activity (described below) may shift the balance from RhoA to Rac1 or Cdc42. Rac1 and Cdc42 stabilize the endothelial barrier by regulating the interactions between α-catenin and cadherins.[58] Rac1 has also been suggested to disrupt the association between β-catenin and IQGAP (another junctional protein), leaving β-catenin free to associate with cadherins and induce the formation of AJs.[59] Rac1 and Cdc42 re-established cell–cell contact by stimulating actin assembly into cytoplasmic protrusions lamellipodia or filopodia,[50,60] which enabled ECs to contact one another and reform IEJs. Upon activation, Cdc42 translocates to the membrane from the cytosol.[61] As VE-cadherin remains bound to β-catenin during internalization, it is possible that activated Cdc42 may reform IEJs by facilitating binding of α-catenin to the β-catenin/VE-cadherin complex.[62] There is delayed activation of Cdc42 in response to thrombin, consistent with the hypothesis that Cdc42 activation is a positive feedback mechanism to promote re-annealing of disrupted junctions.[61] Experiments using a dominant negative form of Cdc42 in ECs stimulated with thrombin showed that Cdc42 was required for proper re-annealing of AJs.[61,62] Likewise, Cdc42 was shown to be required for reversing the increase in lung microvascular permeability post activation of thrombin receptor.[27]

RhoGTPase activity is subject to regulation by guanine nucleotide exchange factors (GEFs), guanine nucleotide dissociation inhibitors (GDIs), or GTPase activating proteins (GAPs). In the context of RhoA activation, studies showed that RhoGDI-1 (RhoGDIα) repressed RhoA activation and thus protected IEJs from disassembly.[63] With stimulation by permeability agonists, RhoGDI-1 was phosphorylated by PKCα, which activated RhoA and enabled the formation of actin stress fibers. RhoA activity can be inhibited through activation of p190RhoGAP.[64] Focal adhesion kinase (FAK) activates p190RhoGAP after thrombin stimulation to inhibit the increase in permeability facilitated by RhoA and to signal for re-assembly of disrupted IEJs.[64] Although much is still a mystery about the role of RhoGTPases in endothelial permeability, therapies that control RhoGTPase function can modulate the increase in endothelial permeability in inflammatory diseases.

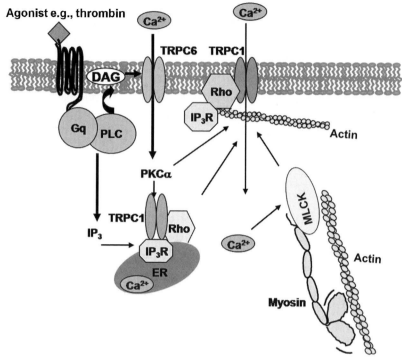

FIGURE 2. Modes of Ca^{2+} entry into ECs. Upon activation by an agonist (such as thrombin), the receptor activates G_q, which binds phospholipase C (PLC). PLC cleaves PIP_2 into IP_3 and DAG. DAG activates receptor-operated channel TRPC6, whereas IP_3 binds the IP_3R on the ER to trigger the release of Ca^{2+} stores from the ER into the cytosol. TRPC6 via $PKC\alpha$ activates RhoA. Upon activation RhoA couples with IP_3R and TRPC1 and facilitates Ca^{2+} entry by inserting TRPC1 in the plasma membrane. $PKC\alpha$ may also phosphorylate TRPC1 to promote RhoA-dependent TRPC1 activity. The increase in Ca^{2+} also activates MLCK. Both RhoA and MLCK induce phosphorylation of MLC and actin–myosin mediated EC contraction. (Modified from Mehta and Malik[1]; used with permission.)

Regulatory Role of Ca^{2+}

A rise in cytosolic Ca^{2+} has been established as the initial pivotal signal that precedes EC contraction and disassembly of IEJs.[22,26] Supporting this notion is the finding that the addition of an intracellular Ca^{2+} chelator prevented the increase in endothelial permeability caused by thrombin.[21] After EC stimulation with permeability-increasing agonists, the intracellular Ca^{2+} concentration increases in two distinct phases (FIG. 2). The initial increase is due to Ca^{2+} release from stores within the endoplasmic reticulum (ER). As shown in FIGURE 2, Ca^{2+} is released from the ER via an inositol triphosphate (IP_3)-dependent mechanism.[65,66] The second phase involves Ca^{2+} entry from extracellular milieu mediated by activation of receptor-operated Ca^{2+} channels (ROC) or store-operated Ca^{2+} channels (SOC) in the plasma membrane.[20,67,68] SOC channels are activated after depletion of ER Ca^{2+} stores when IP_3 associates with the IP_3 receptor (FIG. 2),[20,68,69] whereas ROC channels are activated by diacylglycerol (DAG).[20,68]

Transient receptor potential canonical (TRPC) channels are important in regulating Ca^{2+} concentration and signaling in ECs.[1,51,69,70] TRPC channels can be subdivided into SOCs and ROCs. TRPC1 and TRPC4 form SOCs whereas TRPC3, TRPC6, and TRPC7 form ROCs. ECs express six different isoforms of TRPCs, with TRPC1 and TRPC6 the most abundant forms.[20,69,71] RhoA initiates Ca^{2+} influx through TRPC1 to increase endothelial permeability.[72] The exact mechanism of TRPC1 activation is unknown, although it has been proposed that RhoA-dependent actin reorganization by coupling IP_3 receptor with TRPC1 mediates TRPC1 activation.[66] Studies demonstrate that $PKC\alpha$ phosphorylation of serine/threonine residues on TRPC1 is also required for TRPC1 activation.[73] The expression of TRPC1 determines the extent of the increase in Ca^{2+},[69] implying that the level of TRPC1 expression is important in regulating the extent of the change in permeability. Overexpression of TRPC1 augmented the formation of actin stress fibers caused by thrombin stimulation.[69] In addition, the *in vivo* relevance of TRPC-induced Ca^{2+} entry in regulating microvascular permeability has been demonstrated using $TRPC4^{-/-}$ mice.[74,75] $TRPC4^{-/-}$ mice demonstrated a reduction in the thrombin-induced increase in lung microvessel permeability by $\sim$50%.[75] Microvessel ECs isolated from $TRPC4^{-/-}$ mice also showed an inhibition of Ca^{2+} entry, actin stress fiber formation, and an increase in endothelial monolayer permeability in response to thrombin,[75] consistent with the hypothesis that TRPC1 and TRPC4 expression, and the change in intracellular Ca^{2+}, is an important regulator of the permeability response. Knockdown of TRPC6 (an ROC) inhibited RhoA activation, MLC phosphorylation, and formation of actin stress fibers in response to PKC activators and thrombin,[51] suggesting that TRPC6 activation occurs upstream of RhoA activation. Activation of SOCs did not cause RhoA activation,[51] supporting the hypothesis that TRPC1 activation is downstream of RhoA in ECs. Thus, TRPC6 and TRPC1/4-mediated rise in intracellular Ca^{2+} represent the key mechanism regulating EC contraction and IEJ integrity.

Regulatory Role of PKC Isoforms

PKC isoforms are important regulators of junctional permeability. The PKC activator phorbol ester increases endothelial permeability, and PKC inhibition prevented an increase in permeability.[76] PKCα, a Ca^{2+}- and DAG-dependent isoform, has a crucial role in mediating IEJ disassembly,[20,21,64,73,77,78] although Ca^{2+}-independent isoforms, notably PKCζ, may also be important in IEJ disruption (Minshall, unpublished observations). Agonists increasing endothelial permeability stimulated translocation of PKCα from the cytosol to the IEJ,[21] where it phosphorylated residues on both VE-cadherin[21] and p120.[18] Specifically, PKCα phosphorylated residue S873and caused dephosphorylation of residue S268 on p120[18] to regulate both p120 and cadherin stability.[18,79–82] The latter effect may be the result of the ability of PKCα to activate phosphatases to dephosphorylate critical residues on p120 to destabilize junctions and increase endothelial permeability.[14] Activation of RhoA is regulated by PKCα through phosphorylation of Rho-GDI (as described above)[63] or phosphorylation of p115RhoGEF.[83] This PKCα-RhoA activation pathway may act in concert with PKCα-mediated phosphorylation of AJ proteins to disrupt the barrier.

In addition to PKCα, another Ca^{2+}- and DAG-dependent PKC isoform, PKCβ, regulates endothelial permeability.[84–86] Inhibition of PKCβ decreased endothelial leakage in kidney and retinal microvessels of diabetic animals,[87] and overexpression of PKCβ enhanced the phorbol ester-induced permeability increase.[84] However, PKCβ was also shown to antagonize the thrombin-induced increase in endothelial permeability.[85] In response to thrombin, PKCβ inactivated rising Ca^{2+} levels and MLC contraction.[85]

PKCδ is a "novel" PKC isoform, requiring the presence of DAG but not Ca^{2+}. Studies have shown that PKCδ is required for the permeability increase caused by stimulation with PKC activators or DAG,[88] while others suggest that PKCδ has a barrier-protective function.[78] Overexpression of PKCδ decreases endothelial permeability by increasing focal adhesion contacts.[78] Studies using PKCβ or PKCδ knockout mice models will help to clarify the contribution of these isoforms in mediating changes in endothelial permeability.

Permeability-Increasing Agonists

Activation of ECs results in permeability increases and translocation of transcription factor NF-κB to the nucleus.[89,90] Basal activity of NF-κB regulates homeostatic stability of the endothelium, specifically by regulating TJ barrier proteins.[90] With stimulation of ECs, increased nuclear NF-κB increases expression of inflammatory mediators and permeability-increasing agonists, such as cytokines, chemokines, adhesion molecules, and nitric oxide sythases (NOS).[6,91,92] NF-κB also alters the distribution of TJ proteins to increase permeability, while maintaining the expression level of the proteins.[90,93,94]

Thrombin, a serine protease, binds protease-activated receptor (PAR)-1 on human ECs to cause cell rounding, leukocyte adhesion to ECs, angiogenesis, and increased endothelial permeability.[11,95,96] PAR-1 knockout studies showed that PAR-1 is required for the thrombin-induced permeability increase.[11] Activation of PAR-1 increases cytosolic Ca^{2+} concentration and signals a PKCα-dependent internalization of VE-cadherins[97] (FIG. 3). PAR-1 activation also stimulates RhoA activity, causing cell contraction.[98] PAR-1 activates heterotrimeric G-proteins G_q, $G_{12/13}$, and G_i, all of which are involved in permeability regulation.[96] Activation of G_q mobilizes Ca^{2+} and activates PKCα, RhoA, and EC contraction, resulting in endothelial barrier disruption.[53,96] The initial rise in Ca^{2+} may be due to activation of TRPC6 by a G_q effector molecule.[51] Activation of $G_{12/13}$ increases RhoA activity, which increases permeability through formation of actin stress fibers and inactivation of MLC phosphatases.[96] McLaughlin *et al.* showed that thrombin and PAR-1 peptide are capable of activating different G-proteins.[96] Activation of $G_{12/13}$ requires less thrombin than G_q,[96] suggesting that PAR-1 activation of

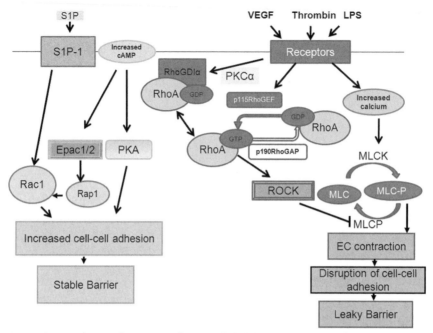

FIGURE 3. Signaling mechanisms regulating endothelial permeability. **(Right)** LPS, thrombin, and VEGF all bind their respective receptors on the EC surface to trigger the disassembly of IEJs and an increase in endothelial permeability. Upon activation PKCα induces GDI-1 and p115RhoGEF phosphorylation, which in turn leads to activation of RhoA. Increased intracellular Ca^{2+} concentration activates MLCK. RhoA and MLCK phosphorylate MLC, allowing EC contraction. **(Left)** Barrier function is enhanced through increased intracellular concentration of cAMP and through binding of S1P to the S1P1 receptor on the cell surface. Increased cAMP activates both PKA and Epac to increase cell–cell adhesion. Epac activates Rap1, which enhances barrier function in a Rac1-dependent manner. S1P via G-proteins Gi (not shown) induces Rac1 activity which increases cell–cell adhesion and thereby endothelial barrier function. (Modified from van Nieuw Amerongen and van Hinsberg.[134])

specific G-proteins determines the downstream effectors mediating the response.

Lipopolysaccharide (LPS) is an endotoxin that resides on the outer membrane of gram-negative bacteria and alerts the immune system of a bacterial infection. LPS binds to Toll-like receptor 4 (TLR4) to signal for translocation of NF-κB to the nucleus to initiate transcription of pro-inflammatory mediators to increase endothelial permeability.[92,98] LPS increases expression of intercellular adhesion molecule 1 (ICAM-1), which also increases permeability.[6,99,100] Cross-linking of ICAM-1 activates RhoA and actin stress fiber formation.[101] ICAM-1 further increases permeability by upregulating RhoA expression in ECs, in what is thought to be a positive feedback mechanism.[101] Interestingly, increased ICAM-2 expression in EC has no effect on RhoA activity or paracellular permeability.[101]

Maintenance of the endothelial barrier requires a basal level of nitric oxide (NO) regulated by endothe-

lial NOS (eNOS).[102] Both reduction in NO production (as in eNOS knockout mice) and increased NO production, secondary to eNOS activation, induced increases in endothelial permeability.[103–106] The permeability response at different concentrations of NO (either high or low) is likely the result of different mechanisms, but both the lack of NO and high NO levels destabilize IEJs.[107–109] During an inflammatory response, production of NO is markedly increased secondary to the expression of inducible NOS (iNOS).[110] iNOS expression activates IP_3, resulting in release of Ca^{2+} from the ER[111] and activation of MLCK. Independent of NOS, generation of NO in response to pressure generated by shear stress also increases permeability.[112] Heparan sulfates expressed on ECs transduce sheer stress into intracellular signals that result in production of NO and reactive oxygen species (ROS), causing barrier dysfunction.[112]

VEGF is a permeability-increasing agonist specific to EC[113] that is essential for angiogenesis during development of the vasculature.[114–117] Tumor cells secrete large amounts of VEGF to increase nutrient exchange with tumor cells and promote metastasis.[116,118] VEGF binds VEGF receptors 1–3 on ECs, which transduce signals that increase permeability.[117,119,120] VEGFR-2 activates Src-family kinases to phosphorylate VE-cadherin, causing its internalization and increasing junctional permeability.[31] VE-cadherin internalization in response to VEGFR-2 activation requires β-arrestin-2, which is recruited after serine phosphorylation of VE-cadherin to facilitate cadherin endocytosis.[31] ROS signaling also plays a role in the response, as it is required for phosphorylation of cadherins by Src-family kinases, downstream of VEGF.[121] Furthermore, extracellular single-stranded RNAs (ssRNAs) have been shown to increase permeability in a VEGFR-2-dependent manner in brain microvascular ECs.[122] ssRNA signals upstream of VEGF and increases permeability through disassembly of TJ. This novel finding suggests that extracellular ssRNA is another regulator of endothelial permeability.[122]

Endothelial Barrier-Protecting Mediators

An increase in vascular permeability is typically reversible, but the mechanisms are less understood than the mechanisms regulating barrier disruption. Understanding how endothelial barrier integrity is restored is important since it may help prevent a potentially fatal increase in permeability. An increase in the concentration of cAMP has been established as a potent second messenger, which prevents increased endothelial permeability in response to several permeability-increasing mediators.[78,123] cAMP induces protein kinase A (PKA) activation, which inhibits RhoA activation and EC contraction in order to decrease paracellular permeability[124] (Fig. 3). PKA inhibition with siRNA showed a dramatic increase in actin rearrangement and interendothelial gap formation, suggesting that PKA is an important stabilizer of endothelial barrier permeability.[125] cAMP can also directly activate Epac, a Rap1GEF,[126] enhancing VE-cadherin junctional integrity and actin reorganization to decrease endothelial permeability.[127] Rap1 decreases basal endothelial permeability by enhancing distribution of junctional proteins for both AJs and TJs[128] through increasing cortical actin, and it antagonizes thrombin-induced increased permeability by inhibiting activation of RhoA.[127,128] Since activation of FAK and Cdc42 also parallels the time course of re-annealing of AJs and endothelial barrier recovery following thrombin challenge,[61,65] it is likely that Rap-1, FAK, and Cdc42 act in concert to downregulate RhoA activity and to promote the re-formation of IEJs.

Sphingosine-1-phosphate (S1P), a biologically active lipid released by platelets and red blood cells (for review, see Ref. 129), has also emerged as an effective barrier-protective agonist in cultured ECs and intact microvessels.[130,131] S1P barrier-protective effects are mediated by S1P receptor (S1P-1). Signaling through S1P-1 activated G_i and caused Rac1-regulated cytoskeletal rearrangement that promoted IEJ reformation[132] (Fig. 3). Other lipid mediators that enhance barrier function include prostaglandins PGE_2 and PGI_2 and Ox-LDL, which help maintain junctional integrity through activation of Rap1, Rac1, and Cdc42.[133]

Conclusions

The endothelium creates a barrier which lines blood and lymphatic vessels and controls the exchange of solutes and fluid. Paracellular or junctional permeability is regulated by the integrity of IEJs. Disruption of IEJs is achieved by increasing intracellular Ca^{2+} to stimulate MLC-dependent cell contraction as well as disassembly of junctional proteins comprising AJs and TJs. Re-annealing of IEJs requires suppression of RhoA activity and actin reorganization mediated by Rac1 and Cdc42. An uncontrolled increase in endothelial permeability underlies pathologies such as atherogenesis, acute lung injury, and metastasis; thus, understanding the subtleties of the signaling mechanisms regulating endothelial permeability will lead to new therapeutic advances.

Conflict of Interest

The authors declare no conflicts of interest.

References

1. MEHTA, D. & A.B. MALIK. 2006. Signaling mechanisms regulating endothelial permeability. Physiol. Rev. **86:** 279–367.

2. LIBBY, P., M. AIKAWA & M.K. JAIN. 2006. Vascular endothelium and atherosclerosis. Handb. Exp. Pharmacol. **176:** 285–306.

3. KIM, J.A. et al. 2006. Reciprocal relationships between insulin resistance and endothelial dysfunction: molecular and pathophysiological mechanisms. Circulation **113:** 1888–1904.

4. MINSHALL, R.D. et al. 2003. Caveolin regulation of endothelial function. Am. J. Physiol. Lung Cell Mol. Physiol. **285:** L1179–L1183.

5. SIMIONESCU, N., M. SIMONIONESCU & G.E. PALADE. 1978. Open junctions in the endothelium of the postcapillary venules of the diaphragm. J. Cell Biol. **79:** 27–44.

6. WYMAN, T.H. *et al*. 2002. A two-insult in vitro model of PMN-mediated pulmonary endothelial damage: requirements for adherence and chemokine release. Am. J. Physiol. Cell Physiol. **283:** C1592–C1603.

7. ORRINGTON-MYERS, J. *et al*. 2006. Regulation of lung neutrophil recruitment by VE-cadherin. Am. J. Physiol. Lung Cell Mol. Physiol. **291:** L764–L771.

8. MINSHALL, R.D. & S.M. VOGEL. 2005. Lung edema and microvascular permeability. *In* Microvascular Research: Biology and Pathology. D. Shepro, Ed.: 491–495. Elsevier Science. Burlington, MA.

9. MINSHALL, R.D., & A.B. MALIK. 2006. Transport across the endothelium: regulation of endothelial permeability. Handb. Exp. Pharmacol. 107–144.

10. FRASER, P.A., A.D. DALLAS & S. DAVIES. 1990. Measurement of filtration coefficient in single cerebral microvessels of the frog. J. Physiol. **423:** 343–361.

11. VOGEL, S.M. *et al*. 2000. Abrogation of thrombin-induced increase in pulmonary microvascular permeability in PAR-1 knockout mice. Physiol. Genomics **4:** 137–145.

12. BALKOVETZ, D.F. 2006. Claudins at the gate: determinants of renal epithelial tight junction paracellular permeability. Am. J. Physiol. Renal Physiol. **290:** F572–F579.

13. LAMPUGNANI, M.G. *et al*. 1995. The molecular organization of endothelial cell to cell junctions: differential association of plakoglobin, beta-catenin, and alpha- catenin with vascular endothelial cadherin (VE-cadherin). J. Cell Biol. **129:** 203–217.

14. XIA, X., D.J. MARINER & A.B. REYNOLDS. 2003. Adhesion-associated and PKC-modulated changes in serine/threonine phosphorylation of p120-catenin. Biochemistry **42:** 9195–9204.

15. YANAGISAWA, M. *et al*. 2004. A novel interaction between kinesin and p120 modulates p120 localization and function. J. Biol. Chem. **279:** 9512–9521.

16. KONDAPALLI, J., A.S. FLOZAK & M.L.C. ALBUQUERQUE. 2004. Laminar shear stress differentially modulates gene expression of p120 catenin, kaiso transcription factor, and vascular endothelial cadherin in human coronary artery endothelial cells. J. Biol. Chem. **279:** 11417–11424.

17. DAVIS, M.A., R.C. IRETON & A.B. REYNOLDS. 2003. A core function for p120-catenin in cadherin turnover. J. Cell Biol. **163:** 525–534.

18. XIAO, K. *et al*. 2003. Cellular levels of p120 catenin function as a set point for cadherin expression levels in microvascular endothelial cells. J. Cell Biol. **163:** 535–545.

19. DUDEK, S.M. & J.G.N. GARCIA. 2001. Cytoskeletal regulation of pulmonary vascular permeability. J. Appl. Physiol. **91:** 1487–1500.

20. TIRUPPATHI, C. *et al*. 2002. Role of Ca^{2+} signaling in the regulation of endothelial permeability. Vascul. Pharmacol. **39:** 173–185.

21. SANDOVAL, R. *et al*. 2001. Ca^{2+} signalling and PKCalpha activate increased endothelial permeability by disassembly of VE-cadherin junctions. J. Physiol. **533:** 433–445.

22. GOECKELER, Z.M. & R.B. WYSOLMERSKI. 1995. Myosin light chain kinase-regulated endothelial cell contraction: the relationship between isometric tension, actin polymerization, and myosin phosphorylation. J. Cell Biol. **130:** 613–627.

23. GARCIA, J., H. DAVIS & C. PATTERSON. 1995. Regulation of endothelial cell gap formation and barrier dysfunction: role of myosin light chain phosphorylation. J. Cell Physiol. **163:** 510–522.

24. NODA, M. *et al*. 1995. Involvement of rho in GTP[gamma]S-induced enhancement of phosphorylation of 20 kDa myosin light chain in vascular smooth muscle cells: inhibition of phosphatase activity. FEBS Letters **367:** 246–250.

25. YOSHIOKA, K. *et al*. 2007. Essential role for class II phosphoinositide 3-kinase {alpha}-isoform in Ca^{2+} induced, Rho- and Rho kinase-dependent regulation of myosin phosphatase and contraction in isolated vascular smooth muscle cells. Mol. Pharmacol. **71:** 912–920.

26. MOY, A.B. *et al*. 1996. Histamine and thrombin modulate endothelial focal adhesion through centripetal and centrifugal forces. J. Clin. Invest. **97:** 1020–1027.

27. PETRACHE, I. *et al*. 2003. The role of the microtubules in tumor necrosis factor-{alpha}-induced endothelial cell permeability. Am. J. Respir. Cell Mol. Biol. **28:** 574–581.

28. VERIN, A.D. *et al*. 2001. Microtubule disassembly increases endothelial cell barrier dysfunction: role of MLC phosphorylation. Am. J. Physiol. Lung Cell Mol. Physiol. **281:** L565–L574.

29. GOROVOY, M. *et al*. 2005. LIM kinase 1 coordinates microtubule stability and actin polymerization in human endothelial cell. J. Biol. Chem. **280:** 26533–26542.

30. KONSTANTOULAKI, M., P. KOUKLIS & A.B. MALIK. 2003. Protein kinase C modifications of VE-cadherin, p120, and {beta}-catenin contribute to endothelial barrier dysregulation induced by thrombin. Am. J. Physiol. Lung Cell Mol. Physiol. **285:** L434–L442.

31. GAVARD, J. & J.S. GUTKIND. 2006. VEGF controls endothelial-cell permeability by promoting the [beta]-arrestin-dependent endocytosis of VE-cadherin. Nat. Cell Biol. **8:** 1223–1234.

32. MARINER, D.J. *et al*. 2001. Identification of Src phosphorylation sites in the catenin p120ctn. J. Biol. Chem. **276:** 28006–28013.

33. NOREN, N.K. *et al*. 2000. p120 Catenin regulates the actin cytoskeleton via Rho family GTPases. J. Cell Biol. **150:** 567–580.

34. LAMPUGNANI, M.G. *et al*. 1997. Cell confluence regulates tyrosine phosphorylation of adherens junction components in endothelial cells. J. Cell Sci. **110:** 2065–2077.

35. BAZZONI, G. 2006. Endothelial tight junctions: permeable barriers of the vessel wall. J. Cell Physiol. **209:** 122–130.

36. BAZZONI, G. & E. DEJANA. 2004. Endothelial cell-to-cell junctions: molecular organization and role in vascular homeostasis. Physiol. Rev. **84:** 869–901.

37. VAN ITALLIE, C.M. & J.M. ANDERSON. 1997. Occludin confers adhesiveness when expressed in fibroblasts. J. Cell Sci. **110:** 1113–1121.

38. MORITA, K. *et al*. 1999. Endothelial claudin: claudin-5/TMVCF constitutes tight junction strands in endothelial cells. J. Cell Biol. **147:** 185–194.

39. CORDENONSI, M. *et al.* 1999. Cingulin contains globular and coiled-coil domains and interacts with ZO-1, ZO-2, ZO-3, and myosin. J. Cell Biol. **147:** 1569–1582.

40. BALDA, M.S. *et al.* 1993. Assembly of the tight junction: the role of diacylglycerol. J. Cell Biol. **123:** 293–302.

41. OJAKIAN, G.K. 1981. Tumor promoter-induced changes in the permeability of epithelial cell tight junctions. Cell **23:** 95–103.

42. WEBER, C., L. FRAEMOHS & E. DEJANA. 2007. The role of junctional adhesion molecules in vascular inflammation. Nat. Rev. Immunol. **7:** 467–477.

43. BAZZONI, G. & E. DEJANA. 2001. Pores in the sieve and channels in the wall: control of paracellular permeability by junctional proteins in endothelial cells. Microcirculation **8:** 143–152.

44. BRUEWER, M. *et al.* 2003. Proinflammatory cytokines disrupt epithelial barrier function by apoptosis-independent mechanisms. J. Immunol. **171:** 6164–6172.

45. MARTINEZ-ESTRADA, O.M. *et al.* 2005. Opposite effects of tumor necrosis factor and soluble fibronectin on junctional adhesion molecule-A in endothelial cells. Am. J. Physiol. Lung Cell Mol. Physiol. **288:** L1081–L1088.

46. ORLOVA, V.V. *et al.* 2006. Junctional adhesion molecule-C regulates vascular endothelial permeability by modulating VE-cadherin-mediated cell-cell contacts. J. Exp. Med. **203:** 2703–2714.

47. AURRAND-LIONS, M. *et al.* 2001. JAM-2, a novel immunoglobulin superfamily molecule, expressed by endothelial and lymphatic cells. J. Biol. Chem. **276:** 2733–2741.

48. KEIPER, T. *et al.* 2005. The role of junctional adhesion molecule-C (JAM-C) in oxidized LDL-mediated leukocyte recruitment. FASEB J. **19:** 2078–2080.

49. OSTERMANN, G. *et al.* 2005. Involvement of JAM-A in mononuclear cell recruitment on inflamed or atherosclerotic endothelium: inhibition by soluble JAM-A. Arterioscler. Thromb. Vasc. Biol. **25:** 729–735.

50. WOJCIAK-STOTHARD, B. & A.J. RIDLEY. 2002. Rho GTPases and the regulation of endothelial permeability. Vasc. Pharmacol. **39:** 187–199.

51. SINGH, I. *et al.* 2007. G{alpha}q-TRPC6-mediated Ca^{2+} entry induces RhoA activation and resultant endothelial cell shape change in response to thrombin. J. Biol. Chem. **282:** 7833–7843.

52. CARBAJAL, J.M. & R.C. SCHAEFFER. 1999. RhoA inactivation enhances endothelial barrier function. Am. J. Physiol. **277:** C955–C964.

53. HIRASE, T. *et al.* 2001. Regulation of tight junction permeability and occludin phosphorylation by RhoA-p160ROCK-dependent and -independent mechanisms. J. Biol. Chem. **276:** 10423–10431.

54. MATSUI, T. *et al.* 1998. Rho-kinase phosphorylates COOH-terminal threonines of ezrin/radixin/moesin (ERM) proteins and regulates their head-to-tail association. J. Cell Biol. **140:** 647–657.

55. TSUKITA, S. & S. YONEMURA. 1999. Cortical actin organization: lessons from ERM (ezrin/radixin/moesin) proteins. J. Biol. Chem. **274:** 34507–34510.

56. KOSS, M. *et al.* 2006. Ezrin/radixin/moesin proteins are phosphorylated by TNF-{alpha} and modulate permeability increases in human pulmonary microvascular endothelial cells. J. Immunol. **176:** 1218–1227.

57. WILDENBERG, G.A. *et al.* 2006. p120-catenin and p190RhoGAP regulate cell-cell adhesion by coordinating antagonism between Rac and Rho. Cell **127:** 1027–1039.

58. KAIBUCHI, K. *et al.* 1999. Regulation of cadherin-mediated cell-cell adhesion by the Rho family GTPases. Curr. Opin. Cell Biol. **11:** 591–596.

59. KURODA, S. *et al.* 1996. Identification of IQGAP as a putative target for the small GTPases, Cdc42 and Rac1. J. Biol. Chem. **271:** 23363–23367.

60. RIDLEY, A. 1999. Rho family proteins and regulation of the actin cytoskeleton. Prog. Mol. Subcell Biol. **22:** 1–22.

61. KOUKLIS,s P. *et al.* 2004. Cdc42 regulates the restoration of endothelial barrier function. Circ. Res. **94:** 159–166.

62. BROMAN, M.T. *et al.* 2006. Cdc42 regulates adherens junction stability and endothelial permeability by inducing alpha-catenin interaction with the vascular endothelial cadherin complex. Circ. Res. **98:** 73–80.

63. GOROVOY, M. *et al.* 2007. RhoGDI-1 modulation of the activity of monomeric RhoGTPase RhoA regulates endothelial barrier function in mouse lungs. Circ. Res. **101:** 50–58.

64. HOLINSTAT, M. *et al.* 2006. Suppression of RhoA activity by focal adhesion kinase-induced activation of p190RhoGAP: role in regulation of endothelial permeability. J. Biol. Chem. **281:** 2296–2305.

65. SCHILLING, W., O. CABELLO & L. RYAN. 1992. Depletion of the inositol 1,4,5-triphosphate-sensitive intracellular Ca^{2+} store in vascular endothelial cells activates the agonist-sensitive Ca(2+)-influx pathway. Biochem. J. **284:** 521–530.

66. MEHTA, D. *et al.* 2003. RhoA interaction with inositol 1,4,5-trisphosphate receptor and transient receptor potential channel-1 regulates Ca^{2+} entry: role in signaling increased endothelial permeability. J. Biol. Chem. **278:** 33492–33500.

67. YAO, X. & C.J. GARLAND. 2005. Recent developments in vascular endothelial cell transient receptor potential channels. Circ. Res. **97:** 853–863.

68. NILIUS, B. & G. DROOGMANS. 2001. Ion channels and their functional role in vascular endothelium. Physiol. Rev. **81:** 1415–1459.

69. PARIA, B.C. *et al.* 2004. Tumor necrosis factor-alpha-induced TRPC1 expression amplifies store-operated Ca^{2+} influx and endothelial permeability. Am. J. Physiol. Lung Cell Mol. Physiol. **287:** L1303–L1313.

70. TIRUPPATHI, C. *et al.* 2006. Ca^{2+} signaling, TRP channels, and endothelial permeability. Microcirculation **13:** 693–708.

71. MOORE, T.M. *et al.* 1998. Store-operated calcium entry promotes shape change in pulmonary endothelial cells expressing Trp1. Am. J. Physiol. Lung Cell Mol. Physiol. **275:** L574–L582.

72. MEHTA, D. *et al.* 2004. Integrated control of lung fluid balance. Am. J. Physiol. Lung Cell Mol. Physiol. **287:** L1081–L1090.

73. AHMMED, G.U. *et al.* 2004. Protein kinase Calpha phosphorylates the TRPC1 channel and regulates store-operated

Ca^{2+} entry in endothelial cells. J. Biol. Chem. **279:** 20941–20949.

74. BIRNBAUMER, L. 2002. TRPC4 knockout mice: the coming of age of TRP channels as gates of calcium entry responsible for cellular responses. Circ. Res. **91:** 1–3.

75. TIRUPPATHI, C. *et al.* 2002. Impairment of store-operated Ca^{2+} entry in TRPC4-/- mice interferes with increase in lung microvascular permeability. Circ. Res. **91:** 70–76.

76. LYNCH, J.J. *et al.* 1990. Increased endothelial albumin permeability mediated by protein kinase C activation. J. Clin. Invest. **85:** 1991–1998.

77. HOLINSTAT, M. *et al.* 2003. Protein kinase C{alpha}-induced p115RhoGEF phosphorylation signals endothelial cytoskeletal rearrangement. J. Biol. Chem. **278:** 28793–28798.

78. HARRINGTON, E.O. *et al.* 2003. Role of protein kinase C isoforms in rat epididymal microvascular endothelial barrier function. Am. J. Respir. Cell Mol. Biol. **28:** 626–636.

79. RATCLIFFE, M.J., L.L. RUBIN & J.M. STADDON. 1997. Dephosphorylation of the cadherin-associated p100/p120 proteins in response to activation of protein kinase C in epithelial cells. J. Biol. Chem. **272:** 31894–31901.

80. RATCLIFFE, M.J., C. SMALES & J.M. STADDON. 1999. Dephosphorylation of the catenins p120 and p100 in endothelial cells in response to inflammatory stimuli. Biochem. J. **338:** 471–478.

81. OHKUBO, T. & M. OZAWA. 1999. p120ctn binds to the membrane-proximal region of the E-cadherin cytoplasmic domain and is involved in modulation of adhesion activity. J. Biol. Chem. **274:** 21409–21415.

82. AONO, S. *et al.* 1999. p120ctn acts as an inhibitory regulator of cadherin function in colon carcinoma cells. J. Cell Biol. **145:** 551–562.

83. MEHTA, D., A. RAHMAN & A.B. MALIK. 2001. Protein kinase C-alpha signals rho-guanine nucleotide dissociation inhibitor phosphorylation and rho activation and regulates the endothelial cell barrier function. J. Biol. Chem. **276:** 22614–22620.

84. NAGPALA, P.G., *et al.* 1996. Protein kinase C beta 1 overexpression augments phorbol ester-induced increase in endothelial permeability. J. Cell Physiol. **166:** 249–255.

85. VUONG, P.T. *et al.* 1998. Protein kinase C beta modulates thrombin-induced Ca2+ signaling and endothelial permeability increase. J. Cell Physiol. **175:** 379–387.

86. DAS EVCIMEN, N. & G.L. KLING. 2007. The role of protein kinase C activation and the vascular complications of diabetes. Pharmacol. Res. **55:** 498–510.

87. YUAN, S.Y. 2002. Protein kinase signaling in the modulation of microvascular permeability. Vascul. Pharmacol. **39:** 213–223.

88. TINSLEY, J.H., N.R. TEASDALE & S.Y. YUAN. 2004. Involvement of PKC{delta} and PKD in pulmonary microvascular endothelial cell hyperpermeability. Am. J. Physiol. Cell Physiol. **286:** C105–C111.

89. COOPER, J.T. *et al.* 1996. A20 blocks endothelial cell activation through a NF-kappa B-dependent mechanism. J. Biol. Chem. **271:** 18068–18073.

90. KISSELEVA, T. *et al.* 2006. NF-{kappa}B regulation of endothelial cell function during LPS-induced toxemia and cancer. J. Clin. Invest. **116:** 2955–2963.

91. ABRAHAM, E. 2003. Nuclear factor-kappaB and its role in sepsis-associated organ failure. J. Infect. Dis. **187**(Suppl 2): S364–S369.

92. OPAL, S.M. 2007. The host response to endotoxin, antilipopolysaccharide strategies, and the management of severe sepsis. Int. J. Med. Microbiol. **297:** 365–377.

93. LEE, H.-S. *et al.* 2004. Hydrogen peroxide-induced alterations of tight junction proteins in bovine brain microvascular endothelial cells. Microvasc. Res. **68:** 231–238.

94. MARTIN, T.A. *et al.* 2006. Synergistic regulation of endothelial tight junctions by antioxidant (Se) and polyunsaturated lipid (GLA) via Claudin-5 modulation. J. Cell Biochem. **98:** 1308–1319.

95. MCLAUGHLIN, J.N. *et al.* 2005. Thrombin modulates the expression of a set of genes including thrombospondin-1 in human microvascular endothelial cells. J. Biol. Chem. **280:** 22172–22180.

96. MCLAUGHLIN, J.N. *et al.* 2005. Functional selectivity of G protein signaling by agonist peptides and thrombin for the protease-activated receptor-1. J. Biol. Chem. **280:** 25048–25059.

97. LUM, H. & A.B. MALIK. 1994. Regulation of vascular endothelial barrier function. Am. J. Physiol. **267:** L223–L241.

98. WU, S.Q. & W.C. AIRD. 2005. Thrombin, TNF-alpha, and LPS exert overlapping but nonidentical effects on gene expression in endothelial cells and vascular smooth muscle cells. Am. J. Physiol. Heart Circ. Physiol. **289:** H873–H885.

99. ZHOU, X. *et al.* 2005. LPS activation of Toll-like receptor 4 signals CD11b/CD18 expression in neutrophils. Am. J. Physiol. Lung Cell Mol. Physiol. **288:** L655–L662.

100. PERO, R.S. *et al.* 2007. Galphai2-mediated signaling events in the endothelium are involved in controlling leukocyte extravasation. Proc. Natl. Acad. Sci. USA **104:** 4371–4376.

101. THOMPSON, P.W., A.M. RANDI & A.J. RIDLEY. 2002. Intercellular adhesion molecule (ICAM)-1, but not ICAM-2, activates RhoA and stimulates c-fos and rhoA transcription in endothelial cells. J. Immunol. **169:** 1007–1013.

102. PREDESCU, D. *et al.* 2005. Constitutive eNOS-derived nitric oxide is a determinant of endothelial junctional integrity. Am. J. Physiol. Lung Cell Mol. Physiol. **289:** L371–L381.

103. KAMINSKI, A. *et al.* 2007. Endothelial nitric oxide synthase mediates protective effects of hypoxic preconditioning in lungs. Respiratory Physiol. Neurobiol. **155:** 280–285.

104. YOU, D. *et al.* 2006. Increase in vascular permeability and vasodilation are critical for proangiogenic effects of stem cell therapy. Circulation **114:** 328–338.

105. BUCCI, M. *et al.* 2005. Endothelial nitric oxide synthase activation is critical for vascular leakage during acute inflammation in vivo. PNAS **102:** 904–908.

106. HATAKEYAMA, T. *et al.* 2006. Endothelial nitric oxide synthase regulates microvascular hyperpermeability in vivo. J. Physiol. **574:** 275–281.

107. MIN, J.-K. *et al.* 2007. Receptor activator of nuclear factor (NF)-{kappa}B ligand (RANKL) increases vascular permeability: impaired permeability and angiogenesis in eNOS-deficient mice. Blood **109:** 1495–1502.

108. SPEYER, C.L. *et al.* 2003. Regulatory effects of iNOS on acute lung inflammatory responses in mice. Am. J. Pathol. **163:** 2319–2328.

109. WORRALL, N. *et al.* 1997. TNFalpha causes reversible in vivo systematic vascular barrier dysfunction via NO-dependent and -independent mechanisms. Am. J. Physiol. **273:** H2565–H2574.

110. RICCIARDOLO, F.L.M., F.P. NIJKAMP & G. FOLKERTS. 2006. Nitric oxide synthase (NOS) as therapeutic target for asthma and chronic obstructive pulmonary disease. Current Drug Targets **7:** 721–735.

111. DAVIDSON, S.M. & M.R. DUCHEN. 2007. Endothelial mitochondria: Contributing to vascular function and disease. Circ. Res. **100:** 1128–1141.

112. DULL, R.O., I. MECHAM & S. MCJAMES. 2007. Heparan sulfates mediate pressure-induced increase in lung endothelial hydraulic conductivity via nitric oxide/reactive oxygen species. Am. J. Physiol. Lung Cell Mol. Physiol. **292:** L1452–L1458.

113. CONNOLLY, D. *et al.* 1989. Tumor vascular permeability factor stimulates endothelial cell growth and angiogenesis. Clin. Invest. **84:** 1470–1478.

114. CARMELIET, P. & D. COLLEN. 2000. Molecular basis of angiogenesis: Role of VEGF and VE-cadherin. Ann. N.Y. Acad. Sci. **902:** 249–264.

115. CROSS, M.J. *et al.* 2003. VEGF-receptor signal transduction. Trends in Biochem. Sci. **28:** 488–494.

116. TAKAHASHI, H. & M. SHIBUYA. 2005. The vascular endothelial growth factor (VEGF)/VEGF receptor system and its role under physiological and pathological conditions. Clin. Sci. **109:** 227–241.

117. TAKAHASHI, T., H. UENO & M. SHIBUYA. 1999. VEGF activates protein kinase C-dependent, but ras-independent raf-MEK-MAP kinase pathway for DNA synthesis in primary endothelial cells. Oncogene **18:** 2221–2230.

118. BERGERS, G. & L. BENJAMIN. 2003. Tumorigenesis and the angiogenic switch. Nat. Rev. Cancer **3:** 401–410.

119. LAMALICE, L. *et al.* 2004. Phosphorylation of tyrosine 1214 on VEGFR2 is required for VEGF-induced activation of cdc42 upstream of SAPK2/p38. Oncogene **23:** 434–445.

120. MAKINEN, T. *et al.* 2001. Isolated lymphatic endothelial cells transduce growth, survival and migratory signals via the VEGF-C/D receptor VEGFR-3. EMBO J. **20:** 4762–4773.

121. USHIO-FUKAI, M. 2007. VEGF signaling through NADPH oxidase-derived ROS. Antioxid. Redox Signal. **9:** 731–739.

122. FISCHER, S. *et al.* 2007. Extracellular RNA mediates endothelial cell permeability via vascular endothelial growth factor. Blood **110:** 2457–2465.

123. SCHMIDT, M. *et al.* 2001. A new phospholipase-C-calcium signalling pathway mediated by cyclic AMP and a Rap GTPase. Nat. Cell Biol. **3:** 1020–1024.

124. QIAO, J., F. HUANG & H. LUM. 2003. PKA inhibits RhoA activation: a protection mechanism against endothelial barrier dysfunction. Am. J. Physiol. Lung Cell Mol. Physiol. **284:** L972–L980.

125. LIU, F. *et al.* 2001. Role of cAMP-dependent protein kinase A activity in endothelial cell cytoskeleton rearrangement. Am. J. Physiol. Lung Cell Mol. Physiol. **280:** L1309–L1317.

126. ROOIJ, J.D. *et al.* 1998. Epac is a Rap1 guanine-nucleotide-exchange factor directly activated by cyclic AMP. Nature **396:** 474–477.

127. KOOISTRA *et al.* 2005. Epac1 regulates integrity of endothelial cell junctions through VE-cadherin. FEBS Letters **579:** 4966–4972.

128. CULLERE, X. *et al.* 2005. Regulation of vascular endothelial barrier function by Epac, a cAMP-activated exchange factor for Rap GTPase. Blood **105:** 1950–1955.

129. HLA, T. 2003. Signaling and biological actions of sphingosine 1-phosphate. Pharmacol. Res. **47:** 401–407.

130. MCVERRY, B.J. & J.G. GARCIA. 2004. Endothelial cell barrier regulation by sphingosine 1-phosphate. J. Cell. Biochem. **92:** 1075–1085.

131. PETERS, S.L.M. & A.E. ALEWIJNSE. 2007. Sphingosine-1-phosphate signaling in the cardiovascular system. Curr. Opin. Pharmacol. **7:** 186–192.

132. MEHTA, D. *et al.* 2005. Sphingosine 1-phosphate-induced mobilization of intracellular Ca^{2+} mediates Rac activation and adherens junction assembly in endothelial cells. J. Biol. Chem. **280:** 17320–17328.

133. BIRUKOVA, A.A. *et al.* 2007. Prostaglandins PGE(2) and PGI(2) promote endothelial barrier enhancement via PKA- and Epac1/Rap1-dependent Rac activation. Exp. Cell Res. **313:** 2504–2520.

134. VAN NIEUW AMEROGEN, G.P. & V.W.M. VAN HINSBERG. 2007. Endogenous RhoA inhibitor protects endothelial barrier. Circ. Res. **101:** 7–9.

Controlling Cardiac Transport and Plaque Formation

ZE'EV ARONIS, SAGI RAZ, ELISHA J.P. MARTINEZ, AND SHMUEL EINAV

Department of Biomedical Engineering, Tel Aviv University, Tel Aviv, Israel

Macro-particles transported in the bloodstream, such as LDL particles and macrophages, are considered to be one of the initiating factors of atherosclerotic plaque development. LDL infiltration from the bloodstream into a blood vessel's wall, whether the coronary, peripheral, or carotid arteries, is considered a major inflammatory factor, recruiting macrophages from the blood flow and leading to the formation of vulnerable atherosclerotic plaques. Infiltration sites are influenced by patterns of blood flow, as regions of lower shear stresses and high oscillations may give rise to higher infiltration rates through the endothelium, exacerbating the growth of a plaque and its tendency to rupture. Previous studies demonstrated a high prevalence of rupture sites proximal to the minimum lumen area, which raised the question of whether the existence of two distinct adjacent plaques, in which the distal plaque is more severe, can give rise to hemodynamic forces that can push the non-stenotic plaque to rupture. Models of the coronary arteries with one and two eccentric and concentric stenotic narrowings were built into a closed flow loop. The single stenosis model had a 75% area reduction narrowing (representing the vunerable atherosclerotic plaque) with relevant elastic properties. The double stenosis model included an additional distal 84% area reduction narrowing. The flow in the area between the two stenoses was recorded and analyzed using continuous doppler particle image velocimetry (CDPIV), together with the hydrostatic pressure acting on the proximal plaque. Results indicated that the combined shear rates and pressure effects in a model with a significant distal stenosis can contribute to the increase in plaque instability by LDL and enhanced macrophage uptake. The highly oscillatory nature of the disturbed flow near the shoulder of the vulnerable atherosclerotic plaque enriches its lipid soft core, and the high hydrostatic pressures acting on the same lesion in this geometry induce high internal maximal stresses that can trigger the rupture of the plaque.

Key words: vulnerable plaque; atherosclerosis; LDL infiltration; continuous doppler particle image velocimetry (CDPIV)

Introduction

Atherosclerosis is a progressive chronic disease of the large arteries and is considered to be the primary cause of heart disease and stroke, the largest causes of mortality in the Western world. Epidemiological studies over the past 50 years have revealed numerous genetic and environmental risk factors for atherosclerosis, including high cholesterol, hypertension, diabetes, family history, high-fat diet, smoking, lack of exercise, and others. The progression of an atheroma is considered to be an inflammatory process[1] that involves the immune system. The process of atherosclerotic plaque formulation is hypothesized to begin with an endothelial dysfunction, in which its permeability and adhesion to leukocytes and platelets are increased. Low-density lipoprotein (LDL) is a major cause of injury to the endothelium and smooth muscle cells (SMC). LDL particles accumulate in the subendothelial matrix and undergo progressive oxidation, forming oxLDL which is internalized by macrophages, producing foam cells. High levels of cholesterol oxides in the cell membrane can trigger foam cell death by apoptosis which in turn may foster the formation of a necrotic lipid core. "Vulnerable atherosclerotic plaque" generally refers to intact lesions that resemble plaques disrupted and compromised by thrombosis. It is assumed that given enough time and/or the right stimulus, the plaques can become disrupted, thus triggering the formation of a thrombus. A large number of vulnerable plaques are relatively uncalcified and non-stenotic; they reduce lumen diameter by less than 50%,[2] have a large lipid core accounting for over 40% of the plaque's total volume, and have a thin fibrous cap of less than 100 μm.[3] A plaque

Address for correspondence: Prof. Shmuel Einav, PhD, Dept. of Biomedical Engineering, Tel Aviv University, Tel Aviv 69978, Israel. Fax: +972-3-6409448.
seinav@sunysb.edu

ruptures when the fibrous cap tears and the necrotic lipid core (which is extremely thrombogenic) is exposed to blood in the arterial lumen. The shoulders of a complex plaque, defined as the longitudinal margins originating at the apex of the plaque up to its connection with the normal vessel wall, are highly prone to rupture and contain macrophages, T lymphocytes, and a paucity of SMC. Plaque rupture and fissuring account for the great majority of thrombi that cause acute coronary syndromes (ACS) with consequent myocardial infarctions (MI). Most occlusive thromboses are clustered within the proximal portion of the major epicardial arteries.[4] From the clinical standpoint, significant stenosis at the proximal left anterior descending (LAD) artery jeopardizes the anterior left ventricular wall and worsens prognosis in coronary artery disease. Eccentric involvement was common in the proximal LAD over a wide spectrum of stenosis severity.[5] Moreover, ruptured plaques show a significantly more eccentric position of the lumen than nonruptured plaques.[6]

High blood pressure is another risk factor influencing the onset of ACS. A measured pressure waveform generally results from coincidental succession of forward and backward wave fronts. Forward wave fronts originate in coronary arteries in the left ventricular cavity, and backward wave fronts originate in small vessels within the myocardium. The nature of wave reflection will depend on the nature of the reflecting site. A positive pressure change reflects a positive pressure change from a closed end but becomes a negative pressure change when reflected from an opening or an expansion. An increase in blood pressure levels due to wave reflections can have a strong effect on the development of atherosclerosis. Increased systolic blood pressure, pulse pressure, and mean blood pressure are all associated with inflammatory markers, such as interleukin-6 (IL-6) and intercellular adhesion molecule-1 (sICAM-1).[7] It has been suggested, therefore, that increased blood pressure may be a stimulus for inflammation, which can explain its role as a risk factor for atherosclerosis.

Localizing rupture-prone plaques proves to be quite a tedious task, as more and more imaging techniques are constantly developed in order to identify them, usually by tracing morphological and histological characteristics that separate these biologically active lesions from other stenotic, stable plaques. While vulnerable plaque ruptures are responsible for most cardiovascular events, the vast presence of these plaques throughout the coronary arterial tree suggests that only a fraction of them tend to rupture, while others stay intact. The reasons for the higher tendency of some plaques to rupture remain quite elusive, and are partially responsible for the difficulties in developing prevention strategies prior to clinical deterioration.

Plaque instability might be caused by pathophysiological processes, such as inflammation, that exert adverse effects throughout the coronary vasculature and therefore result in multiple unstable lesions. Angioscopic studies show that all three major coronary arteries are widely diseased in patients with MI and have multiple nondisrupted plaques.[8] An intravascular ultrasound (IVUS) study of patients with ACS showed that only 37% of plaque ruptures were located on the culprit lesion.[9] Thus, though only a single lesion is clinically the culprit affecting ACS, ACS can be associated with pancoronary destabilization. A high prevalence (54%) of additional plaques was found to be located proximal to the culprit lesion, regardless of whether the clinical situations were acute, evolving MI, or stable coronary syndromes.[10] In patients with acute MI, plaques that were distant from the culprit lesion had a fivefold higher frequency of ulceration, suggesting that the distant plaques may indeed be regarded as vulnerable. An IVUS study conducted on patients with unstable angina, MI, or stable angina, as well as on patients with no syndromes, showed an even higher prevalence (72%) of plaque ruptures that were distinct from the minimum lumen area (MLA).[11] Another study[12] confirmed that most of the ulcerated ruptured plaques in ACS patients were proximal to the minimal lumen site. Even though it can be argued that non-culprit lesions might be an early stage in the development of a stable, highly stenotic plaque, an angiographic study on patients with acute MI showed little change during 6 months of follow-up in those non-culprit complex plaques.[13] According to these observations, most of the ruptures do not occur, surprisingly, in the MLA itself or distal to it, but proximal to the MLA, and in many cases the rupture site and the MLA refer to two distinct plaques. The lower stenosis rate at the rupture site may suggest the occurrence of ruptures at vulnerable plaques, which possess the characteristics of mild narrowing and a high tendency to rupture. This further suggests that the geometry of a mild stenosis with a distal larger narrowing presents a fertile ground for hemodynamic conditions that worsen the stability of the proximal lesions and induce a higher risk of rupture and formation of life-threatening thromboses.

Methodology

A flow loop was designed to mimic the conditions in a blood vessel with two distinct plaques, wherein the proximal stenosis is less narrowed. The flow in the

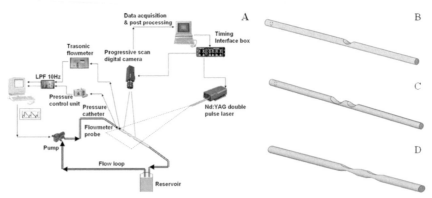

FIGURE 1. Experimental design. A: The experimental setup. B: Eccentric single stenosis (ESS) model. C: Eccentric double stenosis (EDS) model. D: Concentric double stenosis (CDS) model. Direction of flow in models B-D is from bottom-right to upper-left.

region between the two stenoses was recorded and analyzed using continuous digital particle image velocimetry (CDPIV). The CDPIV was designed to measure the hemodynamic parameters of importance, yielding dynamic maps of the shear stresses which act on the proximal plaque and may participate in the evolution of this high-risk plaque. Pressure measurements were used to evaluate the influence of the distal disturbance on the overall resistance of the blood vessel, which may contribute to the normal stresses acting on the vulnerable lesion.

Models of a mild stenosis, both with and without a distal higher narrowing, were employed (FIG. 1B–D). The initial inner diameter of 15 mm corresponded to a scaled up model of a 3 mm diameter LAD artery. The mild stenosis consisted of a 50% diameter stenosis, corresponding to a 75% area stenosis, while the distal narrowing consisted of a 75% diameter stenosis, corresponding to a more severe 84% area stenosis. The single stenosis was constructed as an eccentric single stenosis (ESS) model. A model of eccentric double stenoses (EDS) was also compared to a model of concentric double stenoses (CDS). The models were scaled up, by a diameter factor of 5 in order to improve the spatial resolution, and were made of RTV-630 silicon with a 1-mm wall thickness. The *in vitro* waveform was constructed to mimic the LAD waveform. The apparatus was calibrated with a straight tube without a stenosis, and then replaced by tubes consisting of one and two stenoses. The experimental setup and instrumentation are detailed in an earlier publication,[14] with the exception that the laser used here was a Solo-II 15Hz Nd:Yag laser system (New-Wave Research®; New Wave Research, Inc., Fremont, CA) with a wavelength of 532 nm and energy of 30 mJ.

CDPIV data provide only two components of a three-dimensional vector field. Matlab® (MathWorks Inc., Natick, MA) was used to derive shear strains and stresses from these vectors. Shear stress related parameters, e.g., the oscillatory shear index (OSI), which describes the degree of deviation of the wall shear stress (WSS) from its average direction, were estimated. This index represents a measure of the shear stress acting on the luminal surface due to either cross-flow or reverse-flow velocity components occurring during the pulsatile flow. For purely oscillatory flow, the OSI attains its maximum value at 0.5. The OSI does not take into account the magnitude of the shear stress vectors, merely the directions.

Fluid pressure was measured using a Millar pressure transducer model SPR-524 (Millar Instruments, Houston, TX), with a 2.3 French catheter size and 3.5 French sensor size. The pressure gauge was inserted into the tubular model, and measurements were obtained from different locations along the central tube axis. For the double stenotic model, the measurement locations were on the proximal shoulder of the mild stenosis, its apex, the distal shoulder, the middle of the region between the two stenoses, the proximal shoulder of the severe stenosis, its apex, and its distal shoulder. The one-stenosis model included the same locations excluding those associated with the distal severe stenosis. Measurements were conducted for three different pulsation rates, corresponding to 60 beats per minute (BPM), 80 BPM, and 100 BPM heart rates. All measurements were averaged over five consecutive pulse cycles. Maximal pressure values, as well as pressure drops over the mild stenosis, were analyzed by ANOVA. Differences were considered significant

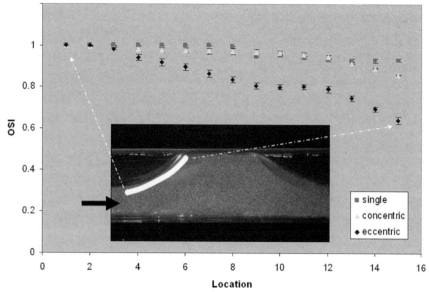

FIGURE 2. OSI comparison at the distal shoulder of the mild-stenosis. *Black arrow* indicates the direction of flow. The abscissa is divided into 15 points, starting at a region closer to the apex of the mild-stenosis, and ending at the end of the stenosis, where it reattaches to the normal straight vessel wall.

at probability levels of $P < 0.05$. The pressure values were compared using Fisher's exact test.

Results

The OSIs on the distal shoulder of the mild stenosis in the three models are compared in FIGURE 2. The measurement path-line is divided into 15 points, starting at a region closer to the apex of the mild stenosis and ending at the end of the stenosis, where it reattaches to the normal straight vessel wall. The OSI for all three models at the region closer to the apex of the stenosis is close to 1, indicating that the flow in that region is unidirectional throughout the whole cycle. An OSI value of 0 represents a unidirectional flow in the opposite direction, while a value of 0.5 represents a purely oscillatory flow. As expected, all three models demonstrate OSI values closer to 0.5 while advancing along the shoulder towards the upper wall, i.e. deeper into the recirculation area. However, both the ESS and CDS models have very similar OSI values, indicating a unidirectional flow for most of the shoulder length. At the very end of the shoulder, where it reconnects with the straight vessel wall (location points 14–15), the OSI values for the CDS model are slightly closer to 0.5 than for the ESS model, indicating that the flow there oscillates a little bit more than in the single stenosis case. However, the EDS model yields OSI values much closer to 0.5 along the same path, reaching an

OSI of 0.64 ± 0.0095, indicating that the presence of a distal disturbance by itself does not necessarily implicate high flow oscillations, and that the eccentricity is an important factor too.

In order to examine the influence of different pulse rates on the shear-related parameters, measurements were conducted at three different heart rates, corresponding to 60 BPM, 80 BPM, and 100 BPM as well as a constant flow. The constant flow was at the average flow rate over a cycle. The comparison in FIGURE 3 is presented for the ESS model. The measurement points correspond to the same locations as in FIGURE 2. All of the different flow rate cases begin with a unidirectional shear stress value of 1, close to the apex of the stenosis. However, advancing into the recirculation zone, the OSI values grow smaller, and the lower the frequency, the closer the oscillatory indicator approaches 0.5. At higher frequencies, the flow within the recirculation zone restores a smaller portion of the core flow direction; therefore it stays for a longer period of the cycle in the same negative direction. This result indicates the favorable effect of the higher flow rates.

Pressure was measured along several locations within the model, starting upstream from the mild stenosis and continuing to the high-grade stenosis downstream. FIGURE 4 shows the maximal pressures of the different models, at three different locations at 60 BPM. Since the proximal mild stenosis is the focus of the current study, only the pressures acting proximally

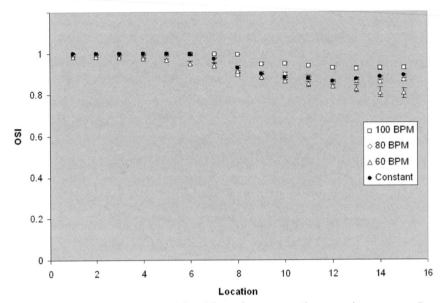

FIGURE 3. OSI comparison for ESS model at different frequencies. Abscissa is the same as in FIGURE 2.

Max. Pressure (@ 60 BPM)

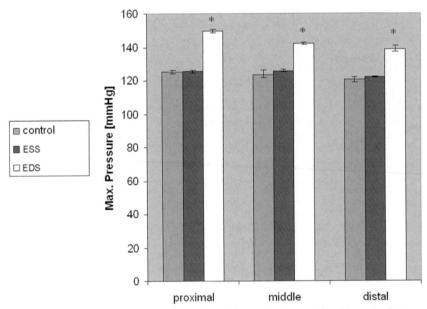

FIGURE 4. Maximal pressure comparison at three different locations (defined in text) relative to the mild stenosis between three model geometries, measured at a flow rate of 60 BPM.

to the narrowing, on its apex, and distally to it, were compared. Thus, the notation "proximal," "middle," and "distal" shown in FIGURE 4 relate to the mild stenosis. There is no significant difference in these locations between the control and the single stenosis. However, a significantly higher pressure ($P < 0.05$) was observed in each of these locations at all flow rates in the EDS model, compared to the control and the ESS models. The pressure at the middle of the mild stenosis at 60 BPM, for example, is 142 ± 0.2 mmHg for the EDS model, versus 125.89 ± 0.035 mmHg for the ESS model. The overall resistance of the tubular vessel to

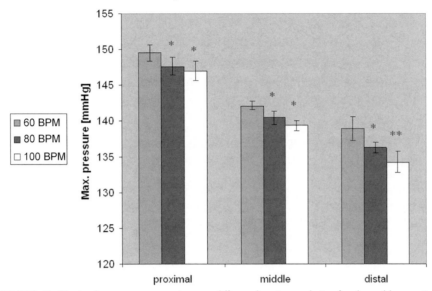

FIGURE 5. Maximal pressure comparison at different locations relative for the mild stenosis, for different flow frequencies. Single asterisk means that results are significantly different from 60 BPM Double asterisk means that results are significantly different from both 60 and 80 BPM.

the flow is higher at the EDS model, and similar pressure differences between the EDS and ESS models are found at the proximal and distal locations. It is also of interest to note that the pressure drop on the stenosis itself, defined as the difference between the maximal distal pressure and the maximal proximal pressure shows no significant difference between the ESS model and the control. The pressure drop is significantly higher for the EDS model, compared to both ESS and control. The pressure drop at 80 BPM in the double stenosis model, for example, is 11.4 ± 1.43 mmHg, compare to 3.75 ± 0.91 mmHg for the single stenosis case. This finding demonstrates that the maximal pressure values are higher for the double stenosis model, and that the pressure drop, which is a measure that doesn't take into account any averaged baseline that could affect the maximal measured pressure, is in fact also higher for this geometry. This effect does not result from the increase in the overall vessel resistance, but is rather a consequence of a reflected pressure wave from the distal narrowing, which adds constructively to the forward pressure wave.

In order to evaluate the effect that different cycle durations have on the value of the measured pressures, a comparison was made between the maximal pressure values at flow rates corresponding to heart rates of 60, 80, and 100 BPM. As presented in FIGURE 5, pressure is significantly lower at both 80 and 100 BPM

than at 60 BPM in the EDS model at all three locations. There is no significant difference between 80 and 100 BPM in the proximal and middle locations. However, at the distal location, which in this case is midway between both stenoses, the measured pressure is also a function of the reflected pressure wave, and at higher frequencies the reflected pressure wave possesses lower peak values that add up to the forward pulse wave. Consequently, the effect of the pressure drop due to higher pulsatility is actually doubled. At this region, the maximal pressure at 100 BPM (which stands on 134.21 ± 1.48 mmHg) is significantly lower than both 60 (138.93 ± 1.68 mmHg) and 80 BPM (136.25 ± 0.72 mmHg).

Discussion

Atherosclerotic plaques are distributed all over the arterial tree in sites which are believed to be non-random. Predilection sites for atherosclerosis occur at sites characterized by low shear stresses and flow reversal. Under such a complex flow behavior, endothelial cells (ECs) were found to have greater monocyte binding[15] compared to ECs exposed to laminar unidirectional flow. It is believed that the ECs that coat the inner surface of the artery respond to the tangential forces imposed upon them by the flowing blood.[16] Variations in the direction of these forces during the cardiac cycle could prevent the cells from adopting a stable config-uration, resulting in a compromised barrier function. The OSI values derived in all three models are close to 1 at the area closer to its apex, indicating a purely unidirectional flow in the negative direction, which has a positive influence on the ECs functioning. However, the OSI gradually decreases with the advance towards the base of the shoulder to close to 0.5, which represents a purely oscillatory flow. The oscillatore OSI values (close to 0.5) in the EDS model, especially at the base of the plaque shoulder, may indicate a longer residence time of particles near the wall,[17] which can be regarded as a hazardous phenomenon, predisposing this geometry to higher risk of more lipid accumulation and anecrotic core enlargement.

The compliance of an arterial wall importantly modulates the consequences of oscillatory flow. If the artery has normal distensibility, the oscillatory flow can actually be protective, and there is some evidence that increasing the frequency of pulsation may be advantageous.[18] The OSI is highly influenced by the pulsation frequency, as lower frequencies caused higher flow oscillations for most of the length of the shoulder area. For all models, a frequency of 80 BPM behaves quite similarly to a constant flow rate, suggesting that this rate value is close to the natural frequency of a tube with the specific geometrical and material properties given in this experiment. The significantly higher OSI values of the EDS model indicate that this is by far the most perilous geometry.

Eccentric plaques at preserved lumen locations are known to experience increased tensile stress at their shoulders, making them prone to fissuring and thrombosis.[19] Since most of the atherosclerotic lesions are eccentric, it is of much interest to know how hazardous this formation is, considering both stress concentration and macromolecule accumulation. OSI values indicate that the CDS model behaves more similarly to the ESS model than to the EDS. This is a surprising finding, considering the fact that the degree of stenosis for both the mild proximal stenosis and the distal high-grade stenosis are the same for both the EDS and CDS models. This finding indicates that a pure concentric orientation of the lesions is enough to compensate almost completely for the harmful flow conditions caused by the presence of a distal high-grade stenosis. Another way to look at it is that a single eccentric mild stenosis is just as "harmful" as two sequential concentric stenoses, from the oscillatory shear stress point of view. This demonstrates the very important role of the degree of eccentricity on the flow conditions, which may harm the normal functioning of the endothelial layer as a barrier to macromolecules infiltration.

The pressure wave is composed of a forward traveling wave generated by the driving flow and a reflected wave from distal obstacles such as arterial narrowing. High blood pressure, such as in patients with hypertension, is known as one of the most important risk factors for atherosclerotic disease, and one possible explanation for this phenomenon is that high pressure levels may stimulate inflammation via expression of cytokines.[7] However, these findings refer to hypertensive disease, which affects the whole systemic vasculature. In the current study the findings refer to a local rise in blood pressure due to a specific geometry in specific arterial segments, and it is not clear whether a localized high pressure can act as a stimulus for local release of cytokines. In patients with several stenoses, especially sequential stenoses, it is important to assess the risk of each lesion to rupture. Various studies have tried to establish clinical evaluation techniques for selecting the most appropriate lesion for treatment procedures, such as pressure measurements in patients with sequential stenoses as a basis of coronary and fractional flow reserve by accounting for stenosis interaction,[20] or by treating arterial pressure as a significant cause of elevated shear stress and accompanying stretch,[21] which are hypothesized to be the primary factors responsible for the topography of atherosclerotic plaques. Convective fluid motion through artery walls aids in the transvascular transport of macromolecules into the arterial wall. It has been shown that an increase in pulsatile pressure creates a significantly increased filtration,[22] which can serve to drive toxic agents, such as oxLDL, deeper into the artery, therefore increasing the injury. High pressure levels may trigger high peak circumferential stress (PCS) within the plaque. Finite-element studies, performed on a series of IVUS-based vulnerable plaque cross-sections on which different pressure levels were applied (45–150 mmHg), showed that a linear relationship exists between the pressure and the PCS on the atherosclerotic cap.[9] Furthermore, reduced cap thickness and decreased core stiffness increase PCS dramatically. Similar results were obtained in a fluid-structure interaction study that also varied these parameters and found that higher pressure, thinner cap thickness, and larger, softer lipid pools were correlated with higher maximum stress.[23] Thus, in the case of a morphologically vulnerable plaque, which has a high tendency to develop high PCS, simple changes in blood pressure can place the plaque in a high-risk, unstable range. Major stress concentrations were found in a finite-element simulation to be located mainly at the edges of plaques,[24] as extremely large stress gradients appear at the interface between tissues that have a substantial difference in their stiffness. Even at a relatively low-grade 30% luminal stenosis, a critical combination of plaque cap thickness and lipid volume that results in high shoulder stress area was established.[25] A model, which attributes plaque rupture to the presence of micro calcified particles, was recently proposed by a group of investigators including one of us (S.E.).[25] These results are consistent with the higher frequency of rupture at the shoulder of the plaques.

In this study, a significant rise in the hydrostatic pressure was measured for the EDS model as compared to the ESS model for all flow rates, while no significant differences were found between the ESS and the control model. This indicates that the existence of a single mild 50% diameter stenosis within the tube does not influence the overall resistance of the vessel, probably because of the compensatory effect of the compliance of the tube. However, the addition of another high-grade 75% diameter stenosis causes a sharp rise in the resistance, as the compliance of the vessel cannot compensate for this significant constriction, therefore the pressure rises sharply by approximately 15 mmHg. A rise in the maximal pressure is visible not only on the highly stenotic distal plaque or proximally to it, but at all locations along the vessel, indicating that the additional narrowing causes a rise in the overall resistance of the tube. Yet, in order to avoid the effect of the rise in total resistance, a pressure difference between two points—just downstream to the mild stenosis and just upstream to it—was also measured. This comparison also shows that there is a greater pressure drop on the mild stenosis in the EDS model, this time due to reflection of a pressure wave from the distal disturbance, adding positively to the forward pressure wave and raising its peak value. The higher normal pressure acting on the mild stenosis may contribute to the internal stresses within the plaque, and if a vulnerable plaque is considered to be the mild stenosis, it can trigger this already unstable lesion toward a higher risk of rupturing. A comparison of the maximal pressures measured at different flow rates demonstrates that the higher flow rate causes a smaller rise in pressure. Combining this beneficial effect of high flowrate with the unidirectional OSI values discussed above, it is clear that higher frequencies have a positive effect over both shear-related qualities and the rise in hydrostatic pressure.

Summary

This study relates to the geometry of a mild stenosis, a model of a possibly vulnerable plaque, which in the presence of a distal high-grade narrowing generates flow patterns, manifested both in shear-related qualities and in hydrostatic pressures, that are hazardous to the stability of the mildly stenotic plaque. This hypothesis was tested by studying the combined role of flow patterns and pressures that could lead to further instability and increase the risk of the mild stenosis plaque rupturing.

The results emphasize the higher potential threat of a specific geometry consisting of two consecutive plaques, wherein the distal plaque is highly stenotic. The combined shear rates and pressure effects in this model yield a highly oscillatory disturbed flow near the shoulder of the mildly stenotic plaque. This effect contributes to an increase in plaque instability and to enhanced uptake of LDL and macrophages, thus enriching its lipid soft core. Higher hydrostatic pressures applied on the same lesion in this geometry can induce higher maximal internal stresses that may trigger the rupture of the plaque. Moreover, the eccentricity of the stenotic plaque is also of importance, as highly eccentric plaques present riskier outcomes, whereas concentricity reduces the negative outcomes to the scale of a single mildly stenotic plaque without any distal disturbance.

Most available diagnostic strategies for the identification of vulnerable plaques are quite limited. Investigators are striving to determine why one plaque is vulnerable and life-threatening while another is resistant and innocuous. The present study offers a partial explanation of the conditions that can influence the stability of an atherosclerotic plaque through shear stress related situations and the effects of the induced internal high stresses on the vulnerable plaques in the specific geometry of two adjacent stenoses.

Conflict of Interest

The authors declare no conflicts of interest.

References

1. KRAMS, R. *et al.* 2003. Inflammation and atherosclerosis: mechanisms underlying vulnerable plaque. J. Interv. Cardiol. **16:** 107–113.
2. BURKE, A.P. *et al.* 1999. Plaque rupture and sudden death related to exertion in men with coronary artery disease. JAMA **281:** 921–926.
3. KOLODGIE, F.D. *et al.* 2001. The thin-cap fibroatheroma: a type of vulnerable plaque: the major precursor lesion to acute coronary syndromes. Curr. Opin. Cardiol. **16:** 285–292.
4. WANG, J.C. *et al.* 2004. Coronary artery spatial distribution of acute myocardial infarction occlusions. Circulation **110:** 278–284.

5. KIMURA, B.J. *et al*. 1996. Atheroma morphology and distribution in proximal left anterior descending coronary artery: in vivo observations. J. Am. Coll. Cardiol. **27:** 825–831.

6. VON BIRGELEN, C. *et al*. 2001. Plaque distribution and vascular remodeling of ruptured and nonruptured coronary plaques in the same vessel: an intravascular ultrasound study in vivo. J. Am. Coll. Cardiol. **37:** 1864–1870.

7. CHAE, C.U. *et al*. 2001. Blood pressure and inflammation in apparently healthy men. Hypertension **38:** 399–403.

8. ASAKURA, M. *et al*. 2001. Extensive development of vulnerable plaques as a pan-coronary process in patients with myocardial infarction: an angioscopic study. J. Am. Coll. Cardiol. **37:** 1284–1288.

9. RIOUFOL, G. *et al*. 2002. [Multiple ruptures of atherosclerotic plaques in acute coronary syndrome. Endocoronary ultrasonography study of three arteries]. Arch. Mal. Coeur. Vaiss. **95:** 157–165.

10. SCHOENHAGEN, P. *et al*. 2003. Coronary plaque morphology and frequency of ulceration distant from culprit lesions in patients with unstable and stable presentation. Arterioscler. Thromb. Vasc. Biol. **23:** 1895–1900.

11. MAEHARA, A. *et al*. 2002. Morphologic and angiographic features of coronary plaque rupture detected by intravascular ultrasound. J. Am. Coll. Cardiol. **40:** 904–910.

12. FUJII, K. *et al*. 2003. Intravascular ultrasound assessment of ulcerated ruptured plaques: a comparison of culprit and nonculprit lesions of patients with acute coronary syndromes and lesions in patients without acute coronary syndromes. Circulation **108:** 2473–2478.

13. LEE, S.G. *et al*. 2004. Change of multiple complex coronary plaques in patients with acute myocardial infarction: a study with coronary angiography. Am. Heart J. **147:** 281–286.

14. RAZ, S. *et al*. 2007. DPIV prediction of flow induced platelet activation-comparison to numerical predictions. Ann. Biomed. Eng. **35:** 493–504.

15. HONDA, H.M. *et al*. 2001. A complex flow pattern of low shear stress and flow reversal promotes monocyte binding to endothelial cells. Atherosclerosis **158:** 385–390.

16. KU, D.N. *et al*. 1985. Pulsatile flow and atherosclerosis in the human carotid bifurcation. Positive correlation between plaque location and low oscillating shear stress. Arteriosclerosis **5:** 293–302.

17. HIMBURG, H.A. *et al*. 2004. Spatial comparison between wall shear stress measures and porcine arterial endothelial permeability. Am. J. Physiol. Heart Circ. Physiol. **286:** H1916–1922.

18. HUTCHESON, I.R. & T.M. GRIFFITH. 1991. Release of endothelium-derived relaxing factor is modulated both by frequency and amplitude of pulsatile flow. Am. J. Physiol. **261:** H257–H262.

19. SLAGER, C.J. *et al*. 2005. The role of shear stress in the generation of rupture-prone vulnerable plaques. Nat. Clin. Pract. Cardiovasc. Med. **2:** 401–407.

20. PIJLS, N.H. *et al*. 2000. Coronary pressure measurement to assess the hemodynamic significance of serial stenoses within one coronary artery: validation in humans. Circulation **102:** 2371–2377.

21. THUBRIKAR, M.J. & F. ROBICSEK. 1995. Pressure-induced arterial wall stress and atherosclerosis. Ann. Thorac. Surg. **59:** 1594–1603.

22. ALBERDING, J.P. *et al*. 2004. Onset of pulsatile pressure causes transiently increased filtration through artery wall. Am. J. Physiol. Heart Circ. Physiol. **286:** H1827–H1835.

23. TANG, D. *et al*. 2004. Effect of a lipid pool on stress/strain distributions in stenotic arteries: 3-D fluid-structure interactions (FSI) models. J. Biomech. Eng. **126:** 363–370.

24. HAYASHI, K. & Y. IMAI. 1997. Tensile property of atheromatous plaque and an analysis of stress in atherosclerotic wall. J. Biomech. **30:** 573–579.

25. VERESS, A.I. *et al*. 2000. Vascular mechanics of the coronary artery. Z. Kardiol. **89**(Suppl 2): 92–100.

Multiscale and Modular Analysis of Cardiac Energy Metabolism

Repairing the Broken Interfaces of Isolated System Components

JOHANNES H.G.M. VAN BEEK

Centre for Integrative BioInformatics, VU University Amsterdam, Centre for Medical Systems Biology, Leiden, Amsterdam, and Rotterdam, and VU University Medical Centre, Amsterdam, the Netherlands

Computational models of large molecular systems can be assembled from modules representing biological function emerging from interactions among a small subset of molecules. Experimental information on isolated molecules can be integrated with the response of the network as a whole to estimate crucial missing parameters. As an example, a "skeleton" model is analyzed for the module regulating dynamic adaptation of myocardial oxidative phosphorylation (OxPhos) to fluctuating cardiac energy demand. The module contains adenine nucleotides, creatine, and phosphate groups. Enzyme kinetic equations for two creatine kinase (CK) isoforms were combined with the response time of OxPhos (t_{mito}; generalized time constant) to steps in the cardiac pacing rate to identify all module parameters. To obtain t_{mito}, the time course of O_2 uptake was measured for the whole heart. An O_2 transport model was used to deconvolute the whole-heart response to the mitochondrial level. By optimizing mitochondrial outer membrane permeability to 21 μm/s the experimental $t_{mito} = 3.7$ s was reproduced. This *in vivo* value is about four times larger, or smaller, respectively, than conflicting values obtained from two different *in vitro* studies. This demonstrates an important rule for multiscale analysis: experimental responses and modeling of the system at the larger scale allow one to estimate essential parameters for the interfaces of components which may have been altered during physical isolation. The model correctly predicts a smaller t_{mito} when CK activity is reduced. The model further predicts a slower response if the muscle CK isoform is overexpressed and a faster response if mitochondrial CK is overexpressed. The CK system is very effective in decreasing maximum levels of ADP during systole and reducing average P_i levels over the whole cardiac cycle.

Key words: mitochondria; membrane permeability; systems biology; oxidative phosphorylation; reverse engineering; creatine kinase; modular modeling

Introduction

Cardiac metabolism represents a crucial function for survival: it energizes the pumping of blood for transport of O_2 and carbon substrates to sustain metabolism throughout the body as well as in the heart itself. Processes in the body are connected across multiple scales.

When approaching the systematic analysis of the whole organism, one often tends to assume that there is a "natural," bottom-up hierarchy of levels: atoms form molecules, molecules, such as nucleic acids, proteins, and small metabolites form the cytosol and organelles, which form the cell. Cells form tissues, and tissues form organs, which form the whole organism. However, causation is not unidirectional, and it does not merely flow upward from the atomic and molecular level. In fact, events at the whole-body level can determine the molecular level. If you often eat a great amount of carbohydrates instead of food with a high fat content, the metabolite levels and turnover in many cells will be different. All scales and functional levels in the organism are important, and causation flows up and down the "hierarchy."[1] Computational modeling of the multiple levels in human physiology and pathology greatly helps us to understand these mutual interdependencies.

Address for correspondence: Hans van Beek, PhD, Dept. Molecular Cell Physiology, FALW, Vrije Universiteit, De Boelelaan 1085, 1081 HV Amsterdam, the Netherlands. Fax: + 31 205987229.

hans.van.beek@falw.vu.nl

Ann. N.Y. Acad. Sci. 1123: 155–168 (2008). © 2008 New York Academy of Sciences.
doi: 10.1196/annals.1420.018

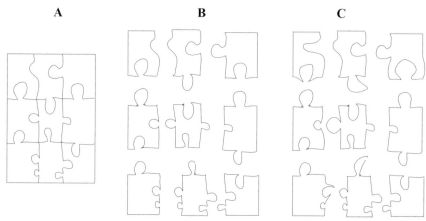

FIGURE 1. (A) A module in the system is illustrated by the rectangle, which is cut in pieces as in a jigsaw puzzle. The puzzle pieces represent the components, which may be enzymes, metabolites, organelles, etc. **(B)** Taking the module apart in its components. **(C)** Often components are distorted during isolation.

Living systems are very complex and cannot be understood and modeled in one big step. To understand how a system works, we tend to decompose the system into components and study their behavior in isolation. A component can be termed a module, defined as an independently operable unit that is part of the total structure. Thus, the response of these modules to changes in their environment and their interaction with each other are studied outside the context of the intact system. It is interesting to note at this point that engineers and computer scientists usually build new complex systems by coupling independently designed modules. Life scientists, on the other hand, analyze existing systems, i.e., they "reverse engineer" the system. The modules can be integrated at a later stage to understand the whole system and predict its behavior and response to interventions.

Here we investigate an example whereby, during the isolation of some components, i.e., modules, of the system, the interface of a component with the rest of the system is altered so that it behaves differently from its natural condition. It is likely that this phenomenon happens often. The change of components when the system is taken apart is illustrated in FIGURE 1. By studying the behavior of the ensemble of components at a higher level, either by observing the behavior of the intact system as a whole or by measuring its response to an experimental challenge, we are able to estimate the correct interaction parameter for the altered component. In this way we can, figuratively speaking, "repair the broken interfaces" which were damaged during isolation from the system.

To illustrate the principles of handling a system piecewise by dividing it into modules, and how this approach is integrated to multiscale modeling, we describe the analysis of a module which, among other things, transfers high-energy phosphoryl groups from the mitochondria to the sites of ATP consumption in the cardiac muscle cells in sarcomeres and ion pumps.

Modeling Cardiac Metabolism

Cardiac carbon substrate and energy metabolism have been studied in great detail under many conditions. Oxidation of fatty acids is usually the main source of energy, but carbohydrates, such as sugar, lactate, and pyruvate, are also metabolized. The regulation of switching between these substrates is therefore important. The mechanism of oxidative phosphorylation (OxPhos) and its regulation in adaptation to an altered energy demand has been modeled often.[2–4] However, despite the large amount of experimental data, it is very difficult to understand all the interactions in the system. Understanding of the cardiac metabolic system is lagging substantially behind the experimental information. Insight may be aided substantially by developing and testing mechanistic computational models for metabolism. While it is possible that some specialists in a particular field have developed a useful model of the system, either implicitly in their mind or in a pictorial representation, these models can often not be effectively utilized by other scientists who work in the same or related fields. Note that nonlinear interactions often yield results that are hard to predict by human reason without resort to computation, even if there are only a few components in a system.

It is not uncommon that experimental scientists criticize computational modeling efforts by stating that the models only reproduce what is put into them,[5] and consequently provide no real new insight. It might be a worthwhile exercise to challenge such scientists to predict the system responses obtained by model simulations just by reasoning. It is very hard to predict the sizes and direction of the metabolic flux responses to enzyme inhibition and degree of damping of oscillations without explicit calculations. Often, the inductive analysis of large collections of data which cover a limited number of experimental conditions establishes correlations rather than causal effects and mechanisms, yielding additional collection of facts, but very limited insight.[6] Insight is more likely obtained by developing and critically testing mechanistic models of the system. This may lead us to predict and identify potential therapeutic interventions in complex molecular networks. The difficulty is that human body function is brought about by the interactions of numerous molecules and our understanding cannot depend on a single hypothesis which must be tested, but rather on a number of detailed hypotheses related to many parallel molecular processes and interactions which are integrated into one system.

One part of cardiac metabolism which has been modeled previously is the creatine kinase (CK) enzyme system which transfers a phosphoryl group from phosphocreatine (PCr) to ADP. This system operates at two time scales. In the fast time scale, it damps ADP oscillations caused by the pulsatile ATP hydrolysis mainly in the systolic phase of the cardiac cycle.[4] However, its dynamic adaptation to heart rate shows changes at the slower time scale. It is shown below that by analyzing the dynamic adaptation of the system as a whole to imposed changes in the pacing frequency of heart rate, an essential parameter of the system, which is usually strongly changed by physical isolation of the component from the system, can be estimated. The parameter in this case is mitochondrial outer membrane permeability, whose value determined *in vitro* is hotly debated.[7]

Tackling the Complexity of the Metabolic System

The large number of enzymes, metabolites, genes, etc., which constitute the metabolic system make it difficult to develop a computational model in one step. In many cases only part of the information necessary

to model a particular metabolic pathway is available. To model all the details, one needs information on the kinetics of several enzymes, specifying the speed with which the enzyme turns over its substrates to products as a function of the concentrations of all substrates, products, and regulating ligands. The effect of post-translational modifications, for instance phosphorylation at specific sites, by other enzymes which influence the function of the investigated enzyme must also be specified.

To tackle this complex system, one can try to describe a metabolic function of a sequence of enzymes, or an organelle, with little detail. For example, one does not describe the individual components of that pathway, or organelle, but represents the overall behavior by an equation determined from measured input–output relations for the pathway as a whole. ATP synthesis by the mitochondria can thus be given by a simple Michaelis–Menten type of equation that describes the dependence of ATP production on ADP and inorganic phosphate (P_i) concentration. Such a summary of experimental data represents the whole organelle as a black box, despite the fact that it is a complex structure containing many interacting proteins which are encapsulated or embedded in two membranes.

The alternative is a detailed white-box model describing the interactions of the internal components. However, the white-box approach is often not sufficiently accurate. For instance, despite the fact that the sequence of amino acids of a protein is known, it is in general difficult to predict substrate affinities or turnover rates of the enzyme accurately, among other reasons due to imperfect energy functions for intramolecular interaction. Consequently, in order to calculate turnover in a metabolic pathway one does not usually use detailed molecular models of the enzymes, but utilizes black-box models in the form of equations which have a relatively crude mechanistic underpinning, but describe kinetic measurements.[8] To determine the parameters, the enzyme is experimentally isolated by controlling and systematically altering the substrates and products of the enzyme in an extract and measuring the rate of formation of the products of the enzyme, often at various levels of regulating molecules.

If the equations of a number of enzymes in a pathway are assembled in series, it is in principle possible to calculate the metabolic flux through the pathway. However, a study of yeast glycolysis shows that model predictions based on such kinetic measurements have rather limited correspondence with fluxes in the intact cell.[8] This may reflect the fact that the assumptions on the coupling of equations for the individual enzymes are not met, or that the enzyme characteristics measured in isolation do not accurately reflect the enzyme function in the intact system. As in the case of predicting the function of the molecule from amino acid sequence, predicting the behavior at the metabolic-network level from information at the enzyme level has limited accuracy. Going from the bottom level upward—from amino acid sequence via the molecular level to the metabolic pathway level—often gives unsatisfactory results at the top level if we do not correct at the intermediate level for potential imperfections in the lower-level description and for incorrect assumptions on the coupling of lower-level components.

In general, it is impossible to exactly reproduce *in vitro* the conditions to which an enzyme is exposed in the intact cell. Concentrations of ions and molecules may differ, protein solutions are often diluted, and subtle interactions with the many other proteins present *in vivo* are disrupted. Furthermore, when coupling enzyme reactions in a model of a metabolic pathway, it is usually assumed that the metabolites diffuse freely and can access the whole cytosol before they react with the next enzyme in the metabolic pathway. If metabolites are channeled directly between enzymes or their accessible diffusion space is restricted by other mechanisms, the calculation of the metabolite concentrations seen by an enzyme is wrong.

Dividing the System into Modules

As stated above, a module is an independently operable part of the system. Rather than assembling large numbers of equations for many enzymes in one great step, it is a better strategy to focus on a series of small separable modules in the metabolic system. The system can then be represented by modules which will be separately analyzed, before they are put together and tested jointly. By testing at the modular level, errors in the characteristics of the components of the module can be identified and repaired.

The idea of studying modules of the system is inspired by engineering and computer programming practices. In object-oriented computer programming, for instance, a module (termed a "class," "object," or "application programming interface") has a well-defined purpose. Its interface consists of a usually limited set of commands to which the program module responds in a prescribed way. This interface is made available to developers of other modules who are not concerned with the inner workings of the software module.[9] In other words, other programmers who connect with this module should not be concerned with,

nor change, the inner workings of such a program module.

A computational model for a module in a metabolic system should be as simple as possible. Albert Einstein is said to have stated: "Any intelligent fool can make things bigger, more complex, and more violent." The metabolic system in the human body is extremely complex in our eyes, although it was not designed by an intelligent fool, but instead arose by evolutionary processes. However, in order for scientists to understand this system, it is advisable to learn from engineers working at complex designs, who are thought to have put forward the K.I.S.S. principle, which stands for "Keep It Simple, Stupid" or "Keep It Short and Simple." The strategy of forward engineering of a new complex system by modularizing it inspires us to apply reverse engineering to the metabolic system. Retaining only the key elements of a module yields a "skeleton model" preserving the essential behavior unclouded by details.

The metabolic system is very densely connected,[10] but the organization of biochemistry textbooks suggest that at the kinetic level the modules can readily be discerned. Metabolic pathways often show high fluxes, which flow from one enzyme to the next in the pathway. If there are side branches where smaller fluxes flow in or out of the main path, it may be fruitful to study the main pathway as a module anyway.

In practical terms a module must show the following characteristics: 1. the module must be experimentally isolatable so that all its inputs can be controlled and its outputs measured; 2. the number of connections between the module and the rest of the system should be low; 3. the components inside the module must be well connected. To be efficiently studied, the module must usually contain a sufficient proportion of well known components whose response characteristics can be well determined. On the other hand, the module may contain one or a few components which are not well known, whose essential characteristics can be estimated from the response of the module as a whole. The term "module" is also used with a more fundamental definition as a group of molecules performing a certain cellular function, and modular design is considered to play a role in evolution.[11] The term "module" is also used for groups of molecules with statistically correlated behavior across data sets. Modules defined in different ways may or may not be similar, or overlap to some extent. Here we concentrate on the definition as an experimentally isolatable, independently operable unit that is part of the system.

There is probably an optimum for efficient analysis: when all components and interactions are well known, the computational model is expected beforehand to behave according to the prediction. However, when too large a proportion of the components are not well known, identification of their correct behavior may not be possible or may require too many complex tests. It is, however, difficult to estimate a priori how well the characterization of an isolated components accurately reflects its function *in vivo*. Some characteristics may be well preserved, but other characteristics, often at the interface of the module, may be very sensitive to damage during isolation.

The module is defined here based on the possibility to experimentally investigate it. Sometimes it may be isolated physically without much functional damage. In other situations, it is feasible to study the module's function in the intact system by minimizing or controlling the effect of other modules on the response of the system.

Can Modules be Designated by Computational Methods?

How can the metabolic system be subdivided in experimentally isolatable and testable modules? The metabolic system is densely connected in many places.[10] However, we have seen that modules may be easier to isolate at the kinetic level than at the structural level. For the identification of modules, metabolic connections with low activity may be neglected in the first approximation to reduce the number of connections of a potential module. To improve isolation of a module, one can experimentally inhibit or switch off one or more pathways that are connected with the module. In the example treated below, the glycolytic ATP production is inhibited so that mitochondrial production is virtually the only ATP source. Mechanistic modules are often found in textbooks. Biochemical pathways and diagrams of signaling networks in textbooks or review articles are a good way to start. Diagrams and maps in databases, such as KEGG (Kyoto Encyclopedia of Genes and Genomes), may also provide a good starting point.

Programs exist that designate modules in biomolecular systems, but these are usually based on clustering high-throughput experimental data.[12] Modules based on correlation of experimental data may not correspond accurately with experimentally isolatable mechanistic modules. This lack of correspondence is caused by high false discovery rates in high-throughput data and difficulties in detecting linear correlations in nonlinear systems. For instance, the intracellular ATP concentration often does not change measurably with ATP turnover in the cell, although both are mechanistically connected. PCr levels in muscle may drop considerably

without showing any measurable correlation with ATP level, although both are connected via the CK reaction and can be part of the same module (see below).

Can the selection of modules be guided by computational methods? Diagrams that visualize the connectivity of components in the system may be very helpful, especially if they are implemented on the computer with navigation and zoom possibilities. The nodes of graphs represent metabolite pools, while the edges are annotated with biochemical reactions. Regulatory pathways may act on the edges. To see modules better at the kinetic level, indications of expected flux sizes may be added to such computer maps. A first attempt to construct a comprehensive model of the human metabolic system is a metabolome-wide model developed by Palsson's group, which contains about 3300 reactions.[13] An important question is whether the selection of modules from such a system is feasible by automated methods. It is conceivable that, given a region of interest as a starting point, an algorithm can be devised to tesselate this region into modules which internally have dense connections, but just a few connections to their environment. Information might be connected to the graph of the metabolic system indicating which metabolites are measurable or controllable, which enzymes have known kinetics, and which pathways are inhibitable. Confidence levels may be attached to the information. An algorithm to suggest a subdivision of the region of interest is conceivable, although investigating the many possible combinations may constitute a computationally hard task. The suggestions resulting from the algorithm need to be critically examined and modified by the investigator.

Integrating Information from Multiple Scales

Information on the parameters of components of a module and the responses of the module as a whole can be combined to improve and test the model. If the module cannot be physically isolated, it may be tested in the context of the system of which it is part. These analyses at multiple levels constitute a multiscale approach. One may envisage that the jigsaw puzzle pieces in FIGURE 1 constitute a model for the molecular level of a metabolic pathway. Likewise, pathways in turn may be put together to form the metabolic system of a cell, as illustrated in FIGURE 2. The metabolic systems of various cell types together form the metabolic system of an organ, constituting the next level. In turn various organs together form the metabolic system of the human body.

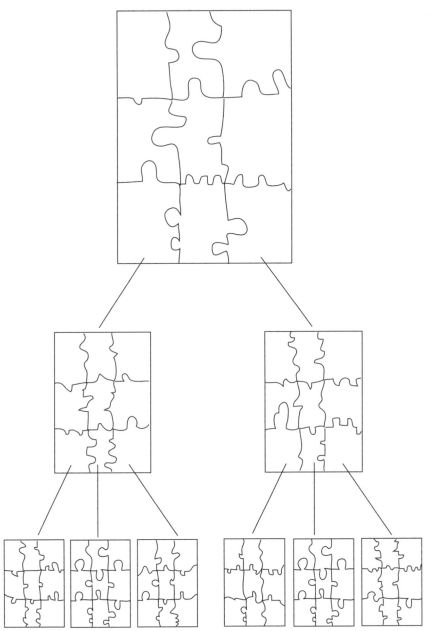

FIGURE 2. Analysis at multiple levels. The bottom layer may, for instance, represent modules consisting of molecules, with the squares representing six molecular pathways, each consisting of nine interacting molecules. The middle layer represents several of these bottom-level modules working together at a higher level. The middle layer may, for instance, represent two cell types, each containing nine molecular pathways. Some molecular pathways are used in multiple cell types, sometimes in slightly modified form. The top layer represents the integration of middle-level modules. This may, for instance, represent an organ consisting of nine cell types.

Below we will demonstrate how several components in the energy transfer pathway, such as CK enzymes and mitochondria in heart muscle cells, were isolated and characterized. In order to estimate the permeability of the mitochondrial outer membrane, which is probably altered during physical isolation of the mitochondria, use was made of the dynamic response time of the system as a whole.

To identify the response of a module embedded in a larger system, in this case the energy transfer system embedded in the cell as a whole, we can measure the response at an even higher level. An example is the estimation of the dynamic response time of O_2 consumption in the mitochondria based on the O_2 uptake of the heart as a whole (see below). Another example might be the measurement of the electrical activity distribution across the heart from measurements of electrical potential distribution at the body surface. In such cases the primary signal of interest is transmitted through the organ or body, and the transfer function of the higher-scale system is used to deconvolute the signal to determine its shape at the source.

Suppose we wish to model the system from the bottom upward. Molecular modeling by accurate calculation of the movement of individual atoms or groups of atoms has limitations, both at the fundamental level and because of the extremely large computation times required. At the next higher level, calculating the performance of a metabolic pathway based on known enzyme kinetics also seems to yield very limited correspondence with whole-pathway performance.[8] Calculation from the molecular bottom level to the whole-system top level seems to yield inaccurate results at the present state of scientific knowledge. However, by testing and improving modules with measured responses of the system at one or more larger scales, we can analyze the system as a whole. Information at a certain scale can be used to predict characteristics both in the upward and downward direction.

To look deep into the metabolic system of the body, transport models are used to describe how dynamic changes in metabolite levels in the blood and tissue relate to the metabolic rates in the cells. One may measure the metabolite content in an organ as a function of time, for instance, with nuclear magnetic resonance (NMR) spectroscopy or positron emission tomography (PET) after introducing a labeled substrate. Metabolic products can accumulate in the cells, and their concentrations depend on the amount transported to the blood in the capillaries, and subsequently out of the organ. Transport models may make it possible to calculate the time course of events in the cells from the time course of blood concentration changes of metabolites in the venous blood flowing out of an organ.

Models of functions of organelles are feasible, for instance, for ATP synthesis in the mitochondria.[2,3] Due to their complexity and the large number of components, it is to be expected that some components will be incorrectly represented. The module will have to be tested experimentally and improved to be combined with other modules. An empirical brief mathematical formula summarizing the organelle's measured response to experimental challenges may represent the organelle's function with an accuracy equaling that of complex mechanistic models. Therefore, it is often necessary and possible to combine black-box models, giving the relation between measured input and measured output for one cluster of components of the system, and white-box models, containing detailed descriptions of mechanisms for other components.

Dynamic Response of Oxidative Phosphorylation

Modeling and testing a module in the energy metabolic system and multiscale analysis will now be illustrated by an example in which we address the first phase of the response of myocardial OxPhos to step changes in heart rate. The demand of the body for blood supply fluctuates constantly, e.g., in response to exercise the heart rate and cardiac output adapt and ATP production in the mitochondria responds correspondingly. In response to a step jump in the heart rate to a sustained higher level, mitochondrial O_2 consumption adapts with a response time (t_{mito}) of about 8 s in rabbit hearts.[14,15] If glycolysis is inhibited and pyruvate is given as a substrate for the tricarboxylic acid (TCA) cycle, t_{mito} is about halved to 3.7 s.[16]

There are two main possibilities to explain the adaptation of mitochondrial OxPhos: substrate effects of ADP and P_i, or increased Ca^{2+} cycling, which follows the step in heart rate and stimulates the mitochondria.[3,14] As shown previously,[17] the response time of OxPhos is identical when either the cardiac end-diastolic volume is changed or the heart rate is changed to increase cardiac workload. Calcium cycling is unchanged in the first situation, whereas it changes immediately in the second. This makes it unlikely that Ca^{2+} is the signal stimulating the dynamic change in O_2 consumption. Indeed, simultaneous measurement of Ca^{2+} in the mitochondria in isolated papillary muscle, using a fluorescent indicator and NADH autofluorescence, showed a slow phase with a time constant of about 25 s for an increase in pacing frequency for the increase of intramitochondrial Ca^{2+} and NADH. For the reverse changes after a decrease in pacing frequency, an even larger time constant of 65 s was found, much slower than the t_{mito} of ~4 s for downregulation of mitochondrial O_2 consumption.[18] The evidence from intact heart and isolated papillary muscle makes it likely that Ca^{2+} is not the direct stimulus for the fast up-

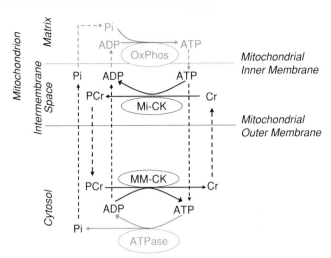

FIGURE 3. Scheme of the adenine nucleotide–creatine–phosphate (ACP) module. The input to the module is the ATP hydrolysis forcing function (ATPase) for ATP splitting. The hydrolyzed ATP is resynthesized in the mitochondrial matrix via oxidative phosphorylation (OxPhos) in response to changes in the concentration of ADP and inorganic phosphate (P_i). The dynamic response to changes in ATP hydrolysis is influenced by the reaction of ATP with creatine (Cr) yielding phosphocreatine (PCr) and ADP, and reverse, catalyzed by isoforms of creatine kinase enzymes (CK): a mitochondrial isoform (Mi-CK) in the intermembrane space and a muscular isoform (MM-CK) in the cytosol.

regulation of OxPhos but rather that ADP and P_i are responsible for the initial fast phase of activation of OxPhos.

During systole Ca^{2+} stimulates contractile activity in the myofibrils, energized by instant hydrolysis of ATP to ADP and P_i. ATP must be resynthesized from ADP and P_i in the mitochondria. At first glance this seems to require that ADP and P_i diffuse to the mitochondria. However, one isoform of CK is located close to the site of ATP hydrolysis and a second isoform in the mitochondrial intermembrane space, close to the site where ADP crosses the mitochondrial inner membrane to enter the mitochondria (Fig. 3). These enzymes catalyze the transfer of the phosphoryl group of PCr to ADP. Thus, $PCr + ADP \Leftrightarrow Cr + ATP$. PCr is present at high concentration in the cytosol of the cardiomyocyte. ADP concentrations in the cytosol are kept low, and ATP is resynthesized close to the site where it is used. Creatine (Cr) may then diffuse freely to the mitochondria where it is rephosphorylated by the mitochondrial isoform of CK, locally forming ADP to stimulate OxPhos. It has even been suggested that the transfer of high-energy phosphate groups by PCr is obligatory, forming a "PCr shuttle" between the sites of ATP synthesis and hydrolysis.[19] The importance of this shuttle has been hotly debated for many years.[20] The permeability to ADP of the mitochondrial outer membrane, which separates the cytosol and mitochondrial intermembrane space, is an important parameter which is varied in the model to study its effect on dynamic responses. Note that P_i is not handled by CK and diffuses directly between the compartments without taking part in the CK reactions.

The system for the transport of high-energy phosphates constitutes the module to be tested. Note that this adenine nucleotide–creatine–phosphate (ACP) module is embedded in the total system by communication with two adjacent modules: ATP hydrolysis during contraction and ATP production by the mitochondria. These communicating modules are rendered in grey in Figure 3. The response of OxPhos is modeled as a mathematical function where ADP and P_i in the intermembrane space are the input variables. This function represents the ATP production module which communicates with the ACP module. The transport processes across the mitochondrial inner membrane are therefore lumped with the enzyme complexes of OxPhos in this single response function. The adenine nucleotide translocator exchanges ADP for ATP, and there is a separate carrier for P_i. The ATP hydrolysis module, on the other hand, is represented by equations which give the time course of ATP hydrolysis derived from measurements. These equations take into account that ATP hydrolysis by the contractile elements and ion pumps is concentrated in the systolic phase of the cardiac cycle.

To isolate the ACP module it is advantageous that PCr and Cr only react in the CK reactions and do not contribute to connections with other modules. Secondly, apart from the dead-end CK reaction, ATP is almost exclusively synthesized by the mitochondria because glycolysis was inhibited in the experiments that were analyzed.

We modeled and tested the small ACP module[21] that is schematically given in FIGURE 3. It is connected to the rest of the system only by ATP, ADP, and P_i. The enzyme kinetic parameters of both isoforms of CK, such as dissociation constants for substrates and products, had been extensively characterized experimentally.[4] The activities of both CK isoforms and the mitochondrial aerobic capacity were measured for the experimental model, the rabbit heart, in which the dynamic response was also measured. Furthermore, the *in vivo* diffusion coefficients of the adenine nucleotides, P_i, Cr, and PCr for muscle are known. OxPhos itself reacts within a few milliseconds to ADP and contributes negligibly to the delay of the response of OxPhos to ATP hydrolysis.[21]

There was one parameter in the module for which no consensus values existed: the permeability of the mitochondrial outer membrane to ADP (P_{ADP}). Values for P_{ADP} varied from 0.16 μm/s[4] to 85 μm/s[21] calculated from measurements on isolated mitochondria or permeabilized muscle fibers. Therefore, all the parameters for the components of the module are known, except one. It turns out that the response time of O_2 consumption to a heart rate step calculated from the model, is very sensitive to P_{ADP}. For $P_{ADP} = 0.16 \mu$m/s, t_{mito} is 14.9 s, which is about four times the experimentally measured value. The calculated ADP level is much higher than experimental values. The very low P_{ADP} estimated for permeabilized muscle fibers is therefore not compatible with experimental data. It is likely that this low permeability derived from experiments in muscle fibers is due to diffusion gradients outside mitochondria.[7]

By systematically varying the permeability, the experimentally determined t_{mito} value of 3.7 s is found for $P_{ADP} = 21 \mu$m/s. This yields the measured response time of the mitochondria in the intact system, (FIG. 4). In this way the parameter values determined from isolated components are complemented with a parameter value derived by optimizing the model response of the whole module to make it correspond with the experimentally determined t_{mito} value. *In vitro* studies of P_{ADP}, resulted in four times higher values derived from measurements on isolated mitochondria[2] and at least four times lower values derived from measurements on permeabilized cardiac muscle fibers.[4] It may be hy-

pothesized that the permeability of the mitochondrial outer membrane is sensitive to isolation or permeabilization procedures and cannot be reliably measured in the isolated state. By estimating this parameter from the response of the system as a whole, this broken interface of an isolated system component has been effectively readjusted.

The determination of P_{ADP} and the testing of a plausible model for regulation of mitochondrial ATP synthesis did not merely establish a good model description of the system, but made it possible to estimate the contribution of the PCr shuttle to high energy phosphoryl group transport, to assess the role of ADP and P_i feedback control in regulation of myocardial mitochondrial ATP synthesis and to establish the function of the CK isoenzyme system (see below).

Exploring the Intact System

The value of t_{mito}, the response time of O_2 consumption in the mitochondria during a step-jump in heart rate, could not be measured directly in the cells, but was derived from time-resolved measurements of the O_2 content in the venous outflow from the heart. To calculate t_{mito} from the venous response, a model for O_2 transport in the heart muscle was developed.[15] After an increase in O_2 consumption in the mitochondria, the local O_2 balance is distorted and the local O_2 concentration around the mitochondrion drops, leading to increased diffusion of O_2 from the capillary network. Consequently, diffusion is part of the transport model. The increased diffusion leads to a lower O_2 concentration in the blood or, in the case of isolated organs, saline solution flowing through the capillary. Convective O_2 transport in the blood vessels is therefore included in the transport model. It turned out that a model for the full time course is not feasible because an accurate distribution function for transport times in the blood vessel system with the mitochondria as starting point cannot be identified. On the other hand, the total amount of O_2 in the whole organ can be calculated from a model for the O_2 concentration gradients in the steady state.[15] An increase in O_2 consumption causes a decrease in the amount of O_2 in the organ. By integrating the global O_2 mass balance, including venous O_2 outflow, for the heart over time, I derived from mathematical equations that dividing the decrease in O_2 amount in the organ by the increase in the O_2 consumption causing this decrease, exactly gives the average transport time for O_2 between mitochondria and the venous site where O_2 concentration was measured.[15] By subtracting this transport

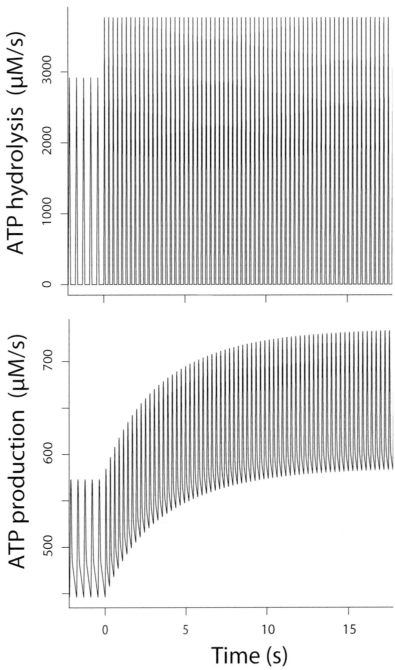

FIGURE 4. Time course of ATP synthesis in the mitochondria in response to a step in ATP hydrolysis caused by a prescribed step in paced heart rate. The pulsatile forcing function, representing ATP hydrolysis during cardiac systole, is plotted in the upper panel. The response of mitochondrial ATP production, calculated from the model, is given in the lower panel. Note the difference in scale for the ordinate between upper and lower panel. This reflects the fact that the large peaks in ATP hydrolysis during systole are damped about 23-fold at the level of mitochondrial ATP production. The mitochondrial outer membrane permeability to ADP is optimized to 21 μm/s to reproduce the experimentally measured response time of O_2 consumption of 3.7 s.

time, about 4–5.5 s, from the response time at the level of the whole organ, the response time at the level of the mitochondria, t_{mito}, was obtained. Note that as a result of the measurements at the whole-organ level, we eventually could estimate the *in vivo* ADP permeability of the mitochondrial outer membrane at the suborganellar level.

One more correction was necessary to obtain the true response time to a step-jump in the workload of the heart. The product of heart rate and developed pressure is a good measure of energy consumption in the heart, determined by its nearly linear relation to O_2 consumption. During a step in the heart rate there is a transient change in the pressure developed by the heart. The time course of the rate-pressure product during a step in the heart rate initially shows an overshoot, probably due to adaptation of calcium cycling during the first beats at an increased pacing rate. The mitochondrial response time was therefore corrected for the overshoot, represented by a negative response time of the rate-pressure product. This measurement at the whole-heart level was used to correct the time course of ATP hydrolysis in the myofibrils.

What Does the Module's Model Explain and Predict?

The reader will have noticed that we made the model fit the experimental response time by adjusting one parameter. Further testing of the model of the module was therefore called for. CK plays an important role in the module, and its role was tested by inhibition. The model predicts that inhibition of CK leads to a very sharp decrease in t_{mito} if the CK activity falls below 3% of its wild-type level. After 98% inhibition of CK, t_{mito} decreased to 2.6 s. The predicted very sharp decrease makes it difficult to establish precise quantitative correspondence between the model prediction and the experimental result, although qualitatively the prediction agrees very well with the experiment. The model can be tested further by investigating several other predictions: for instance, if the muscle isoform of CK is over-expressed, the model predicts a slower response of OxPhos to a changing workload; if the mitochondrial isoform of CK is over-expressed, the model predicts a faster reaction of OxPhos. These and other predictions may form the basis for further testing and refinement of the module.

CK and Cr have been proposed to have several functions, e.g., an emergency energy buffer when O_2 supply fails and serving as a PCr shuttle for transport of high-energy phosphate groups.[19] According to analysis with the ACP module, the fraction of the phosphoryl groups transported as PCr depends strongly on P_{ADP}, the permeability of the mitochondrial outer membrane to ADP. The multiscale model analysis of the dynamic experiments gave a mitochondrial outer membrane permeability of $21\ \mu m/s$. Using this permeability determined *in vivo*, the fraction of phosphoryl groups transported as PCr was calculated from the model to be 32% at modest ATP turnover rates. On the other hand, reduction of the CK activity led to a large increase in the peak value of the ADP oscillation as well as to a clear increase in average P_i value. The model analysis therefore seems to indicate that contribution of Cr to phosphoryl group transfer is modest, but it lowers the maximal ADP and P_i levels quite effectively. ATP synthesis is pulsatile in the beating heart with an oscillation amplitude of roughly 20–25% of the average ATP synthesis, and CK activity buffers large cytosolic ADP oscillations (FIG. 5) and lowers values of P_i, the other breakdown product of ATP hydrolysis. In this way, the multiscale modular analysis of the system revealed the function of the ACP module containing the two isoforms of CK.

Reconstructing the System from Modules

The ACP module is by itself insufficient to describe the behavior of the cardiac energy metabolism *in vivo*. As shown in *ex vivo* experiments,[16] t_{mito} is significantly larger when glycolysis is active. The extent of glycolytic activity *in vivo* will be influenced by mutual inhibition with fatty acid oxidation. Therefore, modules for fatty acid oxidation and glycolysis have to be added if one wants to better understand the dynamic behavior of OxPhos in response to changing cardiac workload and substrate supply. Ca^{2+} cycling will have to be added, too, in order to understand adaptation on the minute time scale. Intracellular signaling pathways which affect the activation states of enzymes are important, and, on a slower time scale, gene expression of the CK isoforms and synthesis of mitochondria are important processes. The need to connect multiple modules to understand the system at different time scales develops naturally. This is an example of the "middle out" approach described by Noble.[1]

So far it was assumed that the system could be described by its average behavior. However, the heart shows very significant spatial heterogeneities in metabolic fluxes, blood supply, enzyme activity, metabolite content, etc.[22] This heterogeneity is not

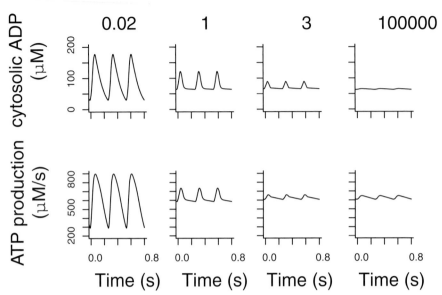

FIGURE 5. Oscillations of ADP concentration and mitochondrial ATP production in response to systolic peaks of ATP hydrolysis in the beating heart (see Fig. 4). The CK activity is given relative to wild-type activity in the rabbit heart. The MM-CK and Mi-CK isoform activity is changed in the same proportion. ATP hydrolysis peaks to 3770 μM/s during systole and is very low during diastole.

characterized by one spatial scale, but shows a fractal pattern of heterogeneity repeated at all scales in a self-similar way.[23] A multifractal model to describe the heterogeneity holds that the blood flow or metabolism in two arbitrary halves of the left ventricle differ by about 20%. The halves can again be subdivided in two halves which differ by 20%. This process of reproducing heterogeneity at all scales in a mathematical model can be repeated until one reaches the level of the capillary bed, which is provided with blood by the smallest arterioles. In the end, the multiscale model must represent the characteristics of the individual organism, perhaps a human patient, and the extensive heterogeneity within her organs. This is found at all levels in the system, probably including gene expression. Analysis of gene expression microarray data[24] shows for instance that local heterogeneity of expression of about 16,000 mRNA's in the joints of rheumatoid arthritis patients, perhaps to a major extent due to blood cell infiltration, is a larger source of variation than the variation between patients and true microarray measurement error.

Modules must repeatedly be tested and modified during the model development process. For their first tests, modules may be embedded in the total system by communication with provisional, simplified surrounding modules. Upon improvement of those provisional modules, reassessment of the embedded module may

be advisable. This modeling process is a necessary development: biology must move from a phase where hunter-gatherers of (molecular) species and genes predominate, via a phase where descriptive models predominate, eventually to a phase where mechanistic models predominate. Physics likewise moved from fact collecting to Kepler's laws and Newtonian mechanics. The human genome project has been considered as an achievement comparable to the Apollo project of man's journey to the Moon. The small step from the spacecraft to the Moon would not have been possible if only piles of loose observations, or even Kepler's laws, had been available. It very likely depended on mechanistic physical models. Likewise, it can be hypothesized that mechanistic models are necessary to make full use of all the information gathered in the genomic era.

Conclusions

Is it true that all the king's horses and all the king's men couldn't put a biomolecular system together again? Perhaps so, but scientists should definitely try. If we approach the system by modularizing it, we may advance a long way *in silico*. The K.I.S.S. principle ("keep it short and simple") is thought to have been invented during the Apollo project with its gigantic engineering effort. Likewise, approaching the large biomolecular system step-by-step by analyzing it module-by-module is a desirable avenue. Multiscale analysis is extremely helpful to put together the pieces and vitalize the interfaces of modules that were broken when the system was taken apart. Our physiological system has a broad range of size and time scales, from the atomic and molecular level, via organelles and cells, to organs and the whole organism. Combining information from multiple scales will enable us to critically test and improve the modules. For instance, enzyme kinetics and diffusivity can be combined with the response speed of the system as a whole, the latter obtained by deconvoluting the response at the organ level with the vascular transport function. This combination enables us to explore the system and optimize unknown or unreliable parameters of system components at smaller scales. Conversely, multiscale models make it possible to demonstrate and predict how the properties at the smaller scales, such as molecular properties and connectivity, influence the systems behavior at the larger scales. This approach yielded surprising results in the analysis of the phosphoryl group system in the heart muscle, represented by the circuit provided by two isoforms of CK. The computational analysis showed that the contribution of the PCr shuttle to high-energy phosphoryl group transfer is very modest. On the other hand, the multiscale computational analysis showed that the CK system is very effective in decreasing maximum levels of ADP during systole and reducing average P_i levels over the whole cardiac cycle. This suggests that buffering of oscillations and lowering P_i levels are important functions of the CK enzymes, and that the high-energy phosphate group transport is a less important function.

Acknowledgment

The Centre for Medical Systems Biology is supported by a Genomics Centre of Excellence grant from the Netherlands Genomics Initiative, which is funded by the Dutch government.

Conflict of Interest

The author declares no conflicts of interest.

References

1. NOBLE, D. 2006. The Music of Life: Biology Beyond the Genome. Oxford University Press. Oxford, UK.

2. BEARD, D.A. 2005. A biophysical model of the mitochondrial respiratory system and oxidative phosphorylation. PLoS Comput Biol. **1:** e36.

3. CORTASSA, S. *et al.* 2003. An integrated model of cardiac mitochondrial energy metabolism and calcium dynamics. Biophys. J. **84:** 2734–2755.

4. VENDELIN, M., O. KONGAS & V. SAKS. 2000. Regulation of mitochondrial respiration in heart cells analyzed by reaction-diffusion model of energy transfer. Am. J. Physiol. Cell Physiol. **278:** C747–C764.

5. PLASTERK, R. 2005. Question and Answer. Curr. Biol. **15:** R861–R862.

6. VAN BEEK, J.H.G.M. 2006. Channeling the data flood: handling large-scale biomolecular measurements in silico. Proc. IEEE **94:** 692–709.

7. KONGAS, O. *et al.* 2002. High K(m) of oxidative phosphorylation for ADP in skinned muscle fibers: where does it stem from? Am. J. Physiol. Cell Physiol. **283:** C743–C751.

8. TEUSINK, B. *et al.* 2000. Can yeast glycolysis be understood in terms of in vitro kinetics of the constituent enzymes? Testing biochemistry. Eur. J. Biochem. **267:** 5313–5329.

9. WEISFELD, M. 2004. The Object-Oriented Thought Process. SAMS Publishing. Indianapolis, In.

10. FELL, D.A. & A. WAGNER. 2000. The small world of metabolism. Nat. Biotechnol. **18:** 1121–1122.

11. HARTWELL, L.H. *et al.* 1999. From molecular to modular cell biology. Nature **402:** C47–C52.

12. TANAY, A. *et al.* 2004. Revealing modularity and organization in the yeast molecular network by integrated analysis of highly heterogeneous genomewide data. Proc. Natl. Acad. Sci. USA **101:** 2981–2986.

13. DUARTE, N.C. *et al.* 2007. Global reconstruction of the human metabolic network based on genomic and bibliomic data. Proc. Natl. Acad. Sci. USA **104:** 1777–1782.

14. VAN BEEK, J.H.G.M. *et al.* 1998. The dynamic regulation of myocardial oxidative phosphorylation: analysis of the response time of oxygen consumption. Mol. Cell Biochem. **184:** 321–344.

15. VAN BEEK, J.H.G.M. & N. WESTERHOF. 1991. Response time of cardiac mitochondrial oxygen consumption to heart rate steps. Am. J. Physiol. **260:** H613–H625.

16. HARRISON, G.J. *et al.* 2003. Glycolytic buffering affects cardiac bioenergetic signaling and contractile reserve similar to creatine kinase. Am. J. Physiol. Heart Circ. Physiol. **285:** H883–H890.

17. HAK, J.B. *et al.* 1992. Influence of temperature on the response time of mitochondrial oxygen consumption in isolated rabbit heart. J. Physiol. **447:** 17–31.

18. BRANDES, R. & D.M. BERS. 2002. Simultaneous measurements of mitochondrial NADH and Ca(2+) during increased work in intact rat heart trabeculae. Biophys. J. **83:** 587–604.

19. BESSMAN, S.P. & P.J. GEIGER. 1981. Transport of energy in muscle: the phosphorylcreatine shuttle. Science **211:** 448–452.

20. MEYER, R.A., H.L. SWEENEY & M.J. KUSHMERICK. 1984. A simple analysis of the "phosphocreatine shuttle". Am. J. Physiol. **246:** C365–C377.

21. VAN BEEK, J.H. 2007. Adenine nucleotide–creatine–phosphate module in myocardial metabolic system

explains fast phase of dynamic regulation of oxidative phosphorylation. Am. J. Physiol. Cell Physiol. **293:** C815–C829.

22. ALDERS, D.J. *et al.* 2004. Myocardial O2 consumption in porcine left ventricle is heterogeneously distributed in parallel to heterogeneous O2 delivery. Am. J. Physiol. Heart Circ. Physiol. **287:** H1353–H1361.

23. VAN BEEK, J.H., S.A. ROGER & J.B. BASSINGTHWAIGHTE. 1989. Regional myocardial flow heterogeneity explained with fractal networks. Am. J. Physiol. **257:** H1670–H1680.

24. VAN DER POUW-KRAAN, T., C. VERWEIJ & J.H.G.M. VAN BEEK. 2007. Analysis of heterogeneity in gene expression. Unpublished observation.

Discovering Regulators of the Drosophila Cardiac Hypoxia Response Using Automated Phenotyping Technology

JACOB D. FEALA,[a] JEFFREY H. OMENS,[a] GIOVANNI PATERNOSTRO,[a,b]
AND ANDREW D. McCULLOCH[a]

[a]Department of Bioengineering, University of California, San Diego, La Jolla,
California, USA

[b]The Burnham Institute of Medical Research, La Jolla, California, USA

Necrosis and apoptosis during acute myocardial infarction result in part from the inability of hypoxic cardiac myocytes to match ATP supply and demand. In contrast, hypoxia-tolerant organisms, such as *Drosophila*, can rapidly regulate cellular metabolism to survive large oxygen fluctuations. A genetic screen of fly heart function during acute hypoxia can be an unbiased way to discover essential enzymes and novel signaling proteins involved in this response. We have developed a prototype to show proof of concept for a genome-scale screen, using computer automation to rapidly gather *in vivo* hypoxic heart data in adult *Drosophila*. Our system automatically anesthetizes flies, deposits them on a microscope slide, and locates the heart organ of each fly. The system then applies a hypoxia stimulus, acquires time-space (M-mode) images of the heart walls, and analyzes heart rate and rhythm. The prototype can produce highly controlled measurements of up to 55 flies per hour, which we demonstrated by characterizing the effect of temperature, oxygen content, and genetic background on the hypoxia response. We discuss the possible applications of a genome-wide cardiac phenotype data set in systems biology analyses of hypoxic metabolism, using genome-scale interaction networks and constraint-based metabolic models.

Key words: cardiac hypoxia; systems biology; automated microscopy; *Drosophila melanogaster*; genomic phenotyping

Introduction

The fruitfly is gaining interest as a model organism for heart research not only for the ease of its handling and short life cycle, as compared to vertebrate models, but also because its well known genome and the array of available genetic tools allows for a genomic and systems biology approach. In contrast with the mouse, in which it takes months to develop new genetic knockout strains, and with the zebrafish, which is a relatively new model organism, *Drosophila* has a genome that is both well understood and easy to manipulate. With its sequence currently in its fourth revision,[1] the fly has one of the best characterized genomes of any multicellular organism. Several publicly available libraries provide strains of single-gene disruptions via deletion, P-element insertion,[2] or transgenic RNA interference.[3]

In particular, transgenic RNAi libraries make it possible to use the binary GAL4/UAS expression system to cause tissue-specific gene inactivation any time during the fly's life span, providing the ability to measure the effects of gene perturbations that would cause lethality in early development or in other cell types.

Flies are a favorite for performing forward genetic screens, but so far there has not been an attempt to screen mutants for heart function under any context. New technology for rapid *in vivo* measurement of adult fly hearts would allow researchers to reverse screen the *Drosophila* genome for novel genetic influences on cardiovascular function. Several systems do exist for measuring rate and contraction of the fly heart based on video microsocopy[4,5] or optical coherence tomography.[6,7] However, the current methods require many steps of labor-intensive manual preparation and data analysis. In order to fully capitalize on the advantages offered by using *Drosophila* as a model (size, fecundity, life cycle, genetic libraries), we have improved on these techniques by automating as many of the steps in the heart measurement process as possible. With computer automation, the fatigue and human error that comes

Address for correspondence: Prof Andrew McCulloch, PhD, Dept of Bioengineering, University of California, San Diego, 9500 Gilman Drive, 0412, La Jolla, CA 92093-0412; Fax: 858-534-5722.
   amcculloch@ucsd.edu

with repeated measurements can be minimized, if not eliminated. Our prototype system can be used for a smaller-scale genetic screen and serves as proof of the concept that a genome-wide screen is feasible with parallel implementation or further engineering efforts.

Screening the Cardiac Hypoxia Response

Since the basics of normal contractile function in cardiac cells are well known in humans, a genetic screen of heart function has more translational benefit under less-understood contexts, such as adaptation to stress or disease states. We have chosen to focus on adaptation to acute hypoxia, but in theory our measurement system could be useful for screening the cardiac effects of drugs, chemicals, aging, environmental stimuli, or most other contexts of interest.

Hypoxia-reoxygenation injury is thought to be the primary cause of cell death in myocardial infarction (MI).[8] Currently the only treatment for acute MI is to restore blood flow as soon as possible, either by mechanical (angioplasty) or chemical (thrombolytic) means. No treatments are currently available for increasing the tolerance of cardiac tissue to hypoxic stress, though some potential lies in drugs that trigger endogenous pathways for ischemic preconditioning.[9,10]

In animals, all cells have intrinsic metabolic and signaling mechanisms for adapting to acute hypoxia. In the ischemic mammalian heart, reduced oxidative metabolism causes a decrease in ATP, which stimulates anaerobic glycolysis, i.e., fermentation of glucose to lactate, through allosteric stimulation of key enzymes.[8] Protein kinases also play a regulatory role in this anaerobic shift, stimulating the breakdown of glycogen into glucose and also regulating pyruvate dehydrogenase (PDH) to alter the flux through the citrate cycle.[8]

Since oxygen fluctuations can manifest in countless ways over the life of a cell, it is likely that the genetic and metabolic core of the cellular hypoxia response evolved very early and is tightly conserved from flies to humans.[11] For example, hypoxia-inducible factor (HIF-1) and AMP-activated protein kinase (AMPK) are important for the hypoxia response in both mammals[12,13] and flies.[14,15] However, the total cellular response is far-reaching and complex, causing an array of changes in metabolic, signaling, and transcriptional networks,[16] and this system-wide orchestration of cellular hypoxia defenses is not well understood. A screen for hypoxia-sensitive responses during cardiac hypoxia would provide a starting point by gathering the list of genes involved.

Hypoxia Tolerance in Drosophila

Fruitflies, like many invertebrates, are extremely tolerant to oxygen fluctuations, surviving total anoxia for 4–6 hours with full recovery.[17] It is thought that this remarkable tolerance stems from the ability to achieve system-wide matching of metabolic supply and demand, especially ATP.[18] The genetic basis for hypoxia tolerance in flies has begun to be explored, e.g., the *Drosophila* gene coding for the enzyme trehalose phosphate synthase increases hypoxia tolerance when overexpressed in flies[19] and when transfected into human cells,[20] which lack the gene. The nitric oxide pathway, known to be important for hypoxia sensing and response in mammalian heart,[21,22] was found to mediate cellular and behavioral responses to hypoxia in fly larvae.[23] Importantly, a genetic screen for hypoxia tolerance[24] discovered that *Drosophila* adenosine deaminase acting on RNA (dADAR), an mRNA editor acting on ion channel transcripts,[25] sensitizes the fly to hypoxia when mutated.

Although *Drosophila* flight muscle metabolism has been well studied, energy metabolism in the fly heart remains completely unknown, partly due to the difficulty of accessing the small heart organ. In flight muscle, carbohydrates are the main source of fuel, and ATP is generated through aerobic glycolysis even during maximum exercise. Under anaerobic conditions, we found that flight muscle produces not only lactate (as in mammalian skeletal and cardiac muscle), but also comparably large amounts of alanine and acetate.[26] It is not known whether the fly heart oxidizes fatty acids and lactate under normal conditions, as mammals do, and we also do not know whether the anaerobic fly heart produces all three end products (lactate, alanine, and acetate) that flight muscle does. Measuring the heart phenotype for targeted perturbations of these enzymes will show which pathways of cardiac metabolism are essential for hypoxia tolerance.

Before *Drosophila* can become an established model organism for studying cardiac hypoxia, there is much to be learned about its metabolism under normal and anaerobic conditions. Targeted gene perturbations can be used with *in vivo* measurement of heart function to test hypotheses for essential pathways and regulators of anaerobic metabolism. Beyond the enzymes and genes of central metabolism, a genome-scale screen of the cardiac hypoxia response in flies may be an unbiased way to uncover unique heart-specific hypoxia-response genes that give flies their extraordinary tolerance.

We have characterized the acute hypoxia response in adult *Drosophila*, both to demonstrate the ability of our new technology to rapidly gather and analyze fly heart data and also to understand limitations and optimize

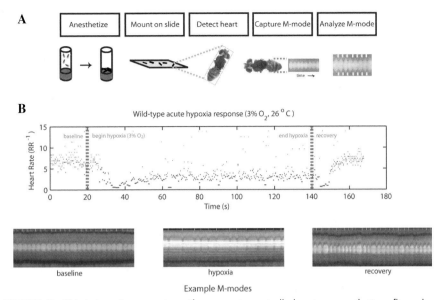

FIGURE 1. (A) Automation overview. The computer-controlled system anesthetizes flies, deposits them on a microscope slide, and detects and focuses on the heart for acquisition of a time-space M-mode image of wall contractions. **(B)** Typical M-mode images and heart rates for base line, 120-s hypoxia stimulus, and recovery in the wild-type fly.

controllable parameters for a future genetic screen. In this report we provide a brief overview of our methods and some characteristic results and then discuss the possible applications of a "genomic phenotype" data set in a systems approach to studying hypoxic myocardium.

Methodology

Automated Measurement

Customized software, written in the Python programming language, was implemented to automate mounting flies on the slide, locating the moving hearts, and measuring heart parameters. The automated measurement process consists of the following steps (FIG. 1A): 1. anesthetize flies and deposit on slide; 2. refine fly position (manually); 3. locate flies; 4. detect heart; 5. position M-mode line; 6. capture time-space M-mode image; 7. detect beats in M-modes.

Upon insertion of the vials into the system, flies are automatically anesthetized with vaporized triethylamine and then deposited on a microscope slide by a network of tubes, pressurized air, and computer-controlled solenoid pinch valves. A robotic stage then manipulates the slide following a series of image-processing algorithms in order to locate and focus on the heart tube of each fly under the microscope. The computer then uses the video feed to construct a time-space (M-mode) representation of a line of pixels across the heart during base line, hypoxia stimulus, and hypoxia recovery. The slide is removed and replaced, and the process is repeated.

Automated Analysis

Heart rate and rhythm can easily be extracted from the M-mode representations using image processing algorithms. When dilated the heart tube is translucent and appears much brighter through the microscope than it does in the contracted state. Therefore, the average brightness of pixels in the M-mode line follows the contractions of the heart. The M-mode is reduced to a one-dimensional intensity signal by averaging the columns of the image. Full 2D M-modes are also saved for future application of heart wall detection algorithms.

Beat detection is more fitting than spectral analysis for analyzing the heart rate in hypoxic flies. As shown under RESULTS, the heart rate slows dramatically under severely hypoxic conditions, which often causes peaks in the spectrum to be masked by low-frequency noise. Also, the heart rhythm is variable during the hypoxic response, resulting in broad or scattered peaks that are difficult to detect and quantify. To overcome these problems, we adapted a beat detection algorithm[27] to locate contractions within the intensity signal.

User Interface

A graphical user interface (GUI) provides the user with full control and monitoring of the process. The GUI allows the user to view intermediate steps of the detection algorithms, and buttons for manual control of the stage allow corrections to be made if the heart is not detected or focused correctly. Clicking on the microscope window centers the M-mode line on the selected pixel, allowing small adjustments to be made during detection and measurement. In addition to the current camera view, the GUI displays the last 2 s of M-mode acquisition so that the user can monitor the raw data, as well as oxygen concentration of the gas mixture.

A separate GUI, written in Matlab, allows fast inspection and analysis of large amounts of M-mode data. This interface allows the user to discard poor data from badly positioned flies, and (blind to the identity of the fly stock) mark beats missed by the analysis algorithm. Also, we built in the capability for displaying the results of heart wall detection and manual marking of systolic and diastolic wall positions in the M-mode images. This leaves our platform open to future work designing image processing to automatically trace heart walls and calculate fractional shortening, a valuable addition to the analysis of heart rate and rhythm presented here.

Results

Characterization of Wild-type Cardiac Hypoxia Response

Our automation allows the highly controlled measurement of hypoxic heartbeats for 25 to 50 flies per hour, depending on the measurement duration. To demonstrate the rapid acquisition of heart measurements, we used our automation technology to characterize the effect of acute hypoxia on heart rate and regularity of rhythm in the Oregon-R wild-type fly for different oxygen percentages, temperatures, ages, sex, and wild-type strains (unpublished data). When nitrogen–oxygen mixtures of less than about 10% O_2 are passed over the fly, the heart rate instantly declines, then partially recovers and maintains a reduced rate. Restoration of atmospheric oxygen causes transient slowing, then a recovery to base line. FIGURE 1 shows example M-modes and heart rate for a single measurement.

FIGURE 2 displays fly heart rates under various controlled conditions. As with other physiological functions in ectotherms such as *Drosophila*, heart rate is dependent on temperature. Similar to previous results,[4]

warmer ambient temperature sped up the heart rate at base line, as well as during the hypoxia stimulus. The strength of hypoxia stimulus (percentage oxygen) changed the hypoxic heart rate as well as the shape of the response. Males and females of the same age did not differ statistically in their heart rate response (not shown), but males aged 40 days had slower heart rates than 3-day-old males over all time points.

Oxygen-sensitive Mutation

The literature describes some known hypoxia-sensitive *Drosophila* genes.[19,23,25] We have used one of these genes, *tps1*, as a positive control for detecting a hypoxia sensitivity with our system. The *tps1* locus codes for the trehalose-phosphate synthase enzyme, which creates the disaccharide trehalose from glucose monomers and causes hypoxia sensitivity when disrupted in flies. Trehalose is important not only as a carbohydrate energy source for anaerobic glycolysis but also due to protective properties when bound to proteins.[19] In one study, the *Drosophila tps1* gene was transfected into human cells, which do not have a copy of the gene, and the cells became more tolerant to hypoxia.[20]

We acquired a heterozygous P-element insertion of *tps1* and outcrossed with a wild-type strain. The *tps1* disrupted flies had similar heart rate in base line and recovery but slightly, though significantly, reduced heart rate during hypoxia when compared with an equivalently outcrossed wild-type strain (FIG. 2).

Discussion

We have developed automation for rapid *in vivo* measurement of heart phenotypes in adult *Drosophila* and demonstrated the speed of measurement by characterizing the wild-type hypoxia response for a wide range of parameters. One major limitation that we discovered was that at oxygen levels below 3%, measurements became difficult due to a reflexive "curling" of the body and wings (unpublished observation). Although 3% O_2 caused significant differences in the hypoxia response of our positive controls (aging flies and the hypoxia-sensitive *tps1* disruption), the differences were small and therefore a genetic screen at that relatively high oxygen level might not cause enough hypoxic stress to expose genes causing minor changes. Nevertheless, if this problem can be overcome, the technology is mature enough to begin a rapid-throughput screen.

Designing a Large-scale Genetic Screen

Considering only the range of parameters tested, using aging and the hypoxia-sensitive mutant *tps1* as

Characterizing the cardiac hypoxia response

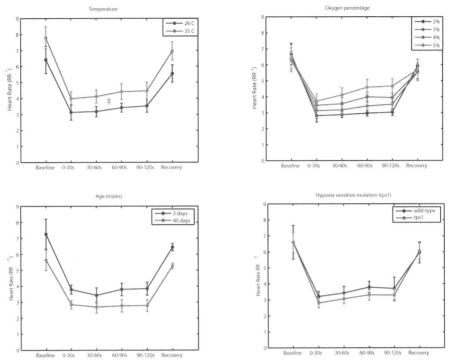

FIGURE 2. Acute hypoxia response over different parameter values such as temperature, percent oxygen, and age. When not varied in the experiments, default parameter values are 26°C, with 3% oxygen over 120-s stimulus, using 3- to 5 day-old flies of both sexes. Heart rate is calculated as the inverse of the average of winned RR intervals. (In color in *Annals* online.)

positive controls, an optimal screen might be performed at 3% oxygen for 90 s at 26°C, with either male or female flies 3- to 5-days old. Supervising our system, one technician working 1 year full-time could feasibly measure the hypoxic heart function for some 2500 mutant stocks under tightly controlled experimental conditions, with minimal human error and fatigue.

Incorporating a Phenotype Screen into a Systems Approach

Although gathering heart data from 2500 mutant *Drosophila* strains would no doubt uncover a handful of novel and interesting genes, the fraction of the genome covered by a screen of that size would be small. Flies have approximately 14,000 identified genes, so even at full-time operation the proposed screen would cover less than one-fifth of the genome per year. Until the methods can be improved further or implemented in parallel, it may be worthwhile to consider strategies for a targeted screen rather than randomly sampling the genome. We previously presented a network-based

strategy for ranking genes more likely to affect life span.[28]

Similarly, it may be possible to build a ranked list of candidate hypoxia-responsive genes based on genome-wide networks. In our lab we use a systems approach to study hypoxic metabolism in *Drosophila* muscle cells.[29] Since the primary effects of oxygen deprivation are metabolic (decreased ATP production, shift in redox potential, pH and ion imbalance), our list of candidates will be generated using our constraint-based network model of hypoxic metabolic regulation.

Constraint-based Metabolic Modeling

A goal of systems biology research is to invent new techniques for integrating different types of high-throughput data into functional network models, which can be simulated to generate testable hypotheses. The constraint-based method is currently the most popular and successful method for modeling genome-wide metabolic networks,[30] although most *in silico* metabolic reconstructions on which constraint-based techniques can be applied have focused on single-cell microbes.

Cardiac phenotypes and single deletion analysis

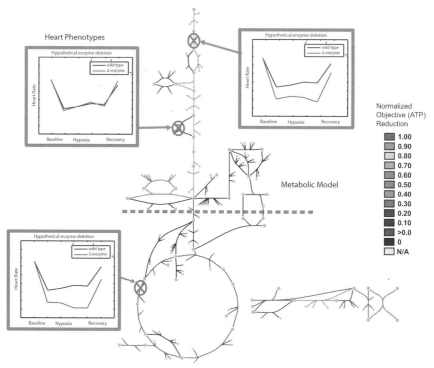

FIGURE 3. A strategy for integrating a phenotype screen with an existing constraint-based model of metabolism. Hypoxic heart rate can be compared to ATP production in the model, for every enzyme with a deletion strain available for screening. This analysis can help improve the model by identifying essential metabolic pathways for cardiac hypoxia tolerance. (In color in *Annals* online.)

However, reconstructions have been performed for the red blood cell[31] and mouse cardiomyocyte,[32] and more recently a multicellular human metabolic network was completed.[33] We are currently expanding and validating our model of central ATP-generating metabolism in *Drosophila* muscle cells using microarrays and metabolomic data.

These genome-wide reconstructions are usually curated manually, starting from a high-throughput data set, such as the annotated genome and/or proteomics data, for the cell type of interest and referring to the literature to confirm putative reactions and enzymes. This method ensures that models start with a complete "parts list" for the system, a cornerstone of the systems biology paradigm.[34] Iterations of high-throughput experimentation and computer simulation are then used to generate and test predictions to progressively increase the accuracy of the model. A genome-wide phenotype screen is a very useful data set for earlier stages of refinement, since the loss of function of particular enzyme gene disruptions identifies the essential reactions to be included in the model. The goal of our lab is

to create a metabolic model that contains all reactions important to hypoxia tolerance in the fly heart, and the genetic screen of cardiac hypoxia will let us identify reactions that manifest at the phenotype level (the "top down" view), which molecular data and modeling can help explain mechanistically (from the "bottom up").

The quantitative nature of our screening method fits well with systems models. Flux–balance simulations of metabolism provide steady-state flux values for all enzymes in the system when optimizing for some objective such as ATP production.[35–37] These intra-system flux values allow the model to be directly validated against fluxomics data derived from metabolomics methods, such as mass spectrometry and nuclear magnetic resonance spectroscopy (as we have done with the *Drosophila* model[26]). Although some assumptions must be made, it is also possible to compare quantitative phenotype values to the ability of the model to meet its objectives. For example, the hypoxic heart rate for a certain mutation can be compared against the amount of ATP produced in the simulated knockout strain (FIG. 3). When performed with a large data

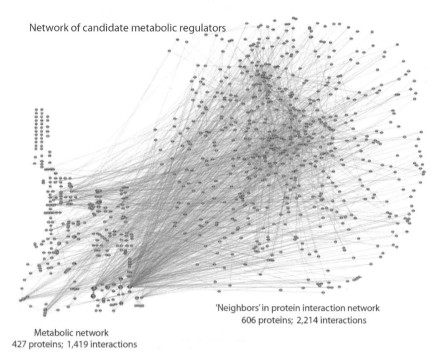

Network of candidate metabolic regulators

'Neighbors' in protein interaction network
606 proteins; 2,214 interactions

Metabolic network
427 proteins; 1,419 interactions

FIGURE 4. A strategy for generating a candidate gene list for a phenotype screen. Rather than screening the entire genome, which is still not feasible, the enzymes in the model and their "first neighbors" in protein interaction networks might be enriched for important hypoxia-responsive metabolic regulators.

set, these comparisons may provide a coarse-grained quantitative view of system operation that goes beyond the "all or none" results of a qualitative screen.

Genome-scale Protein-interaction Networks

One problem with using enzyme deletions is that many are homozygous lethal, especially in the ATP-producing pathways, such as glycolysis, tricarboxylic acid (TCA) cycle, and oxidative phosphorylation. Instead of studying hypoxia while perturbing this core of essential pathways, which we already know to be important under any physiological condition, it would be more interesting to find hypoxia-responsive regulators of metabolism that change in activity during oxygen fluctuations. Although the literature lists some examples, such as glycogen phosphorylase and AMP-activated kinase, we have limited knowledge of the global signaling cascades that mediate the acute shift to anaerobic metabolism (prior to changes in gene expression). It may be possible to take advantage of databases of protein interaction networks, created by high-throughput methods and literature curation, to enrich our list of gene candidates for important hypoxia-responsive regulators.

Although very noisy and incomplete, when combined with other high-throughput data and the right statistical approach, these networks have been successful in providing a coarse-grained view of system-wide activity.[38–40]

As a very simple demonstration, we merged the *Drosophila* protein interaction network with the list of enzyme proteins in our metabolic model, and then extracted the "first neighbors"—all proteins in the global network that have at least one interaction with an enzyme in the model (FIG. 4). The resulting subnetwork of enzyme interactors contains 606 proteins and 2214 interactions, and includes proteins, such as actin, calmodulin, and several protein kinases. This first step toward a list of screen candidates has the advantage of unbiased sampling from the global network, while at the same time being targeted toward likely metabolic regulators. Results from the cardiac phenotype screen can be examined within this network (as node values) and interpreted in contexts, such as vicinity to crucial enzymes in model simulations, clustering with other "first neighbors" of enzymes, and extraction of small modules of highly connected hypoxia-responsive proteins.

Summary

Our system for rapid measurement of cardiac hypoxia phenotypes has the potential to measure upwards of 2500 fly stocks per year, which is of the order of magnitude required for a genome-wide screen. Though valuable on its own, such a screen also has the potential to be integrated into systems analyses, for example with computer simulations of metabolism or in protein interaction networks. We have previously demonstrated the importance of network properties to metabolic function during hypoxic stress. Similarly, the complexity and scope of the myocardial response to acute hypoxia suggests the benefit of studying this system at the genome scale. A reverse genetic screen of cardiac hypoxia phenotypes is the first step toward identifying all of the important components involved.

Conflict of Interest

The authors declare no conflicts of interest.

References

1. ADAMS, M.D. *et al*. 2000. The genome sequence of Drosophila melanogaster. Science **287:** 2185–2195.
2. SPRADLING, A.C. *et al*. 1999. The Berkeley Drosophila Genome Project gene disruption project: single P-element insertions mutating 25% of vital Drosophila genes. Genetics **153:** 135–177.
3. DIETZL, G. *et al*. 2007. A genome-wide transgenic RNAi library for conditional gene inactivation in Drosophila. Nature **448:** 151–156.
4. PATERNOSTRO, G. *et al*. 2001. Age-associated cardiac dysfunction in Drosophila melanogaster. Circ. Res. **88:** 1053–1058.
5. OCORR, K. *et al*. 2007. KCNQ potassium channel mutations cause cardiac arrhythmias in Drosophila that mimic the effects of aging. Proc. Natl. Acad. Sci. USA **104:** 3943–3948.
6. WOLF, M.J. *et al*. 2006. Drosophila as a model for the identification of genes causing adult human heart disease. Proc. Natl. Acad. Sci. USA **103:** 1394–1399.
7. CHOMA, M.A. *et al*. 2006. Images in cardiovascular medicine: in vivo imaging of the adult Drosophila melanogaster heart with real-time optical coherence tomography. Circulation **114:** e35–e36.
8. OPIE, L.H. 1991. The Heart: Physiology and Metabolism. Raven Press. New York.
9. YELLON, D.M. & J.M. DOWNEY. 2003. Preconditioning the myocardium: from cellular physiology to clinical cardiology. Physiol. Rev. **83:** 1113–1151.
10. BUDAS, G.R., E.N. CHURCHILL & D. MOCHLY-ROSEN. 2007. Cardioprotective mechanisms of PKC isozyme-selective activators and inhibitors in the treatment of ischemia-reperfusion injury. Pharmacol. Res. **55:** 523–536.
11. O'FARRELL, P.H. 2001. Conserved responses to oxygen deprivation. J. Clin. Invest. **107:** 671–674.
12. SEMENZA, G.L. 2001. Hypoxia-inducible factor 1: oxygen homeostasis and disease pathophysiology. Trends Mol. Med. **7:** 345–350.
13. HARDIE, D.G. *et al*. 2003. Management of cellular energy by the AMP-activated protein kinase system. FEBS Lett. **546:** 113–120.
14. LAVISTA-LLANOS, S. *et al*. 2002. Control of the hypoxic response in Drosophila melanogaster by the basic helix-loop-helix PAS protein similar. Mol. Cell Biol. **22:** 6842–6853.
15. PAN, D.A. & D.G. HARDIE. 2002. A homologue of AMP-activated protein kinase in Drosophila melanogaster is sensitive to AMP and is activated by ATP depletion. Biochem J. **367:** 179–186.
16. HOCHACHKA, P.W. & G.N. SOMERO. 2002. Biochemical Adaptation: Mechanism and Process in Physiological Evolution. Oxford University Press. New York.
17. HADDAD, G.G. *et al*. 1997. Behavioral and electrophysiologic responses of Drosophila melanogaster to prolonged periods of anoxia. J. Insect. Physiol. **43:** 203–210.
18. HOCHACHKA, P.W. 1980. Living without Oxygen: Closed and Open Systems in Hypoxia Tolerance. Harvard Univ. Press. Cambridge.
19. CHEN, Q. *et al*. 2002. Role of trehalose phosphate synthase in anoxia tolerance and development in Drosophila melanogaster. J. Biol. Chem. **277:** 3274–3279.
20. CHEN, Q. *et al*. 2003. Expression of Drosophila trehalose-phosphate synthase in HEK-293 cells increases hypoxia tolerance. J. Biol. Chem. **278:** 49113–49118.
21. JONES, S.P. & R. BOLLI. 2006. The ubiquitous role of nitric oxide in cardioprotection. J. Mol. Cell Cardiol. **40:** 16–23.
22. SCHULZ, R., M. KELM & G. HEUSCH. 2004. Nitric oxide in myocardial ischemia/reperfusion injury. Cardiovasc. Res. **61:** 402–413.
23. WINGROVE, J.A. & P.H. O'FARRELL. 1999. Nitric oxide contributes to behavioral, cellular, and developmental responses to low oxygen in Drosophila. Cell **98:** 105–114.
24. HADDAD, G.G. *et al*. 1997. Genetic basis of tolerance to O2 deprivation in Drosophila melanogaster. Proc. Natl. Acad. Sci. USA **94:** 10809–10812.
25. MA, E. *et al*. 2001. Mutation in pre-mRNA adenosine deaminase markedly attenuates neuronal tolerance to O2 deprivation in Drosophila melanogaster. J. Clin. Invest. **107:** 685–693.
26. FEALA, J. *et al*. 2007. Flexibility in energy metabolism supports hypoxia tolerance in Drosophila flight muscle: metabolomic and computational systems analysis. Mol. Syst. Biol. **3:** 99. Epub 2007 Apr 17.
27. ABOY, M. *et al*. 2005. An automatic beat detection algorithm for pressure signals. IEEE Trans. Biomed. Eng. **52:** 1662–1670.
28. FERRARINI, L. *et al*. 2005. A more efficient search strategy for aging genes based on connectivity. Bioinformatics **21:** 338–348.
29. FEALA, J. *et al*. 2007. Integrating metabolomics and phenomics with systems models of cardiac hypoxia. Prog. Biophys. Mol. Biol. In press.
30. PRICE, N.D., J.L. REED & B.O. PALSSON. 2004. Genome-scale models of microbial cells: evaluating the consequences of constraints. Nat. Rev. Microbiol. **2:** 886–897.

31. JAMSHIDI, N. *et al*. 2001. Dynamic simulation of the human red blood cell metabolic network. Bioinformatics **17:** 286–287.

32. SHEIKH, K., J. FORSTER & L.K. NIELSEN. 2005. Modeling hybridoma cell metabolism using a generic genome-scale metabolic model of Mus musculus. Biotechnol. Prog. **21:** 112–121.

33. DUARTE, N.C. *et al*. 2007. Global reconstruction of the human metabolic network based on genomic and bibliomic data. Proc. Natl. Acad. Sci. USA **104:** 1777–1782.

34. IDEKER, T., T. GALITSKI & L. HOOD. 2001. A new approach to decoding life: systems biology. Annu. Rev. Genomics Hum. Genet. **2:** 343–372.

35. STEPHANOPOULOS, G. 1999. Metabolic fluxes and metabolic engineering. Metab. Eng. **1:** 1–11.

36. KAUFFMAN, K.J., P. PRAKASH & J.S. EDWARDS. 2003. Advances in flux balance analysis. Curr. Opin. Biotechnol. **14:** 491–496.

37. PALSSON, B. 2006. Systems Biology: Properties of Reconstructed Networks. Cambridge University Press. Cambridge; New York.

38. WORKMAN, C.T. *et al*. 2006. A systems approach to mapping DNA damage response pathways. Science **312:** 1054–1059.

39. HAN, J.D. *et al*. 2004. Evidence for dynamically organized modularity in the yeast protein-protein interaction network. Nature **430:** 88–93.

40. BEGLEY, T.J. *et al*. 2002. Damage recovery pathways in Saccharomyces cerevisiae revealed by genomic phenotyping and interactome mapping. Mol Cancer Res. **1:** 103–112.

Multi-Scale Model of O_2 Transport and Metabolism

Response to Exercise

HAIYING ZHOU,[a] NICOLA LAI,[a] GERALD M. SAIDEL,[a] AND MARCO E. CABRERA[a,b,c]

Departments of [a]Biomedical Engineering, [b]Pediatrics, and [c]Physiology and Biophysics, Case Western Reserve University, Cleveland, Ohio, USA

Regulation of pulmonary oxygen uptake (VO_{2p}) during exercise depends on cellular energy demand, blood flow, ventilation, oxygen exchange across membranes, and oxygen utilization in the contracting skeletal muscle. In human and animal studies of metabolic processes that control cellular respiration in working skeletal muscle, pulmonary VO_2 dynamics is measured at the mouth using indirect calorimetry. To provide information on the dynamic balance between oxygen delivery and oxygen consumption at the microvascular level, muscle oxygenation is measured using near-infrared spectroscopy. A multi-scale computational model that links O_2 transport and cellular metabolism in the skeletal muscle was developed to relate the measurements and gain quantitative understanding of the regulation of VO_2 at the cellular, tissue, and whole-body level. The model incorporates mechanisms of oxygen transport from the airway openings to the cell, as well as the phosphagenic and oxidative pathways of ATP synthesis in the muscle cells.

Key words: modeling; oxygen uptake; mass transport; energy metabolism; muscle exercise

Introduction

In response to an imposed work demand of moderate intensity, muscle power output and the rate of ATP utilization change in less than a second, while pulmonary oxygen uptake (VO_{2p}) increases with a time constant of $\sim$30 s.[1] The corresponding changes in cellular respiration needed to increase energy output are predictably linked to external respiration via the lungs, through the circulation.[2] Factors that contribute to the differences between the dynamic responses of cellular and external respiration are circulatory dynamics,[3] ventilation, oxygen stores in blood and muscle,[4] and oxygen exchange across membranes.

Since direct *in vivo* measurement of muscle oxygen consumption (UO_{2m}) is not feasible during exercise in human and animal experiments, VO_{2p} dynamics measured at the mouth is typically used to investigate the metabolic processes that control cellular respiration in the working skeletal muscles. However, the relationship between VO_{2p} and UO_{2m} kinetics has not been examined quantitatively. At the onset of a constant

work rate exercise of moderate intensity, the step-up responses of VO_{2p} and alveolar oxygen uptake (VO_{2A}) typically show two phases whose slopes are discontinuous: a short ($\sim$20 s) cardio-dynamic rise characterized by a circulatory transit time delay (Phase I), followed by a longer exponential rise to a plateau (Phase II).[1]

The characteristics of these phases depend on a number of factors, among them the training status of the individual, presence of cardiopulmonary disease, or the site in which oxygen uptake is being investigated. For instance, an overshoot may occur within Phase II, as Koppo et al.[5] observed in the VO_{2p} response of trained cyclists exercising at moderate intensity. This phenomenon may be attributed to faster cardiac output dynamics in trained cyclists. In contrast, Grassi et al.[6] determined muscle oxygen uptake kinetics (VO_{2m}) from dynamic measurements of arterial and femoral venous blood and leg blood flow and did not observe an overshoot in Phase II at the muscle or the alveolar level. However, during the transition from light to moderate intensity exercise, they found that the amplitude of the Phase I VO_{2m} response was smaller than that of VO_{2A}, but Phase II response dynamics of VO_{2A} and VO_{2m} did not differ significantly. Assuming no effects from oxygen stores and pulmonary and leg blood flows following a mono-exponential increase towards a steady state, Barstow et al.[7,8] performed simulations

Address for correspondence: Marco E. Cabrera, PhD, Deptartment of Pediatrics, Case Western Reserve University, Cleveland, OH 44106-6011. Voice: +216-844-5085; fax: +216-844-5478.

Marco.Cabrera@cwru.edu

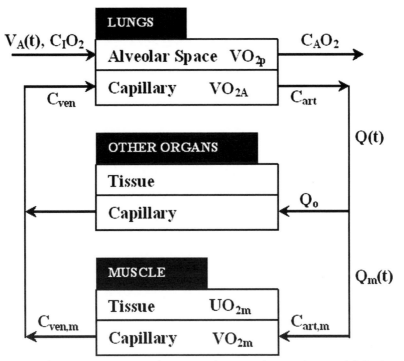

FIGURE 1. Schematic representation of oxygen and transport between lungs and skeletal muscle.

of pulmonary and muscle oxygen uptake in response to a moderate step increase in muscle work. Consistent with the experimental findings of Grassi *et al.*,[6] these simulations showed that VO_{2p} dynamics during Phase II were similar to VO_{2m} dynamics. However, if oxygen stores are taken into consideration in a more general mathematical model of oxygen transport and utilization, then VO_{2p} and VO_{2m} dynamics during Phase II may be different in response to exercise. Indeed, Lai *et al.*[9,10] found the UO_{2m} mean response time to be much shorter than that of VO_{2p} when UO_{2m} dynamics were estimated from muscle oxygen saturation (StO_{2m}) measurements via near-infrared spectroscopy (NIRS) and a computational model of oxygen transport and utilization.

These studies suggest that the relationship between VO_{2p}, VO_{2A}, VO_{2m} and UO_{2m} requires further investigation (a) to establish the physiological basis and interconnections of the dynamics of oxygen uptake at various sites and (b) to gain quantitative understanding of the regulation of VO_2 at the cellular, tissue, and whole-body levels. A critical issue to be examined is the relationship between the mean response times of pulmonary, alveolar, and muscle oxygen uptake during Phase II and that of oxygen utilization in muscle. For this purpose, a multi-scale computational model was developed that links O_2 transport and cellular

metabolism in skeletal muscle.

Methods

A multi-organ model of gas exchange[11] and a skeletal muscle model of oxygen transport and metabolism[10] were combined with experimental measurements of ventilation during exercise to predict muscle oxygenation and pulmonary VO_2 dynamics. Since these models are already available in the literature,[9–11] only a brief description is presented here. The model (FIG. 1) includes three compartments (lungs, muscle, other organs), which are connected by the circulatory system (artery, vein, capillaries). The arterial blood output of the lungs compartment is the input to the tissue compartments, while the venous output of the tissues is the input of the lungs. To simulate O_2 transport between alveolar gas and pulmonary capillary blood, the model represents the alveolar gas phase as a spatially lumped, well-mixed space with a constant average volume. The alveolar mass balance describes the dynamics of gas exchange by ventilation and a transport flux between blood and alveolar space.[11] The rate of pulmonary oxygen uptake is computed as:

$$VO_{2p}(t) = V_A(t)[C_IO_2 - C_AO_2] \tag{1}$$

where $C_I O_2$ is the inspiratory O_2 concentration and $C_A O_2$ is the alveolar (and expiratory) O_2 concentration. The alveolar ventilation, which corresponds to a measured step-up response to exercise, is represented by

$$V_A(t) = V_{A0} + \Delta V_A [1 - e^{(t_0 - t)/\tau_V}] \quad (2)$$

where V_{A0} is the ventilation at warm-up steady-state condition, t_0 is the time at the onset of exercise, and τ_V is the time constant of the ventilatory response.

The spatial and temporal O_2 concentration changes from the arterial to the venous sides in the pulmonary capillary bed are simulated by a one-dimensional convection–dispersion model with a transport flux between the blood and the alveolar space.[11] The rate of oxygen uptake in the alveolar blood is

$$VO_{2A}(t) = Q(t) [C_{art}(t) - C_{ven}(t)] \quad (3)$$

where the cardiac output $Q(t)$ is represented by an exponential equation similar to Equation 2. The O_2 concentration difference between the mixed-venous and arterial blood is given in the brackets. To simulate the transport delay between muscle and lungs and the dispersion in the large vessels of the circulation, the O_2 concentration in the arteries and veins is represented by a one-dimensional convection–dispersion model assuming no mass transport between blood and extra-vascular tissues.

The blood flow and oxygen uptake of other tissues are assumed constant. O_2 concentration dynamics in the capillary blood and tissue cells of the skeletal muscle, are represented by compartmental mass balances.[11] The change in O_2 concentration in the skeletal muscle depends on mass transfer across the blood–tissue membrane and reaction processes (oxidative phosphorylation). The energy demand imposed by exercise is represented by the rate of ATP utilization, which is balanced by ATP production from oxidative phosphorylation and phosphocreatine (PCr) hydrolysis.[10] The contribution of glycolysis to ATP synthesis, however, is negligible when the imposed work rate is of moderate intensity.[12] The rate of oxygen utilization in the tissue can be estimated by

$$UO_{2m}(t) = \phi_{OxPhos} V_{tis} \quad (4)$$

where V_{tis} is the volume of active skeletal muscle and ϕ_{OxPhos}, the oxidative phosphorylation rate, is given by:

$$\phi_{OxPhos} = V_{max} \left[\frac{C_{ADP}}{K_{ADP} + C_{ADP}} \right] \frac{C_{tis}^F}{K_m + C_{tis}^F} \quad (5)$$

where V_{max} is the maximal flux of oxidative phosphorylation, C_{ADP} is the ADP concentration and C_{tis}^F is the free O_2 concentration in muscle cells. The rate of muscle oxygen uptake is computed by

$$VO_{2m}(t) = Q_m(t)(C_{art,m} - C_{ven,m}) \quad (6)$$

where $C_{art,m}$ and $C_{ven,m}$ are the O_2 concentrations in muscle arterial and venous blood, respectively. The exercise response of the muscle blood flow, $Q_m(t)$, has the same form as the cardiac output $Q(t)$

$$Q_m(t) = Q_{m0} + \Delta Q_m [1 - e^{(t_0 - t)/\tau_Q}] \quad (7)$$

where ΔQ_m is the muscle blood flow increase and τ_Q is the time constant of the blood flow response.

The present model, which incorporates these scales through the various subsystems, provides a framework for quantifying the effects of major processes of O_2 delivery from lungs to muscle cells. Using the model, we can simulate and compare the quantitative relationship between cellular, whole-body, alveolar blood, and muscle blood O_2 uptake responses to moderate exercise indicated by UO_{2m}, VO_{2p}, VO_{2A}, and VO_{2m}, respectively. The simulated dynamics of step-up responses can be compared using the mean response time

$$MRT = \frac{\int_{t_0}^{t_1} t \Delta y(t) dt}{\int_{t_0}^{t_1} \Delta y(t) dt} \quad (8)$$

where $\Delta y = y_{max} - y(t)$ represents the amplitude of the dynamic response of VO_{2p}, VO_{2A}, VO_{2m}, or UO_{2m}. t_0 is the onset time of the response from warm-up steady-state condition, and the time to reach the maximum response is t_1.

Simulations were performed with model parameters corresponding to different conditions of three human subjects. Since Phase I and II responses were most evident for VO_{2p} and VO_{2A}, the MRT values were computed for both phases ($t_0 = 0$) and for Phase II alone ($t_0 > 0$) to eliminate the circulatory transit time delay. Using the parameter values of one subject, several key parameters could be examined to determine their effects on oxygen uptake and utilization, specifically, τ_V, τ_Q, and V_{max}. The value of the time constant τ_V can be estimated from a subject's ventilatory response. Also,

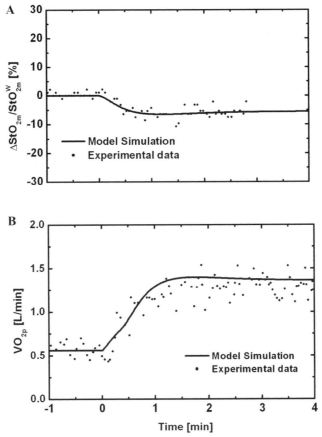

FIGURE 2. Dynamic responses (subject 3) of **(A)** relative oxygen saturation in muscle and **(B)** pulmonary oxygen uptake to moderate exercise.

model simulations of StO_{2m} and VO_{2p} under control conditions can be compared directly with experimental data.

Results

In all the simulations of moderate exercise response from a warm-up base line, values of common model parameters and those specific for three human subjects (i.e., V_{max}, ΔQ_m, and τ_Q) were taken from the literature.[9–11] Model-simulated and experimental responses of the skeletal muscle StO_{2m} and the pulmonary oxygen uptake VO_{2p} are consistent (FIG. 2A and B). Simulated exercise responses corresponding to one subject's oxygen uptake at different sites in the body (VO_{2p}, VO_{2A}, VO_{2m}) and utilization rate (UO_{2m}) are shown in FIGURE 3A. Phases I and II from VO_{2p} and VO_{2A} are more evident in the derivatives of their dynamic responses (FIG. 3B). A comparison

of the derivative responses of VO_{2p} and VO_{2A} shows that signal smoothing occurs between the alveolar level and the airway opening, which tends to increase the mean response time (MRT) (TABLE 1). The MRT of each simulated response is computed based on data from three subjects. For VO_{2p} and VO_{2A}, the MRT is computed also for Phase II, which eliminates the effect of the circulatory transit time delay. In the absence of this delay, it becomes evident that the oxygen uptake rates (VO_{2p}, VO_{2A}, VO_{2m}) have essentially the same MRT, which is larger than the MRT for UO_{2m}. Although not shown, the simulated ATP concentration remains constant, while PCr concentration decreases in response to exercise.

Variations in the maximal flux of oxidative phosphorylation V_{max} resulted in changes of the dynamics of oxygen uptake at different body sites (TABLE 2). With a lower V_{max}, the MRT is greater for the UO_{2m} response and the oxygen uptake everywhere (VO_{2p}, VO_{2A}, VO_{2m}). Under the assumption that ventilation

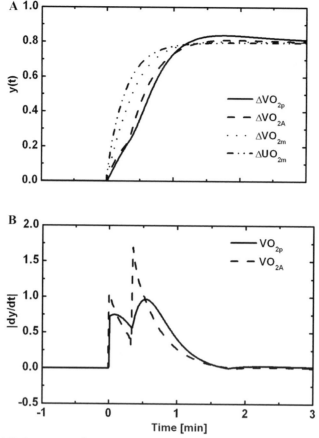

FIGURE 3. (A) Comparison of oxygen uptakes at different scales of the body. **(B)** Derivatives of pulmonary and alveolar blood oxygen uptakes, VO_{2p} and VO_{2A}.

TABLE 1. Mean response time (MRT) of oxygen uptake and utilization responses to moderate exercise. The MRT within parentheses reflects Phase II only

Subject	MRT (s)			
	VO_{2p}	VO_{2A}	VO_{2m}	UO_{2m}
1	26 (18)	21 (13)	14	8
2	31 (23)	29 (22)	23	13
3	24 (18)	22 (17)	19	17
Mean	27 (20)	24 (18)	19	13

TABLE 2. Effects of the maximum oxidative phosphorylation flux (V_{max}) on mean response time (MRT) of oxygen uptake and utilization responses to moderate exercise. The MRT within parentheses reflects Phase II only

V_{max} (mM/min)	MRT (s)			
	VO_{2p}	VO_{2A}	VO_{2m}	UO_{2m}
16.5	38 (35)	41 (40)	39	32
33	24 (18)	22 (17)	19	17
66	22 (15)	19 (13)	13	9

dynamics does not change with V_{max}, the simulation shows that at low V_{max}, the MRT of VO_{2p} is smaller than the MRT of VO_{2A}. At low V_{max}, the effect of Phase I on the MRT is small. The amplitude and duration of Phase I do not change with V_{max} (FIG. 4). With a higher V_{max}, Phase II responses of oxygen uptake produce an overshoot, which is most evident in the VO_{2p} response (FIG. 4).

A change in the time constant of muscle blood flow (τ_Q) from its control value (21 s) cannot affect UO_{2m}, but causes the MRT of VO_{2m} and Phase II of VO_{2p} and VO_{2A} to increase (TABLE 3). When τ_Q is large, an overshoot occurs in all oxygen uptake responses, especially in VO_{2p} (FIG. 5A). With a smaller τ_Q, VO_{2p} and VO_{2A} responses show a greater Phase I amplitude (FIG. 5A and B). When τ_Q decreased from 42 s to 10 s,

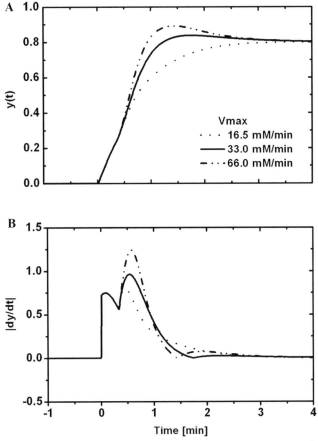

FIGURE 4. Effects of the maximum oxidative phosphorylation flux V_{max} on **(A)** pulmonary oxygen uptake dynamics and **(B)** derivatives of pulmonary oxygen uptake.

TABLE 3. Effects of blood flow dynamics (τ_Q) on mean response time (MRT) of oxygen uptake and utilization responses to moderate exercise. The MRT within parentheses reflects Phase II only

| τ_Q (s) | MRT (s) | | | |
	VO_{2p}	VO_{2A}	VO_{2m}	UO_{2m}
10	25 (22)	28 (27)	25	17
21	24 (18)	22 (17)	19	17
42	28 (21)	26 (18)	21	17

the VO_{2A} Phase I amplitude increased by 100%. The VO_{2m} response (FIG. 5C) shows almost no Phase I and a minimal overshoot in Phase II. With small τ_Q or low V_{max}, the MRT of VO_{2p} can be smaller than that of VO_{2A}. As expected, but not shown, the dynamics of ventilation have negligible effect on either muscle oxygen consumption or muscle oxygen uptake.

Discussion

The multi-scale model presented here combines pulmonary gas exchange[11] and O_2 transport with cellular metabolism in the skeletal muscle.[9,10] Simulated energy demand due to exercise is represented by a rapid increase in ATP utilization. The cellular ATP homeostasis is maintained through oxidative phosphorylation, and the ATPase and creatine kinase reactions. The model also includes explicit relationships between free and bound forms of O_2 that incorporate the effects of hemoglobin and myoglobin in blood and skeletal muscle. To simulate experimental responses of human subjects, the maximal flux of oxidative phosphorylation V_{max} and the blood flow increase ΔQ_m were estimated from measurements of muscle oxygenation during exercise.[10] Simulations with this model show the relationship between VO_{2p}, VO_{2A}, VO_{2m}, and UO_{2m} and the effect of blood flow, ventilation, and V_{max} on oxygen uptake and utilization dynamics.

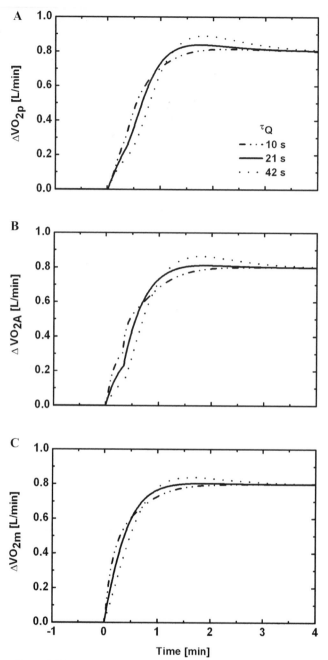

FIGURE 5. Effects of different muscle blood flow dynamics on **(A)** pulmonary oxygen uptake, **(B)** alveolar blood oxygen uptake, and **(C)** muscle oxygen uptake.

Oxygen Uptake and Utilization Dynamics in Response to Exercise

Model simulations of StO_{2m} and VO_{2p} responses to exercise under control conditions correspond closely with experimental data (FIG. 2A and B). The simulated VO_{2p} response shown in FIGURE 2B indicates

two phases, which are evident in FIGURE 3A and B for VO_{2A} also. Phase I reflects the effect of the circulatory transit time delay from skeletal muscle to the lungs. The time delay evident from the model simulation can also be calculated as the ratio of the venous blood volume to blood flow (cardiac output). The Phase I time delays of VO_{2p} and VO_{2A} responses are the same. The VO_{2p} time delay of 22 s simulated in FIGURE 2B is nearly the same as the experimental[1] Phase I duration of 19 s.

As shown in FIGURE 3A, the UO_{2m} on-response to exercise was faster than that of the oxygen uptake (VO_{2m}, VO_{2A}, VO_{2p}), which is reflected in the mean response times (TABLE 1): $MRT(VO_{2p}) > MRT(VO_{2A}) > MRT(VO_{2m}) > MRT(UO_{2m})$. When the effect of circulatory transit time delay (Phase I) was eliminated from the responses of VO_{2p} and VO_{2A}, the MRTs of VO_{2p} and VO_{2A} during Phase II were close to the MRT of VO_{2m} (TABLE 1). The dynamic response of VO_{2m} was slower than that of UO_{2m} because of the processes involving muscle myoglobin and venous O_2 stores.

Effects of Metabolic Rate and Blood Flow Dynamics

The model simulations quantify the relative changes of the oxygen uptake and utilization dynamic responses to exercise with oxygen metabolic rate indicated by V_{max} (TABLE 2). An increase in V_{max}, which represents the fitness level of a subject, generates faster responses (i.e., MRT decreases) in VO_{2p}, VO_{2A}, VO_{2m}, and UO_{2m} (TABLE 2). This is consistent with experimental studies that found trained subjects to have faster VO_{2p} responses than untrained subjects during constant-load exercise.[13] With increased τ_Q and/or V_{max}, a Phase II overshoot can occur in the oxygen uptake response, which is most prominent for VO_{2p} and least prominent for VO_{2m}. No overshoot occurs in the UO_{2m} response. Even with the overshoot of the oxygen uptake response, the MRTs of VO_{2p} and VO_{2A} during Phase II are close to that of VO_{2m}.

With fast blood flow dynamics (small τ_Q) relative to the metabolic rate (low V_{max}), the MRTs of VO_{2p} can be smaller than those of VO_{2A} (TABLES 2 and 3) when the alveolar ventilation response is too slow (large τ_V). Variation of τ_V did not alter the dynamics of VO_{2A}, VO_{2m}, and UO_{2m}. Casaburi *et al.*[14] found in exercise studies with dogs that even a fourfold change of τ_V did not change the Phase II dynamic response of VO_{2A}. While different ventilation dynamics change O_2 concentration in alveolar gas, the O_2 concentration in the blood is essentially constant because of the O_2 associated with hemoglobin. Consequently, the VO_{2A} response, as calculated here in the blood phase, does not change.

Experimentally, Koppo *et al.*[5] reported an overshoot in the VO_{2p} response to moderate-intensity cycle exercise. This behavior was explained as a consequence of a non-constant ATP requirement that is higher at the beginning of exercise. The model simulations, however, suggest that this overshoot can occur in the oxygen uptake responses even with a constant ATP demand, when τ_Q and/or V_{max} are large. A mismatch of τ_Q and V_{max} that produces a large reduction in the muscle O_2 stores at the onset of exercise causes O_2 uptake responses, especially VO_{2p} and VO_{2A}, to increase rapidly to peak values (FIGS. 4A and 5A). An overshoot in the VO_{2p} response of a trained subject, who is expected to have a higher V_{max} than that of an untrained subject, would be more evident.

When τ_Q is increased from 21 s to 42 s, the MRT of VO_{2p} and VO_{2A} increases are primarily associated with Phase I rather than Phase II (TABLE 3). This can be explained by the slower oxygen transport to the lungs that occurs in the earlier part of the response to exercise. However, when τ_Q is decreased from 21 s to 10 s, the MRT of VO_{2p} and VO_{2A} is increased, especially for VO_{2A}. When the circulatory blood flow is faster, the earlier O_2 uptake responses are faster (FIG. 5), but oxygen uptake reaches steady state more slowly. This may be associated with the imbalance of the dynamics of the blood flow and ventilation, since the dynamic response of alveolar ventilation (τ_V) was not changed in these simulations. Experimental studies in dogs[15,16] showed that when blood flow was controlled at a higher steady-state level, it did not change the VO_{2m} dynamics in response to exercise. In human studies,[8,17] faster response of blood flow to exercise led to a slower Phase II response of VO_{2p} (i.e., larger MRT). This may be explained by faster O_2 delivery that leads to a slower dynamic response of the arterial-venous oxygen difference that determines VO_{2A} (Eq. 3).

Summary

The multi-scale mathematical model presented here describes the dynamics of oxygen transport and metabolism from the lungs to skeletal muscle cells and provides a quantitative analysis of the external and internal respiration in response to exercise. Previous single-scale models, based on experiments in isolated mitochondria or muscle preparations, investigated metabolic control in a subsystem in isolation from the whole system. However, the conditions of sufficient oxygen availability or controlled bulk oxygen delivery and fiber recruitment pattern in these early preparations differ from the highly regulated internal environment and centrally controlled physiological responses to exercise, that characterize systems *in vivo*. The integrated model described here mimics the regulated environment of *in vivo* mitochondria and skeletal muscle and elucidates mechanisms of metabolic control and regulation. Our model simulations yield mechanistic explanations of experimentally measured rates of oxygen uptake and utilization under different exercise conditions. The simulated oxygen uptake dynamics that exhibit biphasic responses with overshoot are analyzed relative to local transport and metabolic processes. Furthermore, this integrated model of oxygen transport and metabolism in response to exercise can be extended to comparative quantitative analysis of respiration regulation of healthy subjects and patients suffering from chronic obstructive pulmonary disease, diabetes, or congenital heart disease.

Acknowledgments

This work was supported by grants from the National Aeronautics and Space Administration (NASA–Johnson Space Center NNJ06 HD81G) and the National Institute of General Medical Sciences of the National Institutes of Health (GM–66309).

Conflict of Interest

The authors declare no conflicts of interest.

References

1. WHIPP, B.J. *et al*. 1982. Parameters of ventilatory and gas exchange dynamics during exercise. J. Appl. Physiol. **52:** 1506–1513.

2. WASSERMAN, K. 1994. Coupling of external to cellular respiration during exercise: the wisdom of the body revisited. Am. J. Physiol. **266:** E519–E539.

3. LADOR, F. *et al*. 2006. Simultaneous determination of the kinetics of cardiac output, systemic O2 delivery, and lung O2 uptake at exercise onset in men. Am. J. Physiol. **290:** R1071–R1079.

4. DI PRAMPERO, P.E., U. BOUTELLIER & P. PIETSCH. 1983. Oxygen deficit and stores at onset of muscular exercise in humans. J. Appl. Physiol. **55:** 146–153.

5. KOPPO, K. *et al*. 2004. Overshoot in VO2 following the onset of moderate-intensity cycle exercise in trained cyclists. Eur. J. Appl. Physiol. **93:** 366–373.

6. GRASSI, B. *et al*. 1996. Muscle O2 uptake kinetics in humans: implications for metabolic control. J. Appl. Physiol. **80:** 988–998.

7. BARSTOW, T.J. & P.A. MOLE. 1987. Simulation of pulmonary O2 uptake during exercise transients in humans. J. Appl. Physiol. **63:** 2253–2261.

8. BARSTOW, T.J., N. LAMARRA & B.J. WHIPP. 1990. Modulation of muscle and pulmonary O2 uptakes by circulatory dynamics during exercise. J. Appl. Physiol. **68:** 979–989.

9. LAI, N. *et al*. 2006. Relating pulmonary oxygen uptake to muscle oxygen consumption at exercise onset: in vivo and in silico studies. Eur. J. Appl. Physiol. **97:** 380–394.

10. LAI, N. *et al*. 2007. Linking pulmonary oxygen uptake, muscle oxygen utilization and cellular metabolism during exercise. Ann. Biomed. Eng. **35:** 956–969.

11. ZHOU, H., G.M. SAIDEL & M.E. CABRERA. 2007. Multiorgan system model of O2 and CO2 transport during isocapnic and poikilocapnic hypoxia. Respir. Physiol. Neurobiol. **156:** 320–330.

12. PIIPER, J., P.E. DI PRAMPERO & P. CERRETELLI. 1968. Oxygen debt and high-energy phosphates in gastrocnemius muscle of the dog. Am. J. Physiol. **215:** 523–531.

13. ZHANG, Y. *et al*. 1991. The role of fitness on VO2 and VCO2 kinetics in response to proportional step increases in work rate. Eur. J. Appl. Physiol. **63:** 94–100.

14. CASABURI, R. *et al*. 1979. Determinants of gas exchange kinetics during exercise in the dog. J. Appl. Physiol. **46:** 1054–1060.

15. GRASSI, B. *et al*. 1998. Faster adjustment of O2 delivery does not affect VO2 on-kinetics in isolated in situ canine muscle. J. Appl. Physiol. **85:** 1394–1403.

16. LAI, N. *et al*. 2007. Model of oxygen transport and metabolism predicts effect of hyperoxia on canine muscle oxygen uptake dynamics. J. Appl. Physiol. **103:** 1366–1378.

17. CASABURI, R., S. SPITZER, R. HASKELL & K. WASSERMAN. 1989. Effect of altering heart rate on oxygen uptake at exercise onset. Chest **95:** 6–12.

Mapping Preconditioning's Signaling Pathways

An Engineering Approach

JAMES M. DOWNEY,[a] THOMAS KRIEG,[b] AND MICHAEL V. COHEN[a]

[a]Departments of Physiology and Medicine, University of South Alabama, Mobile, Alabama, USA

[b]Department of Cardiology, Ernst-Moritz-Arndt University Greifswald, Greifswald, Germany

Preconditioning the heart by exposure to brief cycles of ischemia–reperfusion causes it to become very resistant to ischemia-induced infarction. This protection has been shown to depend on a large number of signal transduction components whose arrangements within the cardiomyocyte are unknown. To aid the translation of this phenomenon to the clinical setting, we have attempted to map the signal transduction pathways responsible for this protection. To resolve the signaling order we have injected a signal at an intermediate point in the system transduction pathway and monitored it at a downstream site. System analysis reveals both parallel and series signaling arrangements. Separate trigger and mediator phases could be identified. *The trigger phase* is now well mapped. During the preconditioning ischemia, autacoids—including adenosine, opioids, and bradykinin—are released from the heart. These substances occupy their respective G_i-coupled receptors. Opioid and bradykinin receptors activate phosphatidylinositol 3-kinase (PI3-kinase) which, through phosphoinositide-dependent protein kinase, causes activation of Akt. Opioid couples through transactivation of the epidermal growth factor receptor, while bradykinin's coupling to PI3-kinase is unknown. PI3-kinase causes extracellular signal regulated kinase (ERK)-dependent activation of endothelial nitric oxide synthase. The resulting nitric oxide activates soluble guanylyl cyclase resulting in cyclic C-GMP-dependent protein kinase (PKG) activation through production of cyclic guanosine monophosphate. PKG initiates opening of ATP-sensitive potassium channels on the inner membrane of the mitochondria. Potassium entry into mitochondria causes the generation of free radicals during reperfusion when oxygen is reintroduced. Through redox signaling, these radicals activate protein kinase C (PKC) and put the heart into the protected phenotype that persists for one to two hours. Although adenosine receptors activate PI3-kinase, they also have a second direct coupling to PKC and thus bypass the mitochondrial pathway. *The mediator phase* occurs during the first minutes of reperfusion following the lethal ischemic insult and is still poorly defined. Briefly, PKC somehow potentiates adenosine's ability to activate signaling from low-affinity A_{2b} adenosine receptors. These receptors couple to the survival kinases, Akt and ERK, believed to inhibit the formation of deadly mitochondrial permeability transition pores through the phosphorylation of glycogen synthase kinase-3β. The proposed signaling maps reveal many points at which drugs can trigger the protected phenotype.

Key words: free radicals; ischemia; mitochondria; nitric oxide; permeability transition pores; ATP-sensitive potassium channels; redox signaling; protein kinase C and G; EGF receptor; adenosine receptor

Introduction

Over the past 5 years we have made a concerted effort to map the signal transduction steps involved in the phenomenon of ischemic preconditioning (IPC) whereby exposing the heart to a brief period of ischemia followed by a few minutes of reperfusion, causes myocardium to adapt itself to become highly resistant to infarction from a subsequent ischemic insult. The window of protection lasts for 1–2 h and is the result of the occupancy of receptors during the IPC and the resulting signal transduction pathways to which they are coupled. Prominent among those receptors is the adenosine A_1 subtype. During an IPC, ATP is rapidly broken down to adenosine, which can exit the cell and occupy its surface receptors as an

Address for correspondence: James Downey, PhD, Room 3074, Medical Science Bldg., University of South Alabama, Mobile, AL 36688. Voice: +251-460-6818; fax: +251-460-6464.

jdowney@usouthal.edu

Ann. N.Y. Acad. Sci. 1123: 187–196 (2008). © 2008 New York Academy of Sciences.
doi: 10.1196/annals.1420.022

autacoid. In the 1990s adenosine A_1 receptors,[1] protein kinase C (PKC),[2] and the mitochondrial ATP-sensitive potassium channel (K_{ATP})[3] were identified as necessary steps in IPC's protection. That led to the naïve assumption that A_1 receptor occupancy simply caused PKC to phosphorylate threonine and/or serine groups on sarcolemmal K_{ATP}, causing them to open. It was believed that this would directly protect the heart by reducing calcium entry. Many more components have been identified since then, revealing the true complexity of the system. A map was obviously needed to determine the location of all these components relative to each other. Such a map should be the key to understanding the overall mechanism and should be useful in translating the phenomenon to clinical practice. In the following chapter we outline the overall strategy that we and others have used to make such a map and present the resulting data.

The Strategy

Series Signal Transduction Steps

We have taken advantage of the fact that most signal transduction steps are arranged in series and are unidirectional. Thus, to study these steps we can activate the system at some intermediate point in the signal transduction pathway and monitor the response at some downstream site. Then we can block the system at a point that has already been identified as being part of the pathway. If the block eliminates the response from a trigger of the pathway located at an upstream site, then the site of the block must be between the activation point and the monitoring site. For example, it was recently[4] found that inhibition of either PKC or adenosine receptors at the time of reperfusion would block protection from IPC. One might assume that the proper sequence is that the adenosine receptors activate PKC. However, direct activation of PKC with a 5-min intracoronary infusion of phorbol myristate acetate ester (PMA) at reperfusion mimicked IPC's protection. The PKC inhibitor chelerythrine blocked PMA's protection, confirming that the protection was the result of PKC activation. This established that PKC activation was sufficient to elicit the protected phenotype. Next we infused the adenosine agonist 5'- N-ethylcarboxami doadenosine (NECA) at reperfusion and noted that it too protected the heart. Blocking NECA's protection with an adenosine receptor antagonist[5] confirmed an adenosine receptor-mediated effect. But, when we tested chelerythrine against NECA, we were unable to block protection indicating that the A_{2b} adenosine receptor could not be upstream of PKC.

This was also confirmed when we were able to block PMA's protection with an adenosine receptor blocker. Thus, in the reperfusion phase of IPC, PKC activation somehow triggers occupancy of adenosine receptors, which is the exact opposite of what is normally seen in signal transduction pathways.

The success of the above approach depends on the available tools. For example, some components like phosphatidylinositol 3-kinase (PI3-kinase) cannot be directly activated pharmacologically. Nor do we have any way to directly activate kinases like Akt or ERK, which are physiologically activated by phosphorylation by upstream kinases. Nevertheless, a large number of the components like PKC are amenable to direct activation. Fortunately inhibitors are available for virtually all of the signaling molecules.

Parallel Signal Transduction Steps

Studies in preconditioning have been hampered by the fact that some signal transduction steps are in parallel rather than in series. The usual technique for identifying a critical step in series in the pathway is to inhibit that step and test whether the response has been eliminated. If, however, two or more elements are in parallel at that point, the inhibitor may not eliminate the response and the importance of that element may easily be missed. The best example of parallel signaling is the existence of multiple receptors that participate in triggering IPC. Fortunately serendipity played an important part in the original identification of these receptors. From the beginning, studies of IPC concentrated on eliciting its mechanism. Many of those early studies used many cycles of ischemia to precondition the heart. One investigator used as many as 12 cycles,[6] although a single 5-min cycle seemed to be sufficient to give full protection. Many investigators used 3 or 4 cycles just to be on the safe side while we, being lazy, routinely used just one 5-min cycle. That put us just above the threshold for protection, since a single 2-min ischemia cycle did not protect the rabbit heart.[7] A characteristic of parallel pathways is that inhibition of one pathway should raise the threshold of the stimulus required to activate the downstream target through the remaining pathways. In our case, if one of the three receptors were removed, then a single cycle would no longer be protective.

We had the good fortune to identify the first receptor involved in IPC, the adenosine A_1 receptor. That in itself caused an immediate paradigm shift because at that moment it became clear that IPC was receptor mediated and a product of cell signaling. Other investigators at the time were searching for a pH or a metabolism effect. The discovery of the adenosine receptor was quickly followed by a demonstration that blockade of bradykinin B_2[8] or opioid δ-receptors[9] also eliminated IPC's protection. At first their parallel arrangement was not appreciated, and investigators argued over which receptor was really responsible. But then it was shown that blockade of a single receptor did not eliminate the ability to precondition the heart but merely raised the ischemic threshold required to do so.[10] Protection could be restored in the presence of blockade of a single receptor by simply giving more preconditioning cycles. All three of the above receptors are G_i-coupled, and so far all of the G_i-coupled receptors in the heart that have been tested are capable of triggering the preconditioned state if they are occupied before ischemia.[11]

Temporal Considerations

One of the unique features of IPC is that the temporal sequence of events is complex and critical to its mechanism. Different components do different things at different times. It is convenient to divide the process into a trigger and a mediator phase. In IPC the trigger phase can be considered to comprise the events before the lethal ischemic insult. It is called the trigger phase because transient occupation of the G_i-coupled receptors puts the heart into a preconditioned state that persists even after the ligands are washed off. The actual mechanism of this memory is still not understood, but it may be related to PKC activation and translocation.

The second phase is the mediator. The heart must be reperfused at the end of the index ischemia, and it is at this time when IPC is thought to exert its actual anti-infarct effect. Most of the evidence indicates that IPC prevents the formation of mitochondrial permeability transition pores (mPTP)[12] which destroy the mitochondria in the first minutes of reperfusion, thus eliminating their ATP production. If enough of a cell's mitochondria experience mPTP formation, the cell will not be able to recover metabolically from the energy deficit incurred during ischemia. Loss of ATP required to power membrane ion pumps results in an inability to maintain transmembrane ionic gradients. Sodium accumulation causes myocyte swelling and ultimately membrane rupture with death of the cell. Formation of mPTP is promoted by elevated Ca^{2+}, reactive oxygen species (ROS), and nonspecific ischemic injury [for a review of mPTP see Ref. 13], and all these factors are present at reperfusion. We believe that three populations of cells exist in the reperfused heart: those that were killed by the ischemia, those that are alive and will survive reperfusion, and those that are alive but will be killed by mPTP formation in the first

minutes of reperfusion. IPC would only salvage the latter population.

A unique series of signal transduction steps have been identified in both the trigger and mediator phases. To understand IPC's mechanism it is important to determine in which phase an event is occurring. For example, an animal may be injected with an inhibitor at the beginning of the study and it might be noted that IPC no longer protects this animal. While the experiment shows that the targeted signaling molecule is somehow involved in IPC, it does not reveal whether it was needed in the trigger or the mediator phase. To differentiate between the two phases it is necessary to confine the pharmacological activator or inhibitor to a single phase. Studying the trigger phase requires reversible drugs that can be infused and then washed out prior to the index ischemia. Since removal is obviously a problem with *in situ* preparations, trigger-phase studies are difficult in this model. Fortunately IPC protects against infarction in a buffer-perfused isolated heart just as well as it does in the *in situ* model.[14] In an isolated heart a drug can be started and stopped at will. Also there is the added advantage that drug concentration in the tissue is the same as in the perfusate and thus a selective concentration can be maintained. When a drug is administered intravenously in an *in situ* preparation the determinants of plasma concentration including volume of distribution, protein binding, and rate of metabolism/excretion are seldom known. Mediator phase events can be more easily studied *in situ* by simply not giving the drug until the onset of reperfusion. Maintaining a selective *in situ* concentration, however, is still a challenge. Obviously it is impossible to confine a genetic knockout to a single phase.

Cell versus Whole Heart Models

The current state of signal transduction science is fortunately quite well developed, and new and better tools for working with these pathways appear every day. Pharmacological activators and inhibitors are commercially available for many of the signal transduction molecules. Unfortunately, they are often very expensive. An isolated rabbit heart uses about 50 mL of buffer per minute. Thus if we want to have the drug present during the trigger phase of IPC we would start it 5 min prior to the preconditioning ischemia to load the tissue and continue the infusion for the 5 min of regional ischemia induced by occlusion of a branch of the coronary artery and the initial 5 min of reperfusion. Then during the remaining 5 min of the reperfusion cycle the drug would be omitted from the buffer so it can be washed out of the myocardium. This protocol would require 750 mL of buffer loaded with the inhibitor at an active concentration, which would be economically prohibitive with many of the available agents. *In situ* studies require equally large quantities of drug and are further complicated by the uncertainties of dosing as well as timing. Mouse hearts certainly diminish the need for great quantities of expensive drugs, but hardly solve the problem. One solution is to scale the model down to a cell-based one that uses much less volume.

None of the cell-based models that have been developed to date is in our opinion a really good model of the IPC process. One of the earliest models was developed by Ganote *et al.* in which osmotic fragility was measured in acutely isolated adult cardiomyocytes subjected to simulated ischemia.[15] The suspended cells were placed in 1 mL centrifuge tubes and gently spun into a pellet. The supernatant was then decanted and replaced with a thin layer of mineral oil to exclude oxygen. Every 30 min an aliquot of cells was sucked up with a pipette and resuspended in hypotonic buffer. The percentage of cells that burst from the swelling and stained with Trypan blue was recorded. The longer the ischemia, the more fragile the cells would become, and the rate at which this occurred was appreciably slowed by simulated IPC or adenosine agonists. While this model did respond to many of the known triggers of preconditioning it had one obvious flaw: the end-point was osmotic fragility rather than mPTP formation. Even more problematic was the absence of reoxygenation, meaning that the entire mediator phase, which we now know to be the period when the actual protection is exerted, was not modeled.

An improved model involves cardiomyocytes that are exposed to promoters of mPTP formation. The simplest technique is to expose them to hydrogen peroxide.[16–18] Formation of mPTP is monitored with a voltage-sensitive mitochondrial dye like tetramethyl-rhodamine ester (TMRE). Formation of mPTP causes a loss of mitochondrial membrane potential which results in loss of TMRE fluorescence. The obvious problem is that the cells have not been ischemic and therefore do not have the ischemic injury that predisposes mitochondria to mPTP formation. Remember, reperfusion is innocuous unless the heart has been ischemic long enough to have been injured by it. In the absence of ischemia, cells do not release adenosine, an important component in the mediator pathway. A good cell model for the mediator phase of IPC is still not available.

We were lucky to have a robust model for our study of many of the signals of the trigger pathway. Garlid *et al.* found that the ATP-sensitive potassium channels that are involved in IPC were actually a population located in the inner membrane of the mitochondria (mK_{ATP}) rather than on the sarcolemma.[19] Unlike cells which depend on Na^+ gradients for osmoregulation, mitochondria use K^+. When mK_{ATP} open, the mitochondrial transmembrane electrochemical gradient favors K^+ entry which causes the mitochondria to swell and produce ROS. Pain *et al.*[20] noted that opening of mK_{ATP} acted not as a mediator of protection but rather as a signal transduction event in the trigger phase by causing the mitochondria to produce ROS, which then acted as second messengers for redox signaling. More important, we have shown that virtually all G_i-coupled receptors including opioid and bradykinin receptors trigger protection through mK_{ATP} opening and redox signaling.[11]

The above mentioned discoveries prompted us to design a cellular model in which we could directly measure ROS production in isolated heart cells and use ROS generation as an end-point. We have found that reduced Mitotracker Red is the most sensitive indicator for the small ROS signals. Reduced Mitotracker becomes fluorescent when it is oxidized by ROS, and only the oxidized form is sequestered by the mitochondria. Opening mK_{ATP} increased the fluorescence of cultured adult cardiomyocytes, and the ·OH-selective scavenger N-(2-mercaptoproprionyl) glycine (MPG) blocked that increase. We now had a good model to study most of the trigger pathway for two of the three endogenous agonists involved in IPC and their receptors.[21,22] Interestingly, while adenosine A_1/A_3 receptors do cause mK_{ATP} opening, they also couple directly to PKC, and as a result blocking redox signaling or mK_{ATP} opening does not prevent them from triggering protection.[11]

Monitoring the Cell's Biochemical Signals

One of the most challenging aspects of signal transduction studies has been the measurement of biochemical signals, which are usually very compartmentalized. For example, about 10% of a heart biopsy contains noncardiac muscle tissue, including blood vessels, nerves, connective tissue, blood cells, etc. Thus IPC signals occurring in myocytes in biopsied tissue are contaminated by irrelevant signaling from noncardiac cells. Also there is compartmentalization within the cell. Phospholipid-activated PKC, for example, is known to attach to docking sites.[23] PKC then phosphorylates substrates only in the vicinity of the docking site. Few signals are distributed globally within the cell. Some signals like phosphorylation of Akt and ERK are very robust and easy to measure, while others like activation of PKC are quite difficult. There are more than 10 isoforms of PKC that can be

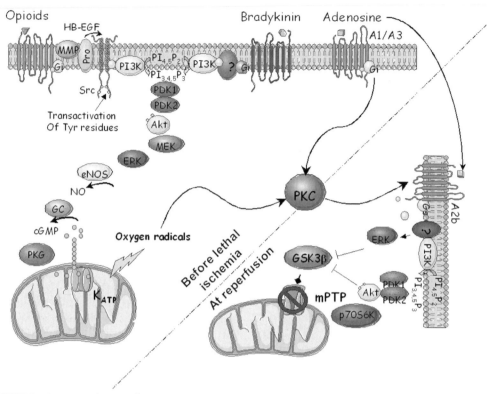

FIGURE 1. A cartoon showing the sequence of signaling events involved in triggering the preconditioned state prior to the ischemic insult (events above the dividing line) and those that mediate protection in the first minutes of reperfusion (events below the dividing line). See text for details. (Reprinted with permission.[54])

independently activated by lipid cofactors. Activated PKC attaches to docking sites within the cell, and measuring this translocation from the soluble to particulate fraction can be subtle and quite challenging.[24]

The next problem was deciding when to measure the signal. Some investigators make biochemical measurements several hours after reperfusion so that the infarct size can be measured in the same heart. It is our opinion that any signals at that time would be irrelevant. Trigger signals should obviously be measured during the trigger phase while mediator signals should be measured in the first minutes of reperfusion. Even then, these signals are often transient, and sampling before or after the signal could yield spurious data. We have used a motorized biopsy tool that allowed us to freeze a transmural biopsy within a couple of seconds of harvest, and we could take serial biopsies from the same heart. We have found it more difficult to make biochemical measurements from isolated adult cardiomyocytes, probably because there is a fair amount of nonmyocyte debris in our cultures. However, others do report success with this method.[25]

Results and Discussion

mK_ATP and ROS

By using the above mentioned strategies we have been able to produce a fairly comprehensive map of the IPC's trigger pathway as shown in FIGURE 1. The complex pathway consists of many steps in both the trigger phase (from receptor occupancy by agonists released by ischemic cells to the release of ROS by mitochondria following opening of mK_{ATP}) and the mediator phase (from PKC activation to phosphorylation of many downstream kinases ending in blockade of mPTP formation). Our first attempt to differentiate between trigger and mediator phases was the demonstration that the kinase activity of PKC could be blocked during the trigger phase with no loss of protection.[26] However, all protection was lost when the reversible PKC inhibitors staurosporine[2,26] or polymyxin[2] were given just before the onset of the index ischemia. Thus PKC was clearly a mediator. Yet a PKC activator acts as a trigger.[2] We believe that PKC activation is the end of the trigger phase and its kinase activity is the fist step in the mediator phase.

mK_{ATP} behave very differently. The critical time for mK_{ATP} opening is during the brief preconditioning cycles. Infusing an mK_{ATP} blocker just prior to the onset of ischemia had no effect on IPC's protection. Furthermore, transient mK_{ATP} opening prior to ischemia mimicked IPC, and that protection could be blocked by co-administration of a free radical scavenger.[20] We have concluded that during the trigger phase the receptor occupancy caused opening of mK_{ATP} which resulted in the mitochondria making ROS. The ROS then acted as second messengers to activate PKC. To test that hypothesis we have examined five G_i-coupled receptor systems in the heart: bradykinin, opioid, α-adrenergic, muscarinic, and adenosine A_1.[11] All five could trigger IPC-like protection if given as a 5-min pulse prior to ischemia. In addition, all except adenosine triggered protection by a pathway that required mK_{ATP} opening and redox signaling.

PI3-kinase and EGFR

Tong *et al.*[27] determined in 2000 that IPC no longer protected rat hearts when preconditioning was performed in the presence of a PI3-kinase inhibitor. PI3 kinase acts to phosphorylate membrane phosphatidyl-inositol in its three positions. The 3-phosphorylated lipid in the membrane then activates phospholipid-dependent protein kinases (PDK) 1 and 2 which in turn phosphorylate a number of target proteins including the protein kinase Akt. PI3-kinase is generally activated by most growth factor receptors. The timing of administration of the inhibitor suggested that PI3-kinase was acting in the trigger phase. Because the PI3-kinase inhibitor did not block protection from a direct PKC activator, Tong *et al.*[27] concluded that PI3-kinase was upstream of PKC. Qin *et al.*[28] tested PI3-kinase inhibitors against pharmacological preconditioning. Wortmannin blocked protection from acetylcholine (ACh), another G_i-protein-coupled receptor, but not from adenosine. The receptor selectivity led Qin to speculate that PI3-kinase was likely coupling mK_{ATP} to the receptors. But how do G_i-coupled receptors activate PI3-kinase? In other cell systems they do it through transactivation of the epidermal growth factor receptor (EGFR), which normally exists as monomers in the membrane. Binding of the EGF ligand to its receptor causes EGFR monomers to dimerize and transphosphorylate each other's tyrosine groups, often referred to as transactivation. The phosphorylated tyrosine groups induce the assembly of a signaling complex that contains, among other things, activated PI3-kinase and Src kinase.

Krieg *et al.*[29] have tested the EGFR hypothesis and showed that ACh caused phosphorylation of Akt (also known as protein kinase B) in rabbit hearts. Akt is a downstream target of PI3-kinase and thus a reliable reporter for PI3-kinase activation. This ACh-mediated phosphorylation of Akt is blocked by a PI3-kinase inhibitor, thus demonstrating a PI3-kinase coupling. Krieg could also block Akt phosphorylation with a tyrosine kinase blocker (EGFR is a tyrosine kinase receptor), a Src kinase, a component of the EGFR signaling complex, and AG1478, a direct inhibitor of EGFR. Additionally, ACh induced EGFR phosphorylation. Unfortunately, AG1478 cannot be used to confirm these observations in the infarct model since it causes ROS generation on its own and is cardioprotective in whole hearts.[30] However, transactivation was confirmed in the intact heart by examining events that lead to EGFR activation. G_i-protein-coupled receptors activate EGFR by cleavage of membrane-bound pro-heparin binding EGF by metalloproteinase. A metalloproteinase inhibitor blocked ACh's protection against infarction.[30] Opioid receptors also triggered protection through the EGFR, but, interestingly, bradykinin did not. None of the interventions known to block signaling by the EGFR affected bradykinin's ability to trigger ROS production in cells or protection in the whole heart.[31]

ROS and Cardiomyocytes

Collagenase perfusion disperses an isolated rabbit heart into millions of individual cardiomyocytes. These myocytes can be kept in a culture medium supplemented with antibiotic for about one week. ROS production can be detected with reduced Mitotracker dye. FIGURE 2 shows typical data from this model.[32] The direct and highly selective mK_{ATP} opener diazoxide increases fluorescence by about 50%. That increase could be blocked by MPG, a scavenger selective for the hydroxyl radical[33] and peroxynitrite,[34] confirming that ROS was responsible for the increased fluorescence. Myxothiazol, a mitochondrial electron transport inhibitor, also blocked the signal, indicating ROS originated from electron transport within the mitochondria. Finally the mK_{ATP} antagonist 5-hydroxydecanoate (5-HD) blocked the signal, confirming diazoxide acted by opening mK_{ATP}. FIGURE 3 shows that ACh mimics diazoxide's increase in fluorescence.[32] Atropine blocked ACh's effect, confirming a muscarinic receptor mechanism, and 5-HD blocked it confirming that ROS was the result of opening of mK_{ATP}.

The targets for ROS have been assumed to be PKC and perhaps phospholipase D, as both have been reported to be directly activated by ROS.[35–37] Both Baines *et al.*[38] and Tritto *et al.*[39] found that they could precondition hearts by direct administration of ROS.

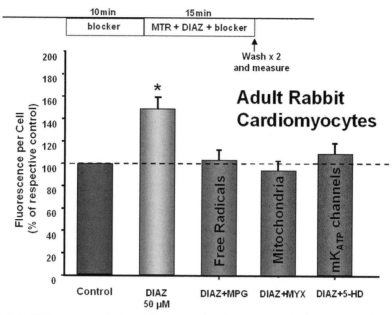

FIGURE 2. ROS generation in isolated heart muscle cells as reported by the increase in fluorescence after 15 min of incubation with reduced Mitotracker Red (MTR). The protocol appears at the top of the figure. The mK_{ATP} opener diazoxide (DIAZ) increased fluorescence by approximately 50%. That increase was blocked by MPG, myxothiazol (MYX), and 5-hydroxydecanoate (5-HD). See text for details. (Reprinted with permission.[32])

Because both studies found that the protection was PKC-dependent, PKC must have been downstream of ROS.

The ROS hypothesis explains one of the perplexing puzzles of IPC. If the preconditioning ischemia causes the release of autacoids that trigger protection, why don't these autacoids protect the non-preconditioned heart? The answer is that the short period of reperfusion prior to the long index ischemia is needed to resupply oxygen to make signaling ROS that only then complete the trigger pathway. ROS will also be generated in the non-preconditioned heart after reintroduction of oxygen, following the release of the coronary occlusion. But we hypothesize that this renewed signaling takes too long to generate the necessary downstream signals that can block mPTP formation in the critical first minutes of reperfusion, thus leading to mitochondrial and cellular disruption. In pharmacological preconditioning, oxygen is abundant when the receptor ligand is given.

Nitric Oxide (NO)

The IPC mimetic ACh triggered protection through EGFR to activate PI3-kinase to open mK_{ATP} to produce ROS. But how does PI3-kinase couple to mK_{ATP}? Oldenburg *et al.*[40] examined bradykinin in the myocyte model and, as expected, it also increased fluo-

rescence by an mK_{ATP}-dependent mechanism. They also tested for NO involvement, a known mediator of many of bradykinin's biological actions. In most systems NO is synthesized by activation of one of the three known isoforms of nitric oxide synthase (NOS). NO acts as a second messenger to activate soluble guanylyl cyclase which makes cyclic GMP (cGMP) from GTP. The diffusible cGMP acts as a second messenger itself to activate the serine/threonine kinase cGMP-dependent kinase (PKG). And PI3-kinase activates endothelial NOS through PDK-1 and PDK-2.[41] Indeed, blockade of NOS with L-NAME (Nw-nitro-L-arginine methyl ester), guanylyl cyclase with ODQ (1-H-[1,2,4] oxadiazole [4,3-9] quinoxalin-1-one), or PKG with Rp-8-Br cGMPS (8-bromoguanosine-3, 5-cyclic monophosphothioate, Rp-isomer) prevented ROS production. Furthermore a direct PKG activator caused ROS generation. In addition, Oldenburg *et al.* found that L-NAME and ODQ could block bradykinin's protection against infarction in whole hearts confirming involvement of NO and cGMP in IPC. Clearly NO was coupling bradykinin to mK_{ATP}. A direct activator of PKG preconditioned the isolated rabbit heart, and that protection was blocked by either 5-HD or MPG.[42] ACh was also found to use the NO-PKG cascade in its coupling to mK_{ATP}.[43] While a peptide PKG blocker could block

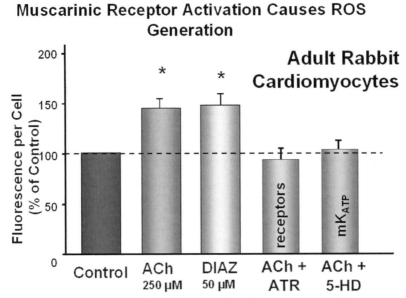

FIGURE 3. ROS generation in isolated heart muscle cells. Acetylcholine (ACh) caused a similar increase in ROS production as the mK$_{ATP}$ opener diazoxide (DIAZ). Atropine (ATR) or 5-hydroxydecanoate (5-HD) both blocked ROS production. See text for details. (Reprinted with permission.[32])

ROS formation from ACh and bradykinin, it could not block ROS production from diazoxide,[44] confirming that PKG was upstream of mK$_{ATP}$. The above data reveal that PI3-kinase activates endothelial NOS to make NO, which then activates guanylyl cyclase to make cGMP which in turn activates PKG.

The reason it took so long to find NO in the pathway is that adenosine receptors bypass it. Isolated hearts trigger primarily through adenosine signaling because the plasma kininogens required for bradykinin synthesis and the innervation required for opioid release are both absent. L-NAME has no effect on IPC in isolated rabbit hearts but blocks IPC's protection *in situ*.[45] Only when pharmacological triggering was done with an mK$_{ATP}$-dependent agonist did NO's involvement become apparent. Because of diversity in the signaling, the true nature of these pathways was only appreciated when each receptor's pathway was studied individually.

PKG Opens mK$_{ATP}$

Costa *et al.*[46] showed that incubating isolated mitochondria in a solution containing cGMP plus PKG caused mK$_{ATP}$ to open. This suggested a phosphorylation site for PKG is present on the mitochondrial outer membrane that can lead to mK$_{ATP}$ opening. It is currently unknown how that phosphorylation site communicates with mK$_{ATP}$ on the inner membrane, but involvement of an intervening step with PKC, probably the ε isoform, was demonstrated.

The Mediator Pathway

While we now have a fairly complete map of the trigger pathway, work on the mediator pathway has just begun. A few components have been identified. Hausenloy *et al.*[47] found that activation of PI3-kinase and p42/p44 MAP kinase (ERK) are required in the first minutes of reperfusion. These pro-survival kinases are thought to inhibit mPTP opening by phosphorylating glycogen synthase kinase-3β (GSK-3β).[48] We have found that at reperfusion, PI3-kinase is under control of adenosine receptors[49] and adenosine receptors are under control of PKC.[4,50] As discussed above, the adenosine receptors are significantly under control of PKC rather than the other way around. The adenosine receptor at reperfusion appears to be the A$_{2b}$ subtype,[4,50] which has a very low affinity for adenosine. In support of that hypothesis, Eckle *et al.*[51] made murine genetic knockout models for all four known adenosine receptors. All but the A$_{2b}$ knockout hearts could be protected by IPC. It would appear that all preconditioning signals converge on the A$_{2b}$ receptor at reperfusion. We have found that activation of PKC increased the sensitivity of the heart to A$_{2b}$ agonists,[4] and that led us to propose that PKC may lower the threshold of these receptors so that endogenous adenosine becomes protective. It is currently not understood why PKC is only turned on at reperfusion in the IPC heart. Nor do we understand how the G$_s$-coupled A$_{2b}$ receptors activate PI3-kinase and ERK. Clearly our understanding of events in the mediator phase is very sketchy and work is currently concentrating on that phase.

The clinical implications of the mediator pathway are enormous, however, because it means that it is still possible to treat the patient right up to the time of reperfusion. This has been borne out by the phenomenon of ischemic postconditioning, in which short cycles of coronary reperfusion/occlusion are given in the first several minutes of reperfusion.[52] We have recently shown that the intermittent ischemia acts to delay mPTP formation by keeping the pH low during early reperfusion which allows enough time for the heart to precondition itself and maintain protection after pH normalizes.[53]

Conflict of Interest

The authors declare no conflicts of interest.

References

1. LIU, G.S. *et al*. 1991. Protection against infarction afforded by preconditioning is mediated by A_1 adenosine receptors in rabbit heart. Circulation **84:** 350–356.
2. YTREHUS, K., Y. LIU & J. M. DOWNEY. 1994. Preconditioning protects ischemic rabbit heart by protein kinase C activation. Am. J. Physiol. **266:** H1145–H1152.
3. GROSS, G.J. & J.A. AUCHAMPACH. 1992. Blockade of ATP-sensitive potassium channels prevents myocardial preconditioning in dogs. Circ. Res. **70:** 223–233.
4. KUNO, A. *et al*. 2007. Protein kinase C protects preconditioned rabbit hearts by increasing sensitivity of adenosine A_{2b}-dependent signaling during early reperfusion. J. Mol. Cell Cardiol. **43:** 262–271.
5. YANG, X.-M. *et al*. 2004. NECA and bradykinin at reperfusion reduce infarction in rabbit hearts by signaling through PI3K, ERK, and NO. J. Mol. Cell. Cardiol. **36:** 411–421.
6. LI, G.C. *et al*. 1990. Myocardial protection with preconditioning. Circulation **82:** 609–619.
7. VAN WINKLE, D.M. *et al*. 1991. The natural history of preconditioning: cardioprotection depends on duration of transient ischemia and time to subsequent ischemia. Coron. Artery Dis. **2:** 613–619.
8. WALL, T.M., R. SHEEHY & J.C. HARTMAN. 1994. Role of bradykinin in myocardial preconditioning. J. Pharmacol. Exp. Ther. **270:** 681–689.
9. SCHULTZ, J.E. J., A.K. HSU & G.J. GROSS. 1996. Morphine mimics the cardioprotective effect of ischemic preconditioning via a glibenclamide-sensitive mechanism in the rat heart. Circ. Res. **78:** 1100–1104.
10. GOTO, M. *et al*. 1995. Role of bradykinin in protection of ischemic preconditioning in rabbit hearts. Circ. Res. **77:** 611–621.
11. COHEN, M.V. *et al*. 2001. Acetylcholine, bradykinin, opioids, and phenylephrine, but not adenosine, trigger preconditioning by generating free radicals and opening mitochondrial K_{ATP} channels. Circ. Res. **89:** 273–278.
12. HAUSENLOY, D.J. *et al*. 2004. Preconditioning protects by inhibiting the mitochondrial permeability transition. Am. J. Physiol. **287:** H841–H849.
13. BERNARDI, P. *et al*. 2006. The mitochondrial permeability transition from in vitro artifact to disease target. FEBS J. **273:** 2077–2099.
14. YTREHUS, K. *et al*. 1994. Rat and rabbit heart infarction: effects of anesthesia, perfusate, risk zone, and method of infarct sizing. Am. J. Physiol. **267:** H2383–H2390.
15. ARMSTRONG, S., J.M. DOWNEY & C.E. GANOTE. 1994. Preconditioning of isolated rabbit cardiomyocytes: induction by metabolic stress and blockade by the adenosine antagonist SPT and calphostin C, a protein kinase C inhibitor. Cardiovasc. Res. **28:** 72–77.
16. AKAO, M. *et al*. 2003. Mechanistically distinct steps in the mitochondrial death pathway triggered by oxidative stress in cardiac myocytes. Circ. Res. **92:** 186–194.
17. XU, Z. *et al*. 2005. Adenosine produces nitric oxide and prevents mitochondrial oxidant damage in rat cardiomyocytes. Cardiovasc. Res. **65:** 803–812.
18. FÖRSTER, K. *et al*. 2006. NECA at reperfusion limits infarction and inhibits formation of the mitochondrial permeability transition pore by activating p70S6 kinase. Basic Res. Cardiol. **101:** 319–326.
19. GARLID, K.D. *et al*. 1997. Cardioprotective effect of diazoxide and its interaction with mitochondrial ATP-sensitive K^+ channels: possible mechanism of cardioprotection. Circ. Res. **81:** 1072–1082.
20. PAIN, T. *et al*. 2000. Opening of mitochondrial K_{ATP} channels triggers the preconditioned state by generating free radicals. Circ. Res. **87:** 460–466.
21. KRENZ, M. *et al*. 2002. Opening of ATP-sensitive potassium channels causes generation of free radicals in vascular smooth muscle cells. Basic Res. Cardiol. **97:** 365–373.
22. OLDENBURG, O. *et al*. 2003. Acetylcholine-induced production of reactive oxygen species in adult rabbit ventricular myocytes is dependent on phosphatidylinositol 3- and Src-kinase activation and mitochondrial K_{ATP} channel opening. J. Mol. Cell. Cardiol. **35:** 653–660.
23. MOCHLY-ROSEN, D., A.I. BASBAUM & D.E. KOSHLAND, JR. 1987. Distinct cellular and regional localization of immunoreactive protein kinase C in rat brain. Proc. Natl. Acad. Sci. **84:** 4660–4664.
24. PING, P. *et al*. 1997. Ischemic preconditioning induces selective translocation of protein kinase C isoforms ε and η in the heart of conscious rabbits without subcellular redistribution of total protein kinase C activity. Circ. Res. **81:** 404–414.
25. XU, Z., X. JI & P.G. BOYSEN. 2004. Exogenous nitric oxide generates ROS and induces cardioprotection: involvement of PKG, mitochondrial K_{ATP} channels, and ERK. Am. J. Physiol. **286:** H1433–H1440.
26. YANG, X.-M. *et al*. 1997. Protection of ischemic preconditioning is dependent upon a critical timing sequence of protein kinase C activation. J. Mol. Cell. Cardiol. **29:** 991–999.
27. TONG, H. *et al*. 2000. Ischemic preconditioning activates phosphatidylinositol-3-kinase upstream of protein kinase C. Circ. Res. **87:** 309–315.

28. QIN, Q., J.M. DOWNEY & M.V. COHEN. 2003. Acetylcholine but not adenosine triggers preconditioning through PI3-kinase and a tyrosine kinase. Am. J. Physiol. **284:** H727–H734.

29. KRIEG, T. *et al.* 2002. ACh and adenosine activate PI3-kinase in rabbit hearts through transactivation of receptor tyrosine kinases. Am. J. Physiol. **283:** H2322–H2330.

30. KRIEG, T. *et al.* 2004. Mitochondrial ROS generation following acetylcholine-induced EGF receptor transactivation requires metalloproteinase cleavage of proHB-EGF. J. Mol. Cell. Cardiol. **36:** 435–443.

31. COHEN, M.V. *et al.* 2007. Preconditioning-mimetics bradykinin and DADLE activate PI3-kinase through divergent pathways. J. Mol. Cell. Cardiol. **42:** 842–851.

32. OLDENBURG, O., M.V. COHEN & J.M. DOWNEY. 2003. Mitochondrial K_{ATP} channels in preconditioning. J. Mol. Cell. Cardiol. **35:** 569–575.

33. BOLLI, R. *et al.* 1989. Marked reduction of free radical generation and contractile dysfunction by antioxidant therapy begun at the time of reperfusion: evidence that myocardial "stunning" is a manifestation of reperfusion injury. Circ. Res. **65:** 607–622.

34. ALTUĞ, S. *et al.* 2000. Evidence for the involvement of peroxynitrite in ischaemic preconditioning in rat isolated hearts. Br. J. Pharmacol. **130:** 125–131.

35. GOPALAKRISHNA, R. & S. JAKEN. 2000. Protein kinase C signaling and oxidative stress. Free Radic. Biol. Med. **28:** 1349–1361.

36. KORICHNEVA, I. *et al.* 2002. Zinc release from protein kinase C as the common event during activation by lipid second messenger or reactive oxygen. J. Biol. Chem. **277:** 44327–44331.

37. NATARAJAN, V. *et al.* 1993. Activation of endothelial cell phospholipase D by hydrogen peroxide and fatty acid hydroperoxide. J. Biol. Chem. **268:** 930–937.

38. BAINES, C.P., M. GOTO & J.M. DOWNEY. 1997. Oxygen radicals released during ischemic preconditioning contribute to cardioprotection in the rabbit myocardium. J. Mol. Cell Cardiol. **29:** 207–216.

39. TRITTO, I. *et al.* 1997. Oxygen radicals can induce preconditioning in rabbit hearts. Circ. Res. **80:** 743–748.

40. OLDENBURG, O. *et al.* 2004. Bradykinin induces mitochondrial ROS generation via NO, cGMP, PKG, and mitoK$_{ATP}$ channel opening and leads to cardioprotection. Am. J. Physiol. **286:** H468–H476.

41. FULTON, D. *et al.* 1999. Regulation of endothelium-derived nitric oxide production by the protein kinase Akt. Nature **399:** 597–601.

42. QIN, Q. *et al.* 2004. Exogenous NO triggers preconditioning via a cGMP- and mitoK$_{ATP}$-dependent mechanism. Am. J. Physiol. **287:** H712–H718.

43. KRIEG, T. *et al.* 2004. Acetylcholine and bradykinin trigger preconditioning in the heart through a pathway that includes Akt and NOS. Am. J. Physiol. **287:** H2606–H2611.

44. KRIEG, T. *et al.* 2005. Peptide blockers of PKG inhibit ROS generation by acetylcholine and bradykinin in cardiomyocytes but fail to block protection in the whole heart. Am. J. Physiol. **288:** H1976–H1981.

45. COHEN, M.V., X.-M. YANG & J.M. DOWNEY. 2006. Nitric oxide is a preconditioning mimetic and cardioprotectant and is the basis of many available infarct-sparing strategies. Cardiovasc. Res. **70:** 231–239.

46. COSTA, A.D.T. *et al.* 2005. Protein kinase G transmits the cardioprotective signal from cytosol to mitochondria. Circ. Res. **97:** 329–336.

47. HAUSENLOY, D.J. *et al.* 2005. Ischemic preconditioning protects by activating prosurvival kinases at reperfusion. Am. J. Physiol. **288:** H971–H976.

48. JUHASZOVA, M. *et al.* 2004. Glycogen synthase kinase-3β mediates convergence of protection signaling to inhibit the mitochondrial permeability transition pore. J. Clin. Invest. **113:** 1535–1549.

49. SOLENKOVA, N.V. *et al.* 2006. Endogenous adenosine protects preconditioned heart during early minutes of reperfusion by activating Akt. Am. J. Physiol. **290:** H441–H449.

50. PHILIPP, S. *et al.* 2006. Postconditioning protects rabbit hearts through a protein kinase C-adenosine A$_{2b}$ receptor cascade. Cardiovasc. Res. **70:** 308–314.

51. ECKLE, T. *et al.* 2007. Cardioprotection by ecto-5′-nucleotidase (CD73) and A$_{2B}$ adenosine receptors. Circulation **115:** 1581–1590.

52. ZHAO, Z.-Q. *et al.* 2003. Inhibition of myocardial injury by ischemic postconditioning during reperfusion: comparison with ischemic preconditioning. Am. J. Physiol. **285:** H579–H588.

53. COHEN, M.V., X.-M. YANG & J. M. DOWNEY. 2007. The pH hypothesis of postconditioning: staccato reperfusion reintroduces oxygen and perpetuates myocardial acidosis. Circulation **115:** 1895–1903.

54. TISSIER, R., M.V. COHEN & J.M. DOWNEY. 2007. Protecting the acutely ischemic myocardium beyond reperfusion therapies: are we any closer to realizing the dream of infarct size elimination? Arch. Mal. Coeur. Vaiss. **100:** 794–802.

The Identity and Regulation of the Mitochondrial Permeability Transition Pore

Where the Known Meets the Unknown

MAGDALENA JUHASZOVA,[a,c] SU WANG,[a,c] DMITRY B. ZOROV,[a,c] H. BRADLEY NUSS,[a] MARC GLEICHMANN,[b] MARK P. MATTSON,[b] AND STEVEN J. SOLLOTT[a]

[a]Laboratory of Cardiovascular Science, Gerontology Research Center, Intramural Research Program, National Institute on Aging, NIH, Baltimore, Maryland, USA

[b]Laboratory of Neurosciences, Gerontology Research Center, Intramural Research Program, National Institute on Aging, NIH, Baltimore, Maryland, USA

The mitochondrial permeability transition (MPT) pore complex is a key participant in the machinery that controls mitochondrial fate and, consequently, cell fate. The quest for the pore identity has been ongoing for several decades and yet the main structure remains unknown. Established "dogma" proposes that the core of the MPT pore is composed of an association of voltage-dependent anion channel (VDAC) and adenine nucleotide translocase (ANT). Recent genetic knockout experiments contradict this commonly accepted interpretation and provide a basis for substantial revision of the MPT pore identity. There is now sufficient evidence to exclude VDAC and ANT as the main pore structural components. Regarding MPT pore regulation, the role of cyclophilin D is confirmed and ANT may still serve some regulatory function, although the involvement of hexokinase II and creatine kinase remains unresolved. When cell protection signaling pathways are activated, we have found that the Bcl-2 family members relay the signal from glycogen synthase kinase-3 beta onto a target at or in close proximity to the pore. Our experimental findings in intact cardiac myocytes and neurons indicate that the current "dogma" related to the role of Ca^{2+} in MPT induction requires reevaluation. Emerging evidence suggests that after injury-producing stresses, reactive oxygen species (but not Ca^{2+}) are largely responsible for the pore induction. In this article we discuss the current state of knowledge and provide new data related to the MPT pore structure and regulation.

Key words: reactive oxygen species; Ca^{2+}; Bcl-2; cyclophilin D; hexokinase II; adenine nucleotide translocase; voltage-dependent anion channel

Introduction

The mitochondrial permeability transition (MPT) is defined as the sudden nonselective increase in the permeability of the inner mitochondrial membrane (IMM) to solutes of molecular mass less than ~1500 Da, and results in loss of mitochondrial membrane potential, mitochondrial swelling, and rupture of the outer mitochondrial membrane (OMM). Hunter *et al.*[1] first reported that the MPT is a consequence of Ca^{2+} induced increased permeability of the IMM that is characterized by simultaneous stimulation of AT-Pase, uncoupling of oxidative phosphorylation, and loss of respiratory control. They later suggested that the MPT is not just an artifact of *in vitro* experimentation, but that it may have a significant physiological relevance and that the mitochondria possess protective mechanisms that guard against its induction. The all-or-nothing character of the MPT led to the conclusion that the MPT is not a consequence of nonspecific mitochondrial membrane damage; it is, instead, the result of the opening of an authentic pore or megachannel.[2,3] From this pivotal work a whole new field in mitochondrial research emerged.

The Quest for MPT Pore Identity

History at a Glance

The last 25 years of MPT research have been characterized by an intensive search for the identity

[c]Contributed equally.

Address for correspondence: Steven J. Sollott, M.D., Laboratory of Cardiovascular Science, Gerontology Research Center, Box 13, Intramural Research Program, National Institute on Aging, NIH, 5600 Nathan Shock Drive, Baltimore, MD 21224-6825. Voice: +410-558-8657; fax: +410-558-8150.

sollotts@mail.nih.gov

Ann. N.Y. Acad. Sci. 1123: 197–212 (2008). © 2008 New York Academy of Sciences.
doi: 10.1196/annals.1420.023

of the MPT megachannel. Naturally, the mitochondrial membrane proteins that have the capability to form channels, such as the adenine nucleotide translocase (ANT; a component of the IMM) and voltage-dependent anion channel (VDAC; known also as porin; the most abundant protein in the OMM) were considered and investigated as the leading candidates. Early on, observations that the MPT is activated by atractyloside and inhibited by ADP and bongkrekic acid implicated ANT as one of the significant players in the formation or modulation of the MPT pore.[4–6] On the other hand, pores formed by VDAC are comparable in size to MPT pores (2–3 nm),[7] suggesting that VDAC might be involved in MPT pore formation.[8] Failure to observe MPT induction by sulfhydryl reagents in mitoplasts led to the conclusion that inner membrane permeability changes require the OMM, and a hypothesis was put forth that local fusion between the inner and outer membranes occurs during MPT induction. It was suggested that under specific conditions this fusion results in the inclusion of VDAC in the inner membrane, near the contacts between the IMM and OMM.[9] The possibility that the role of the contact sites is to provide a physical connection between the VDAC, which binds hexokinase II (HKII) in the OMM and ANT in the IMM, was reported for the first time by Knoll and Brdiczka.[10] Another important advance was made by Kinnally *et al.*,[11] and Szabo and Zoratti,[12] when they found that the mitochondrial megachannel, originally observed in patch-clamp experiments,[13,14] is identical to the benzodiazepine receptor that consists of the VDAC, ANT, and an 18-kDa protein, and that in addition to acting as a benzodiazepine receptor, this complex may form the MPT pore. Experiments with the immunosuppressant, cyclosporin A (CsA), which was found to delay MPT induction,[15,16] in turn, implicated its target, the mitochondrial cyclophilin D (CyP-D; a peptidyl-prolyl isomerase in the mitochondrial matrix) as an important component involved in MPT pore regulation, and it was suggested that the MPT pore might be formed from an interaction between the ANT and CyP-D.[3] At this point, the main players suspected in MPT pore formation for over the past decade had been identified.

The Classical MPT Pore Model

Near the end of the 1990s, an MPT pore model emerged that reflected acquired knowledge and satisfied expectations for the behavior of the pore (e.g., originally reviewed in Crompton,[17] recently in Kroemer *et al.*,[18] and Grimm and Brdiczka[19]). It was proposed that the pore is a dynamic multiprotein complex formed during nonphysiological and pathological conditions that could, depending on circumstances, function as the recruitment and assembling center for other proteins. The pore frame was thought to be formed by the VDAC–ANT–CyP-D complex, which is located at the contact sites between IMM and OMM. MPT pore-like activity was reconstituted into planar bilayers and proteoliposomes from preparations containing complexes made of VDAC, ANT, cytosolic hexokinase, and mitochondrial creatine kinase.[20] This suggested that these kinases are also involved in MPT regulation. It was suspected for a long time that the MPT pore might play an important role in both apoptotic and necrotic cell death. Therefore, it came as no surprise when Bax was found to co-purify with known MPT pore components. Additionally, it was observed that recombinant Bcl-2 or Bcl-x_L augmented the resistance of the reconstituted MPT pore complex to pore opening. This indicated that the MPT pore is under the direct control of anti-apoptotic members of the Bcl-2 family.[21,22] Consequently, Bcl-2 family members were integrated into the MPT pore model as regulatory components. FIGURE 1A shows the proposed MPT pore complex architecture based on these data. This model has become widely accepted and has up until now been recognized by investigators in the field of mitochondrial research. However, there is an increasing volume of recent evidence (based on genetic knockout of individual pore components) that has accumulated against this model (see below and, e.g., Rasola and Bernardi,[23] and Galluzzi and Kroemer[24]).

Reappraisal of MPT Pore Complex Architecture

Discussions about the exact protein composition of the MPT pore still continue. Some alternatives had been proposed, e.g., one suggesting that IMM permeabilization may be the common outcome of several distinct processes involving different proteins or protein complexes, depending on circumstances.[25] While a considerable body of evidence supports the classical MPT pore architecture as it is represented in FIGURE 1A, recent genetic knockout experiments have seemingly contradicted this long-established "dogma" concerning the identity of the main components of the MPT pore. Recent findings provide a basis for a substantial revision of the commonly accepted interpretation of MPT pore molecular structure. Below, we discuss the most important and latest observations related to this topic (summarized in FIG. 1B).

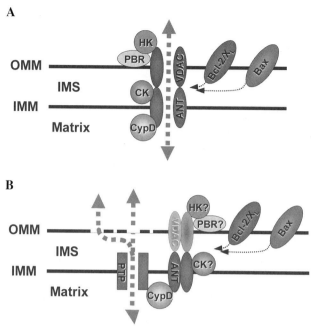

FIGURE 1. Proposed MPT pore complex architecture: **(A)** Classical view. The pore structure is formed by the VDAC–ANT–CyP-D complex, which is located at the contact sites between IMM and OMM. Hexokinase II (HKII) and mitochondrial creatine kinase (CK) are regulatory kinases. Benzodiazepine receptor (PBR) and Bcl-2-family members (Bcl-2, Bcl-x$_L$, and Bax) are included as putative regulatory components. **(B)** Current view, based on reappraisal of the classical model. The elements comprising the pore itself (denoted "PTP" for permeability transition pore) are presently unidentified, but are probably regulated by the adjacent elements as indicated. Note that VDAC, portrayed as a "shadow," is no longer seen as an essential pore component or regulator based on recent genetic evidence. Question mark symbols signify where open questions remain (see text). (In color in *Annals* online.)

Current Standing–VDAC

Three mammalian VDAC isoforms (VDAC1, VDAC2, VDAC3) have been described, and it has been suggested that they may each have a distinct physiological function. VDAC1, the most abundant isoform, is the prototypic version whose properties are highly conserved. Studies on VDAC2-deficient embryonic stem cells suggest that VDAC2 may function as a specific inhibitor of Bak-dependent mitochondrial apoptosis.[26] VDAC2 > VDAC3, but not VDAC1, have been implicated in non-apoptotic cell death induced by an anti-tumor agent, erastin. In these experiments, it was demonstrated that erastin, by direct interaction with VDAC, altered the permeability of the OMM, which resulted in release of oxidative species and, finally, in non-apoptotic, oxidative cell death.[27]

The generation of isoform-specific, VDAC-deficient mice allowed assessment of the role of individual VDAC isoforms in MPT pore structure. Mice missing VDAC1 and VDAC3 were viable but exhibited distinct phenotypes; elimination of VDAC2 resulted in embryonic lethality. MPT properties in mitochondria from VDAC1-null mice were indistinguishable from those of wild-type (WT) mice.[28] Similar conclusions were made in subsequent work by Baines *et al.*,[29] which encompassed all three VDAC isoforms. They found that mitochondria from VDAC1-, VDAC3-, and VDAC1/VDAC3-null mice exhibited Ca^{2+} and oxidative stress–induced MPTs that were indistinguishable from WT mitochondria. Furthermore, they found that fibroblasts lacking VDAC (VDAC1, VDAC2, VDAC3, various combinations of two, or all three VDAC isoforms), showed similar patterns of MPT induction compared to control cells. In addition, their data confirmed the existence of a VDAC-independent model of Bcl-2 family member–mediated cell killing. WT and VDAC-deficient mitochondria and cells responded to activation or over expression of Bax and Bid by equivalent cytochrome c release, caspase cleavage, and cell death. These results make a strong argument against any indispensable role of VDACs in both MPT pore-mediated and Bax–Bak-mediated cell death. The experiments with VDAC-deficient cells and mitochondria strongly cast doubt on the validity of the classical model of MPT pore formation, which assumes that the pore contains VDAC and is formed at the contact sites between mitochondrial membranes.

Current Standing–ANT

Only two ANT isoforms have been definitively identified in mice, as opposed to the three found in humans (ANT1–3). ANTs are co-expressed in tissue-specific patterns. To investigate the role of ANTs in the MPT pore formation, the two isoforms of ANT were genetically inactivated in mouse liver. It was demonstrated that the MPT could still be induced in mitochondria lacking ANT. However, more Ca^{2+} than usual was required to activate the MPT pore, and the pore could no longer be regulated by ANT ligands, including adenine nucleotides. No difference was observed in the sensitivity of hepatocytes deficient in ANTs to various initiators of cell death.[30] Unless the ANT deficiency was compensated for, e.g., by an unforeseen ANT ortholog that is not regulated by conventional ANT ligands, this work provides solid evidence that ANTs are nonessential structural components of the MPT pore, although they may contribute to its regulation. Indeed, we demonstrated that in intact cardiac myocytes, the true specific ANT inhibitor, bongkrekic acid, virtually completely blocked MPT pore induction by reactive oxygen species (ROS).[31] This supports the notion that the ANT may have a regulatory role in controlling MPT pore induction.

Current Position on CyP-D

Recently, genetic studies from four independent groups on CyP-D knockout mice (Ppif$^{-/-}$ mice) confirmed the critical role of CyP-D in regulation of the MPT pore machinery and provided new insight into the role of MPT pore in cell death.[32–35] All four research groups independently provided compelling evidence that while the MPT can still be induced in mitochondria from mice lacking CyP-D, the sensitivity to CsA is lost, proving beyond any doubt that CyP-D is the unique target of CsA. CyP-D-deficient mice show a high level of protection against ischemia/reperfusion injury manifested by an impressive reduction in heart and brain infarct size, while CyP-D-overexpressing mice were sensitized to cell death. Mitochondria from mice lacking CyP-D showed a similar delay in MPT induction as control mitochondria treated with CsA. These observations directly implicate CyP-D as a pore regulator. Similarly to the situation in ANT-deficient mitochondria, Ppif$^{-/-}$ mitochondria were more resistant to Ca^{2+} induced MPT induction and swelling. Interestingly, while ANT protein levels were not changed

in the CyP-D-deficient mice compared to WT, Ppif$^{-/-}$ mitochondria showed increased resistance to the ANT-ligand atractyloside compared to WT mitochondria.[32] In contrast, the ANT-deficient mitochondria have a functional, CsA-inhibitable MPT pore. This might suggest that the CyP-D is a central regulatory component acting in close proximity to MPT pore whereas, if ANT has any regulatory role it is probably executed upstream of CyP-D. Recently, reversing the original view that CyP-D and ANT directly interact,[6] it was proposed that the CyP-D binding protein may not be the ANT,[36] consistent with the emerging view that the ANT is not a requisite structural component of the MPT pore.

These genetic experiments[32–35] also provided critical new insights into fundamental mechanisms of cell death. CyP-D-deficient cells also responded to various apoptotic stimuli in a similar manner as the WT but showed considerable resistance to necrotic cell death, suggesting that CyP-D is not a central component of the apoptotic death pathway. These conclusions lead to a revision of the current dogma and support the idea that the MPT pore is *not* a major participant in apoptotic cell death, but rather plays an important role in necrotic cell death (at least in cardiac myocytes, neurons, and fibroblasts). These ideas represent a significant contribution to the understanding of the role mitochondria play in cell death.

Role of HK II

HK II associates with the mitochondrial surface at the contact sites between OMM and IMM. HKII binding to the VDAC may prevent mitochondrial breakdown during cell stress and injury. On the basis of experimental findings in a reconstituted system, it was suggested that the association between HKII and VDAC correlates with MPT pore closure and vice versa.[20] The interaction between HKII and VDAC might cause a conformational change that favors closure or prevents the interaction between pro-apoptotic Bcl-2 family members and MPT pore components or other regulators.[37] This interaction could be disrupted by a glycogen synthase kinase-3 beta (GSK-3β)-dependent phosphorylation of VDAC and could be promoted by Akt, which inhibits GSK-3β.[38] *In vitro* experiments have demonstrated that the peptide comprising the NH$_2$-terminal 15 amino acids of HKII, which represents the VDAC binding domain, induces detachment of HKII from the mitochondria.[39] For *in vivo* application, a cell-permeable analog was synthesized in which HKII-peptide was fused to the internalization domain of the

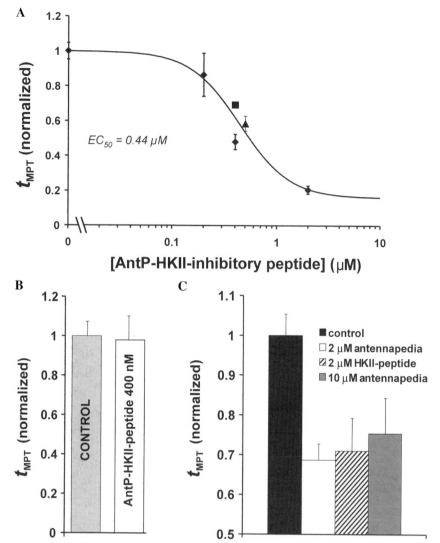

FIGURE 2. Regulation of MPT induction by Antennapedia-Hexokinase II-inhibitory peptide (AntP-HKII) is intrinsic to Antennapedia and not necessarily to HKII. **(A)** MPT susceptibility to ROS (t_{MPT}) in intact cardiac myocytes is regulated by AntP-HKII; dose–response curve. **(B)** Mitochondria in intact hippocampal neurons are relatively insensitive to 400 nM AntP-HKII versus their cardiac myocyte counterparts. **(C)** Antennapedia (AntP); peptides themselves are responsible for an artifact because they regulate the MPT in cardiac myocytes.

Antennapedia homeoprotein that facilitates peptide internalization (AntP-HKII-peptide).[39]

We have previously developed a method for quantifying the susceptibility to induction of the MPT by ROS in individual mitochondria inside live cardiac myocytes.[31] The average time required for the standardized photoproduction of ROS to cause MPT induction (t_{MPT}) is taken as the index of the MPT-induction ROS threshold for that cell. We showed that oxidant stress significantly reduces the ROS thresh-old for MPT induction, while diverse protective agents enhance it.[40] We employed the AntP-HKII-peptide with an aim to study the regulation of MPT pore activity by HKII. In isolated cardiac myocytes, the effect of AntP-HKII-peptide on t_{MPT} was dosedependent ($EC_{50} \sim 400$ nM) (FIG. 2A). Interestingly, hippocampal neurons were relatively insensitive to AntP-HKII-peptide (400 nM): t_{MPT} was comparable in both control and peptide-treated neurons (FIG. 2B). Further investigations also revealed that in cardiac myocytes the

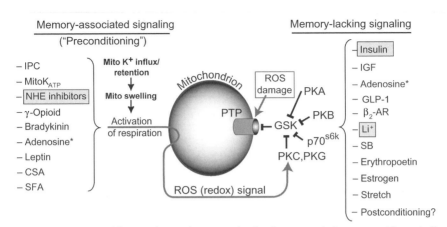

FIGURE 3. Synopsis of the signaling pathways involved in the myocardial protection. The end-effector is the MPT pore complex, which is sensitive to oxidant stress (e.g., pathologically generated ROS resulting from ischemia/reperfusion, etc.) and can determine mitochondrial and cell fate via the consequences of MPT induction. The degree of this sensitivity to ROS can be reduced via activation of a wide spectrum of upstream signaling kinases which converge and inhibit the activity of GSK-3β, which in turn results in protection by increasing the MPT–ROS threshold. At more modest levels, ROS acting locally can also play a role as a redox signal, for example, generated by ischemic preconditioning (IPC), which can activate certain isoforms of PKC and result in protection. The multiplicity of upstream activators of protection signaling arise from two general pathways: one pathway, shown schematically in the left column (of effectors, phenomena, and targets), acts via regulatory increases in mitochondrial volume and respiration which produces the ROS/redox signal critical for preconditioning that can result in protection lasting several hours after the upstream stimulus is removed (i.e., memory-associated signaling). The other pathway (on the right) has little or no memory and produces protection via ROS-independent kinase signaling mechanisms[40,44] (see text and Ref. 40 for additional details).*Adenosine signaling has elements from both major pathways. (From Juhaszova *et al.*, with permission.[44])

internalization sequence of the Antennapedia homeoprotein (AntP) itself had an equivalent effect on t_{MPT} as does AntP-HKII-peptide (FIG. 2C). Thus, we concluded that the experimental results with AntP-HKII-peptide are likely an artifact and thus cannot be interpreted in terms of the function of HKII. These new findings cast doubt on conclusions related to *in vivo* action of HKII derived from experiments using AntP-HKII-peptide, and the role of HKII in MPT pore regulation remains an open question that requires further study.

MPT Pore: End-Effector of Cardioprotection Signaling

The idea that MPT induction is linked to cell damage and death as a result of ischemia/-reperfusion and toxic insult has been recognized for a long time.[41–43] Cell protection, specifically protection of the MPT pore complex against ROS induction, may result from activation of various signaling pathways. Direct measurements of the MPT–ROS threshold showed that diverse upstream protective signaling pathways (e.g.,

acting through kinases including, PKA, PKB, PKC, PKG, and p70s6k) converge on GSK-3β, which serves as a basic point-of-integration. Inhibition of GSK-3β by its phosphorylation relays the protective signal downstream to the end-effector, the MPT pore complex, which results in prevention of MPT induction. FIGURE 3 represents an overview of the signaling pathways involved in cell protection.[40,44] Direct pharmacologic inhibition of GSK-3 reduces infarct size and improves post-ischemic energy metabolism and function in both the brain and heart,[45–48] as well as affording functional and morphological protection to the kidney.[49]

The exact regulatory target(s) of GSK-3β relating to the sensitivity of the MPT pore to induction by ROS have not yet been conclusively identified. Bcl-2 family members have been implicated in MPT pore regulation since the late 1990s (reviewed in Rasola and Bernardi[23]). Indeed, the search for motifs within proteins that are likely to be phosphorylated by the serine/threonine kinase GSK-3β has yielded several candidates: for example, Bcl-2, the Bcl-2–binding protein Bis (also called Bcl-2–binding athanogene-3 [BAG-3]), and the serine/threonine protein phosphatase 2A

(PP2A, subunit B).[40] Expression levels of myeloid cell levkemia-1 (MCL-1), a Bcl-2 family member, have been associated with cell protection. MCL-1 is a labile protein that is stabilized and regulated by GSK-3 inhibition. It was suggested that GSK-3 regulates mitochondrial outer membrane permeabilization and apoptosis by destabilization of MCL-1.[50]

Pastorino *et al.* proposed a direct connection between VDAC and GSK-3β, suggesting that activation of GSK-3β results in VDAC phosphorylation.[38] They proposed that phosphorylation of VDAC prevents HKII binding and consequently its dissociation from the mitochondria. If additional experiments show unambiguously that the binding of HKII to mitochondria is protective (see above), then it would be plausible that GSK-3β–mediated phosphorylation of VDAC (and/or other proteins to be determined) and consequent dissociation of HKII from mitochondria might result in increased sensitivity of mitochondria to MPT induction. On the other hand, since it was shown that VDAC is neither a required nor regulatory element of the MPT pore, the mitochondrial binding partners(s) of HKII through which these actions are exerted remain to be established.

Role of Bcl-2 Family Members in Cell Protection

Bcl-2 has been shown to protect the heart against ischemia/reperfusion injury.[51,52] The BH4 domain of Bcl-x_L was sufficient to provide protection against mitochondrial dysfunction and apoptosis,[53] and hearts perfused with a BH4 peptide derived from Bcl-x_L and conjugated to the protein transduction domain of HIV transactivator of transcription (TAT). TAT-BH4 demonstrated reduced infarct size after ischemia/reperfusion.[54,55] In isolated cardiac myocytes we measured the effect of TAT-BH4 on t_{MPT} to establish the underlying mechanisms. Consistent with the infarction data, TAT-BH4 enhanced the MPT threshold to ROS by about 40% (FIG. 4A). Thus, the BH4 peptide limits MPT pore induction to a degree similar to that seen by GSK-3β–mediated (ischemic preconditioning and pharmacologic) protection.[40]

In the next set of experiments, we employed HA14-1, a small cell-permeable molecule that is thought to inhibit Bcl-2 by binding to the Bcl-2 surface pocket thereby blocking its biological action.[56] HA14-1 inhibited not only the protection afforded by TAT-BH4, but also any protection induced by activation of diverse upstream protective signaling pathways. Specifically, HA14-1 blocked protection induced by: 1) the NHE-inhibitor HOE 694 (HOE) which results in activation of PKCε, 2) activation of Akt and p70s6k down-stream of the insulin receptor via insulin, and 3) direct inhibition of GSK-3β with Li[+] (FIG. 4A, B, and C). Since GSK-3β is the convergence point of the wide variety of diverse upstream protection signaling pathways[40,44] and GSK-3β-mediated resistance to oxidant stress is reversed by HA14-1, our experiments suggest that the target of HA14-1, Bcl-2, is located downstream of GSK-3β, presumably in close proximity to the MPT pore where it regulates pore activity. At high concentrations, HA14-1 can have side effects on mitochondrial respiration. Thus, a structural analog of HA14-1, EM20-25, a compound devoid of these potentially confounding side effects, was also tested. It was reported that similarly to HA14-1, EM20-25, also had deleterious effects on MPT pore induction.[57]

Furthermore, we found that a specific Bcl-2 peptide inhibitor that binds to Bcl-2 with high affinity ($IC_{50} = 130$ nM), derived from the BH3 domain of Bad, completely blocked protection afforded by HOE, insulin, and Li[+], while the negative control peptide analog (in which a single amino acid is switched— Leu151 replaced by Ala), which exhibits nearly a 15-fold reduction in Bcl-2 binding, had no effect on protection (FIG. 4D, E, and F). We conclude that conventionally defined "pro-apoptotic" and "anti-apoptotic" members of the Bcl-2 family (via BH3 and BH4 domains, respectively) are critical mediators of protection signaling downstream of GSK-3β, where they regulate susceptibility of the MPT pore to oxidant stress. It should be pointed out that in this context of limiting MPT pore induction, the actions of the Bcl-2 family described here are effecting changes in cell survival and death via necrosis rather than apoptosis-related pathways, as discussed above. The diagram in FIGURE 5 summarizes our current view of the regulatory role played by Bcl-2 family members in cell protection.

MPT Induction: Paradoxical Role of Ca²⁺

Mostly based on *in vitro* results from experiments on **isolated** mitochondrial suspension, it was concluded and widely accepted that the MPT pore in intact cells is opened by elevated Ca^{2+} levels (see for examples, Halestrap[58] and Kowaltowski *et al.*[59]). This led to the conventional assumption that the opening of the MPT pore by mitochondrial Ca^{2+} loading may be the cause of many types of cell death in the heart and brain, such as, for example, death induced by ischemia/reperfusion injury and chemical toxins. Based on *in vivo* data, however, in **intact** cardiomyocytes and neurons, normal excitation potentially brings Ca^{2+} to levels approaching those believed to induce the MPT.

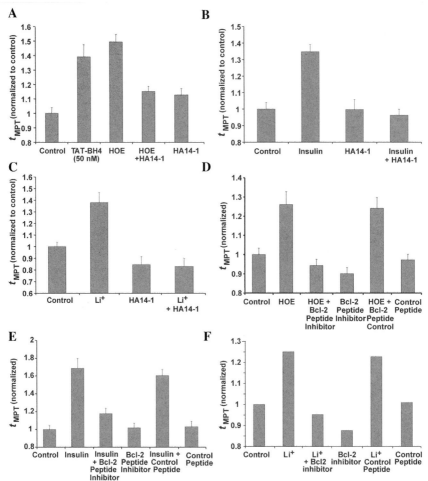

FIGURE 4. Role of Bcl-2 family members in cardioprotection. (**A**) Effect of TAT-BH4 peptide on MPT–ROS threshold (t_{MPT}) in isolated cardiac myocytes. Magnitude of protection afforded by 50 nM TAT-BH4 peptide was comparable to protection induced by the NHE-inhibitor HOE 694 (10 μM). Small, cell-permeable, nonpeptidic Bcl-2 ligand, HA14-1 (10 μM) completely abolished cell protection induced by diverse upstream signaling pathways: HA14-1 blocked protection induced by 10 μM HOE (**A**), 30 nM insulin (**B**), and 3 mM Li^+ (**C**). Cardioprotection afforded by 10 μM HOE (**D**), 30 nM insulin (**E**), and 3 mM Li^+ (**F**) was abolished by 2 μM BH3 peptide inhibitor of Bcl-2, while the negative control peptide analog (in which a single amino acid is switched—Leu151 replaced by Ala) that exhibits a nearly 15-fold reduction in Bcl-2 binding had no effect whatsoever on protection. (See Mothodology section, and Ref. 40 for additional methodological details.)

This suggested that a serious paradox needed to be reconciled: the current "dogma" regarding the role of Ca^{2+} in MPT induction *in vitro* with the empirical evidence of health and longevity of these excitable cells *in vivo*. We hypothesized that mitochondrial Ca^{2+} might be inappropriately "blamed" for cell death after cardiac myocytes and neurons suffer injury-producing stresses such as hypoxia/reoxygenation or excitotoxic stress.

Role of Ca^{2+} in MPT Pore Induction in Cardiac Myocytes

To address this issue we devised a set of experiments aiming to evaluate the effect of intracellular Ca^{2+} concentration on MPT induction in intact cardiac myocytes. The intracellular Ca^{2+} was "clamped" at levels >1 μM for >20 min (using 10 Hz electrical tetanization in 5 mM bathing Ca^{2+} in the presence of thapsigargin or cyclopiazonic acid to disable

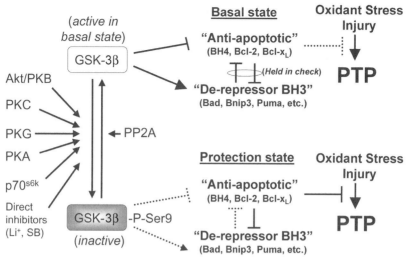

FIGURE 5. Overview of the role of the Bcl-2 family members in the protection signaling. In the basal state, when the GSK-3β is fully active, "anti-apoptotic" and "pro-apoptotic" Bcl-2 family members are held in mutual check. This mutually offsetting balance yields the apparent MPT–ROS threshold "basal state." In the protection state, GSK-3β is phosphorylated at Ser-9, which inactivates GSK-3β and induces cell protection. When protection signaling is activated in this way, the balance between Bcl-2 family members is shifted toward unmasking the activity of "anti-apoptotic" Bcl-2 family members and results in protection of the MPT pore against oxidant stress injury.

the sarcoplasmic reticulum (SR). FIGURE 6A shows the experimental design of the tetanization technique in Indo-1–free acid loaded cardiac myocytes in 1, 2, and 5 mM bathing Ca^{2+}.[60] The MPT–ROS threshold in tetanized cells (with >1 μM cytoplasmic Ca^{2+}) was identical to that of unstimulated cells (with a cytoplasmic Ca^{2+} of ~100 nM) (FIG. 6B). However, once the sarcolemma of these same cardiac myocytes was permeabilized and then maintained in carefully prepared Ca^{2+}/EGTA buffered solution, the outcome was very different: the MPT–ROS threshold in 100 nM Ca^{2+} was the same as in intact cells; however, bathing permeabilized cells in 500 nM Ca^{2+} resulted in a decrease in the MPT–ROS threshold by more than half of that observed in 100 nM Ca^{2+} (FIG. 6B). Thus, we speculate that isolation of cardiac mitochondria from the cytoplasm results in loss of some critical factor(s), and, consequently, the permeability transition pore complex becomes susceptible to elevated Ca^{2+}. On the other hand, mitochondria inside these excitable cells are relatively insensitive to high Ca^{2+}. Furthermore, it has been shown that Sr^{2+} is ineffective (or even inhibitory) in MPT induction in isolated mitochondria (more than an order of magnitude greater levels of Sr^{2+} than Ca^{2+} are required for MPT induction).[1] We took advantage of this convenient property of Sr^{2+} to devise experiments in intact cells that would be essentially free of intracellular Ca^{2+} by replacing

it with a divalent cation that is relatively ineffective in causing MPT induction in isolated mitochondria. Complete equimolar replacement of Ca^{2+} with Sr^{2+} (for 6 hours) in intact cardiac myocytes resulted in the same MPT–ROS threshold as seen in cells with normal Ca^{2+} (FIG. 6C). The MPT–ROS threshold measurements after replacement and restoration of normal Ca^{2+} to these Sr^{2+} treated cells was comparable to the initial controls (FIG. 6C). Thus, it is unlikely that Ca^{2+} plays an important role in *either* direction in mediating MPT induction in intact cardiac myocytes.

FIGURE 6D shows that in cardiac myocytes manipulation of extracellular bathing Ca^{2+} concentrations either by limiting Ca^{2+} influx (in nominally Ca^{2+}-free buffer) or by increasing Ca^{2+} influx by incubating the cell in 3 mM bathing Ca^{2+} does not negatively affect cardiac myocyte death after hypoxia/reoxygenation injury. On the contrary, it might even be beneficial: cell death was significantly lower in the group incubated in 3 mM Ca^{2+}. Furthermore, in myocytes incubated in 1 mM extracellular Ca^{2+} and in nominally Ca^{2+}-free media, buffering intracellular Ca^{2+} with 1,2-*wis*-(o-Aminophenoxy)-ethane-N,N,N′,N′-tetraacetic aci(BAPTA), applying the sarcoplasmic reticulum Ca^{2+}-ATP$_{ase}$ (SERCA) blocker thapsigargin, or treati cells with the mitochondrial calcium uniporter bloc Ru360 did not influence cell death outcome hypoxia/reoxygenation injury, although the r

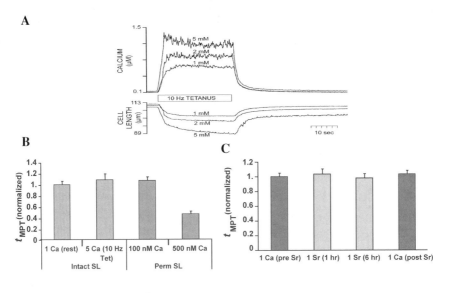

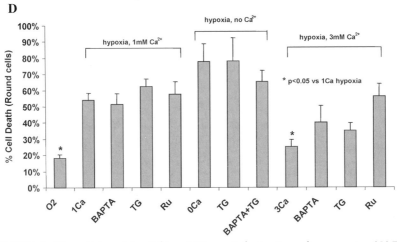

FIGURE 6. Effect of intracellular Ca^{2+} on MPT pore induction in cardiac myocytes. **(A)** Tetanization (thapsigargin pretreatment, 10 Hz electrical stimulation) of indo-1-free acid-loaded cardiac myocytes in 1, 2, and 5 mM bathing Ca^{2+}. **(B)** Role of Ca^{2+} in ROS induction of the MPT in cardiac myocytes with intact sarcolemma (SL) (control cells where intracellular Ca^{2+} concentration is $\sim$100 nM and tetanized myocytes where intracellular Ca^{2+} concentration reaches >1 μM) and in myocytes with a permeabilized SL (maintained in 100 nM and 500 nM Ca^{2+}/EGTA buffered solution). **(C)** Effect of equimolar replacement of intracellular Ca^{2+} for Sr^{2+} (1 mM) on ROS induction of MPT in intact cardiac myocytes. (Modified from Juhaszova *et al.*,[40] with permission). **(D)** Role of Ca^{2+} (1 mM Ca^{2+} versus nominally Ca^{2+}-free versus 3 mM Ca^{2+} bathing solution) in cardiomyocyte death after hypoxia–reoxygenation injury. (Abbreviations: TG, thapsigargin; Ru, Ruthenium 360.)

afforded by 3 mM bathing Ca^{2+} was par-
·d by all three tested agents. Thus, mildly
· influx might actually tend to promote
injury-producing stresses in cardiac
·ertain conditions. Overturning the
·, we can suggest that the function
·ide a death signal through open-

ing of the MPT pore, but potentially to provide a signal that tends to enhance cell survival in the presence of a functional SR. Still, we can not exclude the role of dysfunctional sarcoplasmic reticulum/endoplasmic reticulum (SR/ER) Ca^{2+} regulation in augmentation of myocardial injury during ischemia/reperfusion (reviewed in Tanno and Miura)[61], as our experiments show a trend to reversal of the benefit of increased Ca^{2+} in the presence of SERCA inhibitor thapsigargin, mitochondrial calcium uniporter blocker Ru360, or buffering of intracellular Ca^{2+} with BAPTA.

Additionally, it was reported that manipulation of SR function by changing the levels of SR Ca^{2+} handling proteins might play a role in reperfusion injury. Indeed, reduction of SR Ca^{2+} uptake rate by overexpressing histidine-rich Ca^{2+}-binding SR protein significantly attenuated cell death during ischemia/reperfusion.[62] On the other hand, enhancement of SR Ca^{2+} release and uptake by knocking out phospholamban significantly augmented myocardial injury during ischemia/reperfusion.[63] This effect was gender specific; protection of female hearts against phospholamban ablation was decreased by a NOS inhibitor and mimicked in males by an NO donor. This raises a possibility that the protection related to Ca^{2+} was NOS mediated and supports the notion for the role of NO in cell protection. We have previously shown that an NO donor (SNAP) significantly enhances the MPT–ROS threshold in live cardiac myocytes,[31] which likely explains these ischemia/reperfusion results. Furthermore, it was reported that the endogenous production of NO by eNOS could be induced by stretching cardiac myocytes and this can directly modulate the excitation–contraction coupling (ECC) process by enhancing the capacity to release Ca^{2+} from the SR. The stretch-mediated enhancement of ECC involves PI(3)K-dependent phosphorylation events that include the activation of Akt/PKB and eNOS. This stretch related NO production exerts its effects on Ca^{2+} release probably through S-nitrosylation of the ryanodine receptor. It was proposed that this cardiac PI(3)K-Akt-eNOS axis serves as a physiological sensor of cardiac stretch.[64] Thus, while acute activation of NOS can regulate cell contraction, long-term Ca^{2+} and/or stretch activation of NOS may prolong cell survival, e.g., by inhibition of apoptosis by NO-dependent nitrosation,[65] and a large body of evidence supports the notion that NO may exert a protective role in modulating the severity of ischemia/reperfusion injury in myocardium (reviewed in, e.g., Bolli[66]).

Role of Ca^{2+} in MPT Pore Induction in Neurons

Having demonstrated that Ca^{2+} is not responsible for MPT induction in intact cardiac myocytes (at least under the circumstances tested), we went on to prove that this concept is likely generally valid for a variety of classically excitable cells by performing similar experiments in cultured neurons (FIG. 7). FIGURE 7A demonstrates the experimental line scan protocol to assess the susceptibility of the MPT pore to induction by ROS in mitochondria inside primary hippocampal neurons. This method was adapted from our established protocol for cardiac myocytes.[31] The t_{MPT} is the average time required for the standardized photoproduction of ROS to cause MPT induction, and reflects the ROS threshold in that cell. Similar levels of intracellular Ca^{2+} concentration can be achieved by different stimuli, i.e., glutamate versus KCl depolarization (FIG. 7B). While Ca^{2+} mobilization by both excitotoxic glutamate and KCl depolarization resulted in comparable increases in intracellular Ca^{2+} levels, the outcome as far as the effect on MPT pore induction was diametrically opposed. Treatment with glutamate resulted in a significantly reduced MPT–ROS threshold, and this effect was not prevented by eliminating the rise in intracellular Ca^{2+} through maintenance of cells in nominally Ca^{2+}-free buffer or by attenuating the influx of mitochondrial Ca^{2+} by the Ca^{2+} uniporter blocker Ru360. Mobilization of intracellular Ca^{2+} by KCl depolarization, on the other hand, did not result in significantly altered MPT–ROS threshold (FIG. 7C). Thus, we can conclude that factors other than Ca^{2+}, e.g., ROS and/or metabolic disturbances, are possibly responsible for glutamate-induced reduction in MPT–ROS threshold. Furthermore, since mild alkalosis significantly reduces the MPT–ROS threshold while mild acidosis is protective (FIG. 7D), it is unlikely that, in neurons, the MPT–ROS threshold effect of glutamate, which is accompanied by an intracellular acidification (data not shown), is due to pH disturbances.

Another series of experiments was performed to explore the role of Ca^{2+} and the MPT pore in excitotoxic neuronal death. Compared to normal extracellular Ca^{2+} concentration, cell death was delayed, but not prevented, by keeping the cells in a nominally Ca^{2+}-free extracellular medium. To discriminate between the effect of reduced extracellular Ca^{2+} on the MPT-related (presumably necrotic) and non-MPT-related (most likely apoptotic) cell death,[32,33] we employed CsA and structurally distinct Sanglifehrin A (SfA),[67] both of which bind to CyP-D and desensitize the pore.[68] As gauged by the effects of CsA and SfA, we may conclude that MPT-mediated cell death is only a relatively minor component ($\sim$25–30%) of excitotoxic neuronal death under these conditions. In agreement with our observation that manipulation of intracellular Ca^{2+} level did not influence MPT–ROS threshold (Fig. 7C), both CsA and SfA afforded similar magnitudes of protection, suggesting similar contributions of MPT-dependent mechanisms, regardless of the levels of Ca^{2+} (FIG. 7E).

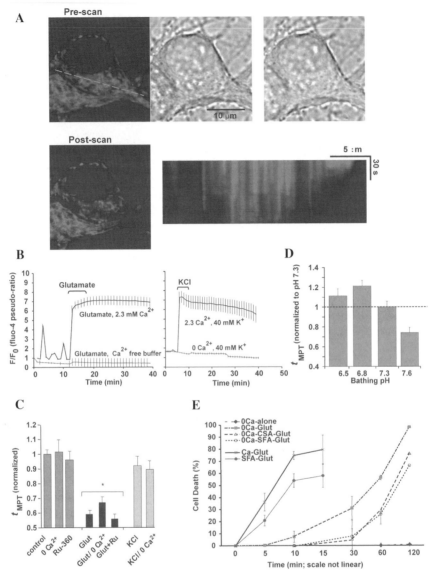

FIGURE 7. Effect of intracellular Ca^{2+} and pH on MPT pore induction in hippocampal neurons. **(A)** Experimental line scan protocol employed to assess the susceptibility of the MPT pore to induction by ROS in mitochondria inside intact neurons. (In color in *Annals* online.) **(B)** Similar levels of intracellular Ca^{2+} concentration were achieved by stimulation with 100 μM glutamate and depolarization induced by 40 mM KCl. **(C)** Effect of intracellular Ca^{2+} concentration on MPT–ROS threshold in intact hippocampal neurons. Glutamate caused a significant decrease in the MPT–ROS threshold regardless of whether or not intracellular Ca^{2+} was allowed to increase. In contrast, KCl depolarization did not affect the MPT–ROS threshold, regardless of whether or not intracellular Ca^{2+} was allowed to increase to levels comparable to that seen with glutamate. **(D)** Influence of acidosis and alkalosis on MPT–ROS threshold in intact hippocampal neurons. Hippocampal neurons were exposed to bathing media at pH 7.3 (control), 6.5, and 6.8, and at the relatively alkaline pH 7.6. The MPT–ROS threshold was measured (as in panel A). Mild alkalosis was found to significantly reduce the MPT–ROS threshold, while mild acidosis was protective, compared to the control at pH 7.3. **(E)** Role of Ca^{2+} and MPT in excitotoxic neuronal death. Preventing the Ca^{2+} rise merely delayed, but did not prevent, the kinetics of processes leading to cell death. The contribution of MPT-dependent mechanisms, as judged by the protective effects of CsA and SfA, is apparently a minor component and of similar magnitude, regardless of the levels of Ca^{2+}. (See text.)

It is well established that, depending on the intensity of glutamate receptor activation (and on other as yet unknown factors), excitotoxic glutamate could induce both apoptosis and necrosis. A shift of this balance towards the apoptotic pathway has been observed, e.g., as a result of caspase-mediated cleavage of glutamate receptor subunits.[69] This, in turn, could lead to a reduction of Ca^{2+} influx and intracellular Ca^{2+} accumulation, which would have a "negative feedback" effect on the progression of certain cell-death mechanisms. One can speculate that delaying excitotoxic cell death by reducing extracellular Ca^{2+} under certain conditions could potentially result in shift towards apoptotic-death pathways through a process that is clearly distinct from (and independent of) the MPT pore mechanisms. Contrary to necrosis, apoptosis is a programmed form of cell death that plays a major role in the development of the nervous system, is regulated by many interactive feed-forward and feedback loops, and can exert local action on the structure and function of growth cones and synapses thus might play a regulatory role in neuronal plasticity.[70]

Methodology

The procedures and materials used here were the same as described earlier.[40]

Materials. All peptides and HA14–1 were obtained from EMD Chemicals, Inc., San Diego, CA. Antennapedia-Hexokinase II-inhibitory peptide (AntP-HKII) (H-RQIKIWFQNRRMKWKK-**MIA-SHLLAYFFTELN**-NH2; the HKII peptide sequence is depicted in bold); Antennapedia peptide (AntP) (H-RQIKIWFQNRRMKWKK-OH); TAT−BH4 peptide (Ac-GRKKRRQRRR-βA-**SNRELVVDFLSYKLSQKGYS**-OH; the conserved N-terminal homology domain (BH4; in bold) of Bcl-x$_L$ (amino acids 4–23) is linked to a carrier peptide, a 10-amino acid HIV-TAT$_{48-57}$ sequence with a β-alanine residue as a spacer; HIV-TAT$_{48-57}$-β-Ala-Bcl-xL BH4$_{4-23}$); BH3 peptide inhibitor of Bcl-2 (Decanoyl-KN^{140}LWAAQRYGRELRRMSDEFEGSFKGL^{165}-OH; a peptide derived from the BH3 domain that binds to Bcl2 with high affinity and the negative control peptide analog in which a single amino acid is switched—Leu151 replaced by Ala). Nonpeptidic Bcl-2 ligand, HA14-1 (10 μM).

Determination of MPT–ROS Threshold and Ca^{2+} Concentration

To assess the susceptibility of the MPT pore to induction by ROS, primary hippocampal neurons were loaded with 30 nM tetramethylrhodamine methyl ester (TMRM) (Molecular Probes Inc., Eugene, OR) for more than 2 h at room temperature and imaged with an LSM 510 inverted confocal microscope with a 63x/1.4 NA oil lens (Carl Zeiss Inc., Jena, Germany). Experimental line scan protocol employed inside intact neurons is shown in FIGURE 7. TMRM fluorescence image is on the top left; corresponding bright field image is in the center, and overlay of both pictures is on the top right (FIG. 7A). Image shows a high magnification view of a neuronal cell body (oval void in TMRM-labeled mitochondria in the overlay of fluorescence and bright field images is the region occupied by the nucleus). Dashed line in the pre-scan image shows the position of the line scan imaging. The post-scan image is shown in the lower right panel. MPT induction in individual mitochondria is seen at the point in time where "columns" of the line scan image suddenly lose TMRM fluorescence intensity and become black. Fluo-4 fluorescence was employed to assess the effect of stimulation with 100 μM glutamate and depolarization induced by 40 mM KCl on intracellular Ca^{2+} concentration (FIG. 7B). Note that nominally Ca^{2+}-free media perfusion immediately prior to the stress blocks any intracellular Ca^{2+} rise in both the glutamate and KCl protocols.

Conclusions

Despite major efforts, the molecular nature of the MPT pore remains enigmatic. Recent genetic evidence has overturned long-standing models: in particular, it is now evident that VDAC is not an essential component of the MPT pore and does not function as a pore regulator. While ANT has been excluded as an essential pore component, it is possible, if not likely, that it serves some regulatory function. The function of HKII as a regulator of the pore, however, is yet unresolved. Only the role of the CyP-D has been clearly established. It was confirmed that CyP-D acts as an important regulator of the MPT pore.

Traditional views about the role of Ca^{2+} in MPT pore induction require revision. The MPT pore in intact cardiomyocytes and neurons is relatively insensitive to increased cytosolic Ca^{2+} (to levels ~10x basal), but it becomes Ca^{2+} sensitive when the mitochondria are separated from the normal cytoplasm. Consequently, we conclude that mitochondrial Ca^{2+} has probably been inappropriately "blamed" for induction of MPT and for the MPT-related intrinsic cell death pathways in cells suffering injury-producing stress, e.g., hypoxia/reoxygenation in cardiac myocytes or excitotoxic stress in neurons. That is not to say that the impairment of other Ca^{2+}-dependent mechanisms does not remain an important factor in promoting cell injury, particularly via MPT-independent and extrinsic cell death mechanisms. Indeed, it has been shown that impairment of SR/ER Ca^{2+} regulation could lead to significant augmentation of cell injury during ischemia/reperfusion.

The function of the MPT pore in cell death has been well established, and critical insights have been developed in relation to the nature of cell death. The fact that the hearts of CyP-D knockout mice are resistant to ischemia/reperfusion injury (similar to CsA treatment), but not to induction of apoptosis by Bcl-2 family member–mediated triggers, suggests that the MPT pore plays a central role in necrosis, but not usually in apoptosis in tissues such as the heart and brain. We can conclude that despite continued uncertainty and debate about the structural identity of the MPT pore, there is no doubt that it plays a central role in cell protection and cell death. Deciphering the molecular structure of the MPT pore will provide additional insight into pathways that are critical for cell survival and cell death, and present a vital target for the development of a novel class of drugs.

Acknowledgment

The work was supported by the Intramural Research Program of the NIH, National Institute on Aging.

Conflict of Interest

The authors declare no conflicts of interest.

References

1. HUNTER, D.R., R.A. HAWORTH & J.H. SOUTHARD. 1976. Relationship between configuration, function, and permeability in calcium-treated mitochondria. J. Biol. Chem. **251:** 5069–5077.
2. HUNTER, D.R. & R.A. HAWORTH. 1979. The Ca2+-induced membrane transition in mitochondria. I. The protective mechanisms. Arch. Biochem. Biophys. **195:** 453–459.
3. CROMPTON, M. & A. COSTI. 1988. Kinetic evidence for a heart mitochondrial pore activated by Ca2+, inorganic phosphate and oxidative stress. A potential mechanism for mitochondrial dysfunction during cellular Ca2+ overload. Eur. J. Biochem. **178:** 489–501.
4. HUNTER, D.R. & R.A. HAWORTH. 1979. The Ca2+-induced membrane transition in mitochondria. II. Nature of the Ca^{2+} trigger site. Arch. Biochem. Biophys, **195:** 460–467.

5. LE-QUOC, K. & D. LE-QUOC. 1988. Involvement of the ADP/ATP carrier in calcium-induced perturbations of the mitochondrial inner membrane permeability: importance of the orientation of the nucleotide binding site. Arch. Biochem. Biophys. **265:** 249–257.
6. HALESTRAP, A.P. & A.M. DAVIDSON. 1990. Inhibition of Ca2(+)-induced large-amplitude swelling of liver and heart mitochondria by cyclosporin is probably caused by the inhibitor binding to mitochondrial-matrix peptidyl-prolyl cis-trans isomerase and preventing it interacting with the adenine nucleotide translocase. Biochem. J. **268:** 153–160.
7. BLACHLY-DYSON, E. *et al.* 1990. Selectivity changes in site-directed mutants of the VDAC ion channel: structural implications. Science **247:** 1233–1236.
8. MORAN O. *et al.* 1990. Electrophysiological characterization of contact sites in brain mitochondria. J. Biol. Chem. **265:** 908–913.
9. LE-QUOC, K. & D. LE-QUOC. 1985. Crucial role of sulfhydryl groups in the mitochondrial inner membrane structure. J. Biol. Chem. **260:** 7422–7428.
10. KNOLL, G. & D. BRDICZKA. 1983. Changes in freeze-fractured mitochondrial membranes correlated to their energetic state. Dynamic interactions of the boundary membranes. Biochim. Biophys. Acta. **733:** 102–110.
11. KINNALLY, K.W. *et al.* 1993. Mitochondrial benzodiazepine receptor linked to inner membrane ion channels by nanomolar actions of ligands. Proc. Natl. Acad. Sci. USA **90:** 1374–1378.
12. SZABO, I. & M. ZORATTI. 1993. The mitochondrial permeability transition pore may comprise VDAC molecules. I. Binary structure and voltage dependence of the pore. FEBS Lett. **330:** 201–205.
13. KINNALLY, K.W., M.L. CAMPO & H. TEDESCHI. 1989. Mitochondrial channel activity studied by patch-clamping mitoplasts. J. Bioenerg. Biomembr. **21:** 497–506.
14. PETRONILLI, V., I. SZABO & M. ZORATTI. 1989. The inner mitochondrial membrane contains ion-conducting channels similar to those found in bacteria. FEBS Lett. **259:** 137–143.
15. CROMPTON, M., H. ELLINGER & A. COSTI. 1988. Inhibition by cyclosporin A of a Ca2+-dependent pore in heart mitochondria activated by inorganic phosphate and oxidative stress. Biochem. J. **255:** 357–360.
16. BROEKEMEIER, K.M., M.E. DEMPSEY & D.R. PFEIFFER. 1989. Cyclosporin A is a potent inhibitor of the inner membrane permeability transition in liver mitochondria. J. Biol. Chem. **264:** 7826–7830.
17. CROMPTON, M. 1999. The mitochondrial permeability transition pore and its role in cell death. Biochem. J. **341:** 233–249.
18. KROEMER, G., L. GALLUZZI & C. BRENNER. 2007. Mitochondrial membrane permeabilization in cell death. Physiol. Rev. **87:** 99–163.
19. GRIMM, S. & D. BRDICZKA. 2007. The permeability transition pore in cell death. Apoptosis **12:** 841–855.
20. BEUTNER, G. *et al.* 1998. Complexes between porin, hexokinase, mitochondrial creatine kinase and adenylate translocator display properties of the permeability transition pore.

Implication for regulation of permeability transition by the kinases. Biochim. Biophys. Acta. **1368:** 7–18.

21. MARZO, I. *et al.* 1998. Bax and adenine nucleotide translocator cooperate in the mitochondrial control of apoptosis. Science **281:** 2027–2031.

22. MARZO, I. *et al.* 1998. The permeability transition pore complex: a target for apoptosis regulation by caspases and bcl-2-related proteins. J. Exp. Med. **187:** 1261–1271.

23. RASOLA, A. & P. BERNARDI. 2007. The mitochondrial permeability transition pore and its involvement in cell death and in disease pathogenesis. Apoptosis **12:** 815–833.

24. GALLUZZI, L. & G. KROEMER. 2007. Mitochondrial apoptosis without VDAC. Nat. Cell Biol. **9:** 487–489.

25. ZORATTI, M., I. SZABO & U. DE MARCHI. 2005. Mitochondrial permeability transitions: how many doors to the house? Biochim. Biophys. Acta **1706:** 40–52.

26. CHENG, E.H. *et al.* 2003. VDAC2 inhibits BAK activation and mitochondrial apoptosis. Science **301:** 513–517.

27. YAGODA, N. *et al.* 2007. RAS-RAF-MEK-dependent oxidative cell death involving voltage-dependent anion channels. Nature **447:** 864–868.

28. KRAUSKOPF, A. *et al.* 2006. Properties of the permeability transition in VDAC1(−/−) mitochondria. Biochim. Biophys. Acta **1757:** 590–595.

29. BAINES, C.P. *et al.* 2007. Voltage-dependent anion channels are dispensable for mitochondrial-dependent cell death. Nat. Cell. Biol. **9:** 550–555.

30. KOKOSZKA, J.E. *et al.* 2004. The ADP/ATP translocator is not essential for the mitochondrial permeability transition pore. Nature **427:** 461–465.

31. ZOROV, D.B. *et al.* 2000. Reactive oxygen species (ROS)-induced ROS release: a new phenomenon accompanying induction of the mitochondrial permeability transition in cardiac myocytes. J. Exp. Med. **192:** 1001–1014.

32. BAINES, C.P. *et al.* 2005. Loss of cyclophilin D reveals a critical role for mitochondrial permeability transition in cell death. Nature **434:** 658–662.

33. NAKAGAWA, T. *et al.* 2005. Cyclophilin D-dependent mitochondrial permeability transition regulates some necrotic but not apoptotic cell death. Nature **434:** 652–658.

34. SCHINZEL, A.C. *et al.* 2005. Cyclophilin D is a component of mitochondrial permeability transition and mediates neuronal cell death after focal cerebral ischemia. Proc. Natl. Acad. Sci. USA **102:** 12005–12010.

35. BASSO, E. *et al.* 2005. Properties of the permeability transition pore in mitochondria devoid of Cyclophilin D. J. Biol. Chem. **280:** 18558–18561.

36. LEUNG, A.W.C. & A.P. HALESTRAP. 2007. The cyclophilin-D binding protein of the mitochondrial permeability transition pore may not be the adenine nucleotide translocase. J. Mol. Cell. Cardiol. **42:** S117.

37. MATHUPALA, S.P., Y.H. KO & P.L. PEDERSEN. 2006. Hexokinase II: cancer's double-edged sword acting as both facilitator and gatekeeper of malignancy when bound to mitochondria. Oncogene **25:** 4777–4786.

38. PASTORINO, J.G., J.B. HOEK & N. SHULGA. 2005. Activation of glycogen synthase kinase 3beta disrupts the binding of hexokinase II to mitochondria by phosphorylating voltage-dependent anion channel and potentiates chemotherapy-induced cytotoxicity. Cancer Res. **65:** 10545–10554.

39. PASTORINO, J.G., N. SHULGA & J.B. HOEK. 2002. Mitochondrial binding of hexokinase II inhibits Bax-induced cytochrome c release and apoptosis. J. Biol. Chem. **277:** 7610–7618.

40. JUHASZOVA, M. *et al.* 2004. Glycogen synthase kinase-3beta mediates convergence of protection signaling to inhibit the mitochondrial permeability transition pore. J. Clin. Invest. **113:** 1535–1549.

41. CROMPTON, M. & A. COSTI. 1990. A heart mitochondrial Ca2+-dependent pore of possible relevance to reperfusion-induced injury. Evidence that ADP facilitates pore interconversion between the closed and open states. Biochem. J. **266:** 33–39.

42. GRIFFITHS, E.J. & A.P. HALESTRAP. 1993. Protection by Cyclosporin A of ischemia/reperfusion-induced damage in isolated rat hearts. J. Mol. Cell. Cardiol. **25:** 1461–1469.

43. GRIFFITHS, E.J. & A.P. HALESTRAP. 1995. Mitochondrial non-specific pores remain closed during cardiac ischaemia, but open upon reperfusion. Biochem. J. **307:** 93–98.

44. JUHASZOVA, M. *et al.* 2005. Protection in the aged heart: preventing the heart-break of old age? Cardiovasc Res. **66:** 233–244.

45. TONG, H. *et al.* 2002. Phosphorylation of glycogen synthase kinase-3beta during preconditioning through a phosphatidylinositol-3-kinase–dependent pathway is cardioprotective. Circ. Res. **90:** 377–379.

46. SASAKI, T. *et al.* 2006. Lithium-induced activation of Akt and CaM kinase II contributes to its neuroprotective action in a rat microsphere embolism model. Brain Res. **1108:** 98–106.

47. MOON, C. *et al.* 2006. Erythropoietin, modified to not stimulate red blood cell production, retains its cardioprotective properties. J. Pharmacol. Exp. Ther. **316:** 999–1005.

48. NISHIHARA, M. *et al.* 2006. Erythropoietin affords additional cardioprotection to preconditioned hearts by enhanced phosphorylation of glycogen synthase kinase-3 beta. Am. J. Physiol. Heart Circ. Physiol. **291:** H748–H755.

49. NELSON, P.J. 2007. Renal ischemia-reperfusion injury: renal dendritic cells loudly sound the alarm. Kidney Int. **71:** 604–605.

50. MAURER, U. *et al.* 2006. Glycogen synthase kinase-3 regulates mitochondrial outer membrane permeabilization and apoptosis by destabilization of MCL-1. Mol. Cell. **21:** 749–760.

51. BROCHERIOU, V. *et al.* 2000. Cardiac functional improvement by a human Bcl-2 transgene in a mouse model of ischemia/reperfusion injury. J. Gene Med. **2:** 326–333.

52. CHEN, Z. *et al.* 2001. Overexpression of Bcl-2 attenuates apoptosis and protects against myocardial I/R injury in transgenic mice. Am. J. Physiol. **280:** H2313–H2320.

53. SHIMIZU, S. *et al.* 2000. BH4 domain of antiapoptotic Bcl-2 family members closes voltage-dependent anion channel and inhibits apoptotic mitochondrial changes and cell death. Proc. Natl. Acad. Sci. USA **97:** 3100–3105.

54. CHEN, M. *et al.* 2002. Calpain and mitochondria in ischemia/reperfusion injury. J. Biol. Chem. **277:** 29181–29186.

55. ONO, M. *et al.* 2005. BH4 peptide derivative from Bcl-xL attenuates ischemia/reperfusion injury thorough

anti-apoptotic mechanism in rat hearts. Eur. J. Cardiothorac. Surg. **27:** 117–121.

56. WANG, J.L. *et al.* 2000. Structure-based discovery of an organic compound that binds Bcl-2 protein and induces apoptosis of tumor cells. Proc. Natl. Acad. Sci. USA. **97:** 7124–7129.

57. MILANESI, E. *et al.* 2006. The mitochondrial effects of small organic ligands of BCL-2: sensitization of BCL-2-overexpressing cells to apoptosis by a pyrimidine-2,4,6-trione derivative. J. Biol. Chem. **281:** 10066–10072.

58. HALESTRAP, A.P. 1999. The mitochondrial permeability transition: its molecular mechanism and role in reperfusion injury. Biochem. Soc. Symp. **66:** 181–203.

59. KOWALTOWSKI, A.J., R.F. CASTILHO & A.E. VERCESI. 2001. Mitochondrial permeability transition and oxidative stress. FEBS Lett. **495:** 12–15.

60. SOLLOTT, S.J., B.D. ZIMAN & E.G. LAKATTA. 1992. Novel technique to load indo-1 free acid into single adult cardiac myocytes to assess cytosolic Ca2+. Am. J. Physiol. **262:** H1941–H1949.

61. TANNO, M. & T. MIURA. 2007. Protecting ischemic hearts by modulation of SR calcium handling. Cardiovasc. Res. **75:** 453–454.

62. ZHOU, X. *et al.* 2007. Overexpression of histidine-rich Ca-binding protein protects against ischemia/reperfusion-induced cardiac injury. Cardiovasc. Res. **75:** 487–497.

63. CROSS, H.R. *et al.* 2003. Ablation of PLB exacerbates ischemic injury to a lesser extent in female than male mice: protective role of NO. Am. J. Physiol. **284:** H683–H690.

64. PETROFF, M.G. *et al.* 2001. Endogenous nitric oxide mechanisms mediate the stretch dependence of Ca2+ release in cardiomyocytes. Nat. Cell. Biol. **3:** 867–873.

65. MANNICK, J.B. *et al.* 1999. Fas-induced caspase denitrosylation. Science **284:** 651–654.

66. BOLLI, R. 2001. Cardioprotective function of inducible nitric oxide synthase and role of nitric oxide in myocardial ischemia and preconditioning: an overview of a decade of research. J. Mol. Cell. Cardiol. **33:** 1897–1918.

67. JAVADOV, S.A. *et al.* 2003. Ischaemic preconditioning inhibits opening of mitochondrial permeability transition pores in the reperfused rat heart. J. Physiol. **549:** 513–524.

68. SORIANO, M.E., L.NICOLOSI & P. BERNARDI. 2004. Desensitization of the permeability transition pore by cyclosporin A prevents activation of the mitochondrial apoptotic pathway and liver damage by tumor necrosis factor-alpha. J. Biol. Chem. **279:** 36803–36808.

69. GLAZNER, G.W. *et al.* 2000. Caspase-mediated degradation of AMPA receptor subunits: a mechanism for preventing excitotoxic necrosis and ensuring apoptosis. J. Neurosci. **20:** 3641–3649.

70. GILMAN, C.P. & M.P. MATTSON. 2002. Do apoptotic mechanisms regulate synaptic plasticity and growth-cone motility? Neuromolecular Med. **2:** 197–214.

Control of Cardiac Rate by "Funny" Channels in Health and Disease

ANDREA BARBUTI AND DARIO DIFRANCESCO

Department of Biomolecular Sciences and Biotechnology, The PaceLab, University of Milan, Milan, Italy

Activation of the "funny" (pacemaker, I_f) current during the diastolic depolarization phase of an action potential is the main mechanism underlying spontaneous, rhythmic activity of cardiac pacemaker cells. In the past three decades, a wealth of evidence elucidating the function of the funny current in the generation and modulation of cardiac pacemaker activity has been gathered. The slope of early diastolic depolarization, and thus the heart rate, is controlled precisely by the degree of I_f activation during diastole. I_f is also accurately and rapidly modulated by changes of the cytosolic concentration of the second messenger cAMP, operated by the autonomous nervous system through β-adrenergic, mainly β₂, and in the opposite way by muscarinic receptor, stimulation. Recently, novel *in vivo* data, both in animal models and humans, have been collected that confirm the key role of I_f in pacemaking. In particular, an inheritable point mutation in the cyclic nucleotide-binding domain of human HCN4, the main hyperpolarization-activated cyclic nucleotide (HCN) isoform contributing to native funny channels of the sinoatrial node, was shown to be associated with sinus bradycardia in a large family. Because of their properties, funny channels have long been a major target of classical pharmacological research and are now target of innovative gene/cell-based therapeutic approaches aimed to exploit their function in cardiac rate control.

Key words: pacemaker; funny current; I_f current; f-channels; heart rate; HCN channels

Introduction

The physiological heart rhythm is generated in the sinoatrial node (SAN), a specialized region of the heart where myocytes spontaneously fire regular and repetitive action potentials, thus determining the rate of contraction of the whole heart. Spontaneous activity relies on the presence of the slow depolarization that develops during diastole and drives the membrane potential to firing threshold. Several mechanisms are involved in the generation of pacemaker activity; among them, a well-established role in initiating the slow diastolic depolarization is played by the pacemaker ("funny") I_f current.[1-4]

The pacemaker current was first described in the late 1970s[5] and named "funny" because its properties were, at the time, unusual in most respects when compared to those of all other known ion channels. Particularly unusual, I_f was found to be a mixed sodium/potassium inward current slowly activating on hyperpolarization below a threshold of about −40 to −45 mV.[6-8] I_f activation appeared therefore to be specifically relevant to generation of a slow depolarizing phase in the pacemaker range of voltages.

Another unusual feature of I_f is its dual activation by voltage and by cyclic nucleotides. Cyclic adenosine monophosphate (cAMP) molecules bind directly to f-channels and increase their open probability.[9,10] cAMP dependence is a particularly relevant physiological property, since it underlies the I_f-dependent autonomic regulation of heart rate.[5,8,11]

In the late 1990s, almost two decades after the first description of I_f, a family of genes encoding the molecular correlates of pacemaker channels was finally sequenced and cloned.[12-18] These genes and the corresponding protein products were named hyperpolarization-activated, cyclic nucleotide-gated (HCN) channels; four isoforms are known in mammals (HCN1–4). Based on sequence similarities, HCN channels were shown to belong to the superfamily of the voltage-gated K^+ (Kv) and the cyclic nucleotide gated (CNG) channels. HCN channels are composed of six transmembrane domains (S1–6), with the positively charged S4 domain acting as the voltage sensor, a pore region, characterized by the GYG sequence typical of K^+-permeable channels, and a

Address for correspondence: Prof. Dario DiFrancesco, Department of Biomolecular Sciences and Biotechnology, University of Milan, via 26 Celoria, 20133 Milan, Italy. Voice: +39-02-50314931; fax: 39-02-50314932.

dario.difrancesco@unimi.it

Ann. N.Y. Acad. Sci. 1123: 213–223 (2008). © 2008 New York Academy of Sciences.
doi: 10.1196/annals.1420.024

cyclic nucleotide binding domain (CNBD), homologous to that of CNG channels, at the C-terminus.

Expression of the various HCN isoforms in heterologous expression systems (human embryonic kidney [HEK] cells, Chinese hamster ovary cells, and *Xenopus* oocytes) elicits currents qualitatively similar to the native pacemaker current but quantitatively different in terms of voltage of activation, time constants, and sensitivity to cAMP. Activation/deactivation kinetics, for example, are fastest for HCN1 and slowest for HCN4, while HCN2 and HCN3 have intermediate activation/deactivation rates. Furthermore, HCN2, HCN3, and HCN4 channels are sensitive to cAMP, while HCN1 channels are nearly insensitive (for reviews, see Refs. 3, 4, 19, 20).

Of the four HCN isoforms, HCN4 is the most highly expressed in the SAN.[21,22] Although the embryologic origin of the cardiac conduction system and in particular of the SAN is far from being fully resolved, some of the steps leading to the development of the cardiac pacemaker region are understood. For example, it has been shown recently that the SAN develops from cardiac progenitors of the second heart field following activation of a specific gene program.[23-25] Among the early genes activated specifically in the cardiac region committed to become the SAN, at embryonic day E7.5, is the HCN4 gene; indeed HCN4 delineates specifically the primary pacemaker region in embryonic as well as in adult tissue.[22,25]

I_f–Modulation of the Cardiac Rate

I_f–Mediated Autonomic Modulation

The properties of I_f make it a central player not only in the generation of cardiac pacemaking but also in the cellular mechanisms underlying autonomic modulation of heart rate.[5,8,11,26,27] Fine modulation of rate relies on fine control of the steepness of the slow diastolic depolarization operated by autonomic stimuli. By activation of β-adrenergic (β$_1$ and β$_2$) and muscarinic M2 receptors, respectively, the sympathetic and parasympathetic neurotransmitters control the cytosolic concentration of the second messenger cAMP.

A major contribution to the autonomic control of rate is provided by the cAMP dependence of f-channels. Binding of cAMP molecules to the C terminus of pacemaker channels increases the probability of f-channel opening via a positively directed shift of the voltage dependence of the activation curve.[9] A reduced intracellular cAMP concentration gives rise to the opposite action, i.e., a negative shift of the activation curve

and a reduction of open probability at any given voltage (FIG. 1A). Sympathetic/parasympathetic control of intracellular cAMP, therefore, induces, through cAMP-induced f-channel modulation, an increase/decrease of the net inward current during diastolic depolarization and a consequent increase/decrease of firing rate, respectively (FIG. 1B).

I_f Is Modulated Preferentially by β$_2$-Adrenergic Receptors

The adaptive function of sympathetic and parasympathetic innervation relies on the ability of the autonomic nervous system to respond effectively to rapidly changing physiological demands; accordingly, the signal transduction pathways leading to rate modulation need to be rapid.

At the cellular level, rapid and effective rate modulation is achieved through the contribution of two distinct processes. First, direct binding of cAMP to pacemaker channels guarantees a quicker and more specific modulation than that achieved by, for example, a cAMP-mediated phosphorylation process. Second, efficient modulation relies on the compartmentation of cAMP signaling.

cAMP is an intracellular second messenger activating several cellular processes, and its concentration is strictly controlled by cellular compartmentation of the biochemical factors responsible for cAMP synthesis and degradation, i.e., adenylyl cyclase and phosphodiesterases, in order to prevent unwanted spreading of cAMP signaling.[28-32] This requires that variations of cAMP occur in close proximity of the final target. We have shown[33,34] that pacemaker channels, and specifically HCN4 subunits, are confined, together with the β$_2$-adrenergic receptors (β$_2$-ARs), to membrane caveolae, cellular microdomains whose function is to keep in close proximity proteins involved in a specific signal transduction pathway.[35-39]

In cardiac cells, the physiological response to sympathetic stimulation is mediated by both β$_1$ and β$_2$ subtypes of β-ARs. In their expression patterns in cardiac cells, these two subtypes differ both in the density and localization. β$_1$-ARs are generally more abundant and widely distributed than β$_2$-ARs in the whole heart; in the SAN, however, β$_2$-ARs are expressed at a much higher level than in the rest of the heart.[40-44] Furthermore, whereas β$_2$-ARs are specifically localized in caveolae, β$_1$-ARs are for the most part excluded from these structures.[36,45,46] We have shown that in rabbit SAN myocytes, β$_2$-ARs co-localize with the HCN4 isoform of pacemaker channels in membrane caveolae.[34] As illustrated in FIGURE 1C and D, specific β$_2$ stimulation (obtained by combined β$_2$ stimulation with

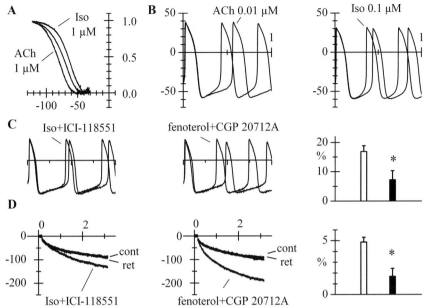

FIGURE 1. Funny channels mediate autonomic modulation of rate. **(A)** I_f activation curve is shifted negative/positive, which decreases/increases current, by muscarinic/β-adrenergic stimulation according to a mechanism involving cAMP-dependent channel modulation. **(B)** A decreased I_f reduces the steepness of diastolic depolarization and slows rate in the presence of ACh (*left*); the opposite occurs in the presence of Iso (*right*). **(C, D)** $β_1$-specific activation, obtained by perfusion with isoproterenol 1 μM + ICI-118551 10 μM, accelerates rate (7.2%; **C,** *filled bar*) and shifts the I_f activation curve (1.7 mV; **D,** *filled bar*) less than $β_2$-specific activation obtained by perfusion with fenoterol 10 μM + CGP20712A 0.3 μM (16.9%; **C,** *white bar,* versus 4.9 mV; **D,** *white bar,* respectively). (Data in **C** and **D** modified from Ref. 34 with permission).

fenoterol and $β_1$ inhibition by CGP 20712A) results in a larger shift of the I_f activation curve (4.9 mV) and stronger rate acceleration (16.9%) than does specific $β_1$ stimulation (obtained by combined $β_1$ stimulation by isoproterenol and $β_2$ inhibition by the specific $β_2$-inverse agonist (ICI-118551): 1.7 mV and 7.2%, respectively).[34] These data show the existence of a preferential $β_2$-mediated modulation of funny channels and control of cardiac chronotropism.

I_f Is Responsible for Rate Control during Vagal Tone

When I_f was first shown to be inhibited by ACh,[47] the general belief was that the mechanism responsible for rate slowing by vagal stimulation was the activation of an ACh-dependent K^+ current ($I_{K,ACh}$[48]). This view was challenged by the findings that inhibition of I_f occurs at 20-fold lower concentrations than activation of $I_{K,ACh}$, and that ACh concentrations able to inhibit I_f slow cardiac rate substantially.[11] The impact of this finding on the interpretation of the parasympathetic regulation of rate was twofold: first, these data showed for the first time that ACh-induced I_f inhibition, rather

than K^+-current activation, is the process underlying slowing of cardiac rate by low ACh doses; second, evidence that both I_f and rate are sensitive to fairly low ACh concentrations introduced the novel concept that rate control during vagal tone is mediated by I_f. Thus, under resting conditions, a mildly inhibited I_f keeps rate to the normal, low resting level of about 70 bpm, and rate acceleration associated with removal of vagal tone, by pharmacological means, for example, is likely caused by removal of resting I_f inhibition.

Functional Relevance of I_f. Genetic Approach

Relevance of I_f to pacemaking is a long debated issue (for reviews, see Refs. 4, 8, 19, 49, 50). Cloning of the HCN channels has represented important progress in the understanding of the functional properties of I_f and has provided the basis for further experimental evidence supporting the role of funny channels in generation of spontaneous activity and rate control. Expression of HCN channels in the heart is high in the conduction system, including the SAN, the atrioventricular node (AVN), and the Purkinje fibres.[2,3,8,51]

The possibility to specifically knock out a gene in mice has provided direct support for the functional role of HCN channels in heart rate control. HCN2-deficient mice display, along with various neuronal disturbances, alteration of sinus rhythm (sinus dysrhythmia), although rate modulation by autonomic neurotransmitters is normal.[52] Homozygous cardiac-specific HCN4-deficient mice (HCN4$^{-/-}$) die early during embryogenesis (between E9.5 and E11.5), and during their short embryonic life display a significantly reduced cardiac rate. Furthermore, HCN4$^{-/-}$ animals show a 75–95% reduction in I_f amplitude and are insensitive to isoproterenol-induced β-adrenergic stimulation.[53]

Taken together, these data indicate that HCN4 is necessary for the correct development of the cardiac pacemaker and is the major contributor to native SAN pacemaker current in mice; HCN4 is also the main mediator of β-adrenergic modulation. HCN2 is important for the correct propagation of electrical stimuli throughout the heart and contributes, even though less than HCN4, to I_f in mice but does not mediate β-adrenergic stimulation.

While transgenic animals have proved important in providing direct evidence for the involvement of HCN channels in the generation and modulation of cardiac rhythm, results highlighting the pacemaking role of HCN in humans have been slower to come. Early data investigating the existence of a link between mutations of the *hHCN4* gene and cardiac dysfunctions were either based on a single-patient report[54] or could not identify specific aspects of a complex arrhythmic behavior associated with a given mutation[55] and did not, therefore, provide compelling evidence for this link.

More recently, however, data showing that mutations of the *hHCN4* gene are indeed associated with specific rhythm disturbances have been obtained. In particular, an inherited mutation of a highly conserved residue in the CNBD of the HCN4 protein (S672R; Fig. 2A and B) was shown to be associated with inherited sinus bradycardia in a large Italian family spanning three generations.[56]

The 15 individuals heterozygous for the mutation had a significantly slower heart rate (mean rate: 52.2 bpm) than the 12 wild-type individuals (mean rate: 73.2 bpm). Heterologous expression in HEK293 cells of the mutated hHCN4 cDNA showed that the S672R mutation causes a change in the biophysical properties of HCN4 channels. Heteromeric wild-type/S672R channels had a more negative activation curve than wild-type channels (−4.9 mV shift) and faster deactivation kinetics, leading to reduced inward current during diastole and slower rhythm.[56] In its action, the mutation mimicked the effect of a low concentration of acetylcholine (about 20 nM) such as that released during mild vagal stimulation.[11] In FIGURE 2C, slowing induced by ACh 0.01 μM (−3.2 mV shift according to Ref. 11) is shown for comparison.

These findings confirmed the relevance of funny channels in the control of cardiac rate and provided novel evidence supporting the view that malfunctioning channels can cause disturbances of normal rhythm.

Pharmacological Approach

The involvement of funny channels in pacemaking makes them ideal targets for pharmacological interventions aimed to specifically modify heart rate. In the last two decades pharmaceutical companies have been actively searching for pure heart-rate-modulating agents, drugs able to change heart chronotropism without altering other parameters that might negatively influence cardiovascular performance (e.g., cardiac inotropism).

The concept of heart-rate reduction as a useful therapeutic intervention for some cardiovascular diseases is validated by several studies showing a tight association between elevated heart rate and mortality in subjects affected by coronary artery disease (CAD) as well as in the general population.[57] Many cardiovascular conditions such as chronic angina, ischemic heart disease, and heart failure are characterized by oxygen imbalance which would greatly benefit from a moderate reduction of heart rate. A lower basal heart rate indeed decreases cellular oxygen demand and improves myocardial perfusion by prolonging diastole.

Effective and widely used pharmacological interventions aimed to reduce heart rate are based on Ca^{2+}-antagonists and β-blockers. These compounds slow the heart efficiently, but have the undesired side effect of reducing cardiac inotropism due to the decreased contractile force developed by the myocardium when Ca^{2+} entry is reduced.

Starting in the 1980s, compounds able to reduce the steepness of the slow diastolic depolarization and slow cardiac rate with limited effects on cardiac inotropism were developed. These drugs were named originally "pure bradycardic agents" (PBAs) and subsequently "heart-rate reducing agents" after their properties (reviewed by Ref. 3), and were shown to exert their action by blocking pacemaker f-channels.[58,59]

f-channel blockers with heart-rate reducing action include alinidine (the N-allyl-derivative of clonidine),

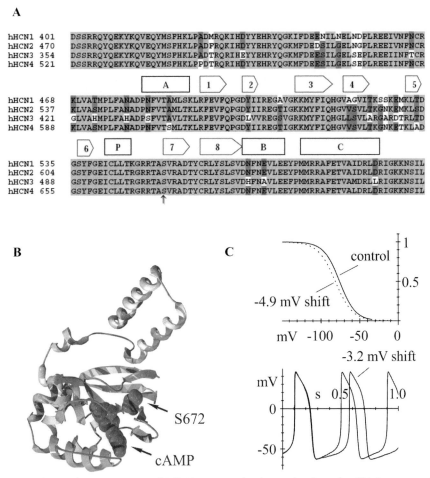

A

```
hHCN1 401  DSSRRQYQEKYKQVEQYMSFHKLPADMRQKIHDYYEHRYQGKIFDEENILNELNDPLREEIVNFNCR
hHCN2 470  DSSRRQYQEKYKQVEQYMSFHKLPADFRQKIHDYYEHRYQGKMFDEDSILGELNGPLREEIVNFNCR
hHCN3 354  DSSRRQYQEKYKQVEQYMSFHKLPADTRQRIHEYYEHRYQGKMFDEESILGELSEPLREEIINFTCR
hHCN4 521  DSSRRQYQEKYKQVEQYMSFHKLPDTRQRIHDYYEHRYQGKMFDEESILGELSEPLREEIINPNCR
```

```
hHCN1 468  KLVATMPLFANADPNFVTAMLSKLRPEVFQPGDYIIREGAVGKKMYFIQHGVAGVITKSSKEMKLTD
hHCN2 537  KLVASMPLFANADPNFVTAMLTKLKPEVFQPGDYIIREGTIGKKMYFIQHGVVSVLTKGNKEMKLSD
hHCN3 421  GLVAHMPLFAHADPSFVTAVLTKLRPEVFQPGDLVVREGSVGRKMYFIQHGLLSVLARGARDTRLTD
hHCN4 588  KLVASMPLFANADPNFVTSMLTKLRPEVFQPGDYIIREGTIGKKMYFIQHGVVSVLTKGNKETKLAD
```

```
hHCN1 535  GSYFGEICLLTKGRRTASVRADTYCRLYSLSVDNFNEVLEEYPMMRRAFETVAIDRLDRIGKKNSIL
hHCN2 604  GSYFGEICLLTRGRRTASVRADTYCRLYSLSVDNFNEVLEEYPMMRRAFETVAIDRLDRIGKKNSIL
hHCN3 488  GSYFGEICLLTRGRRTASVRADTYCRLYSLSVDHFNAVLEEPPMMRRAFETVAMDRLLRIGKKNSIL
hHCN4 655  GSYFGEICLLTRGRRTASVRADTYCRLYSLSVDNFNEVLEEYPMMRRAFETVALDRLDRIGKKNSIL
```

B   **C**

S672

cAMP

control

0.5

-4.9 mV shift

mV -100 -50 0

-3.2 mV shift

mV

s 0.5 1.0

0

-50

FIGURE 2. Single-point mutation of HCN4 associated with sinus bradycardia. **(A)** Sequences of the CNBDs of the four human HCN isoforms (1–4). Flat and arrowed rectangles represent α helices and β sheets, respectively. In the heterozygous mutation found in bradycardic individuals from a large Italian family,[56] a serine is replaced by an arginine (S672R in hHCN4, *arrow*). **(B)** Model 3D ribbon plot of the CNBD of hHCN4 obtained by DeepView-Swiss-PdbViewer homology modeling, based on the published mHCN2 CNBD crystal structure.[89] The mutated residue S672 is located close to the cAMP binding pocket: S672 and a bound cAMP molecule (*solid surface plots*) are indicated by *arrows*. **(C, top)** Relative to wild-type hHCN4 channels expressed in HEK293 cells (*full line*), heterozygous mutant + wild-type channels have a mean activation curve displaced to the left by 4.9 mV (*broken line*). (Data in **C, top,** modified from Ref. 56 with permission). **(C, bottom)** Slowing induced by ACh 0.01 µM, which causes a mean shift of −3.2 mV,[11] shown for comparison.

falipamil (AQ-A39), zatebradine (UL-FS49) and cilobradine (DK-AH26) (compounds originally derived from the Ca^{2+} channel blocker verapamil), and ZD7288. Unfortunately many of these molecules were shown to be potentially proarrhythmic, largely because at the concentration used to inhibit I$_f$ they also affected other cardiac ion channels.[3,60,61]

Ivabradine is a more recently developed f-channel blocker and, to date is the only bradycardic agent

validated and marketed for the treatment of chronic stable angina pectoris. Ivabradine was shown to slow the rate of isolated SAN cells by decreasing specifically the slope of the slow diastolic phase[62]; this effect could be attributed to a specific inhibition of I$_f$, since the T-type and L-type Ca^{2+} current and the delayed K$^+$ currents of SAN cells were unaffected at the same drug concentration (3 µM) blocking 60% of I$_f$.[63]

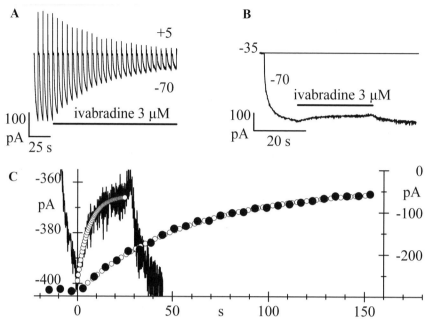

FIGURE 3. Block of If by ivabradine. **(A, B)** Pulsed protocol consisting of hyperpolarizing steps (4.5 s/−70 mV) followed by depolarizing steps (0.5 s/+5 mV), applied every 6 s from a holding potential of −35 mV **(A)** and long hyperpolarizing step to −70 mV **(B)**; application of ivabradine 3 μM results in a much larger block during application of the pulsed protocol than during the long step. **(C)** Time course of block development for the current at −70 mV from the experiments in **A** and **B;** fitting by single exponential decay (*open circles*) yielded steady-state blocks of 85.9% and 8.4% and time constants of 57.87 and 6.08 s for the pulsed protocol in **A** (*filled-circle curve*) and the long step in **B** (*noisy trace*), respectively.

The mechanisms of ivabradine block of native rabbit SAN f-channels and of individual HCN isoforms have been investigated in some detail.[64,65] In native f-channels, the drug acts by accessing the channel vestibule from the intracellular side and blocking the pore in a use-dependent and "current"-dependent way. Ivabradine is able to access its binding site only when channels are in the open state (state-dependence) and its block is favored by depolarization. As shown in FIGURE 3, ivabradine 3 μM applied during a pulsed protocol (activation at −70 mV, deactivation at +5 mV) induced a much larger block (FIG. 3A, about 86%) than that induced by a prolonged channel opening at −70 mV (FIG. 3B, about 8%). This behavior implies that block is strongly favored by depolarization. This voltage-dependent behavior can be shown to depend not so much on voltage itself, but rather by a "current"-dependence of block. In other words, when ions flow through the channel into the cell they tend to displace (kick-off) drug molecules from their binding site, while the opposite occurs in the short time when the current is outward during channel deactivation (kick-in).[64] Since channels open on hyperpolarization,

there appears to be a contradiction between the need to hyperpolarize (in order to open channels) and to depolarize (in order for significant block to occur); on the contrary, however, these features are the basis for a strong "use dependence" of ivabradine action. This is an useful property: during tachycardia, for example, f-channels will undergo high-rate cycling between open and closed states, and the rate-reducing effect of ivabradine will be stronger. It is, however, interesting to notice that high efficiency of block during high-rate cycling between open and closed states is due to a larger degree of block at steady state, but is not associated with a faster onset of block. Surprisingly, block is weaker, but much faster during long hyperpolarizing steps than during pulsed cycling. In FIGURE 3C, for example, the time constant of block onset during hyperpolarization to −70 mV was 6.8 s, about 10-fold faster than that of 57.9 s during the pulsed protocol.

The action of ivabradine has also been investigated in individual HCN isoforms, with the aim of gaining information on the molecular mechanism of drug-channel interactions. A comparative analysis of cilobradine, zatebradine, and ivabradine blocking effect has shown that all these compounds exert a use-dependent block on all HCN isoforms, with half-block values comparable to those found for native f-channels.[60,63,65]

Since HCN4 and HCN1 both contribute to native funny channels in rabbit SAN myocytes, the former being by far the most highly expressed isoform,[21,22,66,67] a detailed study has been performed of ivabradine block of these two HCN isoforms.[65] This study has highlighted important differences in the two block mechanisms. For example, while ivabradine behaves as an "open channel" blocker of HCN4 channels, as on native channels, it behaves as a "closed channel" blocker of HCN1 channels.[65]

Differences between blocking mechanisms of individual isoforms are especially relevant to the issue of specificity of drug action. A small fraction of patients under chronic use of ivabradine, for example, report visual disturbances (phosphenes), which are normally mild and can be reversed upon withdrawal from treatment.[57] These symptoms are attributable to the action of ivabradine on HCN channels expressed in the retina.[68] Since HCN1, but not HCN4, appears to be highly expressed in photoreceptors,[68] knowledge of the molecular basis for block of individual HCN isoforms can be exploited for designing new drugs with improved function and reduced side effects.

It is now recognized that heart rate reduction is an important factor in preventing ventricular remodeling and thus ejection impairment in conditions of chronic heart failure.[69] Specific inhibition of I_f can, therefore, be a valuable tool also for the treatment of patients with stable CAD and chronic heart failure. Two presently running large clinical trials, BEAUTIFUL and SHIFT, aim to evaluate the possible use of ivabradine in reduction of morbidity and mortality in CAD and heart failure patients.[70,71]

Biological Pacemakers

In the past decade, several laboratories have used both gene-based and cellular-based approaches to manipulate cellular processes involved in cardiac pacemaking. The intent of these manipulations was to generate, in normally quiescent substrates, stable rhythmic activity similar to that of native pacemaker myocytes. Gene-based strategies used so far to induce pacemaker activity followed different strategies including: 1) the upregulation of the β-adrenergic pathway[72,73]; 2) the downregulation of the resting potential stabilizing conductance I_{K1}[74]; and 3) the overexpression of pacemaker channels.[75–79]

In relation to the approach involving pacemaker channels, early attempts showed that overexpression of the HCN2 pacemaker channel induced a large increase in rate of spontaneously beating neonatal ventricular myocytes, mostly by steepening the diastolic depolarization.[75] Additional studies later confirmed that *in vivo* injection of adenovirus carrying the HCN2 gene into the left atrium or into the ventricular conduction system of dogs could induce a persistent spontaneous activity which, after sinus rhythm suppression by vagal stimulation, originated from the site of injection. Electrophysiological analysis demonstrated a significant increase of I_f density in cells isolated from the injected regions compared to control animals.[76,77,79]

Following these studies, other groups have proved that *in vivo* overexpression of HCN isoforms other than HCN2 were able to generate spontaneous activity in different substrates (HCN1[80]; HCN4[81]). These data indicate that the expression of pacemaker channels is sufficient to generate repetitive spontaneous activity in normally quiescent cells, in accordance with the established role of these channels in cardiac pacemaking.

Due to limitations in the duration of protein expression driven by adenoviruses and in the safety of adenoviral constructs, researchers have also explored the possibility of generating a biological pacemaker using a cell-based approach in which undifferentiated mesenchymal stem cells were engineered to express high levels of HCN2 channels.[78] This cellular substrate was able to influence, both *in vitro* and *in vivo*, the beating rate of an electrically coupled excitable substrate. It is interesting to note that mesenchymal stem cells overexpressing HCN2 are not pacing cells (in fact, they are not even excitable cells). The mechanism of pacing is based on the hypothesis that in the presence of electrical coupling, the membrane voltage of mesenchymal cells can be made to hyperpolarize to voltages where HCN2 channels are activated, which in turn will induce in the surrounding tissue a depolarization strong enough to drive pacemaker activity.[78]

By showing that overexpression of HCN channels is able to induce pacing, these strategies represent a successful "proof of concept." However, they also have drawbacks, such as 1) the limited duration of HCN expression, especially with the direct adenovirus-mediated infection; 2) the effect of HCN overexpression on ion concentration: in the long term, pacing driven by overexpression of HCN channels could result in cycle-length instability due to internal K^+ deprivation according to model computations[82]; and 3) lack of stability of undifferentiated mesenchymal stem cells, whose *in vivo* behavior is not predictable; indeed these cells have been shown to differentiate *in vitro*

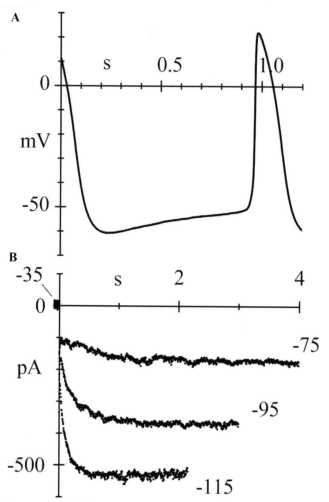

FIGURE 4. Activity recorded from ES cell-derived pacemaker myocytes. **(A)** Spontaneous action potentials. **(B)** I_f traces recorded during steps from a holding potential of -20 mV to the voltages indicated. Recordings were made at 36°C in normal Tyrode solution (see Ref. 34).

and *in vivo* toward several different phenotypes.[83] Differentiated cells could in principle undergo a remodeling process resulting in potentially arrhythmogenic alteration in the expression of HCN and other ion channels.

An alternative approach potentially able to prevent the drawbacks described above would be to use stem cell–derived or engineered cellular substrates with characteristics as close as possible to native pacemaker myocytes. So far, of the different types of stem cells investigated, only embryonic stem cells (ESCs) have been convincingly shown to differentiate into myocytes with a pacemaker phenotype.

Early results showed that murine ESCs can differentiate into cardiac myocytes by following a specific procedure. First, cells are expanded into aggregates called embryoid bodies (EBs), which include a portion with

contractile cells; then, mechanical and enzymatic dissociation of the EB contractile portions leads to isolation of cells with action potentials typical of pacemaker cells and expressing the pacemaker I_f current.[84–86] In FIGURE 4 action potentials and I_f traces recorded from ESC-derived pacemaker myocytes with the above procedure (9 days after EBs plating on gelatin-coated dishes) are shown.

A cellular approach to biological pacemaking has been attempted using human embryonic stem cells (hESC) differentiated into spontaneously contracting EBs.[87,88] The contractile portion of an EB was shown to pace *in vitro* neonatal rat cardiac myocytes and *in vivo* the whole heart of either swine[87] or guinea pig.[88] Although these results were obtained with heterogeneous substrates and the molecular mechanism underlying pacemaking was not directly investigated, it is likely that I_f played there an important role in the generation of rhythmic activity. Evidence for the contribution of I_f derives, for example, from the observation that perfusion of the f-channel blocker ZD7288 was able to slow rate,[88] and that β-adrenergic stimulation by isoproterenol was effective in increasing beating rate.[87,88]

Summary

The role of funny channel activation in the generation of rhythmic activity and control of heart rate has been consolidated through a substantial collection of data since the I_f current was originally identified and described nearly 30 years ago by Brown *et al.*,[5] and is now well established. Recent new applications of the concept of f-channel-based pacemaking have led to important developments of clinical relevance. These include: 1) a pharmacological application to heart-rate reduction therapies for the treatment of angina, achieved by specific f-channel inhibitors like ivabradine; 2) a genetic application to studies of inherited arrhythmias associated with mutations of human HCN4 subunits, such as the bradycardia-inducing S672R mutation; and 3) application to interventional medicine related to development of cell-based biological pacemakers potentially able to eventually replace electronic devices.

The clinical impact of f-channel-mediated pacing is likely to grow in the future in connection with the development of other, more detailed and more refined applicative tools.

Acknowledgments

This study was partially supported by European Union (Normacor) and CARIPLO (2004.1451/10.4878) grants to D.D.

Conflict of Interest

The authors declare no conflicts of interest.

References

1. DiFrancesco, D. 1991. The contribution of the 'pacemaker' current (I_f) to generation of spontaneous activity in rabbit sino-atrial node myocytes. J. Physiol. **434:** 23–40.
2. Accili, E.A. *et al.* 2002. From funny current to HCN channels: 20 years of excitation. NIPS **17:** 32–37.
3. Baruscotti, M., A. Bucchi & D. DiFrancesco. 2005. Physiology and pharmacology of the cardiac pacemaker ("funny") current. Pharmacol. Ther. **107:** 59–79.
4. Barbuti, A., M. Baruscotti & D. DiFrancesco. 2007. The pacemaker current: from basics to the clinics. J. Cardiovasc. Electrophysiol. **18:** 342–347.
5. Brown, H.F., D. DiFrancesco & S.J. Noble. 1979. How does adrenaline accelerate the heart? Nature **280:** 235–236.
6. Brown, H. & D. DiFrancesco. 1980. Voltage-clamp investigations of membrane currents underlying pace-maker activity in rabbit sino-atrial node. J. Physiol. **308:** 331–351.
7. DiFrancesco, D. *et al.* 1986. Properties of the hyperpolarizing-activated current (I_f) in cells isolated from the rabbit sino-atrial node. J. Physiol. **377:** 61–88.
8. DiFrancesco, D. 1993. Pacemaker mechanisms in cardiac tissue. Annu. Rev. Physiol. **55:** 455–472.
9. DiFrancesco, D. & P. Tortora. 1991. Direct activation of cardiac pacemaker channels by intracellular cyclic AMP. Nature **351:** 145–147.
10. DiFrancesco, D. & M. Mangoni. 1994. Modulation of single hyperpolarization-activated channels (I_f) by cAMP in the rabbit sino-atrial node. J. Physiol. **474:** 473–482.
11. DiFrancesco, D., P. Ducouret & R.B. Robinson. 1989. Muscarinic modulation of cardiac rate at low acetylcholine concentrations. Science **243:** 669–671.
12. Santoro, B. *et al.* 1997. Interactive cloning with the SH3 domain of N-src identifies a new brain specific ion channel protein, with homology to eag and cyclic nucleotide-gated channels. Proc. Natl. Acad. Sci. USA **94:** 14815–14820.
13. Gaub, R., R. Seifert & B.U. Kaupp. 1998. Molecular identification of a hyperpolarization-activated channel in sea urchin sperm. Nature **393:** 583–587.
14. Ludwig, A. *et al.* 1998. A family of hyperpolarization-activated mammalian cation channels. Nature **393:** 587–591.
15. Ludwig, A. *et al.* 1999. Two pacemaker channels from human heart with profoundly different activation kinetics. EMBO J. **18:** 2323–2329.
16. Ishii, T.M. *et al.* 1999. Molecular characterization of the hyperpolarization-activated cation channel in rabbit heart sinoatrial node. J. Biol. Chem. **274:** 12835–12839.
17. Seifert, R. *et al.* 1999. Molecular characterization of a slowly gating human hyperpolarization-activated channel predominantly expressed in thalamus, heart, and testis. Proc. Natl. Acad. Sci. USA **96:** 9391–9396.
18. Vaccarti, T. *et al.* 1999. The human gene coding for HCN2, a pacemaker channel of the heart. Biochim. Biophys. Acta **1446:** 419–425.
19. Robinson, R.B. & S.A. Siegelbaum. 2003. Hyperpolarization-activated cation currents: from molecules to physiological function. Annu. Rev. Physiol. **65:** 453–480.
20. Biel, M. *et al.* 1999. Hyperpolarization-activated cation channels: a multi-gene family. Rev. Physiol. Biochem. Pharmacol. **136:** 165–181.
21. Shi, W. *et al.* 1999. Distribution and prevalence of hyperpolarization-activated cation channel (HCN) mRNA expression in cardiac tissues. Circ. Res. **85:** e1–e6.
22. Liu, J. *et al.* 2007. Organisation of the mouse sinoatrial node: structure and expression of HCN channels. Cardiovasc. Res. **73:** 729–738.

23. CHRISTOFFELS, V.M. *et al*. 2006. Formation of the venous pole of the heart from an Nkx2-5-negative precursor population requires Tbx18. Circ. Res. **98:** 1555–1563.

24. MORETTI, A. *et al*. 2006. Multipotent embryonic isl1+ progenitor cells lead to cardiac, smooth muscle, and endothelial cell diversification. Cell **127:** 1151–1165.

25. MOMMERSTEEG, M.T. *et al*. 2007. Molecular pathway for the localized formation of the sinoatrial node. Circ. Res. **100:** 354–362.

26. DIFRANCESCO, D. & C. TROMBA.1988. Inhibition of the hyperpolarization-activated current (I_f) induced by acetylcholine in rabbit sino-atrial node myocytes. J. Physiol. **405:** 477–491.

27. DIFRANCESCO, D. & C. TROMBA. 1988. Muscarinic control of the hyperpolarization-activated current (I_f) in rabbit sino-atrial node myocytes. J. Physiol. **405:** 493–510.

28. JUREVICIUS, J. & R. FISCHMEISTER. 1996. cAMP compartmentation is responsible for a local activation of cardiac Ca^{2+} channels by beta-adrenergic agonists. Proc. Natl. Acad. Sci. USA **93:** 295–299.

29. RICH, T.C. *et al*. 2000. Cyclic nucleotide-gated channels colocalize with adenylyl cyclase in regions of restricted cAMP diffusion. J. Gen. Physiol. **116:** 147–161.

30. RICH, T.C. *et al*. 2001. In vivo assessment of local phosphodiesterase activity using tailored cyclic nucleotide-gated channels as cAMP sensors. J. Gen. Physiol. **118:** 63–78.

31. ZACCOLO, M. & T. POZZAN. 2002. Discrete microdomains with high concentration of cAMP in stimulated rat neonatal cardiac myocytes. Science **295:** 1711–1715.

32. JUREVICIUS, J., V.A. SKEBERDIS & R. FISCHMEISTER. 2003. Role of cyclic nucleotide phosphodiesterase isoforms in cAMP compartmentation following β_2-adrenergic stimulation of ICa,L in frog ventricular myocytes. J. Physiol. **551:** 239–252.

33. BARBUTI, A. *et al*. 2004. Localization of pacemaker channels in lipid rafts regulates channel kinetics. Circ. Res. **94:** 1325–1331.

34. BARBUTI, A. *et al*. 2007. Localization of f-channels to caveolae mediates specific β_2-adrenergic receptor modulation of rate in sinoatrial myocytes. J. Mol. Cell Cardiol. **42:** 71–78.

35. ANDERSON, R.G. 1998. The caveolae membrane system. Annu. Rev. Biochem. **67:** 199–225.

36. RYBIN, V.O. *et al*. 2000. Differential targeting of beta-adrenergic receptor subtypes and adenylyl cyclase to cardiomyocyte caveolae. A mechanism to functionally regulate the cAMP signaling pathway. J. Biol. Chem. **275:** 41447–41457.

37. STEINBERG, S.F. & L.L. BRUNTON. 2001. Compartmentation of G protein-coupled signaling pathways in cardiac myocytes. Annu. Rev. Pharmacol. Toxicol. **4:** 7–7.

38. PARTON, R.G. 2003. Caveolae–from ultrastructure to molecular mechanisms. Nat. Rev. Mol. Cell Biol. **4:** 162–167.

39. DEURS, B. *et al*. 2003. Caveolae: anchored, multifunctional platforms in the lipid ocean. Trends Cell Biol. **13:** 92–100.

40. BRODDE, O.E., F.J. LEIFERT & H.J. KREHL. 1982. Coexistence of beta 1- and beta 2-adrenoceptors in the rabbit heart: quantitative analysis of the regional distribution by (-)-3H-dihydroalprenolol binding. J. Cardiovasc. Pharmacol. **4:** 34–43.

41. DELMONTE, F. *et al*. 1993. Coexistence of functioning β_1- and β_2-adrenoceptors in single myocytes from human ventricle. Circulation **88:** 854–863.

42. LEVY, F.O. *et al*. 1993. Efficacy of β_1-adrenergic receptors is lower than that of β_2-adrenergic receptors. Proc. Natl. Acad. Sci. USA **90:** 10798–10802.

43. RODEFELD, M.D. *et al*. 1996. β-adrenergic and muscarinic cholinergic receptor densities in the human sinoatrial node: identification of a high β_2-adrenergic receptor density. J. Cardiovasc. Electrophysiol. **7:** 1039–1049.

44. BRODDE, O.E. *et al*. 2001. Presence, distribution and physiological function of adrenergic and muscarinic receptor subtypes in the human heart. Basic Res. Cardiol. **96:** 528–538.

45. LI, S. *et al*. 1995, Evidence for a regulated interaction between heterotrimeric G proteins and caveolin. J. Biol. Chem. **270:** 15693–15701.

46. STEINBERG, S.F. 2004. β_2-Adrenergic receptor signaling complexes in cardiomyocyte caveolae/lipid rafts. J. Mol. Cell Cardiol. **37:** 407–415.

47. DIFRANCESCO, D. & C. TROMBA. 1987. Acetylcholine inhibits activation of the cardiac hyperpolarizing- activated current, I_f. Pflügers Arch. **410:** 139–142.

48. SAKMANN, B., A. NOMA & W. TRAUTWEIN. 1983. Acetylcholine activation of single muscarinic K+ channels in isolated pacemaker cells of the mammalian heart. Nature **303:** 250–253.

49. DIFRANCESCO, D. 2006. Funny channels in the control of cardiac rhythm and mode of action of selective blockers. Pharmacol. Res. **53:** 399–406.

50. VINOGRADOVA, T.M. *et al*. 2005. Rhythmic Ca^{2+} oscillations drive sinoatrial nodal cell pacemaker function to make the heart tick. Ann. N.Y. Acad. Sci. **1047:** 138–156.

51. BIEL, M., A. SCHNEIDER & C. WAHL. 2002. Cardiac HCN channels: structure, function, and modulation. Trends Cardiovasc. Med. **12:** 206–212.

52. LUDWIG, A. *et al*. 2003. Absence epilepsy and sinus dysrhythmia in mice lacking the pacemaker channel HCN2. EMBO J. **22:** 216–224.

53. STIEBER, J. *et al*. 2003. The hyperpolarization-activated channel HCN4 is required for the generation of pacemaker action potentials in the embryonic heart. Proc. Natl. Acad. Sci. USA **100:** 15235–15240.

54. SCHULZE-BAHR, E. *et al*. 2003. Pacemaker channel dysfunction in a patient with sinus node disease. J. Clin. Invest. **111:** 1537–1545.

55. UEDA, K. *et al*. 2004. Functional characterization of a trafficking-defective HCN4 mutation, D553N, associated with cardiac arrhythmia. J. Biol. Chem. **279:** 27194–27198.

56. MILANESI, R. *et al*. 2006. Familial sinus bradycardia associated with a mutation in the cardiac pacemaker channel. N. Engl. J. Med. **354:** 151–157.

57. DIFRANCESCO, D. & J.A. CAMM. 2004. Heart rate lowering by specific and selective I_f current inhibition with ivabradine: a new therapeutic perspective in cardiovascular disease. Drugs **64:** 1757–1765.

58. VAN BOGAERT, P.P., M. GOETHALS & C. SIMOENS. 1990. Use- and frequency-dependent blockade by UL-FS 49 of

the I_f pacemaker current in sheep cardiac Purkinje fibres. Eur. J. Pharmacol. **187:** 241–256.

59. DIFRANCESCO, D. 1994. Some properties of the UL-FS 49 block of the hyperpolarization-activated current (I_f) in sino-atrial node myocytes. Pflügers Arch. **427:** 64–70.

60. STIEBER, J. *et al.* 2006. Bradycardic and proarrhythmic properties of sinus node inhibitors. Mol. Pharmacol. **69:** 1328–1337.

61. BUCCHI, A. *et al.* 2007. Heart rate reduction via selective 'funny' channel blockers. Curr. Opin. Pharmacol. **7:** 208–213.

62. THOLLON, C. *et al.* 1994. Electrophysiological effects of S 16257, a novel sino-atrial node modulator, on rabbit and guinea-pig cardiac preparations: comparison with UL-FS 49. Br. J. Pharmacol. **112:** 37–42.

63. BOIS, P. *et al.* 1996. Mode of action of bradycardic agent, S 16257, on ionic currents of rabbit sinoatrial node cells. Br. J. Pharmacol. **118:** 1051–1057.

64. BUCCHI, A., M. BARUSCOTTI & D. DIFRANCESCO. 2002. Current-dependent Block of Rabbit Sino-Atrial Node I_f Channels by Ivabradine. J. Gen. Physiol. **120:** 1–13.

65. BUCCHI, A. *et al.* 2006. Properties of ivabradine-induced block of HCN1 and HCN4 pacemaker channels. J. Physiol. **572:** 335–346.

66. MOOSANG, S. *et al.* 2001. Cellular expression and functional characterization of four hyperpolarization-activated pacemaker channels in cardiac and neuronal tissues. Eur. J. Biochem. **268:** 1646–1652.

67. ALTOMARE, C. *et al.* 2003. Heteromeric HCN1-HCN4 channels: a comparison with native pacemaker channels from the rabbit sinoatrial node. J. Physiol. **549:** 347–359.

68. DEMONTIS, G.C. *et al.* 2002. Functional characterisation and subcellular localisation of HCN1 channels in rabbit retinal rod photoreceptors. J. Physiol. **542:** 89–97.

69. NAGATSU, M. *et al.* 2000. Bradycardia and the role of β-blockade in the amelioration of left ventricular dysfunction. Circulation **101:** 653–659.

70. FOX, K. *et al.* 2006. Rationale and design of a randomized, double-blind, placebo-controlled trial of ivabradine in patients with stable coronary artery disease and left ventricular systolic dysfunction: the morbidity-mortality evaluation of the I_f inhibitor ivabradine in patients with coronary disease and left ventricular dysfunction study. Am. Heart J. **152:** 860–866.

71. TARDIF, J.C. & C. BERRY. 2006. From coronary artery disease to heart failure: potential benefits of ivabradine. Eur. Heart J. **8:** D24–D29.

72. EDELBERG, J.M., W.C. AIRD & R.D. ROSENBERG. 1998. Enhancement of murine cardiac chronotropy by the molecular transfer of the human beta2 adrenergic receptor cDNA. J. Clin. Invest. **101:** 337–343.

73. EDELBERG, J.M. *et al.* 2001. Molecular enhancement of porcine cardiac chronotropy. Heart **86:** 559–562.

74. MIAKE, J., E. MARBAN & H.B. NUSS. 2002. Biological pacemaker created by gene transfer. Nature **419:** 132–133.

75. QU, J. *et al.* 2001. HCN2 overexpression in newborn and adult ventricular myocytes: distinct effects on gating and excitability. Circ. Res. **89:** E8–E14.

76. QU, J. *et al.* 2003. Expression and function of a biological pacemaker in canine heart. Circulation **107:** 1106–1109.

77. PLOTNIKOV, A.N. *et al.* 2004. Biological pacemaker implanted in canine left bundle branch provides ventricular escape rhythms that have physiologically acceptable rates. Circulation **109:** 506–512.

78. POTAPOVA, I. *et al.* 2004. Human mesenchymal stem cells as a gene delivery system to create cardiac pacemakers. Circ. Res. **94:** 952–959.

79. BUCCHI, A. *et al.* 2006. Wild-type and mutant HCN channels in a tandem biological-electronic cardiac pacemaker. Circulation **114:** 992–999.

80. TSE, H.F. *et al.* 2006. Bioartificial sinus node constructed via in vivo gene transfer of an engineered pacemaker HCN Channel reduces the dependence on electronic pacemaker in a sick-sinus syndrome model. Circulation **114:** 1000–1011.

81. CAI, J. *et al.* 2007. Adenoviral gene transfer of HCN4 creates a genetic pacemaker in pigs with complete atrioventricular block. Life Sci. **80:** 1746–1753.

82. VISWANATHAN, P.C. *et al.* 2006. Recreating an artificial biological pacemaker: insights from a theoretical model. Heart Rhythm. **3:** 824–831.

83. LERI, A., J. KAJSTURA & P. ANVERSA. 2005. Cardiac stem cells and mechanisms of myocardial regeneration. Physiol. Rev. **85:** 1373–1416.

84. MALTSEV, V.A. *et al.* 1994. Cardiomyocytes differentiated in vitro from embryonic stem cells developmentally express cardiac-specific genes and ionic currents. Circ. Res. **75:** 233–244.

85. HESCHELER, J. *et al.* 1997. Embryonic stem cells: a model to study structural and functional properties in cardiomyogenesis. Cardiovasc. Res. **36:** 149–162.

86. ABI-GERGES, N. *et al.* 2000. Functional expression and regulation of the hyperpolarization activated non-selective cation current in embryonic stem cell-derived cardiomyocytes. J. Physiol. **523**(Pt 2): 377–389.

87. KEHAT, I. *et al.* 2004. Electromechanical integration of cardiomyocytes derived from human embryonic stem cells. Nat. Biotechnol. **22:** 1282–1289.

88. XUE, T. *et al.* 2005. Functional integration of electrically active cardiac derivatives from genetically engineered human embryonic stem cells with quiescent recipient ventricular cardiomyocytes: insights into the development of cell-based pacemakers. Circulation **111:** 11–20.

89. ZAGOTTA, W.N. *et al.* 2003. Structural basis for modulation and agonist specificity of HCN pacemaker channels. Nature **425:** 200–205.

Experimental Molecular and Stem Cell Therapies in Cardiac Electrophysiology

LIOR GEPSTEIN

The Sohnis Laboratory for Cardiac Electrophysiology and Regenerative Medicine, The Bruce Rappaport Faculty of Medicine, Technion–Israel Institute of Technology, Haifa, Israel

One of the most exciting fields in cardiovascular research today involves the possible use of stem cells, cell and gene therapies, and tissue engineering for the treatment of a variety of cardiovascular disorders. Here, we review on the possible applications of these emerging strategies in the field of cardiac electrophysiology. Initially, the elegant cell and gene therapy approaches proposed for the treatment of bradyarrhythmias are described. These gene therapy approaches are mainly focused on the generation of biological pacemakers either by altering the neurohumoral control of existing pacemaking cells (by overexpressing the β-adrenergic receptor) or by converting quiescent cardiomyocytes into pacemaking cells by shifting the balance between diastolic repolarization and depolarization currents. An alternative approach explores the possibility of grafting pacemaking cells, which were either derived directly during the differentiation of human embryonic stem cells or engineered from mesenchymal stem cells, into the myocardium as a cell therapy strategy for biological pacemaking. We then describe the possible applications of similar strategies for the treatment of common tachyarrhythmias by overexpression of different ion channels, or their modifiers, either directly in host cardiomyocytes or *ex vivo* in cells that will be eventually transplanted into the heart. Next, we discuss the electrophysiological implications of cardiac stem cell therapy for heart failure. Finally, we address the obstacles, challenges, and avenues for further research required to make these novel strategies a clinical reality.

Key words: stem cells; cell therapy; gene therapy; human embryonic stem cells; heart failure; ion channels; arrhythmias; electrophysiology; pacemaker

Introduction

The advances in molecular and stem cell biology have provided researchers and clinicians with exciting tools that may allow the development of novel therapeutic paradigms for a variety of cardiovascular disorders. Consequentially, gene and cell therapies have already been described as experimental strategies for the treatment of ischemic heart disease and heart failure, with some of these strategies already reaching the stage of clinical studies.[1–5] In this review we will focus on the potential applications of these emerging technologies in the field of cardiac electrophysiology. Specifically, we will discuss the possible applications of gene and cell therapy strategies in establishing new anti-arrhythmic paradigms for both bradyarrhythmias

and tachyarrhythmias. In addition, we discuss the electrophysiological implications associated with cardiac stem cell therapy for heart failure in general.

Possible Applications for Treatment of Bradyarrhythmias—Biological Pacemakers

Abnormalities in the generation of the electrical impulse (sinus node dysfunction) or its propagation from the atria to the ventricles (atrioventricular [AV] block) may result in the development of abnormally slow heart rate usually requiring the implantation of an electronic pacemaker. While serving as the treatment of choice for bradyarrhythmias for more than five decades,[6] electronic pacemakers are not without limitations. In recent years, a number of innovative gene and cell therapy approaches have emerged as experimental platforms for the creation of biological pacemakers.[7,8] These strategies were aimed at generating a biological functional unit that will recapitulate some, if not all, of the properties of the native pacemaker of the heart, the sinoatrial node.

Address for correspondence: Prof Lior Gepstein, MD, PhD, The Sohnis Laboratory for Cardiac Electrophysiology and Regenerative Medicine, The Bruce Rappaport Faculty of Medicine, Technion–Israel Institute of Technology, P.O.B. 9649, Haifa 31096, Israel.
Mdlior@tx.technion.ac.il

Ann. N.Y. Acad. Sci. 1123: 224–231 (2008). © 2008 New York Academy of Sciences.
doi: 10.1196/annals.1420.025

Gene Therapy Approaches

The experimental gene therapy approaches for biological pacemaking have focused on altering the function of existing cardiomyocytes; either by modulating (enhancing) the function of existing pacemaking cells by overexpression of the β-adrenergic receptor[9] or by converting quiescent cardiomyocytes into pacemaking cells.[9–12] To achieve the latter, several strategies were suggested that focused on shifting the balance between diastolic repolarization and depolarization currents. To reduce the diastolic repolarization currents, Miake *et al.* utilized a dominant negative approach, to suppress the I_{k1} current (the major diastolic current) in guinea pig ventricular cells.[10,13] Inhibition of this current by adenoviral delivery of a transgene containing a mutated, not active, form of the pore of the Kir2.1 channel (Kir2.1AAA), resulted in the emergence of a new ventricular pacemaking activity.[10]

The alternative strategy of increasing the diastolic depolarization currents consisted of overexpression of the hyperpolarization-activated, cyclic nucleotide-gated (HCN-2) encoded, pacemaker current (I_f) in the existing cardiomyocytes.[11,12] Adenoviral delivery of this transgene into the canine atrial or ventricular tissues resulted in the emergence of an ectopic escape rhythm. More recently, gene transfer of a synthetic pacemaker channel (as an alternative to the HCN-2 channels) was used as a new strategy for biological pacemaking. Kashiwakura *et al.*[14] used site-directed mutagenesis (R447N, L448A, and R453I in S4 and G528S in the pore) to convert a voltage-dependent, potassium-selective channel, Kv1.4, which is not expressed by adult cardiomyocytes, into a hyperpolarization-activated, nonselective channel. Adenoviral gene transfer of the transgene into the ventricular myocytes resulted in the generation of new pacemaker activity both *in vitro* and *in vivo*.

Cell Therapy Approaches

A possible alternative to the gene therapy approaches discussed above, aiming to alter the function of the existing heart cells, may be the use of cell therapy strategies to reconstruct the conduction system. In the case of biological pacemakers, two different strategies have been suggested. The first involves a combined cell and gene therapy approach, in which cells can initially be transfected *ex vivo* to express ionic channels in order to generate an engineered cell with pacemaker-like properties. The second approach proposes to drive the differentiation of stem cells into the desired cardiomyocyte lineage (i.e., to generate cardiomyocyte with stable pacemaking properties).

In both of these strategies, the generated pacemaker-like cells are then transplanted into the heart *in vivo*. To serve as an effective biological pacemaker, these cell grafts have to fulfill the following conditions: (1) they have to survive for a prolonged period of time, (2) they have to continue to express the transgene (in the first strategy) or retain their unique functional pacemaking properties (in the second strategy), (3) they have to functionally couple with host cardiomyocyte, and the degree of coupling should be optimal to allow them to drive the electrical activity of the entire heart, (4) the grafted cells' spontaneous activity should be suppressed by the existence of a faster pacemaker activity (i.e., overdrive suppression), and (5) they should provide a physiological heart rate, which may change based on the metabolic needs of the individual, ideally by reacting to changes in the neurohumoral status.

To demonstrate the feasibility of the first approach, Potapova *et al.*[15] transfected mesenchymal stem cells (MSc) with a plasmid containing the HCN-2 transgene encoding the I_f pacemaker current. One advantage of MSc is their ability to express significant levels of connexin 40 and 43 so that they could form functional gap junctions with host cardiomyocytes. This ability was initially demonstrated *in vitro*[16] in co-culturing experiments by showing dye transfer between the MSc and neighboring cardiomyocytes through gap junctions and by dual-cell patch-clamp recordings showing electrotonic interactions in MSc-cardiomyocyte pairs. Immunostaining studies demonstrated the generation of gap junctions between the MSc and host cardiomyocytes both *in vitro* and *in vivo*.[15,16] The HCN–MSc were then transplanted into the left ventricular epicardium. Following induction of transient AV block[15] the engineered MSc induced a new idioventricular rhythm that could be localized to the site of cell transplantation. A more recent study from the same group demonstrated the long-term ability of the HCN-2 expressing human MSc to pace the canine complete AV block heart model without a need for immunosuppression.[17] The new pacemaker activity required 10–12 days to fully stabilize and continued for 42 days. Interestingly, the new ectopic rhythm was sensitive to catecholamine stimulation and the rate of the new pacemaker activity increased following infusion of epinephrine in some of these animals.

In the alternative cell therapy approach, cardiomyocytes with inherent pacemaker activity are initially generated during *in vitro* differentiation of stem cells. These stem cell–derived cells are then proposed to act as primary pacemakers following *in vivo* transplantation. To demonstrate the feasibility of this concept we recently assessed the ability of human embryonic stem cell (hESC)-derived cardiomyocytes to serve as a biological pacemaker in the swine complete AV block model.[18]

The hESC are pluripotent stem cell lines, derived from human blastocytes, which can be propagated in the undifferentiated state for prolonged periods while retaining the capability to differentiate into cell derivatives of all three germ layers.[19] Using the embryoid body (EB) differentiating system, we and others were recently able to generate a reproducible cardiomyocyte differentiating system from these unique cells.[20–22] Dispersed cells isolated from the spontaneously beating areas within the EBs displayed ultrastructural, gene expression, and functional properties that were consistent with early-stage cardiac phenotype.[20–22] More recently, we have demonstrated that this system is not limited to the generation of isolated cardiomyocytes, but rather a functional cardiomyocyte syncytium is generated with spontaneous pacemaking activity and action potential propagation.[23] Further studies in our laboratory demonstrated a reproducible temporal pattern of proliferation, cell-cycle withdrawal, and ultrastructural maturation of the hESC cardiomyocytes.[24] Electrophysiological studies revealed the presence of typical action potentials and ionic transients.[22,25,26] These studies have also provided mechanistic insights for the basis for spontaneous automaticity and excitability in these cells, namely the presence of a large Na and I_f current in the face of a low density I_{k1} current.[25]

To assess the ability of the hESC-derived cardiomyocytes to serve as a biological pacemaker, we established a swine model of slow heart rate by ablating the AV node.[18] This resulted in complete dissociation between the atrial and ventricular electrical activities, mimicking the clinical scenario of patients suffering from complete AV block. Spontaneously contracting EBs were then injected into the posterolateral left ventricular wall of these animals. A few days following cell grafting, episodes of a new ectopic ventricular rhythm could be detected in 11 out of the 13 animal studied (in six of them it was characterized by sustained and long-term activity). Electrophysiological mapping pinpointed the source of this new ventricular activity to the site of cell grafting, and histological examination confirmed the presence of the hESC-derived cardiomyocytes as well as the formation of gap junctions with host myocytes. More recently, similar results were obtained by others,[27] showing the ability of enhanced green fluorescing protein (eGFP)-expressing hESC-derived cardiomyocytes to pace the isolated guinea pig heart.

As an alternative to the concept of biological pacemaking via electrotonic interaction between grafted and host cells, Cho *et al.*[28] examined the ability to deliver the transgene encoding the pacemaking current into host cardiomyocytes by inducing fusion between the engineered cell and the host cardiomyocytes. In their study, chemically induced fusion between transfected fibroblasts expressing the HCN-1 pacemaker channel and guinea pig ventricular cardiomyocytes both *in vitro* and *in vivo* was successfully performed. The fused myocyte–HCN-1-fibroblast cells exhibited spontaneous automaticity. The firing frequency of these cells increased with β-adrenergic stimulation. Moreover, the heterokaryons generated a source of ectopic pacing at the site of cell delivery.

Possible Applications for Treating Tachyarrhythmias

Gene Therapy Approaches

The same methodologies that were used to increase the local excitability of the myocardium, by creating a biological pacemaker, could theoretically also be utilized to modify the myocardial electrophysiological substrate in an attempt to treat different tachyarrhythmias. The feasibility of the gene therapy approach was evaluated initially by demonstrating that overexpression of different potassium channels can be used to alter the action potential properties in cardiomyocytes *in vitro*[29–31] and even to suppress cardiac hyperexcitability in rabbit ventricular myocytes.[30,32,33] Gene delivery of different potassium channels was suggested as a method to accelerate cardiac repolarization and abbreviate the QT interval for the treatment of the long QT (LQT) syndrome[34] and was also proposed as a way to reverse the downregulation of potassium channels in cardiomyocytes isolated from failing hearts.[31,35]

When targeting a global myocardial disorder, such as the LQT syndrome, gene delivery should be relatively uniform throughout the myocardium. Although electrotonic interactions within the myocardial tissue may theoretically compensate for some inhomogeneities in global myocardial gene delivery, a regional or partial expression of the gene within the myocardium would not only be inefficient in preventing the arrhythmias, but may actually be pro-arrhythmic by inducing significant inhomogeneous electrophysiological properties in the treated area. To this end, Kikuchi *et al.*[36] recently introduced an effective method for uniformly infecting the atrial wall using direct application of adenoviral vectors to the epicardial surface in combination with a poloxamer gel and mild trypsinization to increase virus penetration and incubation time.

The obstacles discussed above may limit the use of gene therapy for applications requiring global transgene cardiac delivery in the near future. Perhaps more appealing targets in the short term may be arrhythmias in which localized manipulation of the electrophysiological substrate may be sufficient to allow effective treatment. Such a localized, *in vivo* application was recently proposed by Donahue *et al.*[37] Using localized viral gene transfer (through the AV nodal artery), these investigators demonstrated that AV nodal conduction properties could be modified by overexpression of Gαi2, and that this approach represents a novel strategy for ventricular rate control for atrial fibrillation, mimicking the effects of β-adrenergic antagonists.

More recently, the same concepts were used to target ventricular tachycardia by manipulating the electrophysiological properties of the myocardial infarct border zone. In this elegant study, Sasano *et al.*[38] used targeted gene delivery to the pig chronic myocardial infarct border zone using the dominant negative version of the KCNH2 potassium channel. This resulted in prolongation of the action potential duration, effective refractory period in the infarct border zone, and elimination of ventricular tachycardia in this model.

The aforementioned pioneering studies established the feasibility of gene delivery to modify the excitable properties of the myocardial tissue but also raised several limitations to the utility of this approach. Clinical application of this approach may be significantly hampered, therefore, by the possible drawbacks of using viral vectors such as the possible inflammatory response, the relatively transient effect achieved by conventional adenoviral gene delivery systems, and the unpredictable level of transgene expression within the tissue.

Using Genetically Modified Cell Grafts

An alternative approach that can overcome some of the aforementioned shortcomings of gene therapy may be the use of genetically modified cell grafts that can be initially transfected *ex vivo* with excellent long-term efficiency and then transplanted to the *in vivo* heart. The general hypothesis underlying this strategy is that the engineered cells would couple with host cardiac tissue and influence the local myocardial electrophysiological properties by way of electrotonic interactions. According to this hypothesis, changes in the local myocardial electrophysiological substrate would depend on both the passive properties (cell capacitance and degree of cell coupling) and active properties (the type of transfected ionic channel) of the grafted and host cells. Thus, based on the type of current expressed, the grafted cell may have either suppressive or excitable effects.

One of the assumptions in using this approach is that the grafted cells will generate functional connections (through gap junctions) with host cells and could therefore modulate their electrophysiological properties through electrotonic interactions. Recent studies have demonstrated that different cell types, such as MSc[8] and fibroblasts,[39–41] are capable of forming gap junctions with host cardiomyocytes and that specific electrotonic interactions can be generated between host and grafted cells.

Based on the above concepts, we have recently examined the feasibility of using genetically engineered fibroblasts, transfected to express the voltage-gated potassium channel Kv1.3, to modify the electrophysiological properties of cardiomyocyte cultures.[42] In this study, a high-resolution multi-electrode array mapping technique was used to assess the electrophysiological and structural properties of primary neonatal rat ventricular cultures. The transfected fibroblasts were demonstrated to significantly alter the electrophysiological properties of the cardiomyocyte cultures. These changes were manifested by a significant reduction in the local extracellular signal amplitude and by the appearance of multiple local conduction blocks. The location of all conduction blocks correlated with the spatial distribution of the transfected fibroblasts and all of the electrophysiological changes were reversed following the application of a specific Kv1.3 blocker, charybdotoxin. More recently, we extended these observations also to the *in vivo* scenario (unpublished data) by demonstrating the ability of the Kv1.3-fibroblasts to extend local refractoriness following transplantation into the rat ventricle.

Myocardial Cell Therapy for Infarct Repair—Electrophysiological Implications

Myocardial cell replacement strategies are emerging as novel experimental therapeutic paradigms for the treatment of post–myocardial infarction heart failure.[3,4] The rationale underlying these strategies is that myocardial function may be improved by repopulating the diseased areas with a new pool of functional cells. Based on this assumption, a variety of cell types have been suggested as potential sources for cell transplantation. Some of the autologous cell sources (skeletal myoblasts and bone marrow–derived progenitor cells) have already reached the clinical stage in ongoing phase I and II clinical trials.

While most efforts in the field have been geared toward the induction of the differentiation of various stem cell types into the cardiac or vascular lineages and to the assessment of the effects of cell transplantation on histology and global myocardial mechanical function, the electrophysiological aspects of these strategies have received only limited attention thus far. In general, the electrophysiological considerations associated with myocardial cell replacement therapy can be grouped into three areas: electrophysiological integration, arrhythmogenic potential of cell therapy, and prevention of arrhythmias, each of which is discussed below.

Electrophysiological Integration

Numerous reports suggest that myocyte transplantation can improve cardiac performance in animal models of myocardial infarction.[3,4] However, it is not entirely clear whether this functional improvement is due to direct contribution to contractility by the transplanted myocytes or by other indirect mechanisms, such as attenuation of the remodeling process, amplification of an endogenous repair process by cardiac resident progenitor cells, or induction of angiogenesis. True systolic augmentation, however, would require structural, electrophysiological, and mechanical coupling of donor and host tissue so that the transplanted cell graft would participate actively in the synchronous contraction of the ventricle. For example, although transplantation of skeletal myoblasts was demonstrated to improve myocardial performance in animal models, gap junctions and functional coupling were not observed between grafted and host tissues. Yet even the presence of such gap junctions between host and donor cardiomyocyte tissues, as observed in some studies, does not guarantee functional integration. For such integration to occur, currents generated in one cell passing through gap junctions must be sufficient to depolarize neighboring cells, resulting in appropriate action potential propagation.

The ability of the transplanted cells to electrically integrate with host tissue has been evaluated in only a handful of studies. We have recently assessed the ability of cardiomyocytes (derived from hESC) to functionally integrate *in vitro* with primary cultures of neonatal rat ventricular myocytes.[18] The contracting areas within the EBs were mechanically dissected and added to rat cardiomyocyte cultures. Interestingly, within 24 hrs post-grafting we could detect synchronous contraction in the co-cultures that persisted for a number of weeks. We next demonstrated, using a microelectrode array mapping technique, that this synchronous contraction is the result of action-potential propagation between

the human and rat tissue. Immunostaining studies demonstrated that this coupling results from the generation of gap junctions—the protein structures responsible for intercellular electrical coupling—between the rat and human cardiomyocytes. In contrast to the above, Abraham *et al.*[43] used optical mapping of co-cultures of rat neonatal cardiomyocytes and skeletal myoblasts and demonstrated the lack of functional integration between the two cell types. This finding is not surprising given the inability of skeletal myoblasts to form gap junctions.

In vivo electrical integration of grafted cells was also evaluated in a small number of studies. As described above we have recently demonstrated the ability of hESC-derived cardiomyocytes to electrically integrate with host cardiac tissue by showing their ability to serve as a biological pacemaker and drive the electrical activity of the entire heart in the pig model of complete AV block.[18] To assess the ability of grafted cells to integrate with host tissue at the single-cell level *in vivo*, Rubart and colleagues[44] created a unique method. This technique uses two-photon microscopy to perform high-resolution imaging of calcium transients at a single-cell resolution in the isolated Langendorff perfused heart. By using transplanted cells that are tagged with a genetic fluorescent marker (eGFP) the investigators can document simultaneously calcium transients in both the transplanted eGFP-expressing cell and neighboring host cardiomyocytes. Using this novel methodology, Rubart *et al.* were able to document electrical integration (by way of intracellular calcium imaging) between transplanted fetal cardiomyocytes and host cells in the intact heart.[44] A similar study by the same group noted the absence of such integration when skeletal myoblasts were used for transplantation.[45]

The key question of electrical integration was recently addressed in an elegant study by Halbach and colleagues.[46] In this study, eGFP-expressing mouse fetal cardiomyocytes were transplanted into the cryoinfarct area and into adjacent healthy myocardium in mice. Using a novel *ex vivo* heart slice preparation,[47] the investigators were able to perform direct microelectrode recordings from the grafted cells. By pacing the preparation simultaneous with the intracellular recordings, the investigators were able to show, unambiguously, cell integration, directly at a single-cell resolution in the intact heart. This integration was highly efficacious, as several of the grafted cells were able to follow host myocardium in a 1:1 conduction airing rapid pacing. The results of the study showed that 82% of transplanted fetal cardiac myocytes surrounded by viable host tissue were electrically integrated with host myocardium. In contrast, transplanted cardiomyocytes surrounded by cryoinjured tissue, while showing spontaneous electrical and contractile activity, were not integrated with host tissue.

Arrhythmogenic Potential of Cell Therapy

Another important obstacle to the successful development of cell therapy strategies for myocardial repair is the possibility of provoking side effects, such as the induction of malignant ventricular arrhythmias. In the initial clinical skeletal myoblast trials, for example, a disturbingly high incidence of ventricular arrhythmias was noted.[48] One possible reason for this arrhythmogenesis may be the lack of formation of gap junctions in the myotubes that may completely uncouple these cells from surrounding ventricular myocytes. To assess this question, Abraham et al.[43] performed optical mapping of co-cultures of skeletal myoblasts and neonatal rat ventricular myocytes. Their results demonstrated that an increase in the percentage of the myoblasts resulted in the formation of anatomical obstacles, increased tissue inhomogeities, slowed conduction, and increased the likelihood of the formation of reentrant arrhythmias. By engineering the skeletal myoblasts to overexpress connexin 43 (the major gap junction protein) they could prevent the initiation of these arrhythmias.

While early-stage cardiomyocytes can form gap junctions and functionally integrate with host tissue as discussed above, they too in theory may have the potential for arrhythmogenesis through a number of mechanisms.[49] Cell transplantation may theoretically increase the likelihood of reentrant arrhythmias by generating slow conducting viable channels within the border zone of the infarct (the anatomical substrate responsible for post-myocardial infarction ventricular tachycardia in patients) or by serving as ectopic foci either due to abnormal automaticity or triggered activity.

Prevention of Arrhythmias

Finally, although cell grafting could theoretically increase the potential for arrhythmias, the opposite may also occur. A number of myocardial pathologies, such as ischemic heart disease, heart failure, and local myocardial autonomic denervation, have been implicated in the pathogenesis of the different cardiac arrhythmias. Somatic gene and cell therapies have been suggested as novel treatment modalities for the aforementioned disorders by targeting the abnormal cardiac structure (promoting angiogenesis or sympathetic reinnervation or preventing restenosis) or function (increasing contractility). Successful application of the above-mentioned strategies for the treatment of these

myocardial pathologies may also result in a secondary favorable effect on the myocardial electrophysiological substrate. This in turn may lead to a decreased tendency to the development of the typical arrhythmias commonly observed during the clinical course of these disease states. For example, if cardiomyocyte transplantation will result in efficient regeneration of the infarct, existing slow conduction pathways within the scar border zone may be eliminated, reducing the risk for the emergence of ventricular tachycardias. In this respect it is interesting to note that in a recent study[50] we were able to improve myocardial performance in the rat chronic infarct model by transplantation of hESC-derived cardiomyocytes. Interestingly, the hESC-derived cardiomyocytes showed superior engraftment properties at the border zone and less so in the center of the infarct.

Summary

Cardiac arrhythmias represent a major cause of worldwide morbidity and mortality. Although marked progress has been achieved in several of the pharmacological and nonpharmacological therapeutic modalities for these rhythm disorders in recent years, there is still a need for the development of new therapeutic paradigms that are more targeted and associated with fewer side effects. Improvement in the understanding of the mechanisms underlying many of these arrhythmias, and the development of molecular and cellular tools suggest a future role for gene and cell therapies for the treatment of these different rhythm disorders. Similarly, as discussed above, understanding the electrophysiological properties of stem cells that are used for myocardial cell replacement therapies as well as their interactions with host tissue at the single-cell, tissue, and whole-organ levels may have important implications to the emerging field of cardiovascular regenerative medicine. Nevertheless, bridging the gap between the proof of concept and the clinical application will require important methodological developments as well as extensive animal experimentation

Acknowledgments

This study was supported in part by the Israel Science Foundation (#520/01) and by the Stephan and Nancy Grand Foundation.

Conflict of Interest

The author declares no conflicts of interest.

References

1. ISNER, J.M. 2002. Myocardial gene therapy. Nature **415:** 234–239.
2. HAJJAR, R.J. *et al*. 2000. Prospects for gene therapy for heart failure. Circ. Res. **86:** 616–621.
3. LAFLAMME, M.A. & C.E. MURRY. 2005. Regenerating the heart. Nat. Biotechnol. **23:** 845–856.
4. DIMMELER S., A.M. ZEIHER & M.D. SCHNEIDER. 2005. Unchain my heart: the scientific foundations of cardiac repair. J. Clin. Invest. **115:** 572–583.
5. REINLIB, L. & L. FIELD. 2000. Cell transplantation as future therapy for cardiovascular disease? A workshop of the National Heart, Lung, and Blood Institute. Circulation **101:** E182–187.
6. KUSUMOTO, F.M. & N. GOLDSCHLAGER. 1996. Cardiac pacing. N. Engl. J. Med. **334:** 89–97.
7. GEPSTEIN, L. 2005. Stem cells as biological heart pacemakers. Expert Opin. Biol. Ther. **5:** 1531–1537.
8. ROSEN, M.R. *et al*. 2004. Genes, stem cells and biological pacemakers. Cardiovasc. Res. **64:** 12–23.
9. EDELBERG, J.M., W.C. AIRD & R.D. ROSENBERG. 1998. Enhancement of murine cardiac chronotropy by the molecular transfer of the human beta2 adrenergic receptor cDNA. J. Clin. Invest. **101:** 337–343.
10. MIAKE, J., E. MARBAN & H.B. NUSS. 2002. Biological pacemaker created by gene transfer. Nature **419:** 132–133.
11. PLOTNIKOV, A.N. *et al*. 2004. Biological pacemaker implanted in canine left bundle branch provides ventricular escape rhythms that have physiologically acceptable rates. Circulation **109:** 506–512.
12. QU, J., A.N. PLOTNIKOV, *et al*. 2003. Expression and function of a biological pacemaker in canine heart. Circulation **107:** 1106–1109.
13. MIAKE, J., E. MARBAN & H.B. NUSS. 2003. Functional role of inward rectifier current in heart probed by Kir2.1 overexpression and dominant-negative suppression. J. Clin. Invest. **111:** 1529–1536.
14. KASHIWAKURA, Y. *et al*. 2006. Gene transfer of a synthetic pacemaker channel into the heart: a novel strategy for biological pacing. Circulation **114:** 1682–1686.
15. POTAPOVA, I. *et al*. 2004. Human mesenchymal stem cells as a gene delivery system to create cardiac pacemakers. Circ. Res. **94:** 952–959.
16. VALIUNAS, V. *et al*. 2004. Human mesenchymal stem cells make cardiac connexins and form functional gap junctions. J. Physiol. **555:** 617–626.
17. PLOTNIKOV, A.N. *et al*. 2007. Xenografted adult human mesenchymal stem cells provide a platform for sustained biological pacemaker function in canine heart. Circulation **116:** 706–713.
18. KEHAT, I. *et al*. 2004. Electromechanical integration of cardiomyocytes derived from human embryonic stem cells. Nat. Biotechnol. **22:** 1282–1289.
19. THOMSON, J.A. *et al*. 1998. Embryonic stem cell lines derived from human blastocysts. Science **282:** 1145–1147.
20. KEHAT, I. *et al*. 2001. Human embryonic stem cells can differentiate into myocytes with structural and functional properties of cardiomyocytes. J. Clin. Invest. **108:** 407–414.
21. MUMMERY, C. *et al*. 2003. Differentiation of human embryonic stem cells to cardiomyocytes: role of coculture with visceral endoderm-like cells. Circulation **107:** 2733–2740.
22. XU, C. *et al*. 2002. Characterization and enrichment of cardiomyocytes derived from human embryonic stem cells. Circ. Res. **91:** 501–508.
23. KEHAT, I. *et al*. 2002. High-resolution electrophysiological assessment of human embryonic stem cell-derived cardiomyocytes: a novel in vitro model for the study of conduction. Circ. Res. **91:** 659–661.
24. SNIR, M. *et al*. 2003. Assessment of the ultrastructural and proliferative properties of human embryonic stem cell-derived cardiomyocytes. Am. J. Physiol. Heart Circ. Physiol. **285:** H2355–2363.
25. SATIN, J. *et al*. 2004. Mechanism of spontaneous excitability in human embryonic stem cell derived cardiomyocytes. J. Physiol. **559:** 479–496.
26. HE, J.Q. *et al*. 2003. Human embryonic stem cells develop into multiple types of cardiac myocytes: action potential characterization. Circ. Res. **93:**32–39.
27. XUE, T. *et al*. 2005. Functional integration of electrically active cardiac derivatives from genetically engineered human embryonic stem cells with quiescent recipient ventricular cardiomyocytes: insights into the development of cell-based pacemakers. Circulation **111:** 11–20.
28. CHO, H.C., Y. KASHIWAKURA & E. MARBAN. 2007. Creation of a biological pacemaker by cell fusion. Circ. Res. **100:** 1112–1115.
29. HOPPE, U.C. *et al*. 1999. Manipulation of cellular excitability by cell fusion: effects of rapid introduction of transient outward K+ current on the guinea pig action potential. Circ. Res. **84:** 964–972.
30. JOHNS, D.C. *et al*. 1995. Adenovirus-mediated expression of a voltage-gated potassium channel in vitro (rat cardiac myocytes) and in vivo (rat liver). A novel strategy for modifying excitability. J. Clin. Invest. **96:** 1152–1158.
31. NUSS, H.B. *et al*. 1996. Reversal of potassium channel deficiency in cells from failing hearts by adenoviral gene transfer: a prototype for gene therapy for disorders of cardiac excitability and contractility. Gene Ther. **3:** 900–912.
32. NUSS, H.B., E. MARBAN & D.C. JOHNS. 1999 Overexpression of a human potassium channel suppresses cardiac hyperexcitability in rabbit ventricular myocytes. J. Clin. Invest. **103:** 889–896.
33. JOHNS, D.C., E. MARBAN & H.B. NUSS. 1999. Virus-mediated modification of cellular excitability. Ann. N.Y. Acad. Sci. **868:** 418–422.
34. MAZHARI, R. *et al*. 2002. Ectopic expression of KCNE3 accelerates cardiac repolarization and abbreviates the QT interval. J. Clin. Invest. **109:** 1083–1090.
35. ENNIS, I.L. *et al*. 2002. Dual gene therapy with SERCA1 and Kir2.1 abbreviates excitation without suppressing contractility. J. Clin. Invest. **109:** 393–400.
36. KIKUCHI, K. *et al*. 2005. Targeted modification of atrial electrophysiology by homogeneous transmural atrial gene transfer. Circulation **111:** 264–270.
37. DONAHUE, J.K. *et al*. 2000. Focal modification of electrical conduction in the heart by viral gene transfer. Nat. Med. **6:** 1395–1398.

38. SASANO, T. *et al*. 2006. Molecular ablation of ventricular tachycardia after myocardial infarction. Nat. Med. **12:** 1256–1258.

39. GAUDESIUS, G. *et al*. 2003. Coupling of cardiac electrical activity over extended distances by fibroblasts of cardiac origin. Circ. Res. **39:** 421–428.

40. ROOK, M.B. *et al*. 1992. Differences in gap junction channels between cardiac myocytes, fibroblasts, and heterologous pairs. Am. J. Physiol. **263:** C959–977.

41. FAST, V.G. *et al*. 1996. Anisotropic activation spread in heart cell monolayers assessed by high- resolution optical mapping. Role of tissue discontinuities. Circ. Res. **79:** 115–127.

42. FELD, Y. *et al*. 2002. Electrophysiological modulation of cardiomyocytic tissue by transfected fibroblasts expressing potassium channels: a novel strategy to manipulate excitability. Circulation **105:** 522–529.

43. ABRAHAM, M.R. *et al*. 2005. Antiarrhythmic engineering of skeletal myoblasts for cardiac transplantation. Circ. Res. **97:** 159–167.

44. RUBART, M. *et al*. 2003. Physiological coupling of donor and host cardiomyocytes after cellular transplantation. Circ. Res. **92:** 1217–1224.

45. RUBART, M. *et al*. 2004. Spontaneous and evoked intracellular calcium transients in donor-derived myocytes following intracardiac myoblast transplantation. J. Clin. Invest. **114:** 775–783.

46. HALBACH, M. *et al*. 2007. Electrophysiological maturation and integration of murine fetal cardiomyocytes after transplantation. Circ. Res. **101:** 484–492.

47. HALBACH, M. *et al*. 2006. Ventricular slices of adult mouse hearts—a new multicellular in vitro model for electrophysiological studies. Cell Physiol. Biochem. **18:** 1–8.

48. SMITS, P.C. *et al*. 2003. Catheter-based intramyocardial injection of autologous skeletal myoblasts as a primary treatment of ischemic heart failure: clinical experience with six-month follow-up. J. Am. Coll. Cardiol. **42:** 2063–2069.

49. ZHANG, Y.M. *et al*. 2002. Stem cell-derived cardiomyocytes demonstrate arrhythmic potential. Circulation **106:** 1294–1299.

50. CASPI, O. *et al*. 2007. Transplantation of human embryonic stem cell-derived cardiomyocytes improves myocardial performance in infarcted rat hearts. J. Am. Coll. Cardiol. **50:** 1884–1893.

Controlling Ischemic Cardiovascular Disease

From Basic Mechanisms to Clinical Management

RAFAEL BEYAR

Rambam Medical Center and Faculties of Medicine and Biomedical Engineering, Technion, Israel Institute of Technology, Haifa, Israel

Progress in cardiovascular disease understanding and management continues at an exponential pace. Our understanding of the molecular basis of disease is enhanced by newer molecular measurement techniques, sophisticated models of physiological protein functions, understanding of the genetic foundation for diseases, and the incorporation of population genetic tools in our clinical analysis. In this review, I discuss prevention and therapy of coronary stenosis impeding coronary flows, prevention of acute and chronic manifestation of coronary flow impairment, and interfering with myocardial manifestation of acute or chronic deprivation of coronary flow. Mechanical heart failure and arrhythmias are common causes of myocardial dysfunction that originate, in part, from the loss of myocardial tissue and function. Techniques for interfering with cardiac function, in order to address the molecular mechanisms associated with restenosis, range from pharmacologic to mechanical procedures including mechanical dilation and scaffolding of coronary stenosis. The use of stents with and without drug coating is leading the clinical world of revascularization side-by-side with cardiac bypass surgery. Other topics that are discussed here include managing myocardial damage and acute and chronic pump failure. Finally, population genetics of cardiac health and the potential for genetic therapeutic guidance in managing ischemic cardiovascular diseases are discussed.

Key words: coronary stenosis; ischemic heart disease; myocardial damage; heart failure

Introduction

Ischemic Heart Disease

Ischemic heart disease (IHD) is primarily a result of the atherosclerotic process and is well known as the leading cause of death in the western world.[1] A major decline in cardiovascular mortality of close to 50% has been observed over the last two decades.[2,3] The reasons for this decline in the United States have been analyzed by Ford *et al.*[3] and is equally attributed to the reduction in risk factors (i.e., high cholesterol, high blood pressure, smoking, and obesity) and therapeutic interventions. Among the therapeutics, secondary preventive therapies after myocardial infarction or revascularization, initial treatments for acute coronary syndrome or myocardial infarction, treatments for heart failure, and revascularization for chronic angina, accounted for 11%, 10%, 9%, and 5% of the reduction in mortality, respectively.

The development of therapeutics for coronary artery disease has focused on revascularization of ischemic regions. In general, surgical methods for revascularization were the first to be developed, but percutaneous methods now dominate this field[4] following the first balloon angioplasty performed by Andreas Grüntzig in 1978.[5] Many techniques have since been developed to open occluded arteries and maintain their patency.

There are three major challenges within the field of percutaneous revascularization. The first is acute success in enlarging the stenotic lumen of the vessel or creating a new channel in total occlusion cases. The second is preventing thrombotic occlusion of the treated segment using local or systemic methods, and the third challenge is prevention of restenosis resulting from an exaggerated proliferative response of the vessel wall to injury.

Mechanical Solutions of Coronary Stenosis

The mechanical solution for enlarging a stenosis (or creating a new lumen for the vessel in case the artery is totally occluded) involves the initial creation of the channel and maximizing the lumen created during the procedure.

Address for correspondence: Professor Rafael Beyar, MD, DSc, Director, Rambam Health Care Campus 8 Ha'Aliyah Street, Haifa 35254, Israel. Voice: +972 4854 2389; fax: +972 4854 2907.

r_beyar@rambam.health.gov.il

Ann. N.Y. Acad. Sci. 1123: 232–236 (2008). © 2008 New York Academy of Sciences.
doi: 10.1196/annals.1420.026

Channel Creation

Channel creation is typically obtained by crossing the lesion with a designated wire and dilating the artery with a balloon that has been sized to the vessel dimensions. The more challenging lesions are total occlusions where there is no practical lumen so that the lumen has to be recreated by crossing it with a wire followed by balloon dilatation.[6] There are many challenges to crossing a total occlusion with a wire, the main one being the semiblind dissection of the plaque with a specialized wire optimizing stiffness, flexibility, and friction, thereby entering the lumen distal to the occlusion.[7] In spite of the methods that aid the cardiologist with imaging and visualization of this procedure, total occlusion remains a major challenge in cardiology today. Techniques that attempt to guide crossing of the occlusions with the wire, vary from optical methods to detect apposition against the vessel wall,[7] using radiofrequency catheters for crossing the occlusion[8] and stiff and inflexible wires that cross the lesion by forceful blunt dissection. Skill of the operator is of primary importance in determining success rates and the frequencies of complication while opening total occlusions.

Maximizing the Lumen

Maximizing the lumen achieved during percutaneous intervention (PCI) can involve balloon dilatation, methods for plaque removal, and methods for scaffolding the arterial wall to prevent internal collapse. Balloon dilatation is the primary method of lumen creation; it creates a larger lumen by shifting the plaque and stretching the vessel wall.[9] Its limitations include the recoil of the vessel upon deflation of the balloon and tears created during the dilatation; the larger the balloon, the higher the chances of arterial rupture. Novel balloons, involving microblades creating controlled dissection, have been developed in attempts to have more predictable results.[10] Balloons are the workhorse of interventional cardiology and are an integral part of almost all of the percutaneous revascularization procedures.

Plaque removal has been a major area of research in IHD since the early days of angioplasty. The methods for plaque removal involve directional atherectomy, cutting and removing plaques,[11] high speed rotational atherectomy,[12] drilling through the lesion, laser evaporation of the plaque,[13,14] and other novel methods of drilling and plaque removal. The various atherectomy and drilling devices are, in general, effective in achieving adequate acute results during the procedure. However, because of procedure complexity, the limitations include the need for increased operator skill and

exposure to higher complication rates, such as perforation and rupture of the vessel wall, distal remobilization, and, most importantly, the long-term proliferative tissue response to injury that causes restenosis. While atherectomy is still a valid approach, it is presently used in a limited way for niche indications. Laser atherectomy has become limited as an appropriate method for coronary revascularization because of the associated high restenosis rates.

Prevention of arterial collapse is typically done with stents.[15] Stent insertion has become the major technique used to obtain a large and predictable lumen during PCI. Most stents are balloon expandable metal stents that are dilated to a predefined diameter, equal or slightly larger than the normal reference vessel wall. Various types of metal have been tried to optimize the biocompatibility of stents. Restenosis and thrombosis have typically been the two major limitations of this procedure. Restenosis occurs from an excessive proliferative response of the vessel wall to the injury cased by stent expansion. It has been shown that restenosis occurs less frequently for stents with thinner struts.[15] Optimal antiplatelet treatment using antiplatelet drugs has reduced the rate of stent thrombosis to less than 1%. Typically, the use of metal stents results in 15–30% restenosis in the vessels,[15] and this is the major limitation of stents that led to the development and introduction of drug-eluting stents (DES).

Molecular Mechanisms Associated with Restenosis and Its Solution

Restenosis is the process of endoluminal tissue proliferation within the artery leading to recurrence of vessel stenosis. A proliferative response following vessel injury, whether generated by balloon angioplasty alone or by stenting, is a natural response of the body and occurs in all patients. It is a multifactorial process dependent on the amount of injury to the arterial wall, various tissue factors that are released from the injured tissue, platelets adhering to the injured surface, and the secretion of various factors that enhance the proliferative response.[16] The proliferative response is typically associated with inflammation[17] resulting from injury or from a response to a foreign body (either the stent metal or polymers that are used for drug delivery). Typically, the proliferative response is limited to a few months and then fades away.

The concept of proliferative response prevention (thus reducing restenosis rates) originated from the well-studied mechanisms of proliferation and involves physical means, pharmacologic approaches, and other biotechnological methods. For example, brachytherapy, which uses various forms of intraluminal radiation, did result in substantial reduction of vascular proliferation following angioplasty.[18] However, the clinical price paid was increased rates of late thrombotic occlusion. It is clear now that the radiated segment of his artery had a markedly altered endothelial function, leading to late consequences of radiation therapy. With the recent introduction of DES, this mode of therapy has been practically abandoned.

Pharmacologic interference with the proliferative and inflammatory process is a common practice in the fields of oncology and transplant immunology. Two major drugs that interfere with various stages of the cell cycle and are widely used in applying stent-mediated local delivery for prevention of restenosis are Rapamycin and Taxol.[19–21] These drugs are slowly released over a 1-month period and effectively reduce restenosis. Other drugs from the same family, but using other mechanisms, are being tested and are at different phases of entering the clinical market. DES have been extremely effective in reducing restenosis, reducing the thickness of the intima inside the stent from roughly $500\,\mu$ to $100–200\,\mu$, and reducing the restenosis rates to less than 10% in most studies.[19] Because of the very high effectiveness of DES, their market penetration has reached 80% of the catheterizations and has been extremely rapid—at a rate not previously known in medicine. However, concerns regarding late (after 1 year) thrombotic occlusion of these stents were recently raised.[21] It is clear that reduction of the proliferative process may result in no coverage of the stent with tissue and to long-term exposure of stent struts to blood-enhancing coagulation. Furthermore, some undegradable stent polymer coatings that are used to carry the drug may also be part of this hazard. The risk of late thrombosis with DES is roughly 0.6% per year, according to various statistical analyses.[21]

The search for stents with the benefits of DES, with a long-term safety profile, and with minimal possible thrombotic events must obviously continue. Other techniques to prevent restenosis and enhance stent coverage using less studied endothelial progenitor cells and other molecular methods are discussed elsewhere.[22]

Myocardial Damage and Pump Dysfunction

Myocardial dysfunction resulting from ischemia and leading to heart failure is one of the major causes of morbidity and mortality and is being intensively investigated. Large clinical studies have consistently shown that the earlier we open an occluded artery during acute myocardial infarction, the less muscle damage is generated.[23] Therefore, the primary challenge in this area is to limit the time to treatment. Treatment methods include intravenous thrombolysis and primary coronary intervention. Both are valid approaches with respect to limiting myocardial injury and are used interchangeably depending on the clinical presentation, local expertise in the catheterization laboratory, and other factors. Opening of an artery supplying an acutely ischemic region is also associated with myocardial damage from reperfusion injury.[24,25] Many factors lead to reperfusion injury and effective methods are being sought for preventing it during primary PCI. Some examples of prevention techniques involve pharmacologic blockages of pathways leading to myocyte death or apoptosis,[26] cooling of the myocardium,[27] as well as cell transplantation during the acute event.[28]

Methods to Manage Acute and Chronic Pump Failure

Acute and chronic pump failures are life-threatening conditions and require aggressive management. Various pharmacologic approaches have been developed and used to alter myocardial contractile function and provide temporary relief for pump failure.[29] Yet, those agents that increase contractility are generally considered ineffective and are therefore used with caution and skepticism. Mechanical devices for pump failure are intra-aortic balloon pumps,[30] surgically implantable ventricular assist devices,[31] and artificial hearts.[32]

Cell therapy for heart failure is currently under intense investigation.[33] It may involve direct injection of cells into the myocardium through an endo-ventricular or a surgical approach, and it can also be done through a direct intracoronary injection.[34] Several studies have shown that the use of bone marrow cells, endocardial progenitor cells, or bone marrow–derived stem cells may result in an improved outcome during acute myocardial infarction. Studies with human embryonic stem cells have also been applied to animal models that show the great potential of these cells to functionally integrate with the myocardium[35,36] and improve ventricular function.[37] Some related topics are discussed in detail elsewhere in this volume.

Population Health and Genetics

While environmental factors are well-known contributors to IHD,[1] the genetic inheritance and risk factors for IHD are well known. IHD clusters in families, suggesting a strong genetic link. Despite extensive exploration of many genes, strong evidence of a molecular genetic association with coronary artery disease or myocardial infarction remains to be obtained. Studies have shown an association between the haptoglobin gene alleles and IHDs in diabetics.[38] In contrast to earlier reports, Ferrari *et al.*[39] and Koch *et al.*[40] did not find an association between stent restenosis and converting enzyme insertion/deletion polymorphism. The recent development of high-density genotyping arrays provides whole-genome assessment of variants associated with common diseases. Samani *et al.*[41] identified and described several genetic loci that affect the risk of developing coronary artery disease and myocardial infarction. Importantly, the finding that a single locus was the strongest predictor of IHD in two separate studies carries promise for clinically relevant progress in our understanding of the genetics of coronary artery disease. Such genetic association studies may be important not only in predicting prognosis but also in guiding therapy.

The Future

Therapies for IDH and interventional cardiology have advanced in the last decade, although there are many more obstacles to cross and summits to conquer. Prevention of restenosis and thrombosis is still required. Tailored antiplatelet therapy, whether local or systemic, may be a solution to the current problem with late thrombosis. Methods of revascularization of total occlusions or complex anatomic lesions still depend on advances in medical technology combined with sophisticated imaging methods. Cell therapy for heart failure may become the leading technique in the future; but major biologic, logistic, and ethical problems have yet to be properly addressed. The role of embryonic stem cells may have a huge impact, but there is still a long way to go before these cells reach the clinical market. Imaging, mechanics, biology, molecular medicine, and genetics must interact to provide us with ever better tools to assure the health and safety of our patients.

Conflicts of Interest

The author declares no conflicts of interest.

References

1. LOPEZ, A.D., C.D. MATHERS, *et al.* 2006. Global and regional burden of disease and risk factors, systematic analysis of population health data. Lancet **367:** 1747–1757.

2. O'FLAHERTY, M.E., E.S. FORD, *et al.* 2007. Coronary heart disease trends in England and Wales from 1984 to 2004: concealed leveling of mortality rates among young adults. Heart. In press.

3. FORD, E.S., U.A. AJANI, J.B. CROFT, *et al.* 2007. Explaining the decrease in U.S. deaths from coronary disease, 1980–2000. N. Engl. J. Med. **356:** 2388–2398.

4. DAEMEN, J., P.W. SERRUYS. 2006. Optimal revascularization strategies for multivessel coronary artery disease. Curr. Opin. Cardiol. **21:** 595–601.

5. ROUBIN, G. & A. GRÜNTZIG. 1986. The coronary artery bypass surgery-angioplasty interface. Cardiology **73:** 269–277.

6. WEISZ, G. & J.W. MOSES. 2007. New percutaneous approaches for chronic total occlusion of coronary arteries. Expert Rev. Cardiovasc. Ther. **5:** 231–241.

7. SURMELY, J.F., E. TSUCHIKANE, *et al.* 2006. New concept for CTO recanalization using controlled antegrade and retrograde subintimal tracking: the CART technique. J. Invasive Cardiol. **18:** 334–338.

8. HOYE, A., P.A. LEMOS & P.W. SERRUYS. 2005. Successful use of a new guidewire with radiofrequency ablation capability for the treatment of chronic total occlusion at the ostium of the left anterior descending artery. J. Invasive Cardiol. **17:** 277–279.

9. TOPOL, E.J. & S.E. NISSEN. 1995. Our preoccupation with coronary luminology. The dissociation between clinical and angiographic findings in ischemic heart disease. Circulation **92:** 2333–2342.

10. TSETIS, D., R. MORGAN & A.M. BELLI. 2006. Cutting balloons for the treatment of vascular stenosis. Eur. Radiol. **16:** 1675–1683.

11. HONDA, Y. & P.J. FITZGERALD. 2001. The renaissance of directional coronary atherectomy: a second look from the inside. J. Invasive Cardiol. **13:** 748–751.

12. REITH, S., P.W. RADKE, O. VOLK, *et al.* 2000. The place of rotablator for treatment of in-stent restenosis. Semin. Interv. Cardiol. **5:** 199–208.

13. KOSTER, R., J. KAHLER, *et al.* 2002. Laser coronary angioplasty: history, present and future. Am. J. Cardiovasc. Drugs **2:** 197–207.

14. KOSTER, R., C.W. HAMM, *et al.* 1999. Laser angioplasty of restenosed coronary stents: results of a multicenter surveillance trial. J. Am. Coll. Cardiol. **34:** 25–32.

15. HOLMES, D.R. JR., J. HIRSHFELD, JR. *et al.* 1998. ACC expert consensus document on coronary artery stents: document of the American College of Cardiology. J. Am. Coll. Cardiol. **32:** 1471–1482.

16. BHODAY, J., S. DE SILVA & Q. XU. 2006. The molecular mechanisms of vascular restenosis: Which genes are crucial? Curr. Vasc. Pharmacol. **4:** 269–275.

17. GASPARDONE, A. & F. VERSACI. 2005. Coronary stenting and inflammation. Am. J. Cardiol. **96:** 65L–70L.

18. NIKOLSKY, E., E. ROSENBLATT, E. GRENADIER, *et al.* 2002. A prospective single-center registry for the use of intracoronary gamma radiation in patients with diffuse in-stent restenosis. Catheter. Cardiovasc. Interv. **56:** 46–52.

19. MORICE, M.C., P.W. SERRUYS, *et al.* 2002. RAVEL Study Group. A randomized comparison of a sirolimus-eluting stent with a standard stent for coronary revascularization. N. Engl. J. Med. **346:** 1773–1780.

20. LEMOS, P.A., A. HOYE, D. GOEDHART, *et al.* 2004. Clinical, angiographic, and procedural predictors of angiographic restenosis after sirolimus-eluting stent implantation in complex patients – An evaluation from the Rapamycin-Eluting Stent evaluated at Rotterdam Cardiology Hospital (RESEARCH) study. Circulation **109:** 1366–1370.

21. DAEMEN, J. & P.W. SERRUYS. 2007. Drug-eluting stent update 2007: part I. A survey of current and future generation drug-eluting stents: meaningful advances or more of the same? Circulation **116:** 316–328.

22. LINDT, R., F. VOGT, I. ASTAFIEVA, *et al.* 2006. A novel drug-eluting stent coated with an integrin-binding cyclic Arg-Gly-Asp peptide inhibits neointimal hyperplasia by recruiting endothelial progenitor cells. J. Am. Coll. Cardiol. **47:** 1786–1795.

23. EELEY, E.C., & L.D. HILLIS. 2007. Primary PCI for myocardial infarction with ST-segment elevation. N. Engl. J. Med. **356:** 47–54.

24. BASSO, C. & G. THIENE. 2006. The pathophysiology of myocardial reperfusion: a pathologist's perspective. Heart **92:** 1559–1562.

25. GROSS, G.J. & J.A. AUCHAMPACH. 2007. Reperfusion injury: does it exist? J. Mol. Cell. Cardiol. **42:** 12–18.

26. DUPLAIN, H. 2006. Salvage of ischemic myocardium: a focus on JNK. Adv. Exp. Med. Biol. **588:** 157–164.

27. LY, H.Q., A. DENAULT, J. DUPUIS, *et al.* 2005. A pilot study: the noninvasive surface cooling thermoregulatory system for mild hypothermia induction in acute myocardial infarction (the NICAMI Study). Am. Heart J. **150:** 933.

28. DOBERT, N., M. BRITTEN, *et al.* 2004. Related transplantation of progenitor cells after reperfused acute myocardial infarction: evaluation of perfusion and myocardial viability with FDG-PET and thallium SPECT. Eur. J. Nucl. Med. Mol. Imaging **31:** 1146–1151.

29. SHIN, D.D., F. BRANDIMARTE, *et al.* 2007. Review of current and investigational pharmacologic agents for acute heart failure syndromes. Am. J. Cardiol. **99:** 4A–23A.

30. APAIOANNOU, T.G. & C. STEFANADIS. 2005. Basic principles of the intraaortic balloon pump and mechanisms affecting its performance. ASAIO J. **51:** 296–300.

31. BIRKS, E.J., P.D. TANSLEY, *et al.* 2006. Left ventricular assist device and drug therapy for the reversal of heart failure. N. Engl. J. Med. **355:** 1873–1884.

32. GRAY, N.A. JR. & C.H. SELZMAN. 2006. Current status of the total artificial heart. Am. Heart J. **152:** 4–10.

33. JOLICOEUR, E.M., C.B. GRANGER, *et al.* 2007. National Heart, Lung, and Blood Institute Cell Therapy Working Group Members. Bringing cardiovascular cell-based therapy to clinical application: perspectives based on a National Heart, Lung, and Blood Institute Cell Therapy Working Group meeting. Am. Heart J. **153:** 732–742.

34. WOLLERT, K.C. & H. DREXLER. 2005. Clinical applications of stem cells for the heart. Circ. Res. **96:** 151–163.

35. KEHAT, I., D. KENYAGIN-KARSENTI, *et al.* 2001. Human embryonic stem cells can differentiate into myocytes with structural and functional properties of cardiomyocytes. J. Clin. Invest. **108:** 407–414.

36. KEHAT, I., L. KHIMOVICH, *et al.* 2004. Electromechanical integration of cardiomyocytes derived from human embryonic stem cells. Nat. Biotechnol. **22:** 1282–1289.

37. CASPI, O., A. LESMAN, *et al.* 2007. Tissue engineering of vascularized cardiac muscle from human embryonic stem cells. Circ. Res. **100:** 263–272.

38. SULEIMAN, M., D. ARONSON, *et al.* 2005. Haptoglobin polymorphism predicts 30-day mortality and heart failure in patients with diabetes and acute myocardial infarction. Diabetes **54:** 2802–2806.

39. FERRARI, M., H. MUDRA, L. GRIP, *et al.* 2002. OPTICUS ACE Substudy. Angiotensin-converting enzyme insertion/deletion polymorphism does not influence the restenosis rate after coronary stent implantation. Cardiology **97:** 29–36.

40. KOCH, W., A. KASTRATI, J. MEHILLI, *et al.* 2000. A. Insertion/deletion polymorphism of the angiotensin I-converting enzyme gene is not associated with restenosis after coronary stent placement. Circulation **102:** 197–202.

41. SAMANI, N.J., J. ERDMANN, A.S. HALL, *et al.* 2007. Genomewide association analysis of coronary artery disease. N. Engl. J. Med. **357:** 443–453.

Index of Contributors